T0140263

Studies in Computational Intelligence

Volume 683

Series editor

Janusz Kacprzyk, Polish Academy of Sciences, Warsaw, Poland
e-mail: kacprzyk@ibspan.waw.pl

About this Series

The series "Studies in Computational Intelligence" (SCI) publishes new developments and advances in the various areas of computational intelligence—quickly and with a high quality. The intent is to cover the theory, applications, and design methods of computational intelligence, as embedded in the fields of engineering, computer science, physics and life sciences, as well as the methodologies behind them. The series contains monographs, lecture notes and edited volumes in computational intelligence spanning the areas of neural networks, connectionist systems, genetic algorithms, evolutionary computation, artificial intelligence, cellular automata, self-organizing systems, soft computing, fuzzy systems, and hybrid intelligent systems. Of particular value to both the contributors and the readership are the short publication timeframe and the worldwide distribution, which enable both wide and rapid dissemination of research output.

More information about this series at http://www.springer.com/series/7092

Vladik Kreinovich

Editor

Uncertainty Modeling

Dedicated to Professor Boris Kovalerchuk
on his Anniversary

 Springer

Editor
Vladik Kreinovich
Department of Computer Science
University of Texas at El Paso
El Paso, TX
USA

ISSN 1860-949X   ISSN 1860-9503   (electronic)
Studies in Computational Intelligence
ISBN 978-3-319-84554-8   ISBN 978-3-319-51052-1   (eBook)
DOI 10.1007/978-3-319-51052-1

Printed on acid-free paper

This Springer imprint is published by Springer Nature
The registered company is Springer International Publishing AG
The registered company address is: Gewerbestrasse 11, 6330 Cham, Switzerland

Preface

This volume is devoted to the 65th birthday of Dr. Boris Kovalerchuk. Dr. Kovalerchuk's results cover many research areas. Many of these areas are reflected in this volume.

In this preface, I would like to emphasize his contributions to research areas which are the closest to my own research: data processing under uncertainty, especially *fuzzy* data processing, when uncertainty comes from the imprecision of expert opinions.

Fuzzy research area: successes and challenges. Fuzzy techniques have many successful practical applications, especially in intelligent control, where expert knowledge—originally formulated in terms of imprecise (fuzzy) words from natural language—is successfully translated into a computer-understandable form and then used in automated decision making.

However, there are still many applications problems (and even whole application areas) where, at present, we are not that successful in formalizing and using imprecise expert knowledge. To be able to use this knowledge, we must overcome several important challenges. In all these challenges, Dr. Kovalerchuk plays an important role as a research leader.

First challenge: need to select appropriate techniques. The first challenge is that, in contrast to (more traditional) probabilistic methods—which are based on solid foundations—many fuzzy techniques are, by nature, heuristic.

There are usually many ways to translate imprecise expert knowledge into precise terms, and the success of an application often depends on selecting the most adequate translation. To be able to select such a translation, we need to have a general description of all possible translations and ways to deal with them. This activity is known as *foundations* of fuzzy techniques.

This is a very complex area of research, an area that requires deep knowledge of mathematics, computer science, foundations and philosophy of science, and—since the ultimate goal is applications—a good understanding of many application areas.

Boris has all these skills, and he has used them successfully in his numerous seminal papers on fuzzy foundations. His papers appeared as chapters in the Springer book series "Studies in Fuzziness and Soft Computing" (see, e.g., [1, 2]; one of the first was his 1994 chapter [1] devoted to the difficult problem of optimization of an uncertain (fuzzy) objective function under uncertain (fuzzy) constraints.

Second challenge: need to combine fuzzy and probabilistic techniques. The second major challenge is related to the fact that, in addition to *subjective* expert knowledge, we also have measurement-based *objective* information about the corresponding system, information usually formulated in probabilistic terms. To solve the corresponding practical problems, we need to adequately combine fuzzy and probabilistic uncertainty. Here, we face two problems:

- a *foundational* problem—which is the best way to combine these two types of uncertainty?—and
- a *communication* problem, caused by the fact that the two communities are not very familiar with each other's research and, as a result, have misunderstandings about the other research areas, misunderstandings that prevent successful collaboration.

Boris is one of the main research leaders in solving both these problems.

He has published several seminal papers on selecting the best way of combining these two types of knowledge; see, e.g., [3, 4]; I would like to specifically mention his 2012 Springer chapter [2].

He has also done a great job of describing probability ideas to fuzzy community and fuzzy ideas to probability researchers, in particular, by showing that—contrary to the widely spread misunderstanding—fuzzy-related techniques do not violate the main idea of probability, and moreover, many such fuzzy techniques can be meaningfully reformulated (and explained) in probabilistic terms.

In particular, he has shown that many real-life applications of fuzzy techniques can actually be reformulated in probabilistic terms—and that the combination of such reformulated terms and traditional probabilistic techniques can enhance the probabilistic approach. He has also shown that a seeming inconsistency between fuzzy methods (based on t-norms) and probabilistic approach can be resolved within a new formalism that Boris called Exact Complete Context Spaces (ECCS). His series of publications starting with his 1994 paper [5], in which he showed that exact complete context spaces link fuzzy logic and probability theory in a new rigorous way. Specifically, he has shown how the use of ECCS can explain numerous successes of fuzzy control in application; this was the main topic of his 1996 paper [6] that was welcomed by Lotfi Zadeh. This work had been expanded in his other publications published in the proceedings of the IEEE World Congresses on Computational Intelligence WCCI'2008–2012, International Conferences on Information Processing and Management of Uncertainty in Knowledge-Based Systems IPMU'2012–2014, World Congress of International Fuzzy Systems

Association IFSA/NAFIPS'2013, and in several seminal Springer book chapters published in 2012 and 2013; see also [7, 8, 9].

Third challenge: dynamic character of human reasoning. The third challenge is that, in contrast to the objective knowledge, which, once established, remains stable, subjective knowledge changes with time, it is dynamic: an expert may learn new things and/or realize that some of his/her previous opinions were imprecise or even incorrect. To make applications of expert knowledge more adequate, we need to take into account the dynamic nature of human reasoning. This is a very difficult task.

In solving this task, Boris was one of the pioneers. With Leonid Perlovsky and Gregory Wheeler, he established a formal mechanism for modeling such dynamic character, a mechanism that they called Dynamic Logic of Phenomena. This is an approach to solve real-world tasks via a dynamic process of synchronous adapting the task and the solution criteria when both are uncertain. Boris started this research under the grant from the US National Research Council (NRC) when he was working at the US Air Force Research laboratory in 2007–2008. His main results are overviewed in his seminal 2012 paper published in a prestigious Journal of Applied Non-Classical Logics [10].

Fourth challenge: dealing with (somewhat) inconsistent expert knowledge. The fourth challenge is that, due to imprecision of expert reasoning, some of the expert statements are, strictly speaking, contradictory to one another. It is desirable to be able to deal with such seemingly inconsistent knowledge. The logic of such inconsistent knowledge bases is known as *paraconsistent logic*. This a very active and a very difficult area of research, so difficult that at present, it has are very few applications to real-life situations, and most of these applications only deal with "crisp" (non-fuzzy) expert statements.

In his pioneer 2006–2010 joint research with Germano Resconi, Boris developed a theory of *irrational* (=inconsistent) *agents*, a theory that combined fuzzy logic, probability theory, and paraconsistent logic into a general techniques for handling both rational and irrational agents [11, 12, 13–20].

Fifth challenge: translating computer results into human-understandable form. The fifth major challenge is related to the fact that, in contrast to fuzzy control where often a decision needs to be made urgently and thus, has to be automated, in many other application areas—e.g., in many cases of medical diagnostics—there is no such hurry. So, it is desirable to first show the resulting computer-generated decision proposal to an expert, to make sure that the automated system properly took all the experts' knowledge into account. To be able to do that, we face a problem which is reverse to the above-mentioned translation problem underlying fuzzy techniques—a problem of how to better translate numerical results of the computer data processing into expert-understandable form. There are two ways we humans get the information:

- in terms of words, and
- in terms of pictures.

Thus, we need to translate the computer results both into words and into pictures. On both tasks, Boris did a pioneer work.

The question of translating computer results into words is handled in Boris's publications on *interpretability* of fuzzy operations. Not only he analyzed this problem theoretically, he also proposed and conducted empirical studies that established the scope of applicability of different "and"-operations (=t-norms) of fuzzy logic. This work was published in Fuzzy Sets and Systems—the main journal of our community—in Elsevier's Journal of General Systems [21], in proceedings of IEEE WCCI'2010–2012 [22], and in many other places (see, e.g., [23]).

In terms of visualization, Boris is a recognized expert in analytical and visual data mining, and in visual analytics. He has published two related books: *Data Mining in Finance* [24] and *Visual and Spatial Analysis* [25]. Most recently (2014) Boris published a series of four conference papers (jointly with his colleague Vladimir Grishin) on lossless visualization of multi-D data in 2-D; see, e.g., [26, 27].

This is an interesting new development, with a potential for a breakthrough in the critical area of big data research. Boris introduced new concepts of collocated paired coordinates and general line coordinates that dramatically expand the scope of lossless multi-D data visualizations [1, 27, 28].

Need for applications. Finally, once all these challenges are resolved, it is important to actively pursue new applications of the corresponding techniques. Boris has many application papers, ranging:

- from applications to medicine, including breast cancer diagnostics [29, 30];
- to finance [24]
- to geospatial analysis—in a series of SPIE publications during the last 10 years; see, e.g., [31–33], and in [34];
- to efficient applications of his new visualization techniques to World Hunger data analysis and the Challenger disaster.

Dr. Kovalerchuk is a world-renowned researcher. All this research activity has made Boris Kovalerchuk a world-renowned expert in systems and uncertainty modeling.

For example, in 2012, he was invited to present a 3-h tutorial on Fuzzy Logic, Probability, and Measurement for Computing with Words at the IEEE World Congress on Computational Intelligence WCCI'2012.

Service to the research community. In addition to doing research, Boris is also very active in the fuzzy research community. He regularly posts short tutorials and opinions on the relation between possibility and probability to the Berkeley

Initiative Soft Computing (BISC) mailing list, often at the explicit invitation of Dr. Zadeh himself.

He also makes an important contribution to conferences. He chaired two Computational Intelligence Conferences [35, 36]. In 2015, he serves as a technical co-chair of the North American Fuzzy Information Processing Society (NAFIPS) Conference to be held in Redmond, Washington (August 2015). At the IEEE Symposium on Computational Intelligence for Security and Defense Applications, CISDA (New York State, May 2015), he organized and mediated a panel of leading experts from multiple organizations including DARPA on Current Challenges of Computational Intelligence in Defense and Security.

Conclusion. Dr. Boris Kovalerchuk is an excellent well-recognized world-level researcher in the area of fuzzy techniques and uncertainty modeling in general, he is one of the leaders in this research area. We wish him happy birthday and many many more interesting research results!

El Paso, Texas, USA

Vladik Kreinovich

References

1. B. Kovalerchuk, "Current situation in foundations of fuzzy optimization", In: M. Delgado, J. Kacprzyk, J. L. Verdegay and M.A. Vila (Eds.), *Fuzzy Optimization: Recent Advances*, Studies in Fuzziness, Physica Verlag (Springer), Heidelberg, New York, 1994, pp. 45–60.
2. B. Kovalerchuk, "Quest for rigorous combining probabilistic and fuzzy logic approaches for computing with words", In: R. Seising, E. Trillas, C. Moraga, and S. Termini (eds.), *On Fuzziness. A Homage to Lotfi A. Zadeh* (Studies in Fuzziness and Soft Computing Vol. 216), Springer Verlag, Berlin, New York, 2012.
3. B. Kovalerchuk, "Probabilistic solution of Zadeh's test problems", In: A. Laurent et al. (Eds.) *Proceedings of IPMU'2014*, 2014, Part II, CCIS Vol. 443, Springer Verlag, pp. 536–545.
4. B. Kovalerchuk and D. Shapiro, "On the relation of the probability theory and the fuzzy sets foundations", *Computers and Artificial Intelligence*, 1988, Vol. 7, pp. 385–396.
5. B. Kovalerchuk, "Advantages of Exact Complete Context for fuzzy control", *Proceedings of the International Joint Conference on Information Science*, Duke University, 1994, pp. 448–449.
6. B. Kovalerchuk, "Second interpolation for fuzzy control", *Proceedings of the Fifth IEEE International Conference on Fuzzy Systems*, New Orleans, Louisiana, September 1996, pp. 150–155.
7. B. Kovalerchuk, "Spaces of linguistic contexts: concepts and examples", *Proceedings of the Second European Congress on Intelligent Techniques and Soft Computing*, Aachen, Germany, 1994, pp. 345–349.
8. B. Kovalerchuk, "Context spaces as necessary frames for correct approximate reasoning", *International Journal of General Systems*, 1996, Vol. 25, No. 1, pp. 61–80.

9. B. Kovalerchuk and G. Klir, "Linguistic context spaces and modal logic for approximate reasoning and fuzzy probability comparison," *Proceedings of the Third International Symposium on Uncertainty Modeling and Analysis and NAFIPS'95*, College Park, Maryland, 1995, pp. A23–A28.

10. B. Kovalerchuk, L. Perlovsky, and G. Wheeler, "Modeling of phenomena and Dynamic Logic of Phenomena," *Journal of Applied Non-classical Logics*, 2012, Vol. 22, No. 1, pp. 51–82.

11. B. Kovalerchuk and G. Resconi, "Logic of uncertainty and irrational agents", *Proceedings of the IEEE International Conference "Integration of Knowledge Intensive Multi-Agent Systems" KIMAS'07*, Waltham, Massachusetts, April 29–May 3, 2007.

12. B. Kovalerchuk and G. Resconi, "Agent-based uncertainty logic network", *Proceedings of the 2010 IEEE World Congress on Computational Intelligence WCCI'2010*, Barcelona, July 18–23, 2010.

13. G. Resconi and B. Kovalerchuk, "The logic of uncertainty with irrational agents", *Proceedings of the Joint International Information Science Conference*, Taiwan, October 2006.

14. G. Resconi and B. Kovalerchuk, "Explanatory model for the break of logic equivalence by irrational agents in Elkan's paradox", *Computer Aided Systems Theory—EUROCAST 2007*, Spinger Lecture Notes in Computer Science, Vol. 4739, 2007, pp. 26–33.

15. G. Resconi and B. Kovalerchuk, "Hierarchy of logics of irrational and conflicting agents", In: N. T. Nguyen et al. (Eds.), *Agent and Multi-agent Systems: Technologies and Applications*, Springer Lecture Notes in Artificial Intelligence, 2007, Vol. 4496, pp. 179–189.

16. G. Resconi and B. Kovalerchuk, "Fusion in agent-based uncertainty theory and neural image of uncertainty", *Proceedings of the 2008 IEEE World Congress on Computational Intelligence WCCI'2008*, Hong Kong, 2008, pp. 3537–3543.

17. G. Resconi and B. Kovalerchuk, "Agents in neural uncertainty", *Proceedings of the 2009 IEEE International Joint Conference on Neural Networks*, Atlanta, Georgia, June 2009, pp. 2448–2455.

18. G. Resconi and B. Kovalerchuk, "Agents' model of uncertainty", *Knowledge and Information Systems Journal*, 2009, Vol. 18, No. 2, pp. 213–229.

19. G. Resconi and B. Kovalerchuk, "Agent uncertainty model and quantum mechanics representation", In: L. C. Jain and N. T. Nguyen (Eds.), *Knowledge Processing and Decision Making in Agent-Based Systems*, Springer-Verlag, 2009, pp. 217–246.

20. G. Resconi and B. Kovalerchuk, "Agents in quantum and neural uncertainty", In: S.-H. Chen and Y. Kambayashi (Eds.), *Multi-Agent Applications for Evolutionary Computation and Biologically Inspired Technologies*, IGI Global, Hershey, New York, 2010, pp. 50–76.

21. B. Kovalerchuk and V. Talianski, "Comparison of empirical and computed values of fuzzy conjunction", *Fuzzy Sets and Systems*, 1992, Vol. 46, pp. 49–53.

22. B. Kovalerchuk, "Interpretable fuzzy systems: analysis of t-norm interpretability", *Proceedings of the 2010 IEEE World Congress on Computational Intelligence WCCI'2010*, Barcelona, July 18–23, 2010.

23. B. Kovalerchuk and B. Dalabaev, "Context for fuzzy operations: t-norms as scales", *Proceedings of the First European Congress on Fuzzy and Intelligent Technologies*, Aachen, Germany, 1993, pp. 1482–1487.

24. B. Kovalerchuk and E. Vityaev, *Data Mining in Finance: Advances in Relational and Hybrid Methods*, Kluwer Academic Publishers, Dordrecht, 2000.

25. B. Kovalerchuk and J. Schwing, *Visual and Spatial Analysis: Advances in Visual Data Mining, Reasoning and Problem Solving*, Springer, 2005.

26. V. Grishin and B. Kovalerchuk, "Multidimensional collaborative lossless visualization: experimental study", *Proceedings of CDVE'2014*, Seattle, Washington, September 2014, Springer Lecture Notes in Computer Science, Vol. 8693, pp. 27–35.

27. B. Kovalerchuk and V. Grishin, "Collaborative lossless visualization of n-D data by collocated paired coordinates", *Proceedings of CDVE'2014*, Seattle, Washington, September 2014, Springer Lecture Notes in Computer Science, Vol. 8693, pp. 19–26.

28. B. Kovalerchuk, "Visualization of multidimensional data with collocated paired coordinates and general line coordinates", *SPIE Visualization and Data Analysis 2014*, Proceedings of SPIE, 2014, Vol. 9017.
29. B. Kovalerchuk, E. Triantaphyllou, J. F. Ruiz, and J. Clayton, "Fuzzy logic in digital mammography: analysis of lobulation", *Proceedings of the Fifth IEEE International Conference on Fuzzy Systems*, New Orleans, Louisiana, September 1996, pp. 1726–1731.
30. B. Kovalerchuk, E. Triantaphyllou, J. F. Ruiz, and J. Clayton, "Fuzzy logic in computer-aided breast cancer diagnosis: analysis of lobulation", *Artificial Intelligence in Medicine*, 1997, No. 11, pp. 75–85.
31. B. Kovalerchuk, "Correlation of partial frames in video matching", In: M. F. Pellechia, R. J. Sorensen, and K. Palaniappan (eds.), *SPIE Proceedings*, 2013, Vol. 8747, Geospatial InfoFusion III, pp. 1–12.
32. B. Kovalerchuk, L. Perlovsky, and M. Kovalerchuk, "Modeling spatial uncertainties in geospatial data fusion and mining", *SPIE Proceedings*, 2012, Vol. 8396–24, pp. 1–10.
33. B. Kovalerchuk, S. Streltsov, and M. Best, "Guidance in feature extraction to resolve uncertainty", In: M. F. Pellechia, R. J. Sorensen, and K. Palaniappan (eds.), *SPIE Proceedings*, 2013, Vol. 8747, Geospatial InfoFusion III.
34. B. Kovalerchuk and L. Perlovsky, "Integration of geometric and topological uncertainties for geospatial data fusion and mining", *Proceedings of the 2011 IEEE Applied Imagery Pattern Recognition Workshop*, 2011.
35. B. Kovalerchuk (ed.), *Proceedings of the 2006 IASTED International Conference on Computational Intelligence*, San Francisco, California, 2006.
36. B. Kovalerchuk (ed.), *Proceedings of the 2009 IASTED International Conference on Computational Intelligence*, Honolulu, Hawaii, August 17–19, 2009.

Contents

MapReduce: From Elementary Circuits to Cloud

Răzvan Andonie, Mihaela Maliţa, and Gheorghe M. Ştefan

Abstract We regard the MapReduce mechanism as a unifying principle in the domain of computer science. Going back to the roots of AI and circuits, we show that the MapReduce mechanism is consistent with the basic mechanisms acting at all the levels, from circuits to Hadoop. At the circuit level, the elementary circuit is the smallest and simplest MapReduce circuit—the elementary multiplexer. On the structural and informational chain, starting from circuits and up to Big Data processing, we have the same behavioral pattern: the MapReduce basic rule. For a unified parallel computing perspective, we propose a novel starting point: Kleene's partial recursive functions model. In this model, the composition rule is a true MapReduce mechanism. The functional forms, in the functional programming paradigm defined by Backus, are also MapReduce type actions. We propose an abstract model for parallel engines which embodies various forms of MapReduce. These engines are represented as a hierarchy of recursive MapReduce modules. Finally, we claim that the MapReduce paradigm is ubiquitous, at all computational levels.

R. Andonie (✉)
Computer Science Department, Central Washington University, Ellensburg,
WA, USA
e-mail: andonie@cwu.edu

R. Andonie
Electronics and Computers Department, Transilvania University of Braşov,
Braşov, Romania

M. Maliţa
Computer Science Department, Saint Anselm College, Manchester, NH, USA
e-mail: mmalita@anselm.edu

G.M. Ştefan
Electronic Devices, Circuits and Architectures Department,
Politehnica University of Bucharest, Bucharest, Romania
e-mail: gstefan@arh.pub.ro

© Springer International Publishing AG 2017
V. Kreinovich (ed.), *Uncertainty Modeling*, Studies in Computational
Intelligence 683, DOI 10.1007/978-3-319-51052-1_1

1

1   Introduction

MapReduce is a programming framework invented by engineers at Google [4] and used to simplify data processing across massive data sets. Beside Googles' implementation, a very popular MapReduce platform is Hadoop [16], used by companies like Yahoo!, Facebook, and New York Times. Other MapReduce implementations are Disco [5], MapReduce-MPI [6], and Phoenix [14].

In MapReduce, users specify a *map* function that processes a key/value pair to generate a set of intermediate key/value pairs, and a *reduce* function that merges all intermediate values associated with the same intermediate key. Computational processing can occur on data stored either in a file system (unstructured) or within a database (structured). MapReduce operates only at a higher level: the programmer thinks in terms of functions of key and value pairs, and the data flow is implicit. The two fundamental functions of a MapReduce query are[1]:

"Map" function: The master node takes the input, chops it up into smaller sub-problems, and distributes those to worker nodes. A worker node may do this again in turn, leading to a multi-level tree structure. The worker node processes that smaller problem, and passes the answer back to its master node.

"Reduce" function: The master node then takes the answers to all the sub-problems and combines them in a way to get the output—the answer to the problem it was originally trying to solve.

Programs written in this functional style are automatically parallelized and executed on a large cluster of commodity machines (e.g., on a computer cloud). The run-time system takes care of the details of partitioning the input data, scheduling the program's execution across a set of machines, handling machine failures, and managing the required inter-machine communication [4]. This allows programmers to easily utilize the resources of a large distributed system.

Why is MapReduce so popular today? One simple reason would be that it comes from Google. Well, this is not enough, and we have to look closer at what the problems are with Big Data storage and analysis. Yes, we know, Big Data is here. It is difficult to measure the total volume of data stored electronically. We can estimate that this data is in the order of one disk drive for every person in the world [16]. The bad news is that we are struggling to access fast Big Data:

- While the storage capacity of hard drives has increased massively over the years, access speed has not kept up. On one typical drive from 1990 you could read all the data from the full drive in about five minutes. Presently, it takes more than two and a half hours to read all the data from a Tera byte drive. Access speed depends on the transfer rate, which corresponds to a disk's bandwidth.
- Seeking time improves more slowly than transfer rate (seeking time characterizes the latency of a disk operation, i.e., the delay introduced by the rotation of the disk to bring the required disk sector).

---

[1]http://www.mapreduce.org/what-is-mapreduce.php.

The problem is that we expect to keep data access time constant (if not to reduce), whereas the data volume is dramatically increasing. MapReduce solves this problem by collocating data with computer code, so data access is fast since it is local. This way, MapReduce tries to avoid the data access bottleneck. MapReduce is a good answer to the execution time concern, being a linearly scalable model. If you also double the size of the input data, a job will run twice as slow, whereas if you double the size of the cluster, a job will run as fast as the original one.

Even if MapReduce looks revolutionary and Google has been granted a patent, there are authors who consider that MapReduce is too similar to existing products.[2] What is perhaps missing in this debate is the connection between the principles of MapReduce and old results from functional and distributed computing. It may be surprising to see how old ideas, put in a new framework (Big Data in this case), can be rediscovered and made practical. Let us consider the following example.

Here are two of the most used functional forms written in Scheme/Lisp:

```
(define (myMap func list)
  (cond ((null? list) ())
        (#t (cons (func (car list))
                  (myMap func (cdr list))))
  )
)


(define (myReduce binaryOp list)
  (cond ((null? (cdr list)) (car list))
        (#t (binaryOp (car list)
                      (myReduce binaryOp (cdr list))))
  )
)
```

The first form maps the function `func` over a list of elements we call `list`. It is a recursive definition which takes, by turn, each element of the list `list` as argument for the function `func`. The selection is done using the Lisp function `car`, which selects the first element from a list. The selected element is extracted from the list using another Lisp function, `cdr`, which returns what remains from `list` after removing its first element. The final result is provided recursively, step by step, using the Lisp function `cons`, which builds the list of results attaching in front of the shortening list of arguments the value of (`func` (`car list`)). The process ends when the Lisp conditional function, `cond`, evaluates `list` to the empty list, (). For example:

$$(myMap\ inc\ '\ (3\ 5\ 8)) \rightarrow (4\ 6\ 9)$$

The second form reduces the list `list` to an atom by repeated application of the binary function `binaryOp`. It is also a recursive application which stops when

---

[2]http://www.dbms2.com/2010/02/11/google-MapReduce-patent/.

`list` contains just one element. The result occurs, step by step, applying by turn the binary operation `binaryOp` to the first element, `(car(list))`, and `binaryOp` applied on the rest, `(cdr(list))`. For example:

$$(myReduce + ' (3 5 8)) -> 16$$

In MapReduce, the combination of these two forms is considered the fundamental programming model for big and complex data processing. This observation gave us the motivation for our work.

Actually, we claim more: MapReduce is a key concept, from elementary digital circuits to the main tasks performed in the cloud, going through the (parallel) computing model proposed by Stephen Kleene, the functional languages envisaged by Alonzo Church (Lisp, the first to be considered), and the one-chip parallel architectures.

We do not aim to question the originality of Google's patent, or the practical value of MapReduce and its implementations. But going back to the roots of AI and circuits, we aim to show that the MapReduce mechanism is consistent with the basic mechanisms acting at all the levels, starting from circuits to Hadoop. On the structural and informational chain, starting from circuits to Big Data processing, we recognize the same MapReduce behavioral pattern. We discover a unifying principle which guarantees a natural structuring process, instead of an *ad hoc* way of scaling up data processing systems. David Patterson offered us a vivid image about what is meant by an *ad hoc* solution, when, expressing serious concerns about one-chip parallel computing, he wrote [10]:

> The semiconductor industry threw the equivalent of a Hail Mary pass when it switched from making microprocessors run faster to putting more of them on a chip – doing so without any clear notion of how such devices would in general be programmed.

We start by looking at the roots of the MapReduce mechanisms in Kleene's model of computation, and describing the main theoretical models related to MapReduce. We will refer both to the recursive function level and to the computer architecture level (circuits and cloud computing). We discover the MapReduce mechanism as a basic module in circuits. We present the abstract model of parallelism using five particular forms of Kleene's composition, and we define the corresponding generic parallel computational engines. These engines will be represented as a hierarchy of recursive MapReduce modules. At the highest level of abstraction, we analyze the MapReduce mechanism in the cloud computing context. In the Concluding Remarks, we synthesize the results of our study.

2 Theoretical Models

Computation starts with *mathematical models of computation*. For sequential computation, Turing's model worked very well because of its simplicity and an almost direct correspondence with the first actual embodiments. Indeed, the distance from Turing Machine to von Neumann's abstract model is short. For parallel computation, we propose a novel starting point: Stephan Kleene's partial recursive functions model [7]. This model is directly supported by circuits and it supports the functional programming style.

2.1   A Mathematical Model of Computation

Kleene's model consists of three basic functions (zero, increment, selection) and three rules (composition, primitive recursivity, minimalization). In [8, 13], we proved that primitive recursivity and minimalization are special forms of composition. Therefore, only the composition rule should be considered.

Composition captures directly the process of computing a number of $p + 1$ functions by:

$$f(x_1, \ldots, x_n) = g(h_1(x_1, \ldots, x_n), \ldots, h_p(x_1, \ldots, x_n))$$

where, at the first level, we have functions $h_1, \ldots, h_p$ and one p-variable reduction function g. Function g (the reduction function) is sometimes implementable as a $(\log p)$-depth binary tree of $p - 1$ functions.

The circuit structure associated to the composition rule is presented in Fig. 1. At the first level, functions h_i ($i = 1, 2\ldots, p$) are processed in parallel. The reduction function is computed at the second level. In the general case, the input variables are sent to all the p cells processing functions h_i, and the resulting vector $\{h_1(x_1, \ldots, x_n), \ldots, h_p(x_1, \ldots, x_n)\}$ is reduced by g to a scalar.

The modules used in this representation can be implemented in various ways, from circuits to programmable structures. The first level implements a *synchronous parallelism*, while between the two levels there is a pipelined, *diachronic parallelism*. In other words, the first level represents the *map* step and the second level the *reduce* step.

Fig. 1 The physical embodiment of the composition rule. The *map* step performs a synchronous parallel computation, while the *reduce* step is diachronic parallel with the *map* step

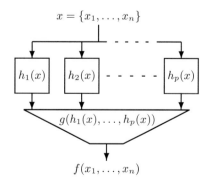

Thus, the mathematical model of computation proposed by Kleene in 1936 is an intrinsic model for parallel computing, and we consider it the starting point for defining the MapReduce parallel computation.

2.2 Circuits

Behind Kleene's model, we have the circuit level. Circuits can be used to implement any h_i or g function. Using Kleene's definition, the three basic functions (zero, increment, projection) can be reduced to one, the elementary selection:

$$s(i, x_1, x_0) = x_i$$

$i \in \{0, 1\}$, because any Boolean function $f : \{0, 1\}^n \to \{0, 1\}$ can be recursively defined by the Boolean expression:

$$f(x_{n-1}, \ldots, x_0) = x_{n-1} \cdot G + x'_{n-1} \cdot H$$

where:

$$G = g(x_{n-2}, \ldots, x_0) = f(1, x_{n-2}, \ldots, x_0)$$

$$H = h(x_{n-2}, \ldots, x_0) = f(0, x_{n-2}, \ldots, x_0)$$

Therefore, any circuit can be seen as a tree of basic circuits corresponding to the elementary `if-then-else` control structure. Indeed, if

$$out = \textbf{if } (sel = 1) \textbf{ then } in1 \textbf{ else } in0$$

is performed for the one bit variables *out, sel, in1, in0*, then the associated circuit is the *elementary multiplexer* expressed in Boolean form as:

$$O = S \cdot I_1 + S' \cdot I_0$$

For $I_0 = 1$ and $I_1 = 0$, the circuit performs the NOT function, i.e., $O = S'$. For $I_0 = 0$, the circuit performs the AND function, i.e., $O = S \cdot I_1$. Thus, we are back to Boolean algebra.

Figure 2a displays the elementary multiplexer. In the world of digital systems, this is the simplest and smallest form of MapReduce.

If we connect the output of this multiplexer to one of the selected inputs, we obtain the basic memory circuit. If $O = I_0$ (see Fig. 2b) and the clock (applied on the selection input) switches from 1 to 0 the value applied on the input I_1 is latched into the circuit, i.e., it is stored, maintained to the output of the circuit as long as the clock stays on 0. In a complementary configuration, if $O = I_1$, the input I_0 is latched when the clock switches from 0 to 1.

Fig. 2 **a** The elementary multiplexer: the first level of ANDs is the *map* step, while the OR circuit makes the reduction. **b** The elementary storage element: the loop closed over the elementary multiplexer provides the 1-bit clocked latch

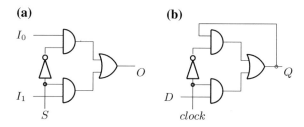

All digital circuits, from the simplest to the most complex programmable systems, are build on the MapReduce elementary structure formally described as an *if-then-else*. Hence, the basic "brick" used in the circuit domain is a MapReduce circuit.

2.3 Programming Style

The lambda calculus was introduced by Church [2] as part of an investigation into the foundations of mathematics. Together with Kleene's partial recursive function model, it has significantly contributed to the development of the Lisp programming language. In our opinion, the most important contribution for the extension of functional programming to parallel systems was provided by Backus [1], much later: the *functional forms*.

Functional forms are based on functions which map objects into objects, where an object can be [1]:

- an atom, x; special atoms are: T (true), F (false), ϕ (empty sequence)
- a sequence of objects, $\langle x_1, \ldots, x_p \rangle$, where x_i are atoms or sequences
- an undefined object $\perp$

The set of functions contains:

- primitive functions which manage:

  - atoms, using functions defined on constant length sequences of atoms, returning constant length sequence of atoms
  - p-length sequences, where p is the number of computational cells

- functional forms for:

  - expanding to sequences the functions defined on atoms
  - defining new functions

- definitions—the programming tool used for developing applications.

A functional form is made of functions which are applied to objects. They are used to define complex functions starting from the set of primitive functions.

We are interested in the following functional forms:

- **Apply to all**: the same function is applied to all elements of the sequence.

$$\alpha f : x \equiv (x = \langle x_1, \ldots, x_p \rangle) \rightarrow \langle f : x_1, \ldots, f : x_p \rangle$$

- **Insert**: the function f has as argument a sequence of objects and returns an object. Its recursive form is:

$$/f : x \equiv ((x = \langle x_1, \ldots, x_p \rangle) \& (p \geq 2)) \rightarrow f : \langle x_1, /f : \langle x_2, \ldots, x_p \rangle \rangle$$

The resulting action looks like a sequential process executed in $O(p)$ cycles, but it can be executed as a reduction function in $O(\log p)$ steps.

- **Construction**: the same argument is used by a sequence of functions.

$$[f_1, \ldots, f_n] : x \equiv \langle f_1 : x, \ldots, f_n : x \rangle$$

- **Threaded construction**: is a special case of construction, when $f_i = g_i \circ i$

$$\theta[f_1, \ldots, f_p] : x \equiv (x = \langle x_1, \ldots, x_p \rangle) \rightarrow \langle g_1 : x_1, \ldots, g_p : x_p \rangle$$

where: $g_i : x_i$ represents an independent thread.

- **Composition**: a pile of functions is applied to a stream of objects. By definition:

$$(f_q \circ f_{q-1} \circ \ldots \circ f_1) : x \equiv f_q : (f_{q-1} : (f_{q-2} : (\ldots : (f_1 : x) \ldots)))$$

This form is a pipelined computation if a ***stream*** of objects, $|x_n, \ldots, x_1|$, is inserted into a pipe of cells, starting with x_1. In this case, each two successive cells will perform:

$$f_i(f_{i-1} : (f_{i-2} : (\ldots : (f_1 : x_j) \ldots)))$$

$$f_{i+1}(f_i : (f_{i-1} : (\ldots : (f_1 : x_{j-1}) \ldots)))$$

The functional forms `apply to all`, `construction`, and `threaded construction` are obviously map-type functions. The function `composition` results by serially connecting map-type functions. The function `insert` is a reduction function. Consequently, Backus's formalism for functional form uses predominantly MapReduce mechanisms.

3   MapReduce Architectures

In this section, we generate abstract models for computing engines and recursive structures, especially for solving Big Data problems. We start our construction from

Kleene's parallel computation model and from Backus' functional forms. We also look at the MapReduce paradigm in the context of cloud computing.

3.1 Abstract Model for Parallel Engines

The computational model defined by Turing was followed in the mid 1940s by the von Neumann [15] and Harvard [3] architectures. Kleene's formalism is also an abstract computational model [7]. Inspired by Kleene's model, in a previous work [9], we have generated five particular composition forms which correspond to the main abstract parallel engines:

1. *Data-parallel*
   If we consider $h_i(x_1, \ldots, x_p) = h(x_i)$ and $g(y_1, \ldots, y_p) = \{y_1, \ldots, y_p\}$, then

   $$f(x_1, \ldots, x_p) = \{h(x_1), \ldots, h(x_p)\}$$

   where $x_i = \{x_{i1}, \ldots, x_{im}\}$ are data sequences. This composition form corresponds to Backus' `apply to all` functional form.
2. *Reduction-parallel*
   If $h_i(x_i) = x_i$, then the general form becomes:

   $$f(x_1, \ldots, x_p) = g(x_1, \ldots, x_p)$$

   which operates a reduction from vector(s) to scalar(s). This corresponds to Backus' `insert` functional form.
3. *Speculative-parallel*
   If the functionally different cells – h_i – receive the same input variable, while the reduction performs the identity function, $g(y_1, \ldots, y_p) = \{y_1, \ldots, y_p\}$, then,

   $$f(x) = \{h_1(x), \ldots, h_p(x)\}$$

   where: x is a sequence of data. This corresponds to Backus' `construction` functional form.
4. *Time-parallel*
   For the special case when $p = 1$, $f(x) = g(h(x))$. Here we have no synchronous parallelism. Only the pipelined, diachronic parallelism is possible, if in each "cycle" we have a new input value. Many applications of type $f(x) = g(h(x))$ result in the m-level *"pipe"* of functions:

   $$f(x) = f_m(f_{m-1}(\ldots f_1(x) \ldots))$$

   where x is an element in the stream of data. This corresponds to the `composition` functional form.

5. *Thread-parallel*

If $h_i(x_1, \ldots, x_n) = h_i(x_i)$ and $g(h_1, \ldots, h_p) = \{h_1, \ldots, h_p\}$, then the general composition form reduces to

$$f(x_1, \ldots, x_p) = \{h_1(x_1), \ldots, h_p(x_p)\}$$

where x_i is a data sequence. Each $h_i(x_i)$ represents a distinct and independent *thread*. This corresponds to the `threaded construction` functional form.

There is a synergic relation between the five abstract parallel engines resulting from Kleene's model and the functional forms proposed by Backus:

Kleene's parallelism $\leftrightarrow$ **Backus' functional forms**

data-parallel $\leftrightarrow$ `apply to all`

reduction-parallel $\leftrightarrow$ `insert`

speculative-parallel $\leftrightarrow$ `construction`

time-parallel $\leftrightarrow$ `composition`

thread-parallel $\leftrightarrow$ `threaded construction`

The MapReduce oriented functional forms find their correspondents in the particular forms of Kleene's MapReduce shaped rule.

3.2 Recursive Structures

The above MapReduce generic parallel engines can be integrated as a recursive hierarchy of MapReduce cells (see Fig. 3), where **cell** consists of **engine** & **memory**. At the lowest level[3] in the hierarchy, for example, we can have:

- **engine**: consists of the **MAP** array of many cells and the **REDUCTION** network loop coupled through CONTROL, with:

  - engine: an execution/processing unit of 8–64 bits
  - memory: a 2–8 KB of static RAM
  - CONTROL: a 32-bit sequential processor

- **memory**: is MEMORY, a 1–4 GB of dynamic RAM (sometimes expanded in a 1–4 TB hard disc)

The above described **cell**, let us call it $cell_1$, can be used to build recursively the next level, $cell_2$, with the same organization but based on $cell_1$ cells. And so on.

In the MAP module, the cellular structure has the simplest interconnections network: a linearly connected network. Each cell contains a local engine and a local memory. The REDUCTION module is a *log*-depth network. At the lowest level, the *log*-depth reduction network contains simple circuits (adders, comparators, logic

---

[3]The lowest level of the generic engine was implemented as BA1024 SoC for HDTV applications [12].

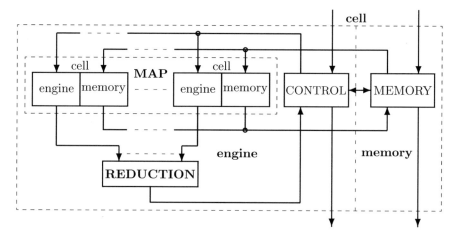

Fig. 3 The recursive definition of the hierarchy of generic MapReduce engines. At each level of this hierarchy, the MAP stage is a linear array of cells, while the REDUCTION stage is a *log*-depth network of circuits

units), while at the highest level, the network is used almost exclusively for communication.

The MapReduce communication pattern allows an easy coherence maintenance of the MEMORY content. The local memory in each $cell_i$ provides the premises for a very high global bandwidth between the storage support and the processing support.

The recursiveness of the structure allows the uniformity of the associated programming environment. Let it be of the form suggested in the first section, where a Lisp-like functional language was used to introduce the map and reduction operations.

3.3 MapReduce and the Cloud

MapReduce programs prove to be very useful for processing big data in parallel. This is performed by dividing the workload across a large number of processing nodes which are not allowed to share data arbitrarily. This feature is one of the explanation of the scalability of a MapReduce application: the communication overhead required to keep the data on the nodes synchronized at all times would prevent the system from performing efficiently on a large scale. In other words, all data elements in MapReduce are immutable, meaning that they cannot be updated. There are several benefits of MapReduce over conventional data processing techniques:

- The model is easy to use, even for programmers without experience with distributed systems, since it hides the implementation details.

- A large variety of problems are easily expressible as MapReduce computations. For example, MapReduce is used for the generation of data for Google's production web search service, for sorting, for data mining, for machine learning, and many other systems.
- MapReduce enables scaling of applications across large clusters of machines comprising thousands of nodes, with fault-tolerance built-in for ultra-fast performance.

The popularity of MapReduce is deeply connected to a very "hot" distributed computer architecture: *the cloud*. MapReduce usually runs on computer clouds and is highly scalable: a typical MapReduce computation processes many Tera bytes of data on thousands of machines. Actually, MapReduce has become the weapon of choice for data-intensive analyses in the cloud and in commodity clusters due to its excellent fault tolerance features, scalability and the ease of use.

In the simplest terms, cloud computing means storing and accessing data and programs over the Internet instead of your computer's hard drive. Cloud computing is the next stage in the Internet's evolution, providing the means through which everything from computing power to computing infrastructure, applications, business processes to personal collaboration can be delivered to you as a service wherever and whenever you need. Cloud computing groups together large numbers of commodity hardware servers and other resources to offer their combined capacity on an on-demand, pay-as-you-go basis. The users of a cloud have no idea where the servers are physically located and can start working with their applications.

Like with MapReduce, the primary concept behind cloud computing is not a new idea. John McCarthy, in the 1960s, imagined that processing amenities is going to be supplied to everyone just like a utility (see [11]). The cloud computing concept is motivated by latest data demands as the data stored on web is increasing drastically in recent times. Combined with MapReduce, the cloud architecture attempts to minimize (and make transparent) the communication bottlenecks between cells and between cells and the system memory.

4   Concluding Remarks

Almost everyone has heard of Google's MapReduce framework which executes on a computer cloud, but few have ever hacked around with the general idea of map and reduce. In our paper, we looked at the fundamental structure of the map and reduce operations from the perspective of recursive functions and computability. This took us back to the foundations of AI and digital circuits.

We can synthesize now the results of our study:

In MapReduce, all is composition We have identified the *MapReduce chain* with six meaningful stages:

- Circuits—for both, combinational and sequential circuits, the basic "brick" is the elementary multiplexer: one of the simplest and smallest MapReduce structure.

- Mathematical model—the composition rule in Kleene's model (the only independent one) is MapReduce.
- Programming style—the functional forms in Backus's approach are MapReduce-type forms.
- Abstract machine model—the five forms of parallelism (data-, reduction-, speculative-, time-, threaded-parallelism) are synergic with the functional forms of Backus and follow the MapReduce mechanism.
- Hierarchical generic parallel structures—represent a direct recursive embodiment of the MapReduce mechanism.
- Cloud computing—is an optimal distributed architecture for MapReduce operations.

MapReduce is the only true parallel paradigm According to our study, this is the only parallel mechanism which solves data synchronization overhead using explicit local buffer management. It is true, MapReduce uses an embarrassingly simple parallelization (a problem that is obviously decomposable into many identical, but separate subtasks is called embarrassingly parallel) and not all problems can be easily parallelized this way.

The programming language matches the structure The composition, embodied in the MapReduce mechanism, generates both the language and the physical structure. It is more "natural" than in the Turing triggered approach, where the language and the engine emerge in two distinct stages: first the engine and then the idea of language as a symbolic entity.

The match between language & structure is maximal at the lowest level in the recursive hierarchy: the one-chip parallel engine level For this reason the power and area performances are in the order of hundred GOPS/Watt[4] and tens GOPS/mm^2 [12].

We claim that the MapReduce chain works as a spinal column in computer science, providing the ultimate coherence for this apparently so heterogeneous domain. We have to observe that the standards for connecting the computer systems and the software needed to make cloud computing work are not fully defined at present time, leaving many companies to define their own cloud computing technologies. Our approach may be useful in this direction.

References

1. J. Backus: "Can Programming Be Liberated from the von Neumann Style? A Functional Style and Its Algebra of Programs". *Communications of the ACM*, August 1978, 613–641.
2. A. Church: "An Unsolvable Problem of Elementary Number Theory", in *American Journal of Mathematics*, vol. 58, p. 345–363, 1936.
3. B. B. Cohen: *Howard Aiken: Portrait of a Computer Pioneer*. Cambridge, MA, USA: MIT Press, 2000.

---

[4]GOPS stands for Giga Operations Per Second.

4. J. Dean, J. Ghemawat: "MapReduce: Simplified Data Processing on Large Clusters", *Proceedings of the 6th Symp. on Operating Systems Design and Implementation*, Dec. 2004.
5. J. Flatow: "What exactly is the Disco distributed computing framework?". http://www.quora.com/What-exactly-is-the-Disco-distributed-computing-framework
6. T. Hoefler, A. Lumsdaine, J. Dongarra: "Towards Efficient MapReduce Using MPI", *Proceedings of the 16th European PVM/MPI Users' Group Meeting on Recent Advances in Parallel Virtual Machine and Message Passing Interface*, 2009, pp. 240–249. http://htor.inf.ethz.ch/publications/img/hoefler-map-reduce-mpi.pdf
7. S. C. Kleene: "General Recursive Functions of Natural Numbers", in *Math. Ann.*, 112, 1936.
8. M. Maliţa, G. M. Ştefan: "On the Many-Processor Paradigm", *Proceedings of the 2008 World Congress in Computer Science, Computer Engineering and Applied Computing (PDPTA'08)*, 2008.
9. M. Maliţa, G. M. Ştefan, D. Thiébaut: "Not multi-, but many-core: designing integral parallel architectures for embedded computation", *SIGARCH Comput. Archit. News*, 35(5):3238, 2007.
10. D. Patterson: "The trouble with multi-core", *Spectrum, IEEE*, 47(7):28–32, 2010.
11. B. T. Rao, L. S. Reddy: "Survey on improved scheduling in Hadoop MapReduce in cloud environments", *CoRR*, 2012.
12. G. M. Ştefan: "One-Chip TeraArchitecture", in *Proceedings of the 8th Applications and Principles of Information Science Conference*, Okinawa, Japan on 11–12 January 2009. http://arh.pub.ro/gstefan/teraArchitecture.pdf
13. G. M. Ştefan, M. Maliţa: "*Can* One-Chip Parallel Computing *Be Liberated From* Ad Hoc Solutions? A Computation Model Based Approach *and Its* Implementation", *18th International Conference on Circuits, Systems, Communications and Computers (CSCC 2014)*, Santorini Island, Greece, July 17–21, 2014, 582–597.
14. J. Talbot, R. M. Yoo, C. Kozyrakis: "Phoenix++: Modular MapReduce for Shared-memory Systems", *Proceedings of the Second International Workshop on MapReduce and Its Applications*, New York, NY, USA, 2011. http://csl.stanford.edu/~christos/publications/2011.phoenixplus.mapreduce.pdf
15. J. von Neumann: "First Draft of a Report on the EDVAC", reprinted in *IEEE Annals of the History of Computing,* Vol. 5, No. 4, 1993.
16. T. White: *Hadoop: The Definitive Guide*, O'Reilly, 2009.

On the Helmholtz Principle for Data Mining

Alexander Balinsky, Helen Balinsky, and Steven Simske

Abstract Keyword and feature extraction is a fundamental problem in text data mining and document processing. A majority of document processing applications directly depend on the quality and speed of keyword extraction algorithms. In this article, an approach, introduced in [1], to rapid change detection in data streams and documents is developed and analysed. It is based on ideas from image processing and especially on the Helmholtz Principle from the Gestalt Theory of human perception. Applied to the problem of keywords extraction, it delivers fast and effective tools to identify meaningful keywords using parameter-free methods. We also define a level of meaningfulness of the keywords which can be used to modify the set of keywords depending on application needs.

1 Introduction

Automatic keyword and feature extraction is a fundamental problem in text data mining, where a majority of document processing applications directly depend on the quality and speed of keyword extraction algorithms. The applications ranging from automatic document classification to information visualization, from automatic filtering to security policy enforcement—all rely on automatically extracted keywords [2]. Keywords are used as basic documents representations and features to perform higher level of analysis. By analogy with low-level image processing, we can consider keywords extraction as low-level document processing.

A. Balinsky (✉)
Cardiff School of Mathematics, Cardiff University, Cardiff CF24 4AG, UK
e-mail: BalinskyA@cardiff.ac.uk

H. Balinsky
Hewlett-Packard Laboratories, Long Down Avenue, Stoke Gifford, Bristol BS34 8QZ, UK
e-mail: Helen.Balinsky@hp.com

S. Simske
Hewlett-Packard Laboratories, 3404 E. Harmony Rd. MS 36, Fort Collins, CO 80528, USA
e-mail: Steven.Simske@hp.com

© Springer International Publishing AG 2017
V. Kreinovich (ed.), *Uncertainty Modeling*, Studies in Computational
Intelligence 683, DOI 10.1007/978-3-319-51052-1_2

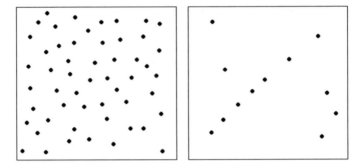

Fig. 1 The Helmholtz principle in human perception

The increasing number of people contributing to the Internet and enterprise intranets, either deliberately or incidentally, has created a huge set of documents that still do not have keywords assigned. Unfortunately, manual assignment of high quality keywords is expensive and time-consuming. This is why many algorithms for automatic keywords extraction have been recently proposed. Since there is no precise scientific definition of the meaning of a document, different algorithms produce different outputs.

The main purpose of this article is to develop novel data mining algorithms based on the Gestalt theory in Computer Vision and human perception. More precisely, we are going to develop Helmholtz principle for mining textual, unstructured or sequential data.

Let us first briefly explain the Helmholtz principle in human perception. According to a basic principle of perception due to Helmholtz [3], an observed geometric structure is perceptually meaningful if it has a very low probability to appear in noise. As a common sense statement, this means that "events that could not happen by chance are immediately perceived". For example, a group of seven aligned dots exists in both images in Fig. 1, but it can hardly be seen on the left-hand side image. Indeed, such a configuration is not exceptional in view of the total number of dots. In the right-hand image we immediately perceive the alignment as a large deviation from randomness that would be unlikely to happen by chance.

In the context of data mining, we shall define the Helmholtz principle as the statement that meaningful features and interesting events appear as large deviations from randomness. In the cases of textual, sequential or unstructured data we derive qualitative measure for such deviations.

Under *unstructured data* we understand data without an explicit *data model*, but with some internal geometrical structure. For example, sets of dots in Fig. 1 are not created by a precise data model, but still have important geometrical structures: nearest neighbours, alignments, concentrations in some regions, etc. A good example is textual data where there are natural structures like files, topics, paragraphs, documents etc. Sequential and temporal data also can be divided into natural blocks like days, months or blocks of several sequential events. In this article, we will assume that data comes packaged into objects, i.e. files, documents or containers. We can also have several layers of such structures; for example, in 20Newsgroups all words are packed into 20 containers (news groups), and each group is divided into individual news. We would like to detect some unusual behaviour in these data and automatically extract some meaningful events and features. To make our explanation more precise, we shall consider mostly textual data, but our analysis is also applicable to any data that generated by some basic set (words, dots, pair of words, measurements, etc.) and divided into some set of containers (documents, regions, etc.), or classified.

The current work introduces a new approach to the problem of automatic keywords extraction based on the following intuitive ideas:

- keywords should be responsible for topics in a data stream or corpus of documents, i.e. keywords should be defined not just by documents themselves, but also by the context of other documents in which they lie;
- topics are signalled by "unusual activity", i.e. a new topic emerges with some features rising sharply in their frequency.

For example, in a book on C++ programming language a sharp rise in the frequency of the words "file", "stream", "pointer", "fopen" and "fclose" could be indicative of the book chapter on "File I/O".

These intuitive ideas have been a source for almost all algorithms in Information Retrieval. One example is the familiar TF-IDF method for representing documents [4, 5]. Despite being one of the most successful and well-tested techniques in Information Retrieval, TF-IDF has its origin in heuristics and it does not have a convincing theoretical basis [5].

Rapid change detection is a very active and important area of research. A seminal paper by Jon Kleinberg [6] develops a formal approach for modelling "bursts" using an infinite-state automation. In [6] bursts appear naturally as state transitions.

The current work proposes to model the above mentioned unusual activity by analysis based on the Gestalt theory in Computer Vision (human perception). The idea of the importance of "sharp changes" is very natural in image processing, where edges are responsible for rapid changes and the information content of images. However, not all local sharp changes correspond to edges, as some can be generated by noise. To represent meaningful objects, rapid changes have to appear in some coherent way. In Computer Vision, the Gestalt Theory addresses how local variations combined together to create perceived objects and shapes.

As mention in [7], the Gestalt Theory is a single substantial scientific attempt to develop principles of visual reconstruction. Gestalt is a German word translatable as "whole", "form", "configuration" or "shape". The first rigorous approach to quantify basic principles of Computer Vision is presented in [7]. In the next section, we develop a similar analysis for the problem of automatic keywords extraction.

The paper is organized as follows. In Sect. 2 we present some results from [1] and further analyse the Helmholtz Principle in the context of document processing and derive qualitative measures of the meaningfulness of words. In Sect. 3 numerical results for State of the Union Addresses from 1790 till 2009 (data set from [8]) are presented and compared with results from [6, Sect. 4]. We also present some preliminary numerical results for the 20Newsgroups data set [9]. Conclusions and future work are discussed in the Sect. 4.

2   The Helmholtz Principle and Meaningful Events

We have defined Helmholtz principle as the statement that meaningful features and interesting events appear as large deviations from randomness. Let us now develop a more rigorous approach to this intuitive statement.

First of all, it is definitely not enough to say that interesting structures are those that have low probability. Let us illustrate it by the following example. Suppose, one unbiased coin is being tossed 100 times in succession, then *any* 100-sequence of heads (ones) and tails (zeros) can be generated with the same equal probability $(1/2)^{100}$. Whilst both sequences

$$s_1 = 10101\ 11010\ 01001\ \ldots\ 00111\ 01000\ 10010$$

$$s_2 = \underbrace{111111111\ldots111111}_{50\ \text{times}}\ \underbrace{000000000\ldots000000}_{50\ \text{times}}$$

are generated with the same probability, the second output is definitely not expected for an unbiased coin. Thus, low probability of an event does not really indicates its deviation from randomness.

To explain why the second output s_2 is unexpected we should explain what an expected output should be. To do this some global observations (random variables) on the generated sequences are to be considered. This is similar to statistical physics where some macro parameters are observed, but not a particular configuration. For example, let μ be a random variable defined as the difference between number of heads in the first and last 50 flips. It is no surprise that the expected value of this random variable (its mean) is equal to zero, which is with high level of accuracy true for s_1. However, for sequence s_2 with 50 heads followed by 50 tails this value is equal to 50 which is very different from the expected value of zero.

Another example can be given by the famous 'Birthday Paradox'. Let us look at a class of 30 students and let us assume that their birthdays are independent and uniformly distributed over the 365 days of the year. We are interested in events that some students have their birthday on a same day. Then the natural random variables will be C_n, $1 \leq n \leq 30$, the number of n-tuples of students in the class having the same birthday. It is not difficult to see that the expectation of the number of pairs of students having the same birthday in a class of 30 is $E(C_2) \approx 1.192$. Similar, $E(C_3) \approx 0.03047$ and $E(C_4) \approx 5.6 \times 10^{-4}$. This means that 'on the average' we can expect to see 1.192 pairs of students with the same birthday in each class. So, if we have found that two students have the same birthday we should not really be surprised. But having tree or even four students with the same birthday would be unusual. If we look in a class with 10 students, then $E(C_2) \approx 0.1232$. This means that having two students with the same birthday in a class of 10 should be considered as unexpected event.

More generally, let Ω be a probability space of all possible outputs. Formally, an output $\omega \in \Omega$ is defined as unexpected with respect to some observation μ, if the value $\mu(\omega)$ is very far from expectation $E(\mu)$ of the random variable μ, i.e. the bigger the difference $|\mu(\omega) - E(\mu)|$ is, the more unexpected outcome ω is. From Markov's inequalities for random variables it can be shown that such outputs ω are indeed very unusual events.

The very important question in such setup is the question of how to select appropriate random variables for a given data. The answer can be given by standard mathematical and statistical physics approach. Any structure can be described by its symmetry group. For example, if we have completely unstructured data, then any permutation of the data is possible. But if we want to preserve a structure, then we can do only transformations that preserve the structure. For example, if we have set of documents, then we can not move words between documents, but can reshuffle words inside each documents. In such a case, the class of suitable random variables are functions which are invariant under the group of symmetry.

2.1 Counting Functions

Let us return to the text data mining. Since we defined keywords as words that correspond with a sharp rise in frequency, then our natural measurements should be counting functions of words in documents or parts of document. To simplify our description let us first derive the formulas for expected values in the simple and ideal situation of N documents or containers of the same length, where the length of a document is the number of words in the document.

Suppose we are given a set of N documents (or containers) $D_1, \ldots, D_N$ of the same length. Let w be some word (or some observation) that present inside one or more of these N documents. Assume that the word w appear K times in all N documents and let us collect all of them into one set $S_w = \{w_1, w_2, \ldots, w_K\}$.

Now we would like to answer the following question: *If the word w appears m times in some document, is this an expected or unexpected event?* For example, the word "*the*" usually has a high frequency, but this is not unexpected. From other hand, in a chapter on how to use definite and indefinite articles in any English grammar book, the word "*the*" usually has much higher frequency and should be detected as unexpected.

Let us denote by C_m a random variable that counts how many times an m-tuple of the elements of S_w appears in the same document. Now we would like to calculate the expected value of the random variable C_m under an assumption that elements from S_w are randomly and independently placed into N containers.

For m different indexes $i_1, i_2, \ldots, i_m$ between 1 and K, i.e. $1 \leq i_1 < i_2 < \ldots < i_m \leq K$, let us introduce a random variable $\chi_{i_1,i_2,\ldots,i_m}$:

$$\begin{cases} 1 & \text{if } w_{i_1}, \ldots, w_{i_m} \text{ are in the same document,} \\ 0 & \text{otherwise.} \end{cases}$$

Then by definition of the function C_m we can see that

$$C_m = \sum_{1 \leq i_1 < i_2 < \ldots < i_m \leq K} \chi_{i_1,i_2,\ldots,i_m},$$

and that the expected value $E(C_m)$ is sum of expected values of all $\chi_{i_1,i_2,\ldots,i_m}$:

$$E(C_m) = \sum_{1 \leq i_1 < i_2 < \ldots < i_m \leq K} E(\chi_{i_1,i_2,\ldots,i_m}).$$

Since $\chi_{i_1,i_2,\ldots,i_m}$ has only values zero and one, the expected value $E(\chi_{i_1,i_2,\ldots,i_m})$ is equal to the probability that all $w_{i_1}, \ldots, w_{i_m}$ belong to the same document, i.e.

$$E(\chi_{i_1,i_2,\ldots,i_m}) = \frac{1}{N^{m-1}}.$$

From the identities above we can see that

$$E(C_m) = \binom{K}{m} \cdot \frac{1}{N^{m-1}}, \tag{1}$$

where $\binom{K}{m} = \frac{K!}{m!(K-m)!}$ is a binomial coefficient.

Now we are ready to answer the previous question:
If in some document the word w appears m times and $E(C_m) < 1$, then this is an unexpected event.

Suppose that the word w appear m or more times in each of several documents. *Is this an expected or or unexpected event?* To answer this question, let us introduce another random variable I_m that counts number of documents with m or more appearances of the word w. It should be stressed that despite some similarity, the random variables C_m and I_m are quite different. For example, C_m can be very large, but I_m is always less or equal N. To calculate the expected value $E(I_m)$ of I_m under an assumption that elements from S_w are randomly and independently placed into N containers let as introduce a random variable $I_{m,i}$, $1 \leq i \leq N$ with

$$I_{m,i} = \begin{cases} 1 & \text{if } D_i \text{ contains } w \text{ at least } m \text{ times,} \\ 0 & \text{otherwise.} \end{cases}$$

Then by definition

$$I_m = \sum_{i=1}^{N} I_{m,i}.$$

Since $I_{m,i}$ has only values zero and one, the expected value $E(I_{m,i})$ is equal to the probability that at least m elements of the set S_w belong to the document D_i, i.e.

$$E(I_{m,i}) = \sum_{j=m}^{K} \binom{K}{j} \left(\frac{1}{N}\right)^j \left(1 - \frac{1}{N}\right)^{K-j}.$$

From the last two identities we have

$$E(I_m) = N \times \sum_{j=m}^{K} \binom{K}{j} \left(\frac{1}{N}\right)^j \left(1 - \frac{1}{N}\right)^{K-j}. \tag{2}$$

We can rewrite (2) as

$$E(I_m) = N \times \mathcal{B}(m, K, p),$$

where $\mathcal{B}(m, K, p) := \sum_{j=m}^{K} \binom{K}{j} p^j (1-p)^{K-j}$ is the *tail of binomial distribution* and $p = 1/N$.

Now, if we have several documents with m or more appearances of the word w and $E(I_m) < 1$, then this is an unexpected event.

Following [7], we will define $E(C_m)$ from (1) as the *number of false alarms* of a m-tuple of the word w and will use notation $NFA_T(m, K, N)$ for the right hand side of (1). The NFA_T of an m-tuple of the word w is the expected number of times such an m-tuple could have arisen just by chance. Similar, we will define $E(I_m)$ from (2) as the number of false alarms of documents with m or more appearances of the word w, and us notation $NFA_D(m, K, N)$ for the right hand side of (2). The NFA_D of an the word w is the expected number of documents with m or more appearances of the word w that could have arisen just by chance.

2.2 Dictionary of Meaningful Words

Let us now describe how to create a dictionary of meaningful words for our set of documents. We will present algorithms for NFA_T. The similar construction is also applicable to NFA_D.

If we observe that the word w appears m times in the same document, then we define this word as *a meaningful word* if and only if its NFA_T is smaller than 1. In other words, if the event of appearing m times has already happened, but the expected number is less than *one*, we have a meaningful event. The set of all meaningful words in the corpus of documents $D_1, \ldots, D_N$ will be defined as a set of keywords.

Let us now summarize how to generate the set of keywords $KW(D_1, \ldots, D_N)$ of a corpus of N documents $D_1, \ldots, D_N$ of the same or approximately same length:

For all words w from $D_1, \ldots, D_N$

1. Count the number of times K the word w appears in $D_1, \ldots, D_N$.
2. For i from 1 to N

   (a) count the number of times m_i the word w appears in the document D_i;
   (b) if $m_i \geq 1$ and

   $$NFA_T(m_i, K, N) < 1, \tag{3}$$

   then add w to the set $KW(D_1, \ldots, D_N)$ and mark w as a meaningful word for D_i.

If the NFA_T is less than ϵ we say that w is ϵ-*meaningful*. We define a set of ϵ-keywords as a set of all words with $NFA_T < \epsilon, \epsilon < 1$. Smaller ϵ corresponds to more important words.

In real life examples we can not always have a corpus of N documents $D_1, \ldots, D_N$ of the same length. Let l_i denote the length of the document D_i. We have three strategies for creating a set of keywords in such a case:

- Subdivide the set $D_1, \ldots, D_N$ into several subsets of approximately equal size documents. Perform analysis above for each subset separately.
- "Scale" each document to common length l of the smallest document. More precisely, for any word w we calculate K as $K = \sum_{i=1}^{N} [m_i/l]$, where $[x]$ denotes an integer part of a number x and m_i counts the number of appearances of the word w in a document D_i. For each document D_i we calculate the NFA_T with this K and the new $m_i \leftarrow [m_i/l]$. All words with $NFA_T < 1$ comprise a set of keywords.
- We can "glue" all documents $D_1, \ldots, D_N$ into one big document and perform analysis for one document as will be described below.

In a case of one document or data stream we can divide it into the sequence of disjoint and equal size blocks and perform analysis like for the documents of equal size. Since such a subdivision can cut topics and is not shift invariant, the better way is to work with a "moving window". More precisely, suppose we are given a document D of the size L and B is a block size. We define N as $[L/B]$. For any word w from D and any windows of consecutive B words let m count number of w in this windows and K count number of w in D. If $NFA_T < 1$, then we add w to a set of keywords and say that w is meaningful in these windows. In the case of one big document that has been subdivided into sub-documents or sections, the size of such parts are natural selection for the size of windows.

If we want to create a set of ϵ-keywords for one document or for documents of different size we should replace the inequality $NFA_T < 1$ by an inequality $NFA_T < \epsilon$.

2.3 Estimating of the Number of False Alarms

In real examples calculating $NFA_T(m, K, N)$ and $NFA_D(m, K, N)$ can be tricky and is not a trivial task. Numbers m, K and N can be very large and NFA_T or NFA_D can be exponentially large or small. Even relatively small changes in m can results in big fluctuations of NFA_T and NFA_D. The correct approach is to work with

$$-\frac{1}{K} \log NFA_T(m, K, N) \tag{4}$$

and

$$-\frac{1}{K} \log NFA_D(m, K, N) \tag{5}$$

In this case the meaningful events can be characterized by $-\frac{1}{K} \log NFA_T(m, K, N) > 0$ or $-\frac{1}{K} \log NFA_D(m, K, N) > 0$.

There are several explanations why we should work with (4) and (5). The first is pure mathematical: there is a unified format for estimations of (4) and (5) (see [7] for precise statements). For large m, K and N there are several famous estimations for large deviations and asymptotic behavior of (5): law of large numbers, large deviation technique and Central Limit Theorem. In [7, Chap. 4, Proposition 4] all such asymptotic estimates are presented in uniform format.

The second explanations why we should work with (4) and (5) can be given by statistical physics of random systems: these quantities represent 'energy per particle' or energy per word in our context. Like in physics where we can compare energy per particle for different systems of different size, there is meaning in comparison of (4) and (5) for different words and documents.

Calculating of (4) usually is not a problem, since NFA_T is a pure product. For (5) there is also possibility of using Monte Carlo method by simulating Bernoulli process with $p = 1/N$, but such calculations are slow for large N and K.

2.4 On TF-IDF

The TF-IDF weight (term frequency—inverse document frequency) is a weight very often used in information retrieval and text mining. If we are given a collection of documents $D_1, \ldots, D_N$ and a word w appears in L documents $D_{i_1}, \ldots, D_{i_L}$ from the collection, then

$$IDF(w) = \log\left(\frac{N}{L}\right).$$

The TF-IDF weight is just 'redistribution' if IDF among $D_{i_1}, \ldots, D_{i_L}$ according to *term frequency* of w inside of $D_{i_1}, \ldots, D_{i_L}$.

The TF-IDF weight demonstrates remarkable performance in many applications, but the IDF part is still remain a mystery. Let us now look at IDF from number of false alarms point of view.

Consider all documents $D_{i_1}, \ldots, D_{i_L}$ containing the word w and combine all of them into one document (*the document about w*) $\widetilde{D} = D_{i_1} + \cdots + D_{i_L}$. For example, if $w =$ 'cow', then $\widetilde{D}$ is all about 'cow'. We now have a *new collection* of documents (containers): $\widetilde{D}, D_{j_1}, \ldots, D_{j_{N-L}}$, where $D_{j_1}, \ldots, D_{j_{N-L}}$ are documents of the original collection $D_1, \ldots, D_N$ that do not contains the word w. In general, $\widetilde{D}, D_{j_1}, \ldots, D_{j_{N-L}}$ are of different sizes. For this new collection $\widetilde{D}, D_{j_1}, \ldots, D_{j_{N-L}}$ the word w appear only in $\widetilde{D}$, so we should calculate number of false alarms or 'energy' ((4) or (5)) per each appearance of w only for $\widetilde{D}$.

Using adaptive window size or 'moving window', (4) and (5) become

$$-\frac{1}{K} \log \left(\binom{K}{K} \frac{1}{\widetilde{N}} \right),$$

i.e.

$$\frac{K-1}{K} \cdot \log \widetilde{N}, \quad \text{where} \quad \widetilde{N} = \frac{\sum_{i=1}^{N} |D_i|}{|\widetilde{D}|}. \tag{6}$$

If all documents $D_1, \ldots, D_N$ are of the same size, then (6) becomes

$$\frac{K-1}{K} \cdot IDF(w),$$

and for large K is almost equal to $IDF(w)$. But for the case of documents of different lengths (which is more realistic) our calculation suggest that more appropriate should be *adaptive IDF*:

$$AIDF(w) := \frac{K-1}{K} \cdot \log \frac{\sum_{i=1}^{N} |D_i|}{|\widetilde{D}|}, \tag{7}$$

where K is term count of the word w in all documents, $|\widetilde{D}|$ is the total length of documents containing w and $\sum_{i=1}^{N} |D_i|$ is the total length of all documents in the collection.

3 Experimental Results

In this section we present some numerical results for State of the Union Addresses from 1790 till 2009 (data set from [8]) and for the famous 20 Newsgroups data set [9].

The performance of the proposed algorithm was studied on a relatively large corpus of documents. To illustrate the results, following [6], we selected the set of all U.S. Presidential State of the Union Addresses, 1790–2009 [8]. This is a very rich data set that can be viewed as a corpus of documents, as a data stream with natural timestamps, or as one big document with many sections.

Fig. 2 Document lengths in hundreds of words is shown by the *solid line*; the document average length is equal to 7602.4 and the sample deviation is 5499.7

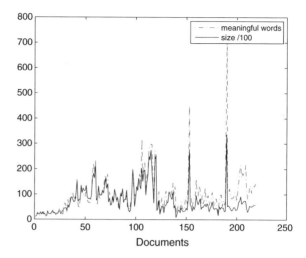

It is important to emphasize that we do not perform any essential pre-processing of documents, such as stop word filtering, lemmatization, part of speech analysis, and others. We simply down-case all words and remove all punctuation characters.

For the first experiment, the data is analyzed as a collection of $N = 219$ individual addresses. The number of words in these documents vary dramatically, as shown in Fig. 2 by the solid line.

As expected, the extraction of meaningful or ϵ-meaningful words using formula (3) from the corpus of different length documents performs well for the near-average length documents. The manual examination of the extracted keywords reveals that

- all stop words have disappeared;
- meaningful words relate to/define the corresponding document topic very well;
- the ten most meaningful words with the smallest NFA follow historical events in union addresses.

For example, five of the most meaningful words extracted from the speeches of the current and former presidents are

Obama, 2009: lending, know, why, plan, restart;
Bush, 2008: iraq, empower, alqaeda, terrorists, extremists;
Clinton, 1993: jobs, deficit, investment, plan, care.

However, the results for the document outliers are not satisfactory. Only a few meaningful words or none are extracted for the small documents. Almost all words are extracted as meaningful for the very large documents. In documents with size more than 19K words even the classical stop word "the" was identified as meaningful.

To address the problem of the variable document length different strategies were applied to the set of all Union Addresses: *moving window*, *scaling to average* and *adapting window size* described in Sect. 2. The results are dramatically improved for outliers in all cases. The best results from our point of view are achieved using an *adaptive window size* for each document, i.e. we calculate (3) for each document with the same K and m_i but with $N = L/|D_i|$ with L being the total size of all documents and $|D_i|$ is the size of the document D_i. The numbers of meaningful words ($\epsilon = 1$) extracted for the corresponding documents are shown by the dashed line in Fig. 2. The remarkable agreement with document sizes is observed.

Our results are consistent with the existing classical algorithm [6]. For example, using a moving window approach, the most meaningful words extracted for The Great Depression period from 1929 till 1933 are: "loan", "stabilize", "reorganization", "banks", "relief" and "democracy", whilst the most important words extracted by [6] are "relief", "depression", "recovery", "banks" and "democracy".

Let us now look at the famous Zipf's law for natural languages. Zipf's law states that given some corpus of documents, the frequency of any word is inversely proportional to some power γ of its rank in the frequency table, i.e. frequency(rank) $\approx$ const/rank$^\gamma$. Zipf's law is mostly easily observed by plotting the data on a *log-log* graph, with the axes being log(rank order) and log(frequency). The data conform to Zipf's law to extend the plot is linear. Usually Zipf's law is valid for the upper portion of the log-log curve and not valid for the tail.

For all words in the Presidential State of the Union Address we plot rank of a word and the total number of the word's occurrences in log-log coordinates, as shown in Fig. 3.

Let us look into Zipf's law for only the meaningful words of this corpus ($\epsilon = 1$). We plot rank of a meaningful word and the total number of the wold's occurrences in log-log coordinates, as shown in Fig. 4. We still can observe the Zipf's law, the curve become smoother and the power γ becomes smaller.

If we increase level of meaningfulness (i.e. decrease the ϵ), then the curve becomes even more smoother and conform to Zipf's law with smaller and smaller γ. This is very much in a line with what we should expect from good feature extraction and dimension reduction: to decrease number of features and to decorrelate data.

Fig. 3 The total number of words in the Presidential State of the Union Addresses as a function of their rank in log-log coordinates

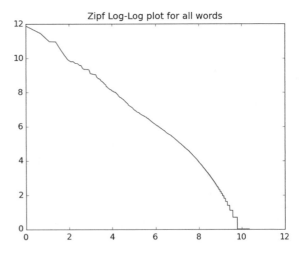

Fig. 4 The total number of meaningful words as a function of their rank in log-log coordinates

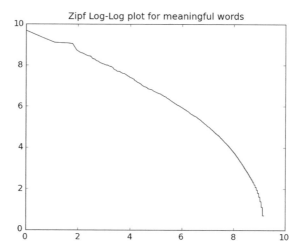

For two sets S_1 and S_2 let us use as a measure of their similarity the number of common elements divided by the number of elements in their union: $W(S_1, S_2) = |S_1 \cap S_2|/|S_1 \cup S_2|$. After extracting meaningful words we can look into similarity of the Union Addresses by calculating similarity W for their sets of keywords. Then, for example, Barack Obama, 2009 speech is mostly similar to George H.W. Bush, 1992 speech with the similarity $W \approx 0.132$ and the following meaningful words in common:

set(['everyone', 'tax', 'tonight', 'i'm', 'down', 'taxpayer', 'reform', 'health', 'you', 'tell', 'economy', 'jobs', 'get', 'plan', 'put', 'wont', 'short-term', 'long-term', 'times', 'chamber', 'asked', 'know']).

George W. Bush, 2008 speech is mostly similar to his 2006 speech (which is very reasonable) with the similarity $W \approx 0.16$ and the following meaningful words in common:

set(['terrorists', 'lebanon', 'al-qaeda', 'fellow', 'tonight', 'americans', 'technology', 'enemies', 'terrorist', 'palestinian', 'fight', 'iraqi', 'iraq', 'terror', 'we', 'iran', 'america', 'attacks', 'iraqis', 'coalition', 'fighting', 'compete']).

From all the Presidential State of the Union Address most similar are William J. Clinton 1997 speech and 1998 speech. Their similarity is $W \approx 0.220339$ and the following meaningful words in common:

set(['help', 'family', 'century', 'move', 'community', 'tonight', 'schools', 'finish', 'college', 'welfare', 'go', 'families', 'education', 'children', 'lifetime', 'row', 'chemical', '21st', 'thank', 'workers', 'off', 'environment', 'start', 'lets', 'nato', 'build', 'internet', 'parents', 'you', 'bipartisan', 'pass', 'across', 'do', 'we', 'global', 'jobs', 'students', 'thousand', 'scientists', 'job', 'leadership', 'every', 'know', 'child', 'communities', 'dont', 'america', 'lady', 'cancer', 'worlds', 'school', 'join', 'vice', 'challenge', 'proud', 'ask', 'together', 'keep', 'balanced', 'chamber', 'teachers', 'lose', 'americans', 'medical', 'first']).

3.1   20 Newsgroups

In this subsection of the article some numerical results for the famous 20 Newsgroup data set [9] will be presented.

This data set consists of 20000 messages taken from 20 newsgroups. Each group contains one thousand Usenet articles. Approximately 4% of the articles are cross-posted. Our only preprocessing was removing words with length ≤ 2. For defining meaningful words we use NFA_T and consider each group as separate container. In Fig. 5, group lengths (total number of words) in tens of words is shown by blue line and number of different words in each group is shown by green line. The highest peek in group lengths correspond to the group 'talk.politics.mideast', and the highest peek in number of different words correspond to the group 'comp.os.ms-windows.misc'.

After creating meaningful words for each group based on NFA_T with $\epsilon = 1$ and removing non-meaningful words from each group, the news group lengths (total number of meaningful words) in tens of words is shown my blue line in Fig. 6. The number of different meaningful words in each group is shown by green line on the same Fig. 6.

Let us now look into the Zipf's law for 20 Newsgroups. We plot rank of a word and total number of the word's occurrences in log-log coordinates, as shown in Fig. 7, and we also plot rank of a meaningful word and total number of the word's occurrences in log-log coordinates, as shown in Fig. 8. As we can see, meaningful word also follow Zipf's law very close.

Similar to the State of the Union Addresses, let us calculate similarity of groups by calculating W for corresponding sets of meaningful words. We will index the groups by integer $i = 0, \ldots, 19$ and denote ith group by Gr[i], for example, Gr[3] = 'comp.sys.ibm.pc.hardware', as shown on the Table 1. The similarity matrix W is 20×20-matrix and is too big to reproduce in the article. So, we show in the Table 1 most similar and most non-similar groups for each group, together with corresponding measure of similarity W. For example, the group 'comp.windows.x'

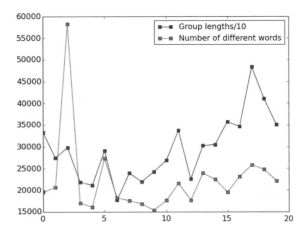

Fig. 5 Group lengths (total number of words) in tens of words and the number of different words in each group

Fig. 6 News group lengths (total number of meaningful words) in tens of words and the number of different meaningful words in each group

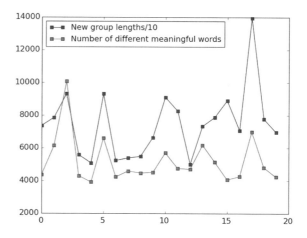

Fig. 7 The total number of words in the 20Newsgroups dataset as a function of their rank in log-log coordinates

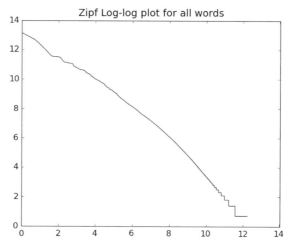

(index = 5) is most similar to the group Gr[1] = 'comp.graphics' with similarity = 0.038, and most non-similar with the group Gr[19] = 'talk.religion.misc' with similarity = 0.0012. As we can see, our feature extraction approach produce very natural measure of similarity for the 20 Newsgroups.

Let us now investigate how sets of meaningful words change with number of articles inside groups. Let us create so called mini-20Newsgroups by selecting randomly 10% of articles in each group. In the mini-20Newsgroups there are 100 articles in each group. We have used for our numerical experiments the mini-20Newsgroups from [9]. After performing meaningful words extraction from the mini-20Newsgroup with NFA_T and $\epsilon = 1$, let us plot together number of meaningful words in each group of original 20Newsgroups, number of meaningful words in each group of mini-20Newsgroup and number of common meaningful words for these two data set. The results are shown in Fig. 9.

Fig. 8 The total number of meaningful words in the 20Newsgroups dataset as a function of their rank in log-log coordinates

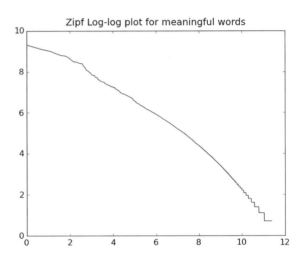

Zipf Log-log plot for meaningful words

Table 1 An example of group similarities

Index	News groups	Highest similarity	Lowest similarity
0	alt.atheism	Gr[19], 0.12	Gr[2], 0.0022
1	comp.graphics	Gr[5], 0.038	Gr[15], 0.0023
2	comp.os.ms-windows.misc	Gr[3], 0.0197	Gr[15], 0.0023
3	comp.sys.ibm.pc.hardware	Gr[4], 0.041	Gr[17], 0.0024
4	comp.sys.mac.hardware	Gr[3], 0.041	Gr[17], 0.0023
5	comp.windows.x	Gr[l], 0.038	Gr[19], 0.0012
6	misc.forsale	Gr[12], 0.03	Gr[0], 0.0024
7	rec.autos	Gr[8], 0.035	Gr[15], 0.0025
8	rec.motorcycles	Gr[7], 0.035	Gr[2], 0.0033
9	rec.sport.baseball	Gr[10], 0.036	Gr[19], 0.0043
10	rec.sport.hockey	Gr[9], 0.036	Gr[15], 0.0028
11	sci.crypt	Gr[16], 0.016	Gr[2], 0.0025
12	sci.electronics	Gr[6], 0.030	Gr[17], 0.0028
13	sci.med	Gr[12], 0.012	Gr[2], 0.0035
14	sci.space	Gr[12], 0.016	Gr[2], 0.0045
15	soc.religion.christian	Gr[19], 0.044	Gr[2], 0.0014
16	talk.politics.guns	Gr[18], 0.042	Gr[2], 0.0021
17	talk.politics.mideast	Gr[18], 0.022	Gr[2], 0.0018
18	talk.politics.misc	Gr[19], 0.043	Gr[5], 0.0017
19	talk.religion.misc	Gr[0], 0.120	Gr[5], 0.0012

As we can see, large part of meaningful words survive when we increase number of articles by ten times, i.e. when we go from mini to full 20Newsgroup data: red and green lines are remarkable coherent.

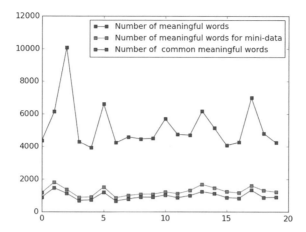

Fig. 9 The number of meaningful words in each news group of the original 20Newsgroups, the number of meaningful words in each group of the mini-20 Newsgroups and the number of common meaningful words for these two data sets

Let us now check how all these meaningful words perform in classification tasks. We would like to have a classifier for finding appropriate newsgroup for a new message. Using 10% of news as training set we have created 20 sets of meaningful words, $MW[i], i = 0, \ldots, 19$. Let us introduce the simplest possible classifier C from messages to the set of 20 Newsgroups. For a message M let us denote by $set(M)$ the set of all different words in M. Then $C(M)$ is a group with largest number of words in $set(M) \bigcap MW[i]$. If there are several groups with the same largest number of words in $set(M) \bigcap MW[i]$, then we select as $C(M)$ a group with smallest index. In the case when all intersections $set(M) \bigcap MW[i]$ are empty, we will mark a message M as "unclassifiable".

The results of applying this classifier to the remaining 90% of 20Newsgroups can be represented by the classification confusion matrix CCM Fig. 10. CCM is a 2020 integer value matrix with CCM(i,j) is the number of messages from ith group classified into jth group. For ideal classifier CCM is a diagonal matrix. For calculating this matrix we used 18000 messages from 20Newsgroups excluding the training set.

REMARK: In each row of the *CCM*, the sum of its elements is equal to 900, which is the number of messages in each group. The exception is the row corresponding to the group "soc.religion.christian", where the sum is equal to 897, because 3 messages from this group remained unclassified, their intersection with the set of meaningful words in each group was empty.

It is also useful to check the classifier performance on the training set itself to validate our approach for selecting meaningful words. The classification confusion matrix for the training set only is shown in Fig. 11.

We now calculate the precision, recall and accuracy of our classifier for each of 20 groups.

639	0	0	1	0	0	2	1	1	0	0	1	1	6	2	88	0	5	3	150
1	743	10	27	8	78	15	5	2	1	1	0	1	1	3	1	1	1	0	2	0
0	159	395	75	12	231	18	4	0	1	0	1	1	0	1	2	0	0	0	0
0	51	10	702	69	38	21	1	0	0	1	1	6	0	0	0	0	0	0	0
0	62	2	113	665	19	30	0	0	0	0	0	6	0	1	0	0	0	1	1
2	139	7	12	1	731	1	0	2	0	0	0	1	0	4	0	0	0	0	0
1	28	1	30	1	8	795	5	0	4	5	1	11	3	5	1	0	1	0	0
0	12	0	8	6	4	81	710	22	11	17	0	13	1	5	0	5	1	3	1
0	8	0	3	2	4	30	20	806	3	7	0	6	0	3	4	1	1	2	0
1	2	0	3	1	0	12	3	2	833	41	0	0	0	1	1	0	0	0	0
0	1	0	1	0	0	4	2	0	15	874	0	0	0	0	1	1	0	1	0
3	60	0	1	0	8	5	0	1	0	0	813	0	2	2	2	2	0	1	0
1	76	2	42	17	14	75	18	1	3	5	12	600	15	16	0	2	1	0	0
1	29	0	3	0	8	24	4	6	4	9	5	7	780	10	1	2	1	6	0
0	32	0	2	0	3	21	3	2	2	4	4	6	9	803	3	0	1	4	1
0	0	0	0	0	0	0	0	0	0	0	0	0	6	0	890	0	1	0	0
1	6	1	0	0	1	0	0	0	1	2	5	1	2	1	6	777	12	59	25
1	1	0	1	0	0	0	2	1	0	1	1	0	0	0	13	2	857	20	0
3	2	0	7	0	1	8	1	1	5	3	2	0	8	12	28	80	73	621	45
113	4	0	8	0	1	6	1	0	1	1	0	0	4	3	174	39	9	88	448

Fig. 10 Classification confusion matrix, where $CCM(i, j)$ is the number of messages from the ith group classified into the jth group

Precision for ith group is defines as

$$P(i) = \frac{CMM(i, i)}{\sum_j CMM(i, j)},$$

Recall for ith group is defines as

$$R(i) = \frac{CMM(i, i)}{\sum_j CMMj, i)},$$

and

Accuracy for ith group is defines as harmonic mean of precision and recall

$$A(i) = \frac{2P(i)R(i)}{P(i) + R(i)}.$$

The results of calculation of precision, recall and accuracy of our classifier for each of 20 groups is shown in the Table 2.

As we can see from the Table 2, this simple classifier performs impressively well for the most of news groups, thus illustrating the success of NFA_T for selecting meaningful features. The smallest accuracy of around 57% has been observed for the group "talk.religion.misc". From the classification confusion matrix *CCM* (Fig. 10), we can see that many articles from "talk.religion.misc" have been classified as belonging to "alt.atheism", "soc.religion.christian" or "talk.politics.misc" groups.

4 Conclusion and Future Work

In this article, the problem of automatic keyword and feature extraction in unstructured data is investigated using image processing ideas and especially the Helmholtz principle. We define a new measure of keywords meaningfulness with good performance on different types of documents. We expect that our approach may not only establish fruitful connections between the fields of Computer Vision, Image Processing and Information Retrieval, but may also assist with the deeper understanding of existing algorithms like TF-IDF.

In TF-IDF it is preferable to create a stop word list, and remove the stop word before computing the vector representation [2]. In our approach, the stop words are removed automatically. It would be very interesting to study the vector model for text

92	0	0	0	0	0	0	0	0	0	0	0	0	0	0	0	0	0	1	7
0	99	0	0	0	1	0	0	0	0	0	0	0	0	0	0	0	0	0	0
0	4	92	2	0	1	1	0	0	0	0	0	0	0	0	0	0	0	0	0
0	0	0	98	1	1	0	0	0	0	0	0	0	0	0	0	0	0	0	0
0	1	0	0	98	0	1	0	0	0	0	0	0	0	0	0	0	0	0	0
0	1	0	0	0	99	0	0	0	0	0	0	0	0	0	0	0	0	0	0
0	0	0	0	0	0	99	0	0	0	0	0	0	1	0	0	0	0	0	0
0	0	0	0	0	0	0	100	0	0	0	0	0	0	0	0	0	0	0	0
0	0	0	0	0	0	1	0	99	0	0	0	0	0	0	0	0	0	0	0
0	0	0	0	0	0	0	0	0	100	0	0	0	0	0	0	0	0	0	0
0	0	0	0	0	0	0	0	0	0	100	0	0	0	0	0	0	0	0	0
0	0	0	0	0	0	0	0	0	0	0	100	0	0	0	0	0	0	0	0
0	0	0	1	0	1	2	0	0	0	0	0	96	0	0	0	0	0	0	0
0	2	0	0	0	0	0	0	0	0	0	0	0	98	0	0	0	0	0	0
0	1	0	0	0	0	0	0	0	0	0	0	0	0	99	0	0	0	0	0
0	0	0	0	0	0	0	0	0	0	0	0	0	0	0	100	0	0	0	0
0	0	0	0	0	0	0	0	0	0	0	1	0	0	0	0	97	0	1	1
0	0	0	0	0	0	0	0	0	0	0	0	0	0	0	0	0	100	0	0
0	0	0	1	0	0	0	0	0	0	0	0	0	0	0	0	3	2	91	3
3	0	0	0	0	0	0	0	0	0	0	0	0	0	0	3	0	0	4	90

Fig. 11 Classification confusion matrix for the training set with data presented as in Fig. 10

Table 2 Precision, recall and accuracy

News groups	Precision	Recall	Accuracy
alt.atheism	0.71	0.8331	0.7667
comp.graphics	0.8256	0.5251	0.6419
comp.os.ms-windows.misc	0.4389	0.9229	0.5949
comp.sys.ibm.pc.hardware	0.78	0.6756	0.7241
comp.sys.mac.hardware	0.7389	0.8504	0.7907
comp.windows.x	0.8122	0.6362	0.7135
misc.forsale	0.8833	0.6925	0.7764
rec.autos	0.7889	0.9103	0.8452
rec.motorcycles	0.8956	0.9516	0.9227
rec.sport.baseball	0.9256	0.9423	0.9339
rec.sport.hockey	0.9711	0.9001	0.9343
sci.crypt	0.9033	0.961	0.9312
sci.electronics	0.6667	0.9091	0.7692
sci.med	0.8667	0.9319	0.8981
sci.space	0.8922	0.9209	0.9063
soc.religion.christian	0.9922	0.7325	0.8428
talk.politics.guns	0.8633	0.852	0.8576
talk, politics, mideast	0.9522	0.8899	0.92
talk.politics.misc	0.69	0.7657	0.7259
talk.religion.misc	0.4977	0.6677	0.5703

mining based with $-\log(NFA)$ as a weighting function. Even the simplest classifier based on meaningful events performs well.

One of the main objectives in [7] is to develop parameter free edge detections based on maximal meaningfulness. Similarly, algorithms in data mining should have as few parameters as possible—ideally none. Developing a similar approach to the keyword and feature extraction, i.e. defining the maximal time or space interval for a word to stay meaningful, is an exciting and important problem. It would also be interesting to understand the relationship between the NFA and [6].

References

1. A. Balinsky, H. Balinsky, and S.Simske, *On Helmholtzs principle for documents processing*, Proc. 10th ACM symposium on Document engineering, Sep. 2010.
2. A. N. Srivastava and M. Sahami (editors), *Text Mining: classification, clustering, and applications*, CRC Press, 2009.
3. D. Lowe, *Perceptual Organization and Visual Recognition*, Amsterdam: Kluwer Academic Publishers, 1985.
4. K. Spärck Jones, *A statistical interpretation of term specificity and its application in retrieval*, Journal of Documentation, vol. 28, no. 1, pp. 1121, 1972.

5. S. Robertson, *Understanding inverse document frequency: On theoretical arguments for idf*, Journal of Documentation, vol. 60, no. 5, pp. 503520, 2004.
6. J. Kleinberg, *Bursty and hierarchical structure in streams*, Proc. 8th ACM SIGKDD Intl. Conf. on Knowledge Discovery and Data Mining, 2002.
7. A. Desolneux, L. Moisan, and J.-M. Morel, *From Gestalt Theory to Image Analysis: A Probabilistic Approach*, ser. Interdisciplinary Applied Mathematics, Springer, 2008, vol.34.
8. Union Adresses. [Online]. Available at http://stateoftheunion.onetwothree.net/
9. 20 Newsgroups Data Set. [Online], 1999. Available at http://kdd.ics.uci.edu/databases/20newsgroups

A Method of Introducing Weights into OWA Operators and Other Symmetric Functions

Gleb Beliakov

Abstract This paper proposes a new way of introducing weights into OWA functions which are popular in fuzzy systems modelling. The proposed method is based on replicating the inputs of OWA the desired number of times (which reflect the importances of the inputs), and then using a pruned n-ary tree construction to calculate the weighted OWA. It is shown that this tree-based construction preserves many useful properties of the OWA, and in fact produces the discrete Choquet integral. A computationally efficient algorithm is provided. The tree-based construction is universal in its applicability to arbitrary symmetric idempotent n-ary functions such as OWA, and transparent in its handling the weighting vectors. It will be a valuable tool for decision making systems in the presence of uncertainty and for weighted compensative logic.

1 Introduction

One of the fundamental aspects in dealing with various forms of uncertainty is operations with uncertainty degrees, for example in the *if-then-else* rule statements. Theories that extend classical logic require extensions of the logical operations, such as conjunction, disjunction, negation and implication. In fuzzy logic in particular, membership degrees from the unit interval are combined by using aggregation functions [4, 14], which in the most general form are monotone increasing functions f with the boundary conditions $f(0, \ldots, 0) = 0$ and $f(1, \ldots, 1) = 1$, so they ensure matching the classical logical operations in the limiting cases.

The first class of prominent aggregation functions are the triangular norms and their duals, the triangular conorms [17]. The product was the first t-norm to appear alongside with the minimum in L. Zadeh's original paper on fuzzy sets [30]. When trying to mimic human decision making empirically, however, it became clear that

G. Beliakov (✉)
Deakin University, 221 Burwood Hwy, Burwood, Australia
e-mail: gleb@deakin.edu.au

© Springer International Publishing AG 2017
V. Kreinovich (ed.), *Uncertainty Modeling*, Studies in Computational
Intelligence 683, DOI 10.1007/978-3-319-51052-1_3

it is a combination of the t-norm and t-conorm, which itself is neither conjunctive nor disjunctive operation, that fits the data best [19, 31]. These functions expressed compensation properties, so that an increase in one input could be compensated by a decrease in another.

Another prominent class of aggregation functions that are in use since the early expert systems MYCIN and PROSPECTOR are the uninorms [4, 7, 26]. These are associative mixed operations that behave as conjunction on one part of their domain and as a disjunction on another. These functions, when scaled to the interval $[-1, 1]$, are useful bi-polar operations, where the information can be interpreted as "positive" or "negative", supporting or negating the conclusion of a logical statement or a hypothesis.

The class of averaging functions is the richest in terms of the number of interesting families. Averaging functions, whose prototypical examples are the arithmetic mean and the median, allow compensation between low values of some inputs and high values of the others. Such functions are also important for building decision models in weighted compensative logic [8], where the concept of Generalized Conjunction/Disjunction (GCD) play a role [10, 12].

Along with many classical means [5], the class of averaging functions contains such constructions as ordered weighted averages (OWA) [24] and fuzzy integrals [15]. The OWA functions in particular became very popular in fuzzy systems community [13, 28, 29]. These are symmetric functions which associate the weights with the magnitude of the inputs rather than with their sources.

The inability of OWA functions to associate weights with the specific arguments in order to model the concepts of importance and reliability of information sources was a stumbling block for their usage in some areas. However the concept of weighted OWA (WOWA) [20] resolved this issue. In WOWA functions, two vectors of weights are used. One vector has the weights associated with the arguments, thus modelling importance of the inputs, whereas the second vector associates weights with the magnitude of the inputs. It was later shown that WOWA are a special class of the discrete Choquet integral [18].

A different way of introducing weights into OWA was presented in [25]. In this paper, the OWA function was applied to the modified arguments, and the modifying function was found by using fuzzy modelling techniques. One such function was a linear combination of the argument and the orness (see Definition 3) of the respective OWA function. This method did not produce idempotent functions except in a few special cases.

In this contribution we present a different approach to introducing weights into OWA, based on n-ary tree construction and recursive application of the base OWA function. Such an approach based in binary trees was recently introduced by Dujmovic [9] in order to incorporate weights into bivariate symmetric means, as well as to extend bivariate means to n variables. More detailed analysis of the properties of this method is in [2, 11].

This paper is structured as follows. After presenting preliminaries in Sect. 2, we describe the construction of WOWA functions by Torra [20] in Sect. 3. In Sect. 4 we describe the method of introducing weights into arbitrary bivariate averaging functions from [9]. Our main contribution is in Sect. 5, where we introduce the n-ary tree construction, outline some of its theoretical properties, and present an efficient computational algorithm. The conclusions are presented in Sect. 6.

2 Preliminaries

Consider now the following definitions adopted from [4, 14]. Let $\mathbb{I} = [0, 1]$, although other intervals can be accommodated easily.

Definition 1 A function $f : \mathbb{I}^n \to \mathbb{R}$ is **monotone** (increasing) if $\forall \mathbf{x}, \mathbf{y} \in \mathbb{I}^n, \mathbf{x} \le \mathbf{y}$ then $f(\mathbf{x}) \le f(\mathbf{y})$, with the vector inequality understood componentwise.

Definition 2 A function $f : \mathbb{I}^n \to \mathbb{I}$ is **idempotent** if for every input $\mathbf{x} = (t, t, ..., t), t \in \mathbb{I}$, the output is $f(\mathbf{x}) = t$.

Definition 3 A function $f : \mathbb{I}^n \to \mathbb{I}$ is a **mean** (or is averaging) if for every $\mathbf{x}$ it is bounded by $\min(\mathbf{x}) \le f(\mathbf{x}) \le \max(\mathbf{x})$.

Averaging functions are idempotent, and monotone increasing idempotent functions are averaging. We consider weighting vectors $\mathbf{w}$ such that $w_i \ge 0$ and $\sum w_i = 1$ of appropriate dimensions.

Definition 4 A function $f : \mathbb{I}^n \to \mathbb{I}$ is **shift-invariant** (stable for translations) if $f(\mathbf{x} + a\mathbf{1}) = f(\mathbf{x}) + a$ whenever $\mathbf{x}, \mathbf{x} + a\mathbf{1} \in \mathbb{I}^n$. A function $f : \mathbb{I}^n \to \mathbb{I}$ is **homogeneous** (of degree 1) if $f(a\mathbf{x}) = af(\mathbf{x})$ whenever $\mathbf{x}, a\mathbf{x} \in \mathbb{I}^n$.

Definition 5 For a given generating function $g : \mathbb{I} \to [-\infty, \infty]$, and a weighting vector $\mathbf{w}$, the **weighted quasi-arithmetic mean** (QAM) is the function

$$M_{\mathbf{w}, g}(\mathbf{x}) = g^{-1}\left(\sum_{i=1}^{n} w_i g(x_i) \right). \tag{1}$$

Definition 6 Let $\varphi : \mathbb{I} \to \mathbb{I}$ be a bijection. The φ-**transform** of a function $f : \mathbb{I}^n \to \mathbb{I}$ is the function $f_\varphi(\mathbf{x}) = \varphi^{-1}(f(\varphi(x_1), \varphi(x_2), ..., \varphi(x_n)))$.

The weighted QAM is a φ-transform of the weighted arithmetic mean with $\varphi = g$.

Definition 7 For a given weighting vector w, $w_i \ge 0$, $\sum w_i = 1$, the **OWA function** is given by

$$OWA_w(x) = \sum_{i=1}^{n} w_i x_{(i)}, \tag{2}$$

where $x_{(i)}$ denotes the i-th largest value of x.

The main properties of OWA are summarised below.

- As with all averaging aggregation functions, OWA are increasing (strictly increasing if all the weights are positive) and idempotent;
- OWA functions are continuous, symmetric, homogeneous and shift-invariant;
- OWA functions are piecewise linear, and the linear pieces are joined together where two or more arguments are equal in value.
- The OWA functions are special cases of the Choquet integral with respect to symmetric fuzzy measures.
- The special case of OWA include the arithmetic mean, the median and the minimum and maximum operators among others.
- The dual of an OWA with respect to the standard negation is the OWA with the weights reversed.

The orness measure allows one to qualify an OWA function as OR-like or AND-like based on whether it behaves more disjunctively or more conjunctively than the arithmetic mean. The expression for the orness measure is given by the following simple formula

$$orness(OWA_w) = \sum_{i=1}^{n} w_i \frac{n-i}{n-1} = OWA_w \left(1, \frac{n-2}{n-1}, \ldots, \frac{1}{n-1}, 0 \right). \quad (3)$$

The OWA functions are OR-like if $orness(OWA_w) \geq \frac{1}{2}$ and AND-like if $orness(OWA_w) \leq \frac{1}{2}$. If the weighting vector is decreasing, i.e., $w_i \geq w_j$ whenever $i < j$, OWA is OR-like and is in fact a convex function. The respective (symmetric) fuzzy measure in this case is sub-modular [3]. The OWA functions with increasing weights are AND-like, concave functions which correspond to the Choquet integral with respect to a super-modular fuzzy measure. OWA with decreasing weighting vectors can be used to define norms [3, 23].

Similarly to quasi-arithmetic means, OWA functions have been generalized with the help of generating functions $g : \mathbb{I} \to [-\infty, \infty]$ as follows.

Definition 8 Let $g : \mathbb{I} \to [-\infty, \infty]$ be a continuous strictly monotone function and let w be a weighting vector. The function

$$GenOWA_{w,g}(x) = g^{-1} \left(\sum_{i=1}^{n} w_i g(x_{(i)}) \right) \quad (4)$$

is called a **generalized OWA**. As for OWA, $x_{(i)}$ denotes the i-th largest value of x.

The generalized OWA is a φ-transform of the OWA function with $\varphi = g$. One special case is the Ordered Weighted Geometric (OWG) function studied in [16, 22]. It is defined by

Definition 9 For a given weighting vector w, the **OWG function** is

$$OWG_w(x) = \prod_{i=1}^{n} x_{(i)}^{w_i}. \tag{5}$$

Similarly to the weighted geometric mean, OWG is a special case of (4) with the generating function $g(t) = \log(t)$. Another special case is the Ordered Weighted Harmonic (OWH) function, where $g(t) = 1/t$.

A large family of generalized OWA functions is based on power functions, similar to weighted power means [27]. Let g_r denote the family of power functions

$$g_r(t) = \begin{cases} t^r, & \text{if } r \neq 0, \\ \log(t), & \text{if } r = 0. \end{cases}$$

Definition 10 For a given weighting vector w, and a value $r \in \mathbb{R}$, the function

$$GenOWA_{w,[r]}(x) = \left(\sum_{i=1}^{n} w_i x_{(i)}^r \right)^{1/r}, \tag{6}$$

if $r \neq 0$, and $GenOWA_{w,[r]}(x) = OWG_w(x)$ if $r = 0$, is called a **power-based generalized OWA**.

3 Weighted OWA

The weights in weighted means and in OWA functions represent different things. In weighted means w_i reflects the importance of the i-th input, whereas in OWA w_i reflects the importance of the i-th largest input. In [20] Torra proposed a generalization of both weighted means and OWA, called WOWA. This aggregation function has two sets of weights w, p. Vector p plays the same role as the weighting vector in weighted means, and w plays the role of the weighting vector in OWA functions.

Consider the following motivation. A robot needs to combine information coming from n different sensors, which provide distances to the obstacles. The reliability of the sensors is known (i.e., we have weights p). However, independent of their reliability, the distances to the nearest obstacles are more important, so irrespective of the reliability of each sensor, their inputs are also weighted according to their numerical value, hence we have another weighting vector w. Thus both factors, the size of the inputs and the reliability of the inputs, need to be taken into account. WOWA provides exactly this type of aggregation function.

WOWA function becomes the weighted arithmetic mean if $w_i = \frac{1}{n}, i = 1, \ldots, n$, and becomes the usual OWA if $p_i = \frac{1}{n}, i = 1, \ldots, n$.

Definition 11 Let w, p be two weighting vectors, $w_i, p_i \geq 0$, $\sum w_i = \sum p_i = 1$. The following function is called **Weighted OWA** function

$$WOWA_{w,p}(x) = \sum_{i=1}^{n} u_i x_{(i)},$$

where $x_{(i)}$ is the i-th largest component of x, and the weights u_i are defined as

$$u_i = g\left(\sum_{j \in H_i} p_j\right) - g\left(\sum_{j \in H_{i-1}} p_j\right),$$

where the set $H_i = \{j | x_j \geq x_i\}$ is the set of indices of i largest elements of x, and g is a monotone non-decreasing function with two properties:

1. $g(i/n) = \sum_{j \leq i} w_j, i = 0, \ldots, n$ (of course $g(0) = 0$);
2. g is linear if all w_i are equal.

Thus computation of WOWA involves a very similar procedure as that of OWA (i.e., sorting components of x and then computing their weighted sum), but the weights u_i are defined by using both vectors w, p, a special monotone function g, and depend on the components of x as well. One can see WOWA as an OWA function with the weights u.

Of course, the weights u also depend on the generating function g. This function can be chosen as a linear spline (i.e., a broken line interpolant), interpolating the points $(i/n, \sum_{j \leq i} w_j)$ (in which case it automatically becomes a linear function if these points are on a straight line), or as a monotone quadratic spline, as was suggested in [20, 21], see also [1] where Schumaker's quadratic spline algorithm was used, which automatically satisfies the straight line condition when needed.

It turns out that WOWA belongs to a more general class of Choquet integral based aggregation functions [18]. It is a piecewise linear function whose linear segments are defined on the simplicial partition of the unit cube $[0, 1]^n$: $\mathscr{S}_i = \{x \in [0, 1]^n | x_{p(j)} \geq x_{p(j+1)}\}$, where p is a permutation of the set $\{1, \ldots, n\}$. Note that there are exactly $n!$ possible permutations, the union of all $\mathscr{S}_i$ is $[0, 1]^n$, and the intersection of the interiors of $\mathscr{S}_i \cap \mathscr{S}_j = \emptyset, i \neq j$.

The next two sections introduce an alternative and generic construction to incorporate weights into any symmetric averaging function. In particular, it will work for OWA and will not have a somewhat unclear issue of selecting the function g in Torra's WOWA.

4   Binary Tree Construction

Consider now a method of incorporating weights into a symmetric bivariate idempotent function f, presented [9] and then extended in [2, 11]. To introduce the weights we use the approach from [6], where each argument x_i is replicated a suitable number of times. To be more precise, we consider an auxiliary vector of arguments

$\mathbf{X} = (x_1, \ldots, x_1, x_2, \ldots, x_2)$, so that x_1 is taken k_1 times and x_2 is taken k_2 times, so that $\frac{k_1}{2^L} \approx w_1$, $\frac{k_2}{2^L} \approx w_2$, and $k_1 + k_2 = 2^L$. Here w_i are the desired weights, and $L \geq 1$ is a specified number of levels of the binary tree shown in Fig. 1. One way of doing so is to take $k_1 = \lfloor w_1 2^L + \frac{1}{2} \rfloor$ and $k_2 = 2^L - k_1$. The vector $\mathbf{X}$ needs to be sorted in the increasing or decreasing order.

Next, let us build a binary tree presented in Fig. 1, where at each node a value is produced by aggregating the values of two children nodes with the given bivariate symmetric averaging function f (denoted by B on the plot and with weights equal to $\frac{1}{2}$). We start from the leaves of the tree which contain the elements of the vector $\mathbf{X}$. In this example we took $w_1 = \frac{5}{8}$ and $w_3 = \frac{3}{8}$. The value y at the root node will be the desired output of the n-variate weighted function.

A straightforward binary tree traversal algorithm for doing so, which starts from the vector $\mathbf{X}$, is as follows:

Aggregation by Levels (ABL) Algorithm

1. Compute $\quad k_1 := \lfloor w_1 2^L + \frac{1}{2} \rfloor, \quad k_2 := 2^L - k_1, \quad$ and $\quad$ create $\quad$ the $\quad$ array $X := (x_1, \ldots, x_1, x_2, \ldots, x_2)$ by taking k_1 copies of x_1 and k_2 copies of x_2;
2. $N := 2^L$;
3. Repeat L times:

   (a) $N := N/2$;
   (b) For $i := 1 \ldots N$ do $X[i] := f(X[2i - 1], X[2i])$;

4. return $X[1]$.

The algorithm is obviously terminating. The runtime of the ABL algorithm is $O(2^L)$, which can make its use prohibitive even for moderate L. Fortunately an efficient algorithm based on pruning the binary tree was presented in [2].

The pruning of the binary tree is done by using the idempotency of f, see Fig. 1, right. Indeed no invocation of f is necessary if both of its arguments are equal. Below

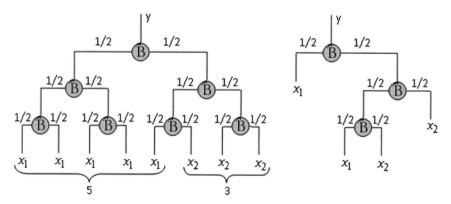

Fig. 1 Representation of a weighted arithmetic mean in a binary tree construction. The tree on the *right* is pruned by using idempotency

we present the pruned tree algorithm whose worst case complexity is $O(L)$, which makes it practically applicable for larger L.

The algorithm is recursive depth-first traversing of the binary tree. A branch is pruned if it is clear that all its leaves have exactly the same value, and by idempotency this is the value of the root node of that branch.

Pruned Tree Aggregation (PTA) Algorithm
function $node(m, N, K, x)$

1. If $N[K] \geq 2^m$ then do:

    (a) $N[K] := N[K] - 2^m$;
    (b) $y := x[K]$;
    (c) If $N[K] = 0$ then $K := K + 1$;
    (d) return y;

    else
2. return $f(node(m - 1, N, K, x), node(m - 1, N, K, x))$.

    function $f\_n(w, x, L)$

1. create the array $N := (k_1, k_2)$ by using
    $k_1 := \lfloor w_1 2^L + \frac{1}{2} \rfloor$, and $k_2 := 2^L - k_1$;
2. $K := 1$;
3. return $node(L, N, K, x)$.

In this algorithm, the array N serves as a counter of how many copies of each of $x[K]$ remains. If there are more than 2^m copies, they belong to a branch that can be pruned, so the function $node$ just returns $x[K]$ and never visits the nodes of that branch. If $N[K] = 1$ then the last remaining copy of $x[K]$ is returned and the value of K is incremented. Every time a branch whose leaves contain identical arguments is encountered (which is detected by the counter $N[K] \geq 2^m$), this branch is pruned.

As an example, consider the binary tree in Fig. 1. Here $L = 3$, $k_1 = 5$ and $k_2 = 3$. In the first call to function $node$ instruction passes to step 2 where $node$ is called recursively twice. In the first recursive call $N[1] = 5 \geq 2^2$ at step 1, hence x_1 is returned and $N[1]$ is set to 1. In the second call to $node$ the instruction goes to step 2, where $node$ is called recursively twice. In the first of those calls the recursion continues until $m = 0$, at which point x_1 is returned and K is incremented. In the subsequent call to $node$ x_2 is returned and then $f(x_1, x_2)$ is computed and returned (the bottom level in the pruned tree in Fig. 1, right). At this point $N[2]$ becomes 2, and in the subsequent call to $node$ step 1 is executed, x_2 is returned and subsequently aggregated in $f(f(x_1, x_2), x_2)$ (middle level of the tree in Fig. 1, right). That last output is aggregated with x_1 at the top level of the tree, and the recursive algorithm terminates, producing the output $y = f(x_1, f(f(x_1, x_2), x_2))$.

To see the complexity of this algorithm note that f is never executed (nor the corresponding node of the tree is visited) if its arguments are the same. There is exactly one node at each level of the tree where the child nodes contain distinct arguments, hence f is executed exactly L times. Also note that both N and K are input-output parameters, so that the two arguments of f at step 2 are different as N and K change from one invocation of the function *node* to another, however the order of execution of the calls to *node* does not matter as the lists of formal parameters are identical.

The ABL and PTL algorithms produce identical outputs but differ in computational complexity. For this reason it may be convenient to formulate (or prove) the results in terms of the complete tree processed by algorithm ABL.

Several useful properties of the binary tree construction were presented in [2]. In particular, the weighted function f_w inherits many properties of the base aggregator f, such as idempotency, monotonicity, continuity, convexity (concavity), homogeneity and shift-invariance, due to preservation of these properties in function composition. Furthermore, when the weights are given in a finite binary representation (as is always the case in machine arithmetic), the sequence of the outputs of the ABL (and hence PTA) algorithm with increasing $L = 2, 3, \ldots$ converges to a weighted mean with the specified weights, and in fact L needs not exceed the number of bits in the mantissa of the weights w_i to match these weights exactly. Finally, when f is a quasi-arithmetic mean, f_w is a weighted quasi-arithmetic mean with the same generator.

Another contribution made in [2, 11] is the extension of the symmetric bivariate means to weighted n-variate means by using essentially the same approach, i.e., by replicating the n inputs a suitable number of times and constructing a binary tree with the desired numbed of levels L. The ABL and PTA algorithms in fact remain the same, safe the definition of the array of multiplicities N which is now n-dimensional.

The big advantage of the binary tree construction is its universality and transparency. It is applicable to any bivariate idempotent function f without modification, and the role of the weights as the respective multiplicities of the arguments as argued in [6] is very clear. The availability of a fast and uncomplicated algorithm for computing the output makes this method immediately applicable.

However the binary tree construction is not suitable for introducing weights into OWA functions, as here we already start with an n-variate function. The approach presented in the next section is an adaptation of the binary tree approach to n-variate OWA functions.

5 Importance Weights in OWA Functions

Our goal here is to incorporate a vector p of non-negative weights (which add to one) into a symmetric n-variate function, by replicating the arguments a suitable number of times. As in the binary tree construction we build an n-ary tree with L levels, as shown in Fig. 2. As the base symmetric aggregator f we take an OWA function OWA_w with specified weights w (although the origins of f are not important for the algorithm).

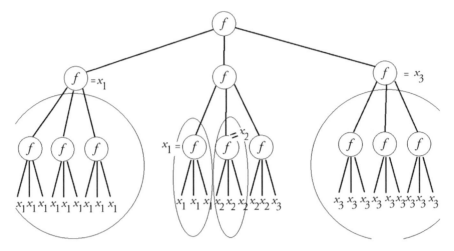

Fig. 2 Representation of a weighted tri-variate function f in a ternary tree construction. The weights are chosen as $\mathbf{p} = (\frac{12}{27}, \frac{5}{27}, \frac{10}{27})$ and $L = 3$. The *circled branches* are pruned by the algorithm

Now, let us create an auxiliary vector $\mathbf{X} = (x_1, \ldots, x_1, x_2, \ldots, x_2, \ldots, x_n, \ldots, x_n)$, so that x_1 is taken k_1 times, x_2 is taken k_2 times, and so on, and $\frac{k_1}{n^L} \approx p_1$, $\frac{k_2}{n^L} \approx p_2$, $\ldots$, and $\sum k_i = n^L$, where $L \geq 1$ is a specified number of levels of the tree shown in Fig. 2. One way of doing so is to take $k_i = \lfloor p_i n^L + \frac{1}{n} \rfloor$, $i = 1, \ldots, n - 1$ and $k_n = n^L - k_1 - k_2 - \cdots - k_{n-1}$.

Pruned n-Tree Aggregation (PnTA) Algorithm
function $node(n, m, N, K, x)$

1. If $N[K] \geq n^m$ then do:

   (a) $N[K] := N[K] - n^m$;
   (b) $y := x[K]$;
   (c) If $N[K] = 0$ then $K := K + 1$;
   (d) return y;

   else
2. for $i := 1, \ldots, n$ do
   $z[i] := node(n, m - 1, N, K, x)$
3. return $f(z)$.

   function $f\_n(n, x, w, p, L)$

1. create the array $N := (k_1, k_2, \ldots, k_n)$ by using
   $k_i := \lfloor p_i n^L + \frac{1}{n} \rfloor$, $i = 1, \ldots, n - 1$, and $k_n := n^L - k_1 - \cdots - k_{n-1}$;
2. $K := 1$;
3. return $node(n, L, N, K, x)$.

The algorithm PnTA works in the same way as the PTA algorithm for binary trees. The vector of counters N helps determine whether there are more than n^m identical elements of the auxiliary array X, in which case they are the leaves of a branch of the tree with m levels. This branch is pruned. The function f is executed only when some of its arguments are distinct, and since the elements of X are ordered, there are at most $n - 1$ such possibilities at each level of the tree, hence the complexity of the algorithm is $O((n - 1)L)$.

Note that the complexity is linear in terms of L, as that of the PTA algorithm, which means that the dimension of the base aggregator f does not matter in this respect. Of course, nominally the n-ary tree is larger than the binary tree, but since we only track the multiplicities of the arguments, never creating the array $\mathbf{X}$ explicitly, memorywise the complexity of the PnTA algorithm is the same as that of PTA.

We also reiterate that the vector $\mathbf{X}$ needs to be sorted, which is equivalent to sorting the inputs $\mathbf{x}$ jointly with the multiplicities of the inputs N (i.e., using the components of $\mathbf{x}$ as the key).

Let us list some useful properties of the function f_p generated by the PnTA algorithm.

Theorem 1 (The Inheritance Theorem) *The weighted extension f_p of a function f by the PnTA algorithm preserves the intrinsic properties of the parent function f as follows:*

1. *f_p idempotent since f is idempotent;*
2. *if f is monotone increasing then f_p is monotone increasing;*
3. *if f is continuous then f_p is continuous;*
4. *if f is convex (resp. concave) then f_p is convex (resp. concave);*
5. *if f is homogeneous then f_p is homogeneous;*
6. *if f is shift-invariant then f_p is shift-invariant;*
7. *f_p has the same absorbing element as f (if any);*
8. *if f generates f_p then a φ-transform of f generates the corresponding φ-transform of f_p.*

Proof The proof easily follows from the properties of composition of the respective functions and idempotency of f. For the φ-transform notice that at each inner level of the tree the composition $\varphi^{-1} \circ \varphi = Id$, while φ is applied to the leaves of the tree and φ^{-1} is applied to the root. □

Now we focus on the OWA functions as the base aggregator f. Here we can show the following.

Theorem 2 *Let $f = OWA_w$. Then the algorithm PnTA generates the weighted function f_p which is the discrete Choquet integral (and is hence homogeneous and shift-invariant).*

Proof The Choquet integral is a piecewise linear continuous aggregation function where the linear pieces are joined together at the intersections of the canonical simplices $\mathscr{S}_i$, i.e., where two or more components of $\mathbf{x}$ are equal. Since OWA_w is continuous and piecewise linear, so is f_p, by the properties of function composition. Now, let us show that the function f_p is not differentiable only on the sets where some of the components of the input vector $\mathbf{x}$ are equal, which will imply that the result is the discrete Choquet integral. Indeed at each node in the n-ary tree there is a function (OWA) not differentiable when some of inputs are equal. At the L-th level of the tree these are the points where the components of $\mathbf{x}$ are equal.

At the level $L - 1$, the arguments of the OWA function are equal if and only if the arguments of the child nodes are equal, because the smallest argument of the left child node is no smaller than the largest argument of the right node (we recall that $\mathbf{X}$ is sorted). Continuing this recursion, we end up with the root node where the resulting function is not differentiable if and only if some of the arguments of the nodes at the bottom level L are equal, which are exactly the components of $\mathbf{x}$. Hence f_p is a piecewise linear, continuous aggregation function, and the linear pieces are joined together at the intersections of $\mathscr{S}_i$, so f_p is the discrete Choquet integral. □

As the special cases of Choquet integral we have the following results.

Theorem 3 *Let $f = OWA_w$. Then the algorithm PnTA generates the weighted function f_p with the following properties:*

1. *for the weights $w_i = \frac{1}{n}$, f_p is the weighted arithmetic mean with the weights $\mathbf{p}$;*
2. *for the weights $p_i = \frac{1}{n}$, f_p is OWA_w;*
3. *when $f = OWA_w = \min$ (or $= \max$) and $p_i > 0$ for all i, f_p is also $\min$ (respectively, $\max$);*
4. *when $f = OWA_w = $ median and n is odd, f_p is the weighted median;*
5. *if OWA_w generates f_p, then the dual OWA_w^d generates the dual f_p^d, and in particular an OWA with the reverse weights generates the respective weighted OWA with the reverse weights.*

Proof 1. In this case OWA_w is the arithmetic mean, and hence f_p is the weighted arithmetic mean with the respective weights.

2. In this case, each argument x_i is repeated exactly n^L times, hence the inputs to each node of the n-ary tree (except the root node) are all equal, and by idempotency the tree is pruned to just one level, and hence delivers the original OWA_w function.

3. The function at each node in the tree returns the minimum (maximum) of its arguments, hence the result does not depend of the weights if they are strictly positive. However, when the weights p_i could be 0, the result is not true, as the smallest (largest) component of $\mathbf{x}$ can be excluded from the calculation of the minimum (maximum). Note that f_p is not a weighted minimum or weighted maximum functions, as those are the instances of the Sugeno and not Choquet integral.

4. The weighted median can be written as the median of a vector with the components repeated the relevant number of times, i.e., $median(\mathbf{X})$ [4], p. 121. While in general the median of medians of subsets of inputs is not the median of the whole set of inputs, for an odd n at each level of the tree the median is the value of the central child node, which in turn is the value of its central child node, and so on until

the bottom level where we get the centralof $\mathbf{X}$; see Fig. 2 for an illustration. This statement does not work for an even n if we consider the lower (respectively, upper) medians, because now we need to take the value of the right middle child node at every level, but the lower median of $\mathbf{X}$ sorted in the decreasing order corresponds to $X_{n^L/2+1}$, which happens to be the left child of its parent node. This is clearly seen in the case of $n = 2$ where the lower median of the bivariate function f is the minimum of its arguments, hence $f_p(\mathbf{x}) = \min(\mathbf{X})$ which clearly does not always coincide with the median of $\mathbf{X}$. For example, in the binary tree in Fig. 1 $median(\mathbf{X}) = X_5 = x_1$ whereas $f_p(x_1, x_2) = X_8 = x_2$.

5. Follows from the preservation of φ-transform. $\qquad\qquad\square$

Theorem 4 *Let* $f = OWA_w$ *and let the weighting vector be decreasing (increasing). Then the algorithm PnTA generates a Choquet integral with respect to a submodular (supermodular) fuzzy measure.*

Proof An OWA function with decreasing weights is convex (respectively concave for increasing weights), and hence is a special case of the Choquet integral with respect to a submodular (supermodular) fuzzy measure [3]. Since the convexity (concavity) is preserved in the n-ary tree construction as per Theorem 1, the resulting weighted function f_p is also convex (concave), and hence it is a Choquet integral with respect to a submodular (supermodular) fuzzy measure [3]. $\qquad\square$

This result is useful when constructing weighted norms from OWA with decreasing weights, see [3, 23].

On the technical size we note that we do not need to sort the arguments in each OWA function in the n-ary tree, as the vector $\mathbf{x}$ is already sorted, hence only one sort operation for the inputs is required. Another note is that when the weights $\mathbf{p}$ are specified to m digits in base n, $L = m$ levels of the n-ary tree is sufficient to match these weighs exactly. For example if $\mathbf{p}$ are specified to 3 decimal places and $n = 10$, we only need to take $L = 3$. Therefore to match the weights to machine precision (e.g., 53 bits for data type `double`) n^L need not exceed the largest 64-bit integer, and hence the algorithm PnTA can be implemented with 64-bit data types. The source code in C++ is presented in Fig. 3.

Finally by using Definition 8 we can introduce weights into generalized OWA functions in the same was as for OWA functions, by using the n-ary tree construction. This can be done in two ways: a) by using $GenOWA_{\mathbf{w},g}$ function as f, or b) by using a φ-transform of a weighted OWA with $\varphi = g$, that is, by applying g and g^{-1} only to the leaves and to the root node of the tree, relying on the preservation of φ-transforms. The second method is computationally more efficient as functions g and g^{-1} need not be used in the middle of the tree, where they cancel each other.

This way we also obtain the special cases of weighted OWG and weighted power based generalized OWA functions.

```
double OWA(int n, double x[],double w[])
{ /* no sorting is needed when used in the tree */
    double z=0;
    for(int i=0;i<n;i++) z+=x[i]*w[i];
    return z;
}
double node(int n,double x[],long int N[],long int C,int & k,
      double w[],double(*F)(int,double [],double[]),double* z)
{
    /* recursive function in the n-ary tree processing
    Parameters: x input vector, N vector of multiplicities of x
    m current level of recursion counted from root node L to 0
    k input-output parameter, the index of x being processed */
    if(N[k]==0) k++;
    if(N[k]>= C) {  /* we use idempotency to prune the tree */
        N[k] -= C;
        if(N[k]<=0) return x[k++]; else return x[k];
    }
     C /= n;
    /* tree not pruned, process the children nodes */
    for(int i=0;i<n;i++) z[i]=node(n,x,N,C,k,w,F,z+n);
    return F(n,z,w);
}

double weightedf(double x[], double p[], double w[], int n,
         double(*F)(int, double[],double[]), int L)
/*
 Function F is the symmetric base aggregator.
 p[] = array of weights of inputs x[],
 w[] = array of weights for OWA, n = the dimension of x, p, w.
 the weights must add to one and be non-negative.
 L = number of binary tree levels. Run time = O[(n-1)L]   */
{
    long int t=0, C=1;
    int k=0;
    for(int i=0;i<L;i++) C*=n;   /* C=n^L */
    sortpairs(x, x+n, p);
    long int N[n];/* multiplicities of x based on the weights*/
    for(int i=0;i<n-1;i++)  { N[i]=p[i]*C+1./n;   t+=N[i]; }
    N[n-1]=C-t;
    double z[n*L]; /* working memory */
    return node(n,x,N,C,k,w,F,z);
}
/* example: calling the function */
int n=4, L=4;
double x[4]={0.2,0.2,0.4,0.8};
double w[4]={0.1,0.2,0.3,0.4};
double p[4]={0.3,0.2,0.1,0.4};
double y=weightedf(x,p,w,n,&OWA,L);
```

Fig. 3 A C++ implementation of the pruned n-ary tree algorithm PnTA. The function `sortpairs` (not shown) implements sorting of an array of pairs (x_i, p_i) in the order of decreasing x_i

6 Conclusions

The proposed method of introducing weights into n-ary symmetric functions has several advantages. Firstly, it is a generic method universally applicable to any symmetric idempotent function, in particular to OWA functions. Secondly, the handling of the weights is transparent and intuitive: the weights correspond to the multiplicities of the arguments. Thirdly, many important properties of the base symmetric aggregator are preserved in the n-ary tree construction, which is very useful as these properties need to be verified only for the base aggregator. Finally, the pruned n-ary tree algorithm delivers a numerically efficient way of calculating weighed averages, among them the weighted OWA. This algorithm has complexity linear in n and the number of levels of the tree L, and L is bounded by the desired accuracy of the weights. We believe the n-ary tree algorithm constitutes a very competitive alternative to the existing weighted OWA approaches.

References

1. G. Beliakov. Shape preserving splines in constructing WOWA operators: Comment on paper by V. Torra in Fuzzy Sets and Systems 113 (2000) 389–396. *Fuzzy Sets and Systems*, 121:549–550, 2001.
2. G. Beliakov and J.J. Dujmovic. Extension of bivariate means to weighted means of several arguments by using binary trees. *Information Sciences* 331:137–147, 2016.
3. G. Beliakov, S. James, and G. Li. Learning Choquet-integral-based metrics for semisupervised clustering. *IEEE Trans. on Fuzzy Systems*, 19:562–574, 2011.
4. G. Beliakov, A. Pradera, and T. Calvo. *Aggregation Functions: A Guide for Practitioners*, volume 221 of *Studies in Fuzziness and Soft Computing*. Springer-Verlag, Berlin, 2007.
5. P.S. Bullen. *Handbook of Means and Their Inequalities*. Kluwer, Dordrecht, 2003.
6. T. Calvo, R. Mesiar, and R.R. Yager. Quantitative weights and aggregation. *IEEE Trans. on Fuzzy Systems*, 12:62–69, 2004.
7. B. De Baets and J. Fodor. Van Melle's combining function in MYCIN is a representable uninorm: An alternative proof. *Fuzzy Sets and Systems*, 104:133–136, 1999.
8. J.J. Dujmovic. Continuous preference logic for system evaluation. *IEEE Trans. on Fuzzy Systems*, 15:1082–1099, 2007.
9. J.J. Dujmovic. An efficient algorithm for general weighted aggregation. In *Proc. of the 8th AGOP Summer School*, Katowice, Poland, 2015.
10. J.J. Dujmovic. Weighted compensative logic with adjustable threshold andness and orness. *IEEE Trans. on Fuzzy Systems*, 23:270–290, 2015.
11. J.J. Dujmovic and G. Beliakov. Idempotent weighted aggregation based on binary aggregation trees. *International Journal of Intelligent Systems*, 32:31–50, 2017.
12. J.J. Dujmovic and H.L. Larsen. Generalized conjunction/disjunction. *Int. J. Approx. Reasoning*, 46:423–446, 2007.
13. A. Emrouznejad and M. Marra. Ordered weighted averaging operators 1988-2014: A citation-based literature survey. *Int. J. Intelligent Systems*, 29:994–1014, 2014.
14. M. Grabisch, J.-L. Marichal, R. Mesiar, and E. Pap. *Aggregation Functions*. Encyclopedia of Mathematics and Its Foundations. Cambridge University Press, 2009.
15. M. Grabisch, T. Murofushi, and M. Sugeno, editors. *Fuzzy Measures and Integrals. Theory and Applications*. Physica – Verlag, Heidelberg, 2000.

16. F. Herrera and E. Herrera-Viedma. A study of the origin and uses of the Ordered Weighted Geometric operator in multicriteria decision making. *Int. J. Intelligent Systems*, 18:689–707, 2003.
17. E.P. Klement, R. Mesiar, and E. Pap. *Triangular Norms*. Kluwer, Dordrecht, 2000.
18. Y. Narukawa and V. Torra. Fuzzy measure and probability distributions: Distorted probabilities. *IEEE Trans. on Fuzzy Systems*, 13:617–629, 2005.
19. U. Thole, H.-J. Zimmermann, and P. Zysno. On the suitability of minimum and product operators for the intersection of fuzzy sets. *Fuzzy Sets and Systems*, 2:167–180, 1979.
20. V. Torra. The weighted OWA operator. *Int. J. Intelligent Systems*, 12:153–166, 1997.
21. V. Torra. The WOWA operator and the interpolation function W*: Chen and Otto's interpolation revisited. *Fuzzy Sets and Systems*, 113:389–396, 2000.
22. Z.S. Xu and Q.L. Da. The ordered weighted geometric averaging operator. *Int. J. Intelligent Systems*, 17:709–716, 2002.
23. R. R. Yager. Norms induced from OWA operators. *IEEE Trans. on Fuzzy Systems*, 18(1):57–66, 2010.
24. R.R. Yager. On ordered weighted averaging aggregation operators in multicriteria decision making. *IEEE Trans. on Systems, Man and Cybernetics*, 18:183–190, 1988.
25. R.R. Yager. Including importances in OWA aggregations using fuzzy systems modeling. *IEEE Trans. on Fuzzy Systems*, 6:286–294, 1998.
26. R.R. Yager. Uninorms in fuzzy systems modeling. *Fuzzy Sets and Systems*, 122:167–175, 2001.
27. R.R. Yager. Generalized OWA aggregation operators. *Fuzzy Optimization and Decision Making*, 3:93–107, 2004.
28. R.R. Yager and J. Kacprzyk, editors. *The Ordered Weighted Averaging Operators. Theory and Applications*. Kluwer, Boston, 1997.
29. R.R. Yager, J. Kacprzyk, and G. Beliakov, editors. *Recent Developments in the Ordered Weighted Averaging Operators: Theory and Practice*. Springer, Berlin, New York, 2011.
30. L. Zadeh. Fuzzy sets. *Information and Control*, 8:338–353, 1965.
31. H.-J. Zimmermann and P. Zysno. Latent connectives in human decision making. *Fuzzy Sets and Systems*, 4:37–51, 1980.

Uncertainty Management: Probability, Possibility, Entropy, and Other Paradigms

Bernadette Bouchon-Meunier

Abstract Uncertainty modeling is a domain explored by researchers for centuries and it is difficult to bring stones to this long history of thought. Boris Kovalerchuk has been concerned for many years with specificities, relations and complementarities of the main paradigms enabling the construction of automatic systems able to handle imperfect information in a real-world environment, mainly probability theory and fuzzy set theory. We point out the necessity to cope with several aspects of uncertainty apparent in complex systems, mainly due to the complexity of natural phenomena and, more recently, to the size and diversity of artifacts, in addition to the necessity to take observers of the phenomena into account.

Keywords Uncertainty · Fuzzy sets · Probability · Complexity · Entropy · Perception · Subjectivity · Imprecision · Incomplete information

1 Introduction

Uncertainty modeling is a domain explored by researchers for centuries and it is difficult to bring stones to this long history of thought. After the preeminence of probability theory until the 1960s, the necessity to built automatic systems able to handle imperfect information in a real-world environment drove to the emergence of new paradigms, among which fuzzy set and possibility theories have led the field, mainly because of their efficiency to cope with the complexity of real-world problems. These theories are surrounded by evidence theory, imprecise probabilities, interval methods, non-classical logics, to name but a few among the most important paradigms to represent uncertainty and imperfect information. Boris Kovalerchuk has been concerned for many years with specificities of several of these paradigms, their relations and their complementarities and he has keenly explored solutions

B. Bouchon-Meunier (✉)
Sorbonne Universités, UPMC Univ Paris 06, UMR 7606, LIP6, 75005 Paris, France
e-mail: bernadette.bouchon-meunier@lip6.fr

B. Bouchon-Meunier
CNRS, UMR 7606, LIP6, 75005 Paris, France

© Springer International Publishing AG 2017
V. Kreinovich (ed.), *Uncertainty Modeling*, Studies in Computational
Intelligence 683, DOI 10.1007/978-3-319-51052-1_4

to make them collaborate, in particular in the Computing With Words approach [2, 13, 14]. In this paper, we point out the necessity to cope with several aspects of uncertainty underlying complex systems, mainly due to the complexity of natural phenomena and, more recently, to the size and diversity of artifacts, in addition to the necessity to take observers of the phenomena into account. We can consider that the imperfection of available real-world information takes three main forms bringing uncertainty, the first one being a doubt on the outcome of an experiment or in the forecasting of a system state. The second one is incompleteness of information, which entails uncertainty on unknown aspects of the studied phenomenon. The third one is imprecision and vagueness resulting either from the observer, due to natural ability of humans to cope with imprecise concepts and values or to inaccuracies of observation tools; imprecision and vagueness also exist in nature and provide uncertainty on precise values or states.

In Sect. 2, we insert the question of uncertainty management in approaches to complexity, mainly proposed by E. Morin and J.-L. Le Moigne. In Sect. 3, we review concepts of entropy as methods to deal with uncertainty, showing that several approaches are available, according to the level of complexity of the observed system we take into account. In Sect. 4, we propose to differentiate several dimensions of uncertainty to show that they must work together and adapt to fit the requirements of any given real-world problem.

2   Complexity Intelligence

In a restricted view, complexity intelligence could be regarded as an approach to deal with the present complexity of available data, due to the large amount and the heterogeneity of information available in a digital environment in the so-called big data paradigm. This aspect of complexity intelligence corresponds to the use of artificial intelligence approaches and, in particular, the efficiency of computational intelligence methods, to deal with large amounts of data, to explore them, to extract relevant information, to discover patterns, to detect exceptional elements, outliers or weak signals, in order to support decisions and to predict risk. This aspect is certainly a major one in the modern environment, but it does not cover the whole concept of complexity intelligence advocated by E. Morin [16] and J.-L. Le Moigne [15].

They consider that, in its general form, complexity intelligence refers to the impossibility to separate the observed world from the observer, and the necessity to take into account perception and context in the analysis of complex phenomena. Natural intelligence is then involved in the handling of complexity inherent in all systems, in the construction of models and the communication of knowledge. According to E. Morin, consciousness of complexity is necessary to understand a phenomenon and to generate science. Progresses made by humans on scientific certainties induce progresses on their uncertainties and recognition of their ignorance associated with their knowledge. E. Morin claims that principles of incompleteness and uncertainty are at the heart of complexity intelligence, even though it is based on a kind of multidimensional thinking tending to take into account all aspects of the observed phenomenon.

In [15], the authors point out the necessity for Science to be conscious of its own complexity, impossible to avoid because of relations between humans and the observed world, be it natural or artifactual. They recommend that science of complexity produces concepts and theories supporting intelligibility and being in the realm of the possible, rather than the necessary. They claim that, before being conscious of the need to approach complexity through intelligence, perception and understandability, science was based on four "pillars of certainty" tending to eliminate complexity and to introduce simplicity in order to grasp complex phenomena. E. Morin [15] considers that the first pillar of certainty is the absolute order, considering that a strict ordering governs everything in the universe and disorder can only be the consequence of a lack of knowledge. The second pillar of certainty is the separability principle, associated with the capability to split any problem into simple elements, forgetting the whole and the existing relations between parts of the system. The third pillar of certainty is the reduction principle, reducing knowledge to measurable quantities. The fourth pillar of certainty is inductive/deductive logic, denying concepts like creation, abduction and hypothesis-based reasoning.

E. Morin insists on the need to take into account qualitative, as well as quantitative effects and to accept contradictions in observations. Uncertainty is unavoidable, as is complexity inherent in the real world, and it takes various forms, such as "logical uncertainty", "empirical uncertainty", or "cognitive uncertainty" referring to human mental categories. E. Morin and Le Moigne's vision is embedded in system science, the foundations of which have been laid by Ludwig von Bertalanffy in the fifties, at the same time as cybernetics was emerging, disseminating the concept of entropy. It is worth noting that their work is related to the so-called second-order cybernetics, invented by H. von Foerster [1, 17] and regarded as the cybernetics of observing systems, as opposed to the cybernetics of observed systems. The observer is clearly involved in the analysis of complexity and the management of uncertainty, involving perception and taking into account context and environment of the system.

Such an approach leads to search for solutions to deal with complexity through the representation of incomplete information, the preservation of possible situations even in the case where they seem incompatible and the capacity to revise and update information in non-classical logics. Furthermore, it appears that the world complexity cannot be grasped by means of the only utilization of probabilities to cope with so diverse uncertainties. We can think of L.A. Zadeh's search for the involvement of perception in the automated management of information, and in particular his perception-based theory of probabilistic reasoning [20] or a computational theory of perception [19] adding to classic probability theory or predicate logic the capacity to compute and reason with perception-based information.

More generally, a fuzzy set-based knowledge representation following L.A. Zadeh's seminal work and the associated possibility theory provide solutions to escape the four above-mentioned pillars of certainty. Among fuzzy relations, fuzzy orders, similarity and indistinguishability relations provide methods to avoid absolute orders. Partial membership, fuzzy partitions, overlapping classes enable to weaken the separability principle, according to human capabilities of considering unsharp categories and to accept contradictions to some extent. Fuzzy sets, providing a numerical/symbolic interface, are source of expressiveness and linguistic summarization of numerical data can be achieved, as an alternative proposal to reduction principle. Finally, approximate reasoning, general modus ponens, fuzzy inductive reasoning, fuzzy abduction and case-based reasoning, are solutions to avoid the rigidity of deduction in classical logic, when suitable.

3 Entropy

After means of representing uncertainty reviewed in Sect. 2, we can consider means of evaluating uncertainty, which are mainly studied in information theory. We can establish a parallel between these two paradigms and their history.

The previous proposals to deal with complexity go beyond traditional knowledge representation methods, by means of the involvement of the observer and his perceptions in the representation of uncertainty in complex systems. They can be compared to novel information theory that also appeared in the 1960s, after the first concept of entropy proposed by C. Shannon in coding theory and N. Wiener at the origin of cybernetics, in 1948. This concept was dedicated to communication and information transmission and was considered as syntactical, with no semantic value. This is the reason why a more general theory of information was proposed by J. Kampé de Fériet and B. Forte [10] as a theoretical approach to the concept of information, based on axioms [8] not necessarily dependent on probabilities and taking an observer into account. We see again, as we did in E. Morin's approach to uncertain information representation, that observers take part in the evaluation of uncertain knowledge, with a role assigned to a "headquarter" [11] in the case where there exist several observers. The objective of the authors is to accept subjectivity in information.

Another extension of the original concept of entropy has been simultaneously proposed in [3, 9] to take into account qualitative aspects of information related to its utility for the fulfillment of a goal, considering again the context of the observed phenomena. A fresh look at the concept of entropy was taken by De Luca and Termini [7] in their definition of a non-probabilistic entropy regarded as a measure of a quantity of information not necessarily related to random experiments, provided by a different kind of uncertainty based on a fuzzy set-based knowledge representation.

Such works were the presages of a long list of extensions of the concept of entropy [12] supposed to go along with various uncertain knowledge representation, such as intuitionistic entropy [6], or entropy in a mathematical theory of evidence [18]. The reasons why the introduced quantities are called entropy, as well as their properties, are diverse [5], but in any case they are assumed to evaluate a degree of imperfection taking a part in some uncertainty inherent in an observed system, whatever the chosen means of representing uncertainty are.

4 Dimensions of Uncertainty

Introduced and developed in the 17th century by B. Pascal, P. de Fermat and J. Bernoulli, probability has been the main mathematical concept to represent uncertainty during three centuries. Subjective probabilities were a first attempt to soften this concept in order to take into account the observer, to some extent, and they were proposed by F. Ramsey and B. de Finetti in the 1930s. Another alternative to classic probabilities was introduced by A. Dempster in 1967 to encompass a higher form of uncertainty, regarded as uncertainty on degrees of uncertainty, by means of upper and lower probabilities, then developed by G. Shafer in 1976 as evidence theory. All these works were addressing the problem of *doubt and uncertainty about the outcome* of an experiment or the state of a phenomenon [2], which can be regarded as the first level of uncertainty in complex systems. Examples of such intrinsic uncertainty are "Peter may attend the meeting today" or "Mr. X will probably be elected by the assembly".

A second level of uncertainty corresponds to a *doubt resulting from an imprecision, an approximation, or from incomplete knowledge*. For instance, "I will be at the railway station around noon" is based on imprecise information and entails an uncertainty on my exact arrival time and the fact that I can catch a train leaving at noon. This level of uncertainty was first tackled by L.A. Zadeh in 1965 in his seminal paper on fuzzy sets, and also by Ramon Moore in 1966 in his book on interval analysis addressing a narrower aspect of imprecise information. It is only in the framework of fuzzy set theory that the induced uncertainty itself is represented by means of possibilities, introduced by L.A. Zadeh in 1978.

A third level of uncertainty can be identified, corresponding to a *doubt due to a subjective appreciation or a judgment* expressed by an individual. For instance, "I don't believe that Peter is in Paris" or "I am not sure that Mr. X will be elected by the assembly". Such appreciation can autonomously rate the doubt on a fact or be regarded as some meta-uncertainty added to an uncertainty of the first or second level. In addition to possibilistic logic developed in the environment of possibility theory, extensive work has been published on non-classical logics such as modal logics since the 1960s, nonmonotonic reasoning to cope with evolving information and its revision since 1980, Truth Maintenance Systems introduced by J. Doyle in 1979 to enable the revision of beliefs, autoepistemic logics introduced by Robert Moore in 1983 to take into account partial information with semantical considerations, probabilistic logic introduced by N.J. Nilsson in 1986.

This list is not exhaustive, as the development of non-classical logics has been intense, but our purpose is to point out the existence of solutions to cope with uncertainty, answering E. Morin and J.-L. Le Moigne's concerns and accepting uncertain information instead of eliminating it, considering the need to revise beliefs and to take context and observer into account in the representation of complex situations.

This typology of levels of uncertainty can be completed by a view of uncertain information representation methods along three dimensions according to the nature of the addressed uncertainty [2], with the aim of choosing an appropriate method when facing a real-world problem.

The first dimension refers to the distinction between *numerical and symbolic information*: probabilities or masses of assignment in evidence theory are examples of methods to deal with numerical information ("Mr. X is expected to be elected with 62% of the votes"), while modal logics manage symbolic uncertain information ("I believe that Mr. X will be elected"). Fuzzy set-based methods and the associated possibility theory are intermediate between these kinds of information, coping with numerical data by means of a symbolic representation, while equipping knowledge representation with capabilities of interpretability and expressiveness ("Mr. X is expected to be elected with a large majority of the votes").

The second dimension of uncertain information corresponds to the distinction between *intrinsic and extrinsic uncertainty*, the first one being attached to the real world phenomenon, the second one being due to the process of observation itself. Interval analysis can be used to handle extrinsic uncertainty deriving for instance from measurement errors ("within 10% of the measured value"), while fuzzy classes can also address various cases of intrinsic uncertainty, such as the contours of regions in a digital image or categories with blurred boundaries like "young" and "old".

The third dimension of uncertain information in a complex world indicates a gradual degree of *subjectivity*, from an objective measurement to a subjective evaluation [4], from physical properties of an object to its perceived properties, for instance from the numerical representation of colors from red ("wavelength between 620–700 nm", "RGB: 255, 36, 0") in a digital image to the perception of a dominant color by a human agent ("red"), and furthermore to a subjective appreciation, such as a feeling or an emotion ("passion", "energy").

It is worth noting that these dimensions of uncertain information do not generally appear independently and representation methods can be regarded in this three dimensional space. We give a few examples of elements in this space.

The most classic uncertainty representation regards numerical intrinsic uncertainty, dealt with by means of objective or subjective probabilities, according to the importance attached to subjectivity.

Examples of objective symbolic intrinsic uncertainty can be associated with imprecise descriptions of variables managed through fuzzy sets in the case of objective and measured data. For instance [2], in the case of descriptions of spots on a mammography, qualified as "round" by a medical doctor, an uncertainty comes from the imprecise description, since "round" is only a linguistic approximation of the mathematical characterization "circular", and the membership function is obtained in a machine learning-based process from measurements of various criteria such as convexity or elongation, performed during the image processing.

Objective numerical extrinsic uncertainty is present in sensor imprecision or in estimations. Interval analysis, fuzzy sets, or confidence intervals can be used to deal with it.

A subjective intrinsic symbolic uncertainty can be identified in felt probabilities (such as "highly probable"), which can be represented by fuzzy values, whose membership functions can be obtained by means of a psychometric approach.

Subjective extrinsic numerical uncertainty is identified in web-based information quality scoring, and it can be managed by means of possibility theory or evidence theory, for instance.

A symbolic extrinsic source of uncertainty is related to the difficulty to characterize a given complex phenomenon, solved by means of linguistic expressions. An example of objective such uncertainty appears when it is impossible to obtain precise values from an observer, for instance evaluating a distance ("far from the house"), this information being easily represented through fuzzy sets. The case of a subjective symbolic extrinsic uncertainty is observed when the observer expresses a doubt on the validity of data ("I believe these news"), and modal logic is one of the candidates to cope with it.

In real-world applications, the selection of an uncertainty modeling technique relies on the nature of uncertainty present in the problem to solve or the phenomenon to observe. It is also oriented by the necessity to obtain an expressible result and/or a numerical coefficient or mark expressed by users. Let us remark that, in some situations, it may be interesting to preserve the uncertainty associated with the outcomes of a system and to enable the user to use his/her expertise to make a final decision. In other words, it may be important to know that two different events may occur and to be prepared to both of them, which can be achieved by means of possibility theory or fuzzy logic, rather than to look for a deterministic decision. In addition, it is often interesting to use several uncertainty modeling techniques simultaneously in a given environment, for instance probabilities and fuzzy sets, which is possible in the so-called Soft Computing paradigm.

5 Conclusion

We have presented a prism to study uncertainty in an operational environment, with the aim of helping users to appropriately choose a knowledge representation method. We have situated this study in the analysis of complex systems, in which uncertainty is fundamentally inherent. We have shown three levels and three dimensions of uncertainty and we claim that it is impossible to reduce one form of uncertainty to another one to deal with complex systems.

References

1. Andreewsky E., Delorme, R. (eds.), *Seconded cybernétique et complexité – Rencontres avec Heinz von Foerster*, L'Harmattan, Paris (2006)
2. Beliakov, G., Bouchon-Meunier, B., Kacprzyk, J., Kovalerchuk, B., Kreinovich, V., Mendel, J., Computing With Words (CWW): role of fuzzy, probability and measurement concepts, and operations, in: *Mathware and Soft Computing*, 19–2, 27–45 (2012)
3. Belis, M., S. Guiasu, A quantitative-qualitative measure of information in cybernetic systems (Corresp.), *IEEE Transations on Information Theory*, 14(4), 593–594 (1968)

4. Bouchon-Meunier, B., Lesot, M.-J., Marsala, C., Modeling and management of subjective information in a fuzzy setting, *International Journal of General Systems*, vol. 42(1), pp. 3–19 (2013)
5. Bernadette Bouchon-Meunier, Christophe Marsala. Entropy measures and views of information. in J. Kacprzyk, D. Filev, G. Beliakov (eds.) Granular, Soft and Fuzzy Approaches for Intelligent Systems, 344, Springer, 2016, Studies in Fuzziness and Soft Computing, pp. 47–63
6. Burillo, P., Bustince, H.: Entropy on intuitionistic fuzzy sets and on interval-valued fuzzy sets, *Fuzzy Sets and Systems* 78, 305–316 (1996)
7. De Luca, A., Termini, S.: A definition of a nonprobabilistic entropy in the setting of fuzzy sets theory. *Information and Control* 20, 301–312 (1972)
8. Forte, B., Measures of information: the general axiomatic theory. *Revue française d'informatique et de recherche opérationnelle, série rouge*, 3(2), 63–89 (1969)
9. Guiasu, S.: Weighted entropy, *Reports on Mathematical Physics* 2(3), 165–179 (1971)
10. Kampé de Fériet, J., Forte, B. (1967). Information et probabilité. *Comptes Rendus Hebdomadaires des Séances de l'Académie des Sciences, Série A*, 265(12), 350, Gauthier-Villars (1967)
11. Kampé de Fériet, J., La théorie généralisée de l'information et la mesure subjective de l'information. In *Theories de l'information*, J. Kampé de Fériet, C.F. Picard (eds.) 1–35, Springer-Verlag (1974)
12. Klir, G., Wierman, M.J., *Uncertainty-Based Information. Elements of Generalized Information Theory*. Studies in Fuzziness and Soft Computing. Springer-Verlag (1998)
13. Kovalerchuk B., Quest for Rigorous Combining Probabilistic and Fuzzy Logic Approaches for Computing with Words, In: R. Seising, E. Trillas, C. Moraga, S. Termini (eds.): *On Fuzziness. A Homage to Lotfi A. Zadeh* (Studies in Fuzziness and Soft Computing Vol. 216), Berlin, New York: Springer 2013. Vol. 1. pp. 333–344
14. Kovalerchuk, B., Probabilistic Solution of Zadeh's test problems, In: A. Laurent et al. (Eds.): *Proc. IPMU 2014*, Part II, CCIS 443, pp. 536–545, 2014, Springer
15. Morin, E., Le Moigne, J. L., *L'intelligence de la complexité*, L'Harmattan, Paris (1999)
16. Morin E., *Science avec conscience*, Librairie Arthème Fayard, 1982; Paris: Seuil (1990)
17. Von Foerster, H., *Understanding Understanding, Essays on Cybernetics and Cognition*, Springer (2003)
18. Yager, R.R.: Entropy and specificity in a mathematical theory of evidence. In: R. Yager, L. Liu, (eds.), *Classic Works of the Dempster-Shafer Theory of Belief Functions*, Studies in Fuzziness and Soft Computing, vol. 219, pp. 291–310. Springer Berlin Heidelberg (2008)
19. Zadeh, L.A., A New Direction in AI, Toward a Computational Theory of Perceptions, *AI Magazine*, Volume 22, Number 1 pp. 73–84 (2001)
20. Zadeh, L.A., Toward a perception-based theory of probabilistic reasoning with imprecise probabilities, In: *Intelligent Systems for Information Processing, from representation to applications*, B. Bouchon-Meunier, L. Foulloy, R.R. Yager (eds.), Elsevier, pp. 3–34 (2003)

Relationships Between Fuzziness, Partial Truth, and Probability in the Case of Repetitive Events

Jozo Dujmović

Abstract Frequency-based probabilistic models are suitable for the quantitative characterization and analysis of repetitive events. Quantitative models based on the concept of fuzzy set can be applied both in the case of repetitive events and in cases where frequency-based probabilistic models are not appropriate. In the case of repetitive events there is a possibility of comparison of relationships between probabilistic and fuzziness-based interpretations of the same physical or perceptual reality. In this paper we analyze these relationships using three characteristic examples and show that in the case of repetitive events, fuzziness, probability and partial truth are three coexisting compatible interpretations of the same reality, i.e. practically equivalent concepts.

1 Introduction

Relationships between fuzziness and probability have a long history of controversial opinions and hot discussions. The latest contribution to this area can be found in [6, 8]. In spring 2014, BISC community (Berkeley Initiative in Soft Computing) was a forum for exchanging a variety of opinions about fuzziness, probability, possibility and partial truth, as formalisms for dealing with uncertainty. This paper includes material that was initially used as author's contribution to that discussion.

Repetitive events are events that have predecessors and successors. Any form of life creates repetitive events. Some sequences of events have the first and the last event in the sequence, but the nature of sequence is still repetitive. Even the first event in the sequence can frequently be interpreted as a repetition of the first event in some similar previous sequence that consisted of similar events. Since the repetitive events are defined using predecessors and successors, singular events should be defined as events without predecessor and successor. Taking into account that predecessors and successors only need to be sufficiently similar to an analyzed event, it seems

J. Dujmović (✉)
Department of Computer Science, San Francisco State University,
1600 Holloway Avenue, San Francisco, CA 94132, USA
e-mail: jozo@sfsu.edu

© Springer International Publishing AG 2017

V. Kreinovich (ed.), *Uncertainty Modeling*, Studies in Computational Intelligence 683, DOI 10.1007/978-3-319-51052-1_5

to be extremely difficult to find truly singular events. Even if such events can be identified, it is very easy to show that they are a negligible minority compared to repetitive events. Therefore, whenever in this paper we speak about probability and probabilistic models we always assume frequency-based probability and frequency-based probabilistic models derived from observation of repetitive events.

Most of human experiences are repetitive. For example, human work in any profession is essentially a repetitive experience that consists of many days spent working with similar people, offering similar services, and solving similar problems. Regular use of machines and tools (e.g. cars and computers), by both individuals and corporate users creates repetitive events where some experiences are more positive (machines provided satisfactory service) and some are less positive (performance of machines was not sufficiently satisfactory). Human physical and other properties (e.g. weight, height) are also repetitive and can be analyzed using data about selected populations of human subjects.

It is useful to note that some events are only seemingly singular, i.e. they are interpreted with intention to be classified as singular. A favorite recent example is the possible election of a female president in a country that had a long sequence of male presidents. Of course, one could argue that this is a singular event, but it would be much easier to see that the election of female president follows the same rules as the election of all previous male presidents, and that such a "singular event" already occurred many times in many countries that had a sequence of male presidents followed by a female president. Interpreting events as repetitive is much easier than interpreting them as singular.

Generally, repetitive experiences are neither identical nor certain—they always vary in a specific range and create human perceptions of satisfaction, suitability, quality, value, etc. In such cases some properties are objectively measurable (e.g. the height of a car driver), but much more frequently we have to deal with human percepts that are not measurable. Of course, human perceptions can be modeled using formal models, and such models create the areas of computing with words [9], perceptual computing [7], and aggregation logic [1, 5].

Uncertainty is a human property that primarily reflects a lack of information and human inability to accurately predict future events. Formalisms for dealing with uncertainty are developed with general intention to reduce uncertainty and help in precision of meaning and better describing perceptions in a natural language.

The frequency-based probabilistic approach to reducing uncertainty consists of observing and recording frequencies of past events, and then using them to predict the likelihood of similar future events (e.g. to determine, with a given degree of confidence, intervals where events are likely to occur). Fuzzy approach is based on reducing uncertainty by using fuzzy set membership functions where high degree of membership reflects a high certainty that an object satisfies conditions that define members of a specific fuzzy set. Logic approach to reducing uncertainty consists of computing a degree of truth of assertion that an object has specific properties [2]. If the degree of truth is high, then the certainty of assertion is also high, contributing to precision and reducing the uncertainty. In this paper, our goal is to use characteristic examples to show that in the case of repetitive events fuzzy approach, probabilistic approach, and soft computing logic approach are similar and very frequently equivalent.

2 The Case of Tall Drivers

Let us consider a set of n licensed car drivers (in the USA $n > 200 \times 10^6$). Let h denote the height of a car driver (written in each driver license), and let us sort car drivers according to their height:

$$h_{min} = h_1 < h_2 < \cdots < h_n = h_{max}.$$

The heights are real numbers, and for simplicity we assume that all heights are different. The function $P : [h_{min}, h_{max}] \rightarrow [0, 1]$ denotes the fraction of remaining $n - 1$ drivers that are shorter than the i-th driver:

$$P(h_i) = (i - 1)/(n - 1), \quad i = 1, \ldots, n$$

$$P(h_1) = 0, \quad P(h_n) = 1$$

A typical shape of this function (for large n) is shown in Fig. 1.

Let us now consider a fuzzy set of tall drivers. The membership function of such a fuzzy set $\mu_{tall}(H)$ could be defined in many different ways using various arbitrarily selected intervals of height H. Since the distribution in Fig. 1 represents the objective reality, it is very reasonable to claim that only the tallest of all n drivers has the distinct privilege to be the "full member" of the tall driver fuzzy set, i.e., $\mu_{tall}(h_n) = 1$. Once we decided the status of the tallest driver, we must do the same with the shortest driver: we declare the shortest driver to be the one and only full member of the fuzzy set of short drivers, i.e. $\mu_{short}(h_1) = 1$. Furthermore, $\mu_{tall}(H) + \mu_{short}(H) = 1$.

Another self-evident decision is related to the driver whose membership satisfies $\mu_{tall}(H) = \mu_{short}(H) = 1/2$. The most natural way to select such a driver is by using the median of all heights $H = h_{med}$; assuming that n is odd, the median driver has the height $h_{med} = \mu_{short}^{-1}(1/2) = \mu_{tall}^{-1}(1/2)$, i.e., approximately 50% of all drivers are

Fig. 1 A characteristic shape of the $P(H)$ function

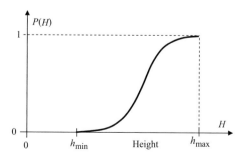

taller than the driver *med* and 50% of all drivers are shorter than the driver *med*. If this reasoning is acceptable, then the most realistic membership functions are the following:

$$\mu_{\text{tall}}(h_i) = (i - 1)/(n - 1)$$

$$\mu_{\text{short}}(h_i) = (n - i)/(n - 1)$$

Consequently, $\mu_{\text{tall}}(H)$ is exactly the function shown in Fig. 1.

Let us now investigate an arbitrary driver D whose height is h_m, and let us find the degree of truth of the statement "D is a tall driver." Again, it seems impossible to find reasons why this degree of truth should differ from $\mu_{\text{tall}}(h_m) = (m - 1)/(n - 1)$, which is the fraction of drivers who are shorter than D. In other words, the degree of truth for "D is a tall driver" is the same as the degree of membership in the fuzzy set of tall drivers, and the probability that a randomly selected driver is shorter then D.

The height of a driver is obviously a repetitive event that occurs with measurable frequency. The number of drivers having the height $h \in [a, b]$ is $n[P(b) - P(a)]$. If we measure frequencies of drivers $f_1, \ldots, f_k$ in k equidistant subintervals of $[h_{\min}, h_{\max}]$, and normalize the frequencies, $p_i = f_i/(f_1 + \ldots + f_k), i = 1, \ldots, k$, then, for large k and n we can consider that h is a continuous random variable, and we get the probability density function of driver height $p(H), 0 < h_{\min} \leq H \leq h_{\max}$ and the (cumulative) probability distribution function $P(H) = \int_{h_{\min}}^{H} p(t)\,dt$. Unsurprisingly, for all practical purposes related to modeling human reasoning, the function shown in Fig. 1 is also the probability distribution function for the height of a driver, i.e. $\Pr[h \leq H] = P(H)$ and $\Pr[a < h \leq b] = P(b) - P(a)$. The mean height of drivers is the following:

$$\overline{h} = n^{-1}(h_1 + \ldots + h_n) = \int_{h_{\min}}^{h_{\max}} t p(t)\,dt =$$

$$\int_0^{h_{\max}} [1 - P(t)]\,dt = h_{\max} - \int_{h_{\min}}^{h_{\max}} P(t)\,dt.$$

Therefore, in the repetitive case of a measurable physical property (height) of a large number of drivers, we have that the fuzzy set membership function coincides with the probability distribution function. In addition, the degree of fuzzy set membership of an object can be naturally interpreted as the degree of truth of a statement claiming that the object is a full member of the set.

3   The Case of Job Satisfaction

Job satisfaction is not an objective physical property. However, it is a well defined perceptual variable. Each worker has a clear percept of the degree of job satisfaction, and such percept can easily be quantified. For example, consider the statement "I am

asked to indicate, on a scale from 0 to 1, the degree to which I like my job. My answer
is: 0.7." This is a value statement (it reports the result of an intuitive evaluation).
Obviously, this statement indicates that the degree of truth of the statement "I have a
perfect job" is 0.7. Following are two interesting questions related to this statement:

(a) Is this statement a consequence of the repetitive nature of job satisfaction?
(b) Is this statement yielding equivalent fuzzy, probabilistic, and truth value inter-
    pretations?

Job satisfaction is a compound perception. It is affected by many factors, including
monetary compensation, fringe benefits, opportunities for professional growth, social
recognition, degree of job-related stress, relationships with coworkers and managers,
etc. There is no doubt that each job creates repetitive experiences every day spent
at work. Job satisfaction can be interpreted as a quantifiable perception that can be
created at the end of each working day. The degree to which a worker likes a job can
be interpreted as an overall perception of job satisfaction obtained by averaging all
repetitive daily job satisfaction degrees.

A typical normalized relative frequency distribution of a daily job satisfaction perception s is shown in Fig. 2. So, $\int_{s_{\min}}^{1} f_s(x)\,dx = 1$ and the corresponding probability
distribution function of job satisfaction is $F_s(x) = \Pr[s \le x] = \int_{s_{\min}}^{x} f_s(t)\,dt$. The
average daily job satisfaction is $S = \bar{s} = \int_{s_{\min}}^{1} x f_s(x)\,dx = 1 - \int_{s_{\min}}^{1} F_s(x)\,dx$ and it
can also be interpreted as the degree of truth of the statement "I have a perfect job".
So, S is a cumulative perception of job satisfaction that in the above example was
reported to be $S = 0.7$. This is also a degree of membership of the current job in the
fuzzy set of "worker's ideal jobs," as well as a degree of likelihood that a randomly
selected day at work will completely satisfy worker's expectations. On the other
hand, the probability distribution function $F_s(x)$ shows the probability $\Pr[s \le x]$,
and it can be interpreted as the membership function of the daily job satisfaction
perception x in a fuzzy set of "high job satisfaction days."

Note that all these interpretations hold only for one specific individual. In the
next iteration we can analyze a group of n related individuals (e.g. medical doctors, or corn farmers, or software engineers) with respect to their perceptions of
the average job satisfaction $S_1, \ldots, S_n$. Here we can directly apply the same reasoning used for the fuzzy set of tall drivers. For example, we can sort the average
job satisfaction of all medical doctors, $S_{\min} = S_1 < \cdots < S_n = S_{\max}$, and then the

Fig. 2 A typical shape of
the daily job satisfaction
distribution for a specific
worker

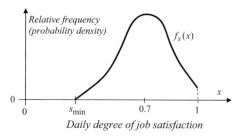

Daily degree of job satisfaction

membership in the fuzzy set of job-satisfied medical doctors is the same as the probability distribution function of the average MD job satisfaction: $P_{MD}(S_i) = \Pr[S \leq S_i] = (i-1)/(n-1) = \mu_{MD}(S_i)$. The mean value of this probability distribution $\overline{S} = s_{max} - \int_{S_{min}}^{S_{max}} P_{MD}(x)\,dx$ can be used as the degree of truth of the statement "medical doctors are completely satisfied with their jobs."

Therefore, this example demonstrates that job satisfaction experiences are repetitive both at the level of individual worker, and at he level of a professional group. In both cases, probabilistic, fuzzy and truth value interpretations are equivalent: the job satisfaction probability distribution is the same as the corresponding membership function in the fuzzy set of satisfied workers, and the mean degree of satisfaction can be interpreted as the degree of truth of the statement claiming a complete satisfaction.

4 The Case of Computer Selection

The percept of value is one of the most frequent human percepts. It is not an objectively measurable property of evaluated object. Our final example investigates computer evaluation and selection, i.e. a decision process of predicting the value of a computer system for a specific stakeholder. The stakeholder is a company (or an individual) who buys a computer system in order to attain specific goals. The computer selection process consists of computing the overall suitability of each competitive computer system and selecting the computer that has the maximum suitability.

Each computer has many quantitative and qualitative parameters and there is no simple probabilistic model for computing the probability of satisfaction of specific stakeholder. Similar difficulties are encountered in defining a precise overall membership function of each competitive computer in the fuzzy set of computers that completely satisfy the stakeholder. What can be done, however, is to use the soft computing evaluation logic methodology and compute the degree of truth of the statement claiming that a given computer completely satisfies justifiable requirements of the stakeholder. That degree of truth is a function of degrees of truth of statements that significant computer attributes satisfy stakeholder's requirements. The computation of the overall degree of truth can be based on a propositional calculus presented in [2, 3]; that calculus is the main component of the Logic Scoring of Preference (LSP) evaluation method [4].

If the evaluation is based on the LSP criterion model, then the first step consists of identifying a set of attributes $a_1, a_2, \ldots, a_n$ that affect the capability of computer system to attain stakeholder's goals. For complex computer systems typical value of n is from 80 to 120. For each attribute it is necessary to define an attribute criterion that specifies the attribute degree of suitability. For example, if a_i denotes the memory capacity, then a simple form of attribute criterion for computing the memory capacity suitability degree x_i might be the following:

$$x_i = g_i(a_i) = \max\left[0, \min\left(1, \frac{a_i - M_{min}}{M_{max} - M_{min}}\right)\right], \quad a_i > 0, \quad 0 \leq x_i \leq 1$$

If $a_i \leq M_{min}$, then $x_i = 0$ and if $a_i \geq M_{max}$ then $x_i = 1$. Between M_{min} and M_{max}, the degree of suitability is approximated as a linear function of a_i. Therefore, the available memory must be greater than the threshold value M_{min}; similarly, M_{max} denotes the maximum necessary memory (buying more would not improve performance but would decrease affordability). The selections of both the attributes and the parameters of their criteria are based on experiences with previous (and/or similar) computer systems, i.e. on repetitive events of satisfaction/dissatisfaction with attributes of previously owned computers running stakeholder's workload. The suitability degree x_i is interpreted as the degree of truth of assertion that the available memory perfectly (completely) satisfies stakeholder's requirements. It can also be interpreted as the degree of membership in the fuzzy set of fully satisfied memory capacity requirements.

After the specification of all attribute criteria and attribute evaluation we have n attribute suitability degrees $x_1, x_2, \ldots, x_n$ that are continuous logic variables affecting the overall degree of suitability x. The aggregation of $x_1, x_2, \ldots, x_n$ is a logic process because these values must satisfy various logic conditions: some groups must be simultaneously satisfied, some attributes can replace each other, some are mandatory, some are optional, etc. Similarly to attribute criteria, these conditions are also derived from experiences generated using previous stakeholder's computer systems. The result of logic aggregation is the overall suitability

$$x = L(x_1, \ldots, x_n) = L(g_1(a_1), \ldots, g_n(a_n)), \ \ 0 \leq x \leq 1.$$

A more detailed description of the logic aggregation process based on the Logic Scoring of Preference aggregators can be found in [2, 4], and discussion of relationships of mathematical models of aggregation and observable properties of human reasoning can be found in [1].

In agreement with the interpretation of attribute suitability degrees, the overall suitability degree x is interpreted as the degree of truth of assertion that the evaluated computer system perfectly (i.e. completely) satisfies all stakeholder's requirements. Similarly to previous examples, it can also be interpreted as the degree of membership of the analyzed computer in the fuzzy set of perfectly suitable computers. In addition, x can also be used as a predictor of the likelihood (or probability) that at any time in the future the analyzed computer will perfectly satisfy stakeholder's needs. In other words, if we have the equivalence between the overall degree of truth, fuzzy membership, and probability, then any one of them can be used to determine or approximate the other two. Compared to previous examples, the case of computer selection has the least visible repetitive properties. Regardless a generally modest number of previously owned computers, the experiences that are necessary to specify evaluation criteria (the selection, structure and parameters of attribute criteria, and the structure and parameters of the suitability aggregation models) can also be obtained from similar users having similar computers and running similar workload, as well as from domain expert knowledge and professional literature. In other words, wherever decisions are made using professional experiences of stakeholders, evaluators, and domain experts, such knowledge can only come from previous repetitive experiences accumulated during education, training, practical work and/or solving similar problems.

5 Discussion and Conclusions

Presented examples show that in the case of repetitive events we can identify an observable object and select, collect, and analyze one or more of its measurable attributes. Each selected attribute can be an objective physical property or a perceptual variable. In all cases we can assume the existence of a number of measurable observations of each selected attribute. In such cases the attribute probability distribution function, the fuzzy set membership function, and the degree of partial truth of the statement claiming the full satisfaction of membership requirements are *equivalent interpretations* of the same physical reality. The prerequisite for the equivalency is the repetitive character of the analyzed variable (e.g. the height of driver) and an appropriate and justifiable interpretation of the concepts of fuzziness, probability, and partial truth.

Generally, for any observable repetitive attribute a that has the probability distribution function $\Pr[a \leq X] = P_a(X), a_{\min} \leq a \leq a_{\max}$ where $P_a(X) = 0$ for $X \leq a_{\min}$ and $P_a(X) = 1$ for $X \geq a_{\max}$ we can define a fuzzy set of "objects that have a large value of attribute a" (i.e. objects whose value of a is close to $a_{\max}$). The membership function in such a fuzzy set for an object having $a = x$ is $\mu_{\text{large } a}(x) = P_a(x)$. The degree of truth of the statement "the object having $a = x$ is the largest" is $\mu_{\text{large } a}(x)$.

It is important to note that a very large number of cases fit in this model. In particular, all professional evaluation problems fall in this category. Indeed, all evaluation criteria are based on evaluator's previous experiences and reflect repetitive character of the evaluation process. It is not possible to differentiate good and bad values of attributes unless the evaluator has knowledge derived from repetitively performing similar evaluations. If a justifiable evaluation model generates for object Ω a resulting degree of suitability $S \in [0, 1]$, then S can be equivalently interpreted in three ways:

(1) S is a degree of truth of the statement "Ω is perfectly satisfying all requirements."
(2) S is the degree of membership of Ω in a fuzzy set of perfect objects.
(3) S is a degree of likelihood (probability) that Ω will deliver a perfect performance.

Let us emphasize that in evaluation problems S is a quantitative estimate (or predictor) of the human percept of the overall (compound) value/quality/suitability. Thus, there is no objective measurable value that could be compared with S. On the other hand, everybody understands that there are stakeholders and their goals and requirements, and that these goals and requirements can regularly be only partially satisfied. The likelihood of full satisfaction, whatever label is used for it (probability, or fuzzy membership, or partial truth), is going to be predicted using the only indicator we have, the overall suitability S. In this situation the concepts of probability, fuzzy membership, and partial truth become equivalent interpretations of a clearly observable but objectively nonmeasurable human percept. Of course, the equivalence of a probability distribution function and a fuzzy set membership function is *not* a general mathematical property based on fundamental assumptions or axiomatic origins of these concepts. However, it is a rather frequent consequence of *interpretations and use* of those concepts in the context of observable human reasoning.

The repetitive nature of human experiences is frequently underestimated or neglected. In many discussions that compare fuzzy and probabilistic approach some authors use the concept of singular events, as unpredictable events that happen without predecessors and consequently their probability cannot be determined using frequency-based probability theory. However, it is extremely difficult, if not impossible, to find events without predecessors, and those that might (partially) qualify for that status usually deserve significantly less attention than the huge majority that has predecessors and can be characterized as repetitive. Since the observable repetitiveness in finite populations (clearly visible in the case of tall drivers) yields probability distribution function that can also serve as a fuzzy membership function, as well as a degree of truth of a related assertion, it seems that the situations where the concepts of probability, fuzzy membership, and partial truth are very close or fully equivalent significantly dominate the situations where that is not the case.

References

1. J. Dujmović, "Aggregation operators and observable properties of human reasoning", in: H. Bustince, J. Fernandez, R. Mesiar, and T. Calvo (eds.), *Aggregation Functions in Theory and in Practice*, Springer, 2013, pp. 5–16.
2. J. Dujmović, "Preference logic for system evaluation", *IEEE Transactions on Fuzzy Systems*, 2007, Vol. 15, No. 6, pp. 1082–1099.
3. J. Dujmović, "Weighted Compensative Logic With Adjustable Threshold Andness and Orness", *IEEE Transactions on Fuzzy Systems*, 2015, Vol. 23, No. 2, pp. 270–290.
4. J. Dujmović and H. Nagashima, "LSP method and its use for evaluation of Java IDE's", *International Journal of Approximate Reasoning*, 2006, Vol. 41, No. 1, pp. 3–22.
5. M. Grabisch, J.-L. Marichal, R. Mesiar, and E. Pap, *Aggregation Functions*, Cambridge University Press, 2009.
6. B. Kovalerchuk, "Quest for Rigorous Combining Probabilistic and Fuzzy Logic Approaches for Computing with Words", In: R. Seising, E. Trillas, C. Moraga, and S. Termini (eds.), *On Fuzziness. A Homage to Lotfi A. Zadeh* (Studies in Fuzziness and Soft Computing Vol. 216), Springer Verlag, 2013, Vol. 1, pp. 333–344.
7. J. M. Mendel and D. Wu, *Perceptual Computing*, Wiley-IEEE, Hoboken, NJ, USA, 2010.
8. L. A. Zadeh, "A note on similarity-based definitions of possibility and probability", *Information Sciences*, 2014, Vol. 267, pp. 334–336.
9. L. A. Zadeh, "Outline of a new approach to the analysis of complex systems and decision processes", *IEEE Transactions on Systems, Man and Cybernetics*, 1973, Vol. 3, pp. 28–44.

Using Extended Tree Kernel to Recognize Metalanguage in Text

Boris A. Galitsky

Abstract The problem of classifying text with respect to metalanguage and language-object patters is formulated and its application areas are proposed. We extend parse tree kernels from the level of individual sentences towards the level of paragraphs to classify texts at a high level of abstraction. The method targets the text classification tasks where keyword statistics is insufficient for text classification tasks. We build a set of extended trees for a paragraph of text from the individual parse trees for sentences. Conventional parse trees are extended across sentences based on anaphora and rhetoric structure relations between the phrases in different sentences. Tree kernel learning is applied to extended trees to take advantage of additional discourse-related information. We evaluate our approach in the security-related domain of the design documents. These are the documents which contain a formal well-structured presentation on how a system is built. Design documents need to be differentiated from product requirements, architectural, general design notes, templates, research results and other types of documents, which can share the same keywords. We also evaluate classification in the literature domain, classifying text in Kafka's novel "The Trial" as metalanguage versus novel's description in scholarly studies as a mixture of metalanguage and language-object.

Keywords Tree kernel · Formalizing discourse · Language-object and metalanguage · Document analysis

1 Introduction

In the majority of text classification problems, keywords statistics is sufficient to determine a class. Keywords are sufficient information to determine a topic of a text or document, such as software vs hardware, or pop rock vs punk. However, there are classification problems where distinct classes share the same keywords, and document phrasing, style and other kinds of text structure information needs

B.A. Galitsky (✉)
Knowledge Trail Inc, San Jose, CA, USA
e-mail: bgalitsky@hotmail.com

© Springer International Publishing AG 2017
V. Kreinovich (ed.), *Uncertainty Modeling*, Studies in Computational
Intelligence 683, DOI 10.1007/978-3-319-51052-1_6

to be taken into account. To perform text classification in such domain, discourse information such as anaphora and rhetoric structure needs to be taken into account.

We are interested in classifying a text belonging to metalanguage or language-object. If a text tells us how to do things, or how something has been done, we classify this text as a language-object. If a text is saying how to write a document which explains how to do things, we classify it as metalanguage. Metalanguage is a language or symbolic system used to discuss, describe, or analyze another language or symbolic system. In theorem proving, metalanguage is a language in which proofs are manipulated and tactics are programmed, as opposed to the logic itself (the object-language). In logic, it is a language in which the truth of statements in another language is being discussed.

Obviously, using just keyword information would be insufficient to differentiate between texts in metalanguage and language-object. Use of parse trees [11] would give us specific phrases in use by texts in metalanguage, but still it will not be sufficient for systematic exploration of metalanguage-related linguistic features. It is hard to identify these features unless one can analyze the discourse structure, including anaphora, rhetoric relations, and interaction scenarios by means of communicative language [28]. Furthermore, to systematically learn these discourse features associated with metalanguage, we need a unified approach to classify graph structures at the level of paragraphs [5, 8, 14].

The design of syntactic features for automated learning of syntactic structures for classification is still an art nowadays. One of the approaches to systematically treat these syntactic features is the set kernels built over syntactic parse trees. Convolution tree kernel [2] defines a feature space consisting of all subtree types of parse trees and counts the number of common subtrees as the syntactic similarity between two parse trees. Tree kernels have found applications in a number of NLP tasks, including syntactic parsing re-ranking, relation extraction, named entity recognition [4] and Semantic Role Labeling [23, 32], relation extraction, pronoun resolution [18], question classification and machine translation [29, 30].

The kernel's ability to generate large feature sets is useful to assure we have enough linguistic features to differentiate between the classes, to quickly model new and not-well-understood linguistic phenomena in learning machines. However, it is often possible to manually design features for linear kernels that produce high accuracy and fast computation time, whereas the complexity of tree kernels may prevent their application in real scenarios. Support Vector Machines (SVM [31]) can work directly with kernels by replacing the dot product with a particular kernel function. This useful property of kernel methods, that implicitly calculates the dot product in a high-dimensional space over the original representations of objects such as sentences, has made kernel methods an effective solution to modeling sentence-level structures in natural language processing (NLP).

An approach to build a kernel based on more than a single parse tree has been proposed, however for a different purpose than treating multi-sentence portions of text. To perform classification based on additional discourse features, we form a single tree from a tree forest for a sequence of sentences in a paragraph of text. Currently, kernel methods tackle individual sentences. For example, in question answering, when a query is a single sentence and an answer is a single sentence, these methods work fairly well. However, in learning settings where texts include multiple sentences, we need to represent structures which include paragraph-level information such as discourse.

A number of NLP tasks such as classification require computing semantic features over paragraphs of text containing multiple sentences. Doing it at the level of individual sentences and then summing up the score for sentences will not always work. In the complex classification tasks where classes are defined in an abstract way, the difference between them may lay at the paragraph level and not at the level of individual sentences. In the case where classes are defined not via topics but instead via writing style, discourse structure signals become essential. Moreover, some information about entities can be distributed across sentences, and classification approach needs to be independent of this distribution [12–14]. We will demonstrate the contribution of paragraph-level approach versus the sentence level in our evaluation.

1.1 Design Documents Versus Design Meta-Documents

We define design document as a document which contains a thorough and well-structured description of how to build a particular engineering system. In this respect a design document according to our model follows the reproducibility criteria of a patent or research publication; however format is different from them. What we exclude is a document which contains meta-level information relatively to the design of engineering system, such as *how to write design docs* manuals, standards design docs should adhere to, tutorials on how to improve design documents, and others.

We need to differentiate design documents from the classes of documents which can be viewed as ones containing meta-language, whereas the genuine design document consists of the language-object. Below we enumerate such classed of *meta-documents*:

(1) design requirements, project requirement document, requirement analysis, operational requirements
(2) construction documentation, project planning, technical services review
(3) design guidelines, design guides, tutorials
(4) design templates (template for technical design document)
(5) research papers on system design
(6) general design-related notes
(7) educational materials on system design
(8) the description of the company, which owns design documents
(9) resume of a design professional
(10) specifications for civil engineering
(11) functional specifications
(12) 'best design practices' description
(13) project proposals

Naturally, design documents are different from similar kinds of documents on the same topic in terms of style and phrasing. To extract these features, rhetoric relations are essential.

Notice that meta-documents can contain object-level text, such as design examples. Object-level documents (genuine design docs) can contain some author reflections on the system design process (which are written in metalanguage). Hence the boundary between classes does not strictly separate metalanguage and language-object. We use statistical language learning to optimize such boundary, having supplied it with a rich set of linguistic features up to the discourse structures. In the design document domain, we will differentiate between texts in mostly meta-language and the ones mostly in language-object.

1.2   Novel in Metalanguage Versus Novel in Language-Object

A mixture of object-language and metalanguage descriptions can be found in literature. Describing the nature, a historical event, an encounter between people, an author uses a language-object. Describing thoughts, beliefs, desires and knowledge of characters about the nature, events and interactions between people, an author uses a metalanguage. The entities/relations of such metalanguage range over the expressions (phrases) of the language-object. In other words, the physical world is usually described in language-object, and the mental world (theory of mind, the world of thoughts) typically combines both levels.

One of the purest examples of use of metalanguage in literature is Franz Kafka's novel "The Trial". According to our model, the whole plot is described in metalanguage, and object-level representation is absent. This is unlike a typical work of literature, where both levels are employed. In "The Trial" a reader learns the main character Joseph K. is being prosecuted, his thoughts are described, meeting with various people related to the trial are presented. However, no information is available about a reason for the trial, the charge, the circumstances of the deed. The novel is a pure example of the presence of meta-theory and absence of object-level theory, from the standpoint of logic. The reader is expected to form the object–level theory herself to avoid ambiguity in interpretation of the novel.

Exploration of "The Trial" would help to understand the linguistic properties of metalanguage and language-object. For example, it is easy to differentiate between a mental and a physical words, just relying on keywords. However, to distinguish meta-language from language-object in text, one need to consider different discourse structures, which we will automatically learn from text.

The following paragraph of text can be viewed as a fragment of an algorithm for how to solve an abstract problem of acquittal. Since it suggests a domain-independent approach (it does not matter what an accused did), it can be considered as a meta-algorithm.

'There are three possibilities: absolute acquittal, apparent acquittal and deferment. Absolute acquittal is the best, but there is nothing I could do to get that sort of outcome.

I don't think there's anyone at all who could do anything to get an absolute acquittal. Probably the only thing that could do that is if the accused is innocent. As you are innocent it could actually be possible and you could depend on your innocence alone. In that case you will not need me or any other kind of help.'

In some sense this algorithm follows along the lines of a 'vanilla' interpreter in Prolog, a typical example of a meta-program:

achieve_acquittal (true).
achieve_acquittal ((A,B)):- achieve_acquittal (A), achieve_acquittal (B).
achieve_acquittal (A):- clause(A, B), achieve_acquittal (B).

where the novel enumerates various *clauses*, but never ground terms expressing the details of a hypothetical crime (no instances of *A* or *B*). *clause(A, B)* is expression of the format A :- B, where A is a term being defined (a clause head) and B is a sequence of defining terms (a body of this clause). This interpreter shows multiple possibilities a term can be proved, similarly to multiple possibilities of acquittal spelled out by Kafka.

We hypothesize that a text expressing such a meta-program, Kafka's text, should have specific sequences of rhetoric relation, infrequent in other texts. We will attempt to find distinct discourse patterns associated with metalanguage and differentiate it with other texts.

In the literature domain, we will attempt to draw a boundary between the pure metalanguage (peculiar works of literature) and a mixed level text (a typical work of literature).

2   Extending Tree Kernel Towards Discourse

To Why can sentence-level tree kernels be insufficient for classification? Important phrases can be distributed through different sentences. So we want to combine/merge parse trees to make sure we cover the phrase of interest.

For the following text:

This document describes the design of back end processor. Its requirements are enumerated below.

From the first sentence, it looks like we got the design document. To process the second sentence, we need to disambiguate the preposition 'its'. As a result, we conclude from the second sentence that it is a requirements document (not a design document).

2.1   Leveraging Structural Information for Classification

How can a sentence structural information be indicative of the class?

The idea of measuring similarity between the question-answer pairs for question answering instead of the question-answer similarity turned out to be fruitful

[23]. The classifier for correct vs incorrect answers processes two pairs at a time, $<q_1, a_1>$ and $<q_2, a_2>$, and compare q_1 with q_2 and a_1 with a_2, producing a combined similarity score. Such a comparison allows to determine whether an unknown question/answer pair contains a correct answer or not by assessing its distance from another question/answer pair with a known label. In particular, an unlabeled pair $<q_2, a_2>$ will be processed so that rather than "guessing" correctness based on words or structures shared by q_2 and a_2, both q_2 and a_2 will be compared to their correspondent components q_1 and a_1 of the labeled pair $<q_2, \quad a_2>$ on the grounds of such words or structures. Since this approach targets a domain-independent classification of answer, only the structural cohesiveness between a question and answer is leveraged, not 'meanings' of an answers.

We take this idea further and consider an arbitrary sequence of sentences instead of question-sentence and answer-sentence pair for text classification. Our positive training paragraphs are "plausible" sequences of sentences for our class, and our negative training paragraphs are "implausible" sequences, irrespectively of the domain-specific keywords in these sentences.

In our opinion, for candidate answer selection task, such structural information is important but insufficient. At the same time, for the text classification tasks just structure analysis can suffice for proper classification.

Given a positive sequence and its parse trees linked by RST relations:

'A hardware system contains classes such as GUI for user interface, IO for importing and exporting data between the emulator and environment, and Emulator for the actual process control. Furthermore, a class Modules is required which contains all instances of modules in use by emulation process.' (Fig. 1).

And a negative sequence and its linked parse trees:

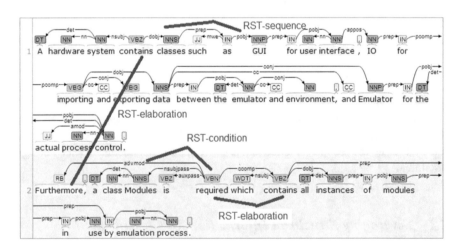

Fig. 1 A sequence of parse trees and RST relations for a positive example

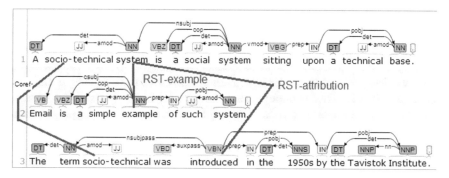

Fig. 2 A sequence of parse trees and RST relations for a negative example

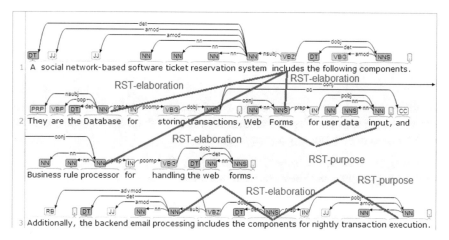

Fig. 3 A sequence of parse trees and RST relations for a text to be classified

'A socio-technical system is a social system sitting upon a technical base. Email is a simple example of such system. The term socio-technical was introduced in the 1950s by the Tavistok Institute.' (Fig. 2).

We want to classify the paragraph

'A social network-based software ticket reservation system includes the following components. They are the Database for storing transactions, Web Forms for user data input, and Business rule processor for handling the web forms. Additionally, the backend email processing includes the components for nightly transaction execution.' (Fig. 3).

One can see that this paragraph follows the rhetoric structure of the top (positive) training set element, although it shares more common keywords with the bottom (negative) element. Hence we classify it as a design document text, since it describes the system rather than introduces a terms (as the negative element does).

To illustrate the similar point in the question answering domain, we use a simple query example. If q_1 is '*What is plutocracy?*' and the candidate answers are $a_1 =$ '*Plutocracy may be defined as a state where ...*' versus $a_0 =$ '*Plutocracy affects the wills of people ...*', comparison with the correct pair formed by $q_2 =$ '*What is a source control software?*' and $a_2 =$ '*A source control software can be defined as a...*' will induce the kernel method to prefer a_1 to a_0. One can see that a_1 has a similar wording and structure to a_2, hence $< q_1, a_1 >$ will get a higher score than $< q_1, a_0 >$ using the kernel method. In contrast, the opposite case would occur using a similarity score matching q_1 with a_1 as compared with matching q_1 with a_0, since both a_1 and a_0 contain keywords *plutocracy* from q_1. This explains why even a bag-of-words kernel adjusting its weights on question/answer pairs has a better chance to produce good results than a bag-of-words question/answer similarity.

2.2 *Anaphora and Rhetoric Relations for Sentiments Topicality Classification*

I sentiment analysis, we classify sentences and documents with respect to sentiments they contain. Let us consider the sentiment classes for the following sentences. We are interested in both polarity and topicality classes:

'I would not let my dog stay in this hotel' (Fig. 4)
'Our dog would have never stayed in this hotel'
'Filthy dirty hotel, like a dog house'
'They would not let my dog stay in this hotel'
'The hotel management did not let me in when I was with my dog'
'We were not allowed to stay there with our dog'

What one observes is that polarity = negative in both cases, whereas topics are totally different. Topic1 = '*hotel is dirty*', and
topic2 = '*dogs are not allowed*'.

This is rather difficult task for keyword-based text classification problems because both classes share the same keywords.

Notice these classes are different from the topic3 = '*hotels intended for dogs*', polarity = '*neutral*'

If you have never been to a dog hotel, now is the time.

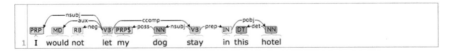

Fig. 4 A parse tree for an individual sentence

Fig. 5 A fragment of text
with a coreference relation

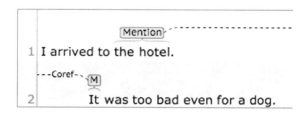

It is even harder to perform classification, when information about 'staying in a hotel and having a dog' is spread through different sentences. An easier case is anaphora:

'I arrived to the hotel. It was too bad even for a dog' (Fig. 5).

The hardest case is when rhetoric structure is needed to link information about a hotel and a dog:

'I was traveling for business. My partner invited me to stay at his place, however it looked like a house for dogs.'

'I was traveling with my dog for business. I was not going to stay at a hotel but at my partner's place, however he turned out to be allergic to dogs. Sadly, the hotel did not let us in.' (Fig. 6).

In the above cases, the parts of the parse trees (sub-trees) essential to determine the meanings occur in different sentences, so needs to be connected. Anaphora is a natural way to do that, but is not always sufficient. Hence we need rhetoric relations to link '*travel, dog owner, hotel*' and permission relationships.

2.3   *Anaphora and Rhetoric Relations for Classification Tasks*

We introduce a domain where a pair-wise comparison of sentences is insufficient to properly learn certain semantic features of texts. This is due to the variability of ways information can be communicated in multiple sentences, and variations in possible discourse structures of text which needs to be taken into account.

We consider an example of text classification problem, where short portions of text belong to two classes:

- Tax liability of a landlord renting office to a business.
- Tax liability of a business owner renting an office from landlord.

Coreference:

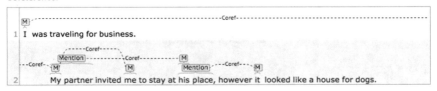

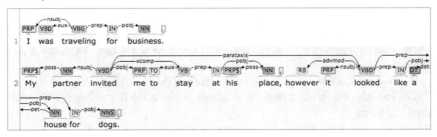

Fig. 6 The parse tree and respective coreference arcs for two sentences

'I rent an office space. This office is for my business. I can deduct office rental expense from my business profit to calculate net income.

To run my business, I have to rent an office. The net business profit is calculated as follows. Rental expense needs to be subtracted from revenue.

To store goods for my retail business I rent some space. When I calculate the net income, I take revenue and subtract business expenses such as office rent.

I rent out a first floor unit of my house to a travel business. I need to add the rental income to my profit. However, when I repair my house, I can deduct the repair expense from my rental income.

I receive rental income from my office. I have to claim it as a profit in my tax forms. I need to add my rental income to my profits, but subtract rental expenses such as repair from it.

I advertised my property as a business rental. Advertisement and repair expenses can be subtracted from the rental income. Remaining rental income needs to be added to my profit and be reported as taxable profit.'

Note that keyword-based analysis does not help to separate the first three paragraph and the second three paragraphs. They all share the same keywords *rental/office/income/profit/add/subtract*. Phrase-based analysis does not help, since both sets of paragraphs share similar phrases.

Secondly, pair-wise sentence comparison does not solve the problem either.

Anaphora resolution is helpful but insufficient. All these sentences include 'I' and its mention, but other links between words or phrases in different sentences need to be used.

Rhetoric structures need to come into play to provide additional links between sentences. The structure to distinguish between

'renting for yourself and deducting from total income' and

'renting to someone and adding to income' embraces multiple sentences. The second clause about *'adding/subtracting incomes'* is linked by means of the rhetoric relation of *elaboration* with the first clause for *landlord/tenant*. This rhetoric relation may link discourse units within a sentence, between consecutive sentences and even between first and third sentence in a paragraph. Other rhetoric relations can play similar role for forming essential links for text classification.

Which representations for these paragraphs of text would produce such common sub-structure between the structures of these paragraphs? We believe that extended trees, which include the first, second, and third sentence for each paragraph together can serve as a structure to differentiate the two above classes.

The dependency parse trees for the first text in our set and its coreferences are shown in Fig. 7. There are multiple ways the nodes from parse trees of different sentences can be connected: we choose the rhetoric relation of elaboration which links the same entity office and helps us to form the structure *rent-office-space— for-my-business—deduct-rental-expense* which is the base for our classification. We used Stanford Core NLP, coreferences resolution [19] and its visualization to form Figs. 1 and 2.

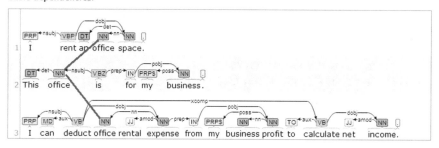

Fig. 7 Coreferences and the set of dependency trees for the first text

Fig. 8 Extended tree which
includes 3 sentences

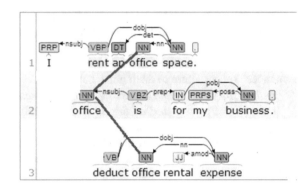

Figure 8 shows the resultant extended tree with the root 'I' from the first sentence.
It includes the whole first sentence, a verb phrase from the second sentence and a
verb phrase from the third sentence according to rhetoric relation of elaboration.
Notice that this extended tree can be intuitively viewed as representing the 'main
idea' of this text compared to other texts in our set. All extended trees need to be
formed for a text and then compared with that of the other texts, since we don't
know in advance which extended tree is essential. From the standpoint of tree kernel
learning, extended trees are learned the same way as regular parse trees.

3   Building Extended Trees

For every arc which connects two parse trees, we derive the extension of these trees,
extending branches according to the arc (Fig. 9).

In this approach, for a given parse tree, we will obtain a set of its extension,
so the elements of kernel will be computed for many extensions, instead of just a
single tree. The problem here is that we need to find common sub-trees for a much
higher number of trees than the number of sentences in text, however by subsumption
(sub-tree relation) the number of common sub-trees will be substantially reduced.

If we have two parse trees P_1 and P_2 for two sentences in a paragraph, and a relation
$R_{12} : P_{1i} \rightarrow P_{2j}$ between the nodes P_{1i} and P_{2j}, we form the pair of extended trees
$P_1 * P_2$:

$\ldots, P_{1i-2}, P_{1i-1}, P_{1i}, P_{2j}, P_{2j+1}, P_{2j+2}, \ldots$

$\ldots, P_{2j-2}, P_{2j-1}, P_{2j}, P_{1i}, P_{1i+1}, P_{2i+2}, \ldots,$

which would form the feature set for tree kernel learning in addition to the original
trees P_1 and P_2. Notice that the original order of nodes of parse trees are retained
under operation '*'.

The algorithm for building an extended tree for a set of parse trees T is presented
below:

Fig. 9 An arc which connects two parse trees for two sentences in a text (on the *top*) and the derived set of extended trees (on the *bottom*)

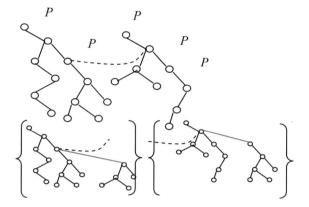

Input:
    (1) Set of parse trees T.
    (2) Set of relations R, which includes relations R_{ijk} between the nodes of T_i and $T_j : T_i \in T, T_j \in T, R_{ijk} \in R$. We use index k to range over multiple relations between the nodes of parse tree for a pair of sentences.
Output: the exhaustive set of extended trees E.
Set $E = \emptyset$;
For each tree $i = 1 : |T|$
    For each relation $R_{ijk}, k = 1 : |R|$
    Obtain T_j
    Form the pair of extended trees $T_i * T_j$;
    Verify that each of the extended trees do not have a super-tree in E
    If verified, add to E;
Return E.

Notice that the resultant trees are not the proper parse trees for a sentence, but nevertheless form an adequate feature space for tree kernel learning.

To obtain the inter-sentence links, we employed coreferences from Stanford NLP [19, 24]. Rhetoric relation extractor based on our rule-based approach to finding relations between elementary discourse units [7, 9, 14]. We combined manual rules with automatically learned rules derived from the available discourse corpus by means of syntactic generalization.

Rhetorical Structure Theory (RST [20]) is one of the most popular approach to model extra-sentence as well as intra-sentence discourse. RST represents texts by labeled hierarchical structures, called Discourse Trees (DTs). The leaves of a DT correspond to contiguous Elementary Discourse Units (EDUs). Adjacent EDUs are connected by rhetorical relations (e.g., Elaboration, Contrast), forming larger discourse units (represented by internal nodes), which in turn are also subject to this relation linking. Discourse units linked by a rhetorical relation are further distinguished based on their relative importance in the text: nucleus being the central part, whereas satellite being the peripheral one. Discourse analysis in RST involves two subtasks: discourse segmentation is the task of identifying the EDUs, and discourse parsing is the task of linking the discourse units into a labeled tree.

3.1 Kernel Methods for Parse Trees

Kernel methods are a large class of learning algorithms based on inner product vector spaces. Support

Vector machines (SVMs) are mostly well-known algorithms. The main idea behind SVMs is to learn a hyperplane,

$$H(\vec{x}) = \vec{w} \cdot \vec{x} + b = 0, \tag{1}$$

where $\vec{x}$ is the representation of a classifying object o as a feature vector, while $\vec{w} \in \Re^n$ (indicating that $\vec{w}$ belongs to a vector space of n dimensions built on real numbers) and $b \in \Re$ are parameters learned from training examples by applying the Structural Risk Minimization principle [31]. Object o is mapped into $\vec{x}$ via a feature function

$$\phi : \mathcal{O} \to \Re^n,$$

where $\mathcal{O}$ is the set of objects; o is categorized in the target class only if

$$H(\vec{x}) \geq 0.$$

The decision hyperplane can be rewritten as:

$$H(\vec{x}) = \left(\sum_{i=1,\dots,l} y_i \alpha_i \vec{x}_i \right) \cdot \vec{x} + b = \sum_{i=1,\dots,l} y_i \alpha_i \vec{x}_i \cdot \vec{x} + b = \sum_{i=1,\dots,l} y_i \alpha_i \phi(o)_i \cdot \phi(o) + b, \tag{2}$$

where y_i is equal to 1 for positive examples and to -1 for negative examples

$$\alpha_i \in \Re \text{ (with } \alpha_i \geq 0, o_i \, \forall i \in \{1, \dots, l\})$$

are the training instances and

$$K(o_i, o) = \langle \phi(o_i) \cdot \phi(o) \rangle$$

is the kernel function associated with the mapping $\phi \cdot$.

Convolution kernels as a measure of similarity between trees compute the common sub-trees between two trees T_1 and T_2. Convolution kernel does not have to compute the whole space of tree fragments. Let the set $\mathcal{T} = \{t_1, t_2, ..., t_{|\mathcal{T}|}\}$ be the set of sub-trees of an extended parse tree, and $\chi_i(n)$ be an indicator function which is equal to 1 if the subtree t_i is rooted at a node n, and is equal to 0 otherwise. A tree kernel function over trees T_1 and T_2 is

$$TK(T_1, T_2) = \sum_{n_1 \in N_{T_1}} \sum_{n_1 \in N_{T_2}} \Delta(n_1, n_2), \tag{3}$$

where N_{T1} and N_{T2} are the sets of T_1's and T_2's nodes, respectively and

$$\Delta(n_1, n_2) = \sum_{i=1}^{|\mathcal{T}|} \chi_i(n_1)\chi_i(n_2). \tag{4}$$

(4) calculates the number of common fragments with the roots in n_1 and n_2 nodes.

3.2 Learning System Architecture

The architecture of learning system is shown in Fig. 10. Once Stanford NLP performs parsing and identifies coreferences, VerbNet components obtains verb signatures, and Stanford NLP also builds anaphora relations, we proceed to finding the same-entity and sub-entity links. After that, we perform segmentation into elementary discourse units and find RST relations, using our own templates (also, third party RST parsers can be incorporated).

As a result, we obtain parse thicket for a text as a sequence of parse trees with additional discourse level arcs linking parse trees for different sentences. For a parse

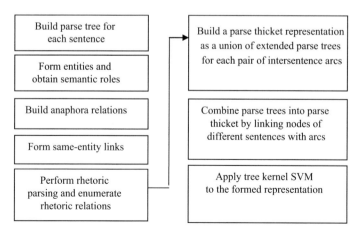

Fig. 10 The architecture of learning system

thicket, we form the set of all phrases and then derive the set of extended phrases according to inter-sentence arcs. Once we build the totality of extended phrases, we form a representation for SVM tree kernel learning, which inputs the set of extended parse trees.

4   Metalanguage and Works of Literature

4.1   Frantz Kafka's Novel "The Trial"

For the example use of metalanguage we consider Frantz Kafka's (1883–1924) novel "The Trial". This novel, written in 1915, has been puzzling many generations of literature critics. Kafka's novels belong to the modernist trends in the literature, which uses various philosophical concepts and new ways of depicting reality. In his works, Franz Kafka went largely beyond other modernists, since his novel structure and language feature very sophisticated artistic expression. The language of Kafka, apparently devoid of any revelations and interpretations, contains an inexhaustible material for linguistic theories. Through scientific analysis of the novel "The Trial", one can see that the writer's use of language is a pointer to an understanding of the underlying aspects of his work. Novel "The Trial" consists of sixteen chapters and an appendix titled "Dream". During his life Franz Kafka could not complete this work. Max Brod, Kafka's closest friend and executor helped publish this novel and gave it the title.

Undoubtedly, many unfinished manuscripts of Kafka makes treating his work with even greater interest, since the absence of ties and open ending always contribute to the construction of the set of inconsistent theories.

"The Trial" does not represent a complex, multi-passage structure, at the first glance. The story presented by Kafka is fairly simple. Joseph K. is the alter ego of the writer. From the first lines of the novel he became involved in a lawsuit. The reader does not get any information about the reason and meaning of this trial, instead the reader is given the common description of its flow and the suspects' encounters with a series of law enforcement officials. It turns out that Joseph K. himself does not know why he ended up in this situation. There are no clues and hints as to why the hero suddenly became defendants and for what crime he may be judged.

4.2   Use of Metalanguage in the Novel

Incomprehensibility of what is happening in the novel makes us scrutinize the peculiarities of language used by Franz Kafka to describe the events in "The Trial".

To comprehend the meaning of the novel as a whole, we need to ascend to the certain level of abstraction. To systematically treat the plot, we consider the text of the novel as written in meta-language, and the remaining part about the reason and the subject of the trial as hypothetically represented in the language-object. Hence what seems to be most interesting and informative to the reader is the theory-object (which is absent), and what is available is the meta-theory (what is explicitly described in the novel) [10].

A metalanguage includes the relations over the expressions of language-object, such as the features of the trial flow. On the contrary, a traditional literature style relies on language-object only, which includes the relations between subjects such as objects of physical and mental world. In most works of literature, metalanguage might be present but its role is secondary and its purpose is to present some generalizations of characters and observations of a writer. However, in "The Trial" the language-object level is totally absent.

Kafka describes all events in metalanguage only, relatively to the subject of the trial. It is easy to verify this statement, having taken an arbitrary paragraph. It describes some discussion of a detail of the trial with some individual involved in law enforcement, such as the following: Object-level details are absent here.

'If you want an apparent acquittal I'll write down an assertion of your innocence on a piece of paper. The text for an assertion of this sort was passed down to me from my father and it's quite unassailable. I take this assertion round to the judges I know. So I'll start off with the one I'm currently painting, and put the assertion to him when he comes for his sitting this evening. I'll lay the assertion in front of him, explain that you're innocent and give him my personal guarantee of it. And that's not just a superficial guarantee, it's a real one and it's binding.'

We attempt to represent the novel "The Trial" at the meta-level only. One can observe that the key character meets with other characters to discuss the details of the trial. All these characters are aware of what is happening, the plot is fully focused on the trial procedures, but the knowledge of the process itself (the object-level) is absent. Franz Kafka naturally used his peculiar writing style and way of narration working on a novel. Many of Kafka's novels are full with mystery and deep analysis of characters' mental states, and the author often uses meta-language as a means of expression. What are the reasons for it?

In the novel the use of metalanguage symbolizes the impossibility to come up with a term for the whole trial. The author is so appalled by what he is describing that he is unable to name it (this name would belong to language-object). Possibly, the described trial is not correlated with the socio-historical context of the 1910s. Metalanguage describes all the specific information about Joseph K., for example, the reader learns from the first pages where Joseph K. lives and what he receives for breakfast and when. But we cannot say for sure that Kafka describes the physical reality, not a human condition between sleeping and waking. In other words, the writer enhances the effect of involvement of the reader in the phantasmagoric and absurd world with the blurred boundaries between dream and reality.

For Franz Kafka's representation of sleep and awakening, it is an inexhaustible source for the exploration of the world and its characters. It was after a restless sleep (troubled dreams) that Gregor Samsa has turned into an insect in the novel "The Metamorphosis" of 1912. On the contrary, a partial detachment from the real world and the abundance of household items in "The Trial" gives the author an opportunity to create a completely different reader perception, subject to human comprehension to a even smaller degree.

4.3   Use of Metalanguage and the Author's Attachment to the Novel

Kafka's scholars consider his biography for better comprehension of his novel. It is important for "The Trial" since the details of Kafka's life are similar to Joseph K.'s. They both occupy an office position (Joseph K. was a bank's vice president). Kafka is familiar with bureaucratic structures which includes even cathedrals. All venues in the novel change as if they are theater decorations, dwellings are instantly turned into a courtroom. Other characters also belong to the same common bureau-cratic/mechanical system, as its indispensable parts. Many characters in the novel are not bright personalities at all, for example, the officers who arrested Joseph K. The writer aims to convincingly capture the whole bureaucratic world, relying on the expressiveness of metalanguage instead of satirical motifs. Apparently, the elements of satire in "The Trial" were deliberately rejected by the author to make the novel sound as a parable.

Creating a novel in which the metalanguage is the only means of descriptions of what is happening, Kafka continues the tradition of historical writings and legends. "The Trial" is a novel where ancient cultural traditions are deeply intertwined with the subconscious mind of the author. In the story of Josef K., described in the meta-language, there is a lot of social terms: the court, the law, the judge, the arrest, the process of accusation. These terms could have been used from ancient times to the modern era of social development. For example, the process of Joseph K. may well be a reference to the Last Judgment and his sentence denotes the divine punishment. In the novel "The Trial" much can be seen from the perspective of theology and spirituality.

For sure, Franz Kafka believed in an idea, not in a reality, being a modernist. His idea came from his own subconscious, which can be described by metalanguage means only [1, 25] to represent processes in which only the metalanguage matters. All moral reflection of the author, his spiritual studies and innermost thoughts reflect his special method of expression.

To some extent, the novel can be attributed to the detective genre. The events occur in the environment Joseph K. is familiar with, and all the characters are described by the author in a fairly routine manner. Yet "The Trial" does not belong to the classical detective because its logical constructions cannot be assessed as a truth. This is because the facts are represented in metalanguage and the object-level facts are missing, so a reader cannot appreciate a solution to a problem, an inherent part of a detective novel. Initiation of "The Trial" occurs in the object level outside of text and cannot be used to accuse the main character.

Kafka's use of metalanguage has a significant impact on the perception of the whole process as something absurd, not subject to any expected rules. All the actions of Joseph K. cannot be viewed from the standpoint of a legal system and are deliberately fairly generic in nature. The writer presents the Court as an omnipotent organization, but in fact this court makes no investigation of the main character, at least from what we know from the novel. The investigation itself cannot exist in this novel, given the style of the description of the whole process to the metalanguage.

The lack of investigation and of disclosure of secrets do not prevent Kafka from maintaining the suspense until the last pages of the novel. One gets the impression that the final sentence of Joseph K. in all circumstances of the case may not be an acquittal. The very first cause of any judicial process lies in the fact of charges being brought. Metalanguage only emphasizes the bureaucratic proceedings, which is the foundation and skeleton of the novel.

Endless absurdity of the described process reminds us about the famous novel by Charles Dickens "Bleak House". Completed in 1853, "Bleak House" was forerunner of a new narrative paradigm. It is important to note that the plot of a Dickens novel in many respects anticipates the history of "The Trial". All events of "Bleak House" take place during an endless litigation process in which the Court of Chancery determines the fate of the characters of the novel.

Dickens conducted to a certain extent an artistic experiment, outlining the events in a very detached way, but not depriving them of a secret meaning. Kafka and Dickens are united in their desire to express in new linguistic forms the meaninglessness and injustice of what is happening in the courts. Both writers were looking for the truth, which is mired in unpleasant courtrooms and stuffy offices. Franz Kafka, perfectly familiar with the works of Dickens, partially enriched "The Trial" by the narrative technique of "Bleak House." But the metalanguage of Kafka does not express deep personal experiences of the character. Instead, it is limited to the unattached vision of distant events.

Commenting and clarification of events by the author is absent in "The Trial". Interestingly, the nature of the novel is close to the authors' perception of what is happening in the society. However, the writer expresses his attitude about the society implicitly, in the metalanguage. The absence of any background and causes of the trial makes the reader focus primarily on the fate of Josef K. Perhaps Kafka strongly felt connection with his main character and considered supplementing the text with his subjective emotions unnecessary. In addition to the above, Kafka's metalanguage hides his deep personal alienation from the system of justice. By the very use of metalanguage the author emphasizes that any litigation is a priori senseless. Also, any man is doomed to become a victim of the system, where bureaucratic principles dominate.

"The Trial" is both illogical and ordered through the use of metalanguage. Joseph K. was not ready for the trial, but at the same time it is clear that without the involvement of a trial this character would be no longer interesting for the reader. Kafka carefully hides mentions of the past of the main character, and also does not give us any opinion about him by the other novel characters. It becomes difficult for the reader to determine whether Josef K. is guilty or not guilty. Metalanguage does not provide any clues about previously occurred events in the novel. The metalanguage of "The Trial" prevents its unequivocal interpretation and leaves the ending of the novel open and ambiguous.

The expectation of the novel outcome is rather strong. The metalanguage expressive means reinforce the impression of the reader via the seemingly meaningless wandering of the main character through the courtrooms. Here the author attempts to create a hostile atmosphere, where the usual course of time is replaced by eternal awaiting of a sentencing. The conflict is manifested not in the struggle of the main character for his honor and dignity, but in the endless discussion of details. Having ruled out by the use of the metalanguage, any information about the reasons for the charges of Josef K., the writer depicts the stages of the judicial process in detail. Even the passage of time in the novel is completely subordinated to the trial. Any decision depends on the unknown higher powers, and the activity of the main character cannot affect the situation.

Metalanguage of "The Trial" is not only a way to describe events, but it also gives this novel an artistic value, independent of the historical and philosophical concepts of the era. Undoubtedly, Franz Kafka does not give any answer in the novel, instead, this novel only raises questions. Writer's method manifests itself in the complete absence of any reasons and plot ties, so that the boundaries of the reader's imagination are unlimited. Thanks to the metalanguage, the writer creates an artistic world where thoughts and speculations become the only possible form of the perception of the novel. Through the analysis of "The Trial", it becomes clear that any conclusions on the product may not be final. Metalanguage creates a paradoxical situation where the lack of basic information makes it possible for any interpretations of the text. By applying a metalanguage in the novel, Franz Kafka included it in the list of works of literature appealing for new generations of readers.

5 Evaluation

5.1 Baseline Classification Approaches

TF*IDF Nearest Neighbor approach finds a document in the training set which is the closest to the given one being recognized. Nearest Neighbor feature is implemented via the search in inverse index of Lucene [3] where the search result score is computed based on TF*IDF model [27]. The query is formed from each sentence of the documents being classified as a disjunctive query including all words except stop-words. The resultant classes along with their TF*IDF scores are weighted and aggregated on the basis of a majority vote algorithm such as [22].

A Naive Bayes classifier is a simple probabilistic classifier based on applying Bayes' theorem (from Bayesian statistics) with strong (naive) independence assumptions. This classifier assumes that the presence (or absence) of a particular feature of a class is unrelated to the presence (or absence) of any other feature. For example, a fruit may be considered to be an apple if it is red, round, and about 4 inches in diameter. Even if these features depend on each other or upon the existence of the other features, a naive Bayes classifier considers all of these properties to independently contribute to the probability that this fruit is an apple. Depending on the precise nature of the probability model, naive Bayes classifiers can be trained very efficiently in a supervised learning setting. In many practical applications, parameter estimation for naive Bayes models uses the method of maximum likelihood. In this study, we use a Naïve Bayes classifier from WEKA package based on [16].

5.2 Forming Training Datasets

For design documents, we built a web mining utility which searched for public design documents on the web in a number of engineering and science domains. We use the following keywords to add to a query for *design document*: *material, technical, software, pharmaceutical, bio, biotech, civil engineering, construction, microprocessor, C++, python, java, hardware, processor, architectural, creative, web*. As a result we formed a set of 1200 documents, it turned out we had 90 % of non-design engineering documents of the classes we want to exclude (meta-documents) and 10% of genuine design documents.

For the literature domain, we collected 200 paragraphs from Kafka's novel "The Trial" describing interaction with people related to the court, as a training set of meta-documents. As a set of object-level documents we manually selected 100 paragraphs of text in the same domain (scholarly articles about "The Trial"). A good example of such language-object documents is a Wikipedia article on the novel, which is a language-object/mixed paragraph (it described the actions of a person):

'K. visits the lawyer several times. The lawyer tells him incessantly how dire his situation is and tells many stories of other hopeless clients and of his behind-the-scenes efforts on behalf of these clients, and brags about his many connections. The brief is never complete. K.'s work at the bank deteriorates as he is consumed with worry about his case.'

On the contrary, a paragraph in metalanguage (an abstract note on the procedure of *apparent acquittal*) looks like this paragraph which we already sited earlier:

'If you want an apparent acquittal I'll write down an assertion of your innocence on a piece of paper. The text for an assertion of this sort was passed down to me from my father and it's quite unassailable. I take this assertion round to the judges I know. So I'll start off with the one I'm currently painting, and put the assertion to him when he comes for his sitting this evening. I'll lay the assertion in front of him, explain that you're innocent and give him my personal guarantee of it. And that's not just a superficial guarantee, it's a real one and it's binding.'

We split the data into five subsets for training/evaluation portions [17]. For the design documents, evaluation results were assessed by quality assurance personnel. For the literature domain, the evaluation was done by the author.

Table 1 Classifying text into metalanguage and language-object

Method	Design document %				Literature %			
	Precision	Recall	F-measure	Standard deviation	Precision	Recall	F-measure	Standard deviation
Nearest neighbor classifier—tf*idf based	55.8	61.4	58.47	2.1	51.7	54.0	52.82	3.4
Naïve Bayesian classifier (WEKA)	57.4	59.2	58.29	3.2	52.4	50.4	51.38	4.2
Manual keyword-based rule selection	93.1	97.5	95.25	1.3	88.1	90.3	89.19	1.3
Manual parse-tree based rules	95.3	97.8	96.53	1.2	N/A	N/A	N/A	1.2
Tree kernel—regular parse trees	73.4	77.6	75.44	2.8	65.7	67.3	66.49	3.8
Tree kernel SVM—extended trees for anaphora	77.0	79.3	78.13	3.1	67.1	67.8	67.45	4.4
Tree kernel SVM—extended trees for RST	78.3	81.5	79.87	2.6	69.3	72.4	70.82	3.7
Tree kernel SVM—extended trees for both anaphora and RST	82.1	85.2	83.62	2.7	70.8	73.0	71.88	4.2

5.3 Evaluation Results

We report the standard deviation of the recognition accuracy expressed as F-measure over five folds achieved by different methods. Table 1 shows evaluation results for the both domains, Design document and Literature. Each row shows the results of a particular classification method.

Keyword statistic-based methods, including Nearest-Neighbor classification and Naïve Bayes, produced rather poor results. Conversely, a manual rule-based system produces a very high accuracy result, especially when manually formed rules go beyond the keywords/phrases and take into account part-of-speech information.

In addition to automated learning, we relied on manual rules for classes. An increase in accuracy by a few percent is achieved in design documents by using manually collected cases for of expressions indicating a use of metalanguage. Also, the rules for writing styles associated with meta-documents have been compiled. These rules also included regular expressions relying on specific document formatting, including a table of content and structure of sections. In the literature domain, that was not possible. Manual rule performance is shown by grayed rows.

Performance of the tree kernel based methods improves as the sources of linguistic properties become richer. For both domains, there is a few percent improvement by using RST relations compared with baseline tree kernel SVM which relies on parse trees only. For the literature domain, the role of anaphora was rather low.

6 Conclusions

In our previous papers we showed how employing a richer set of linguistic information such as syntactic relations between words assists relevance tasks [6, 12, 13]. To take advantage of semantic discourse information, we introduced parse thicket representation and proposed the way to compute similarity between texts based on generalization of parse thickets [8]. We built the framework for generalizing PTs as sets of phrases on one hand, and generalizing PTs as graphs via maximal common subgraphs, on the other hand [14].

In this study we focused on how discourse information can help with a fairly abstract text classification tasks by means of statistical learning. We selected the domain where the only difference between classes lays in phrasing and discourse structures and demonstrated that both are learnable. We compared two sets of linguistic features:

- The baseline, parse trees for individual sentences,
- Parse trees and discourse information,

and demonstrated that the enriched set of features indeed improves the classification accuracy, having the learning framework fixed. We demonstrated that the

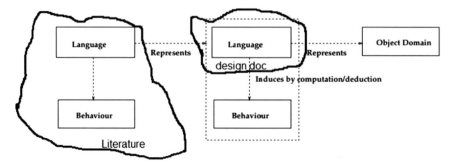

Fig. 11 Meta-reasoning chart: mutual relationships between major classes of our interest

baseline text classification approaches perform rather poorly in the chosen classification domain. Also, kernel-based learning was unable to reach the performance of manually structure-based rules, and we hypothesize that a vast amount of discourse information is not employed in the proposed learning framework.

Meta-reasoning addresses a question of how to give a system its own representation to manipulate. Meta-reasoning needs both levels for both languages and domain behavior. We depict out two main classes of interest in Fig. 11.

[26] outlines a general approach to meta-reasoning in the sense of providing a basis for selecting and justifying computational actions. Addressing the problem of resource-bounded rationality, the authors provide a means for analyzing and generating optimal computational strategies. Because reasoning about a computation without doing it necessarily involves uncertainty as to its outcome, probability and decision theory were selected as main tools.

A system needs to implement metalanguage to impress peers of being human-like and intelligent, needs to be capable of thinking about one's own thinking. Traditionally within cognitive science and artificial intelligence, thinking or reasoning has been cast as a decision cycle within an action-perception loop [21]. An intelligent agent perceives some stimuli from the environment and behaves rationally to achieve its goals by selecting some action from its set of competencies. The result of these actions at the ground level is subsequently perceived at the object level and the cycle continues. Meta-reasoning is the process of reasoning about this reasoning cycle. It consists of both the meta-level control of computational activities and the introspective monitoring of reasoning. In this study we focused on linguistic issues of text which describes such cognitive architecture. It turns out that there is a correlation between a cognitive architecture and a discourse structure used to express it in text. Relying on this correlation, it is possible to automatically classify texts with respect to metalanguage they contain.

In our previous studies we considered the following sources of relations between words in sentences: coreferences, taxonomic relations such as sub-entity, partial case, predicate for subject etc., rhetoric structure relations and speech acts [8]. We demonstrated that a number of NLP tasks including search relevance can be improved if search results are subject to confirmation by parse thicket generalization, when answers occur in multiple sentences. In this study we employed coreferences and rhetoric relation only to identify correlation with the occurrence of metalanguage in text.

Traditionally, machine learning of linguistic structures is limited to keyword forms and frequencies. At the same time, most theories of discourse are not computational, they model a particular set of relations between consecutive states. In this work we attempted to achieve the best of both worlds: learn complete parse tree information augmented with an adjustment of discourse theory allowing computational treatment.

In this paper, we used extended parse trees instead of regular ones, leveraging available discourse information, for text classification. This work describes one of the first applications of tree kernel to industrial scale NLP tasks. The advantage of this approach is that the manual thorough analysis of text can be avoided for complex text classification tasks where the classes are rather abstract. The feasibility of suggested approach to classification lays in the robustness of statistical learning algorithms to unrelated and unreliable features inherent in NLP.

The experimental environment, multi-sentence queries and the evaluation framework is available at https://github.com/bgalitsky/relevance-based-on-parse-trees.

References

1. Aurora, V. 2001. Freudian metaphor and Surrealist metalanguage; in Michel Leiris: The Unconscious and the Sea LittéRéalité, Vol. XIII.
2. Collins, M., and Duffy, N. 2002. Convolution kernels for natural language. In Proceedings of NIPS, 625–32.
3. Croft, B., Metzler, D., Strohman, T. 2009. Search Engines - Information Retrieval in Practice. Pearson Education. North America.
4. Cumby, C. and Roth D. 2003. On Kernel Methods for Relational Learning. ICML, pp. 107–14.
5. Galitsky, B. 2003. Natural Language Question Answering System: Technique of Semantic Headers. Advanced Knowledge International, Adelaide, Australia.
6. Galitsky, B. 2012. Machine Learning of Syntactic Parse Trees for Search and Classification of Text. Engineering Application of AI. 26(3), 1072–91.
7. Galitsky, B. 2013. Transfer learning of syntactic structures for building taxonomies for search engines. Engineering Applications of Artificial Intelligence. Volume 26 Issue 10, pp. 2504–2515.
8. Galitsky, B. 2014. Learning parse structure of paragraphs and its applications in search. Engineering Applications of Artificial Intelligence. 32, 160-84.
9. Galitsky, B., Kuznetsov S. 2008. Learning communicative actions of conflicting human agents. J. Exp. Theor. Artif. Intell. 20(4): 277–317.
10. Galitsky, B., Josep-Lluis de la Rosa, and Boris Kovalerchuk. 2011. Assessing plausibility of explanation and meta-explanation in inter-human conflict. *Engineering Application of AI*, 24(8), 1472–1486.
11. Galitsky, B., de la Rosa JL, Dobrocsi, G. 2012. Inferring the semantic properties of sentences by mining syntactic parse trees. Data & Knowledge Engineering. 81–82, 21–45.
12. Galitsky, B., Gabor Dobrocsi, Josep Lluis de la Rosa 2012. Inferring the semantic properties of sentences by mining syntactic parse trees. Data & Knowledge Engineering http://dx.doi.org/10.1016/j.datak.2012.07.003.

13. Galitsky, B., Usikov, D., and Kuznetsov S.O. 2013. Parse Thicket Representations for Answering Multi-sentence questions. 20th International Conference on Conceptual Structures, ICCS 2013.
14. Galitsky, B., Ilvovsky, D., Kuznetsov SO, and Strok, F. 2013. Improving Text Retrieval Efficiency with Pattern Structures on Parse Thickets, in Workshop Formal Concept Analysis meets Information Retrieval at ECIR 2013, Moscow, Russia.
15. Haussler, D. 1999. Convolution kernels on discrete structures. UCSB Technical report.
16. John, G.H. and Langley, P. 1995. Estimating Continuous Distributions in Bayesian Classifiers. In Eleventh Conference on Uncertainty in Artificial Intelligence, San Mateo, 338–45.
17. Kohavi, R. 1995. A Study of Cross-Validation and Bootstrap for Accuracy Estimation and Model Selection. International Joint Conference on Artificial Intelligence. 1137–43.
18. Kong, F. and Zhou, G. 2011. Improve Tree Kernel-Based Event Pronoun Resolution with Competitive Information. Proceedings of the Twenty-Second International Joint Conference on Artificial Intelligence, 3 1814–19.
19. Lee, H., Chang, A., Peirsman, Y., Chambers, N., Surdeanu, M. and Jurafsky, D. 2013. Deterministic coreference resolution based on entity-centric, precision-ranked rules. Computational Linguistics 39(4), 885–916.
20. Mann, W., Matthiessen, C. and Thompson, S. 1992. Rhetorical Structure Theory and Text Analysis. Discourse Description: Diverse linguistic analyses of a fund-raising text. ed. by Mann, W. and Thompson, S.; Amsterdam, John Benjamins. pp. 39–78.
21. Michael, T., Cox, M.T., and Anita Raja. 2007. Metareasoning: A manifesto.
22. Moore, J.S., and Boyer, R.S. 1991. MJRTY - A Fast Majority Vote Algorithm, In R.S. Boyer (ed.), Automated Reasoning: Essays in Honor of Woody Bledsoe, Automated Reasoning Series, Kluwer Academic Publishers, Dordrecht, The Netherlands, 1991, pp. 105–17.
23. Moschitti, A. 2006. Efficient Convolution Kernels for Dependency and Constituent Syntactic Trees. 2006. In Proceedings of the 17th European Conference on Machine Learning, Berlin, Germany.
24. Recasens, M., de Marneffe M-C, and Potts, C. 2013. The Life and Death of Discourse Entities: Identifying Singleton Mentions. In Proceedings of NAACL.
25. Ricoeur, P. 1975. The Rule of Metaphor: The Creation of Meaning in Language. University of Toronto Press, Toronto.
26. Russell, S., Wefald, E., Karnaugh, M., Karp, R., McAllester, D., Subramanian, D., Wellman, M. 1991. Principles of Metareasoning, Artificial Intelligence, pp. 400–411, Morgan Kaufmann.
27. Salton, G. and Buckley, C. 1988. Term-weighting approaches in automatic text retrieval. Information Processing & Management 24(5): 513—23.
28. Searle, 1969. Speech acts: An essay in the philosophy of language. Cambridge, England: Cambridge University.
29. Sun, J., Zhang, M., and Tan, C. 2010. Exploring syntactic structural features for sub-tree alignment using bilingual tree kernels. In Proceedings of ACL, 306–315.
30. Sun, J., Zhang, M., and Tan. C.L. 2011. Tree Sequence Kernel for Natural Language. AAAI-25.
31. Vapnik, V. 1995. The Nature of Statistical Learning Theory, Springer-Verlag.
32. Zhang, M., Che, W., Zhou, G., Aw, A., Tan, C., Liu, T., and Li, S. 2008. Semantic role labeling using a grammar-driven convolution tree kernel. IEEE transactions on audio, speech, and language processing. 16(7):1315–29.

Relationships Between Probability and Possibility Theories

Boris Kovalerchuk

Abstract The goal of a new area of Computing with Words (CWW) is solving computationally tasks formulated in a natural language (NL). The extreme uncertainty of NL is the major challenge to meet this ambitious goal requiring computational approaches to handle NL uncertainties. Attempts to solve various CWW tasks lead to the methodological questions about rigorous and heuristic formulations and solutions of the CWW tasks. These attempts immediately reincarnated the long-time discussion about different methodologies to model uncertainty, namely: Probability Theory, Multi-valued logic, Fuzzy Sets, Fuzzy Logic, and the Possibility theory. The main forum of the recent discussion was an on-line Berkeley Initiative on Soft Computing group in 2014. Zadeh claims that some computing with words tasks are in the province of the fuzzy logic and possibility theory, and probabilistic solutions are not appropriate for these tasks. In this work we propose a useful constructive probabilistic approach for CWW based on sets of specialized K-Measure (Kolmogorov's measure) probability spaces that differs from the random sets. This work clarifies the relationships between probability and possibility theories in the context of CWW.

1 Introduction

Computing with Words (CWW) is a new area [1, 2] that intends to solve computationally the tasks formulated in a natural language (NL). Examples of these tasks are: "What is the sum of (about 5) + (about 10)?" and "What is the possibility or probability that Mary is middle-aged, if she is 43 and is married for about 15 years?"

In these problems we are given *information I* in the form of membership functions (MFs) such as $\mu_{about-5}$, $\mu_{about-15-years}$, μ_{young}, $\mu_{middle-aged}$, and the respective probability density function (pdfs), e.g., age pdf p_{age}. Similarly for other tasks information *I* can include a membership function μ_{tall} (Height(X)), where X is a person (e.g., John) and a *probability density function* of height p_H (Height(John)) in

B. Kovalerchuk (✉)
Department of Computer Science, Central Washington University,
400 E. University Way, Ellensburg, WA 98926, USA
e-mail: borisk@cwu.edu

© Springer International Publishing AG 2017
V. Kreinovich (ed.), *Uncertainty Modeling*, Studies in Computational
Intelligence 683, DOI 10.1007/978-3-319-51052-1_7

the height interval $U = [u_1, u_n]$, e.g., $u_1 = 160$ cm and $u_n = 200$ cm. The probability space for P_H contains a set of elementary events e_i = "Height of John is u_i": e_1 = "Height of John is u_1", e_2 = "Height of John is u_2",..., e_n = "Height of John is u_n".

Attempts to solve various CWW tasks lead to the methodological questions about rigorous and heuristic formulations and solutions of the CWW tasks which have a tremendous level of uncertainty in both formulations and solutions [3, 4].

These attempts immediately reincarnated the long-time discussion about different methodologies to model uncertainty [5], namely: probability theory (PrT), multi-valued logic (MVL), Fuzzy Sets (FS), Fuzzy Logic (FL), and the Possibility theory. The main forum of the recent discussion was an on-line Berkeley Initiative on Soft Computing (BISC) group in 2014.

Zadeh claims that some computing with words tasks are in the province of the fuzzy logic, and probabilistic solutions are not appropriate for these tasks. "Representation of grade of membership of probability has no validity" [6].

He uses this clam for producing a much wider claim of fundamental limitation of the probability theory for computing with words. "Existing logical systems—other than fuzzy logic—cannot reason and compute with information which is described in natural language [6].

In this paper, we show that this and similar claims may be true if we only consider simple probabilistic models and techniques, like the ones used in routine engineering applications. However, modern probability theory contains more sophisticated concepts and ideas that go way beyond these techniques. We show that, by using such more sophisticated concepts and ideas, we can come up with a probabilistic interpretation of fuzzy sets.

To be more precise, such interpretations have been proposed in the past, e.g., the interpretation using random sets [7–10]. Unfortunately, the random sets interpretation is quite complicated, requires mathematical sophistication from users, and often a lot of data that is not always realistic in many practical applications to provide an efficient computational tool. In contrast, our interpretation that uses a set of specialized Kolmogorov-type probability measures (K-measures, for short), is intuitively understandable and computationally efficient.

In this discussion Zadeh did not specify his term "no validity". This led to the difficulty for an independent observer to test his opinion. To move from an opinion level to a scientific level we need the independently verifiable criteria to test validity of an approach that we propose below for the probabilistic solutions.

Any solution of the task to be a *valid probabilistic solution* must meet criteria of the Axiomatic Probability Theory (APT):

(1) The K-measure spaces (probability spaces) must be created.
(2) All computations must be in accordance with the APT.

In the course of BISC discussion and as result of it we proposed probabilistic formulations and solutions that met criteria (1) and (2) for several CWW test tasks listed below:

(1) "Tall John" (Zadeh's task) [11],

(2) "Hiring Vera, Jane, and Mary" that is based on their age and height (Zadeh's task), [12],
(3) "Degree of circularity of the shape" [11],
(4) "Probably Very is middle-aged" [11, 12],
(5) "Sum of two fuzzy sets" (Zadeh's task) [4].

The proposed K-measure based solutions are at least of the same usefulness as a solution with the fuzzy logic min/max operation. In [4] we proposed several alternative explications of task 5 and their valid solutions within APT.

This work focuses on three problems:

P1: Establishing relations between concepts of possibility and probability;
P2: Establishing relations between grades of membership of a fuzzy set and probabilities;
P3: Outlining future rigorous possibility theory.

The paper is organized as follows: problem P1 is considered in Sect. 2, problem P2 is considered in Sect. 3, problem P3 is considered in Sects. 4 and 5 provides a conclusion.

Within problem P1 we discuss the following topics:

- Possibilistic Semantics,
- Relationship between possibility and probability in natural language (NL),
- Is NL possibility easier than NL probability for people to use?
- Relation between NL words "possibility" and "probability" and the formal math probability concept,
- NL possibility versus modal and non-modal probability,
- Can the sum of possibilities be greater than 1?
- Context of probability spaces for modeling possibility.

Within problem P2 we discuss:

- Are fuzzy sets and possibility distributions equivalent?
- Can we model degrees of membership probabilistically by Kolmogorov's measures spaces (K-spaces)?
- What is a more general concept fuzzy set or probability?
- Is interpreting grade of membership as probability meaningful?
- Relationship among unsharpness, randomness and axiomatic probability?

Within problem P3 we discuss:

- Possibility as upper probability,
- Should evaluation structure be limited by [0, 1] or lattice?
- Exact numbers vs. intervals to evaluate uncertainty.

2 Relations Between Concepts of Possibility and Probability

2.1 Possibilistic Semantics

The literature on the Possibility Theory is quite extensive [8, 9, 13–22]. These studies mostly concentrate on conceptual developments. The important aspects of empirical justification of these developments via psychological and linguistic experiments are still in a very earlier stage. As a result it is difficult to judge whether the current possibility concepts are descriptive or prescriptive. A descriptive theory should reproduce the actual use of the concept of possibility in NL. A prescriptive theory should set up rules for how we should reason about possibilities. Both theories require identification of the semantics of the concept of possibility.

Below we outline semantics of possibility in the natural language based on review in [17]:

- An object/event A that did not yet occur (have not yet been seen) can be a possible event, $Pos(A) \geq 0$.
- The object/event A is possible (not prohibited) if $0 < Pos(A) \leq 1$.
- The possibility of the observed object A is a unitary possibility, $Pos(A) = 1$.
- If to characterize the object/event A somebody selects the statement "A is possible" over a statement "A is probable" then the chances for A to occur are lower than if the second statement would be selected. This is consistent with the common English phrases "X is possible, but not probable.", i.e., $Pos(X) \geq Pr(X)$.
- Types of possible events:

  Type 1: *Merely possible events* are events that may not occur (earthquake) while occurred in the past, $Pos(A)$ is unknown.
  Type 2: *Eventual events* are events that are guaranteed to occur sooner or later (rain), $Pos(A) = 1$

In the Sect. 4.2 we discuss possible deviations from this semantics for some complex situations.

Other aspects of possibilistic semantics are discussed in [14] with four meanings of the word "*possibility*" identified:

- *Feasibility, ease* of achievement, the solution of a problem, satisfying some constraints with linguistic expressions such as "it is possible to solve this problem".
- *Plausibility*, the propensity of events to occur with linguistic expressions such as "it is possible that the train arrives on time".
- *Epistemic*, logical consistency with available information, a proposition is possible, meaning that it does not *contradict* this information. It is an all-or-nothing version of plausibility.
- *Deontic*, possible means *allowed,* permitted by the law.

The expressions below illustrate the semantic difference between NL concepts of possibility and probability where it seems that the expressions with word "probable" are equivalent, but the expressions with word "possible" are not equivalent:

It is *not probable* that "not A" vs. It is probable that A
It is *not possible* that "not A" vs. It is possible that A.

2.2 Is Easiness an Argument for Possibility Theory?

Below we analyze arguments for and against the possibility theory based on [6]. [12] argued for the possibility theory by stating: "*Humans find it difficult to estimate probabilities. Humans find it easy to estimate possibilities.*" To support this claim he presents "Vera's task" that we reconstruct below from his sketchy description.

Vera is middle-aged and middle-aged can be defined as a probability distribution or a possibility distribution. Age is discrete. It is assumed that neither probability nor possibility distributions are given and a person must use his/her judgment to identify them by answering questions for specific ages, e.g.,

(Q1) What is your estimate of the *probability* that Vera is 43?
(Q2) What is your estimate of the *possibility* that Vera is 43?

Zadeh stated that if he would be asked Q1 he "could not come up with an answer. Nor would *anyone* else be able to come up with an answer based on one's perception of middle-aged."

Accordingly answering Q2 must be easier than answering Q1. In fact he equates *possibility* with a *grade of membership*. "Possibility is *numerically equal to grade of membership*. 0.7 is the degree to which 43 fits my perception of middle-aged, that is, it is the grade of membership of 43 in middle-aged." Thus, he is answering question Q3:

(Q3) What is your estimate of the *grade of membership* that middle-aged Vera is 43?

This question is a result of *interpretation* of possibility as a grade of membership. In other words, Zadeh's statement is not about easiness of possibility versus probability for humans, but about *ability or inability to interpret grades of membership as probabilities or possibilities*. We had shown the ability to interpret grades of membership as probabilities in [12] for Zadeh's examples "Mary is middle-aged" and "John is tall". It is done by reinterpreting probabilistically Zadeh's own solution [23] using a set of K-Measure spaces outlined in this paper in Sect. 3.2. For the current example "Vera's age" the reinterpretation is the same. Thus to test easiness of possibility versus probability for humans other arguments are needed.

2.3 Words Possibility and Probability Versus Formal Math Probability

In the BISC discussion Zadeh made several statements:

- "Probabilistic information is *not derivable* from possibilistic information and vice versa" [6].
- "A basic reason is that count and boundary are distinct, *underivable* concepts." [6].
- "0.7 is the possibility that Vera is 43 and has *no relation to the probability* of middle-aged" [6].
- "The possibility that Hans can eat n eggs for breakfast is *independent* of the probability that Hans eats n eggs for breakfast, with the understanding that if the possibility is zero then so is the probability" [6].

Below we show deficiencies of the first three claims and explain independence in the last claim in a way that is consistent with probabilistic interpretation of the concept of possibility. Zadeh bases these statements on claims that, in large measure, the probability theory is "*count*-oriented", fuzzy logic is "fuzzy-*boundary*-oriented", and the possibility theory is "*boundary*-oriented".

Making these statements, Zadeh references Chang's claim [24] that possibility relates to: "Can it happen?" and probability relates to: "How often?" [25] commented on these statements, noting that Zadeh has accepted for his claims the *frequency interpretation* of probability, while there exist five or more, interpretations of probability [26] and in the same way likely multiple interpretations of possibility can be offered depending on some features such as the number of alternatives.

Below we list four well-known different meanings of probability and different people are not equal in using these meanings:

(a) probability as it is known in a *natural language* (NL) without any relation to any type of probability theory,
(b) probability as it is known in the *frequency-based* probability theory,
(c) probability as it is known in the *subjective* probability theory [27],
(d) probability as it is known in the formal *axiomatic* mathematical probability theory (Kolmogorov's theory).

We show deficiencies of Zadeh's statement for meanings (b)–(d) in this section and for meaning (a) later in this paper. It is sufficient to give just a single example of relations for each (b)–(d) that would interpret the possibility that Vera is 43 as a respective probability (b)–(d) of middle-aged.

Consider two events e_1 = {Vera is middle-aged} and e_2 = {Vera is not middle-aged}. In the formal mathematical theory of probability, events are objects of any nature. Therefore, we can interpret each of these two events simply as a sequence of words or symbols/labels, e.g., we can assign the number 0.7 to e_1 and the number 0.3 to e_2. In this very formal way we fully satisfy the requirement of (d) for 0.7 and 0.3 to be called probabilities in the Kolmogorov's axiomatic probability theory [28].

In addition meanings (b) and (c) allow getting 0.7 and 0.3 meaningfully by asking people questions similar to questions like "What is your subjective belief about event e?". An interpretation (b) is demonstrated in depth by Jozo Dujmovic in this volume [29].

To counter Zadeh's claim about the example with n eggs it is sufficient to define two probabilities:

p(Hans *eats* n eggs for breakfast) and p(Hans *can eat* n eggs for breakfast).

The last probability is, in fact, the possibility that Hans eats n eggs for breakfast as the Webster dictionary defines the word "possibility" and we discuss in the next section in details. The idea of that probability (probability of a modal statement with the word "can") was suggested by Cheeseman a long time ago [30]. Thus, we simply have two different probabilities, because we have different sets of elementary events {eat, not eat} and {can eat, cannot eat}. Nobody claims and expects that two probabilities from two different probability spaces must be equal. In the same way, we have two different probabilities, when we evaluate probabilities of rain with two sets of elementary events {daytime rain, no daytime rain} and {nighttime rain, no nighttime rain}.

2.4   Possibility and Probability in Natural Language

The Webster dictionary provides meanings of the words "possibility" and "probability" in the Natural Language (NL). Table 1 contrasts Webster NL definitions of these words. It shows that the major difference is in "will happen", "is happened" associated with word "probability" and "might happen" associated with word "possibility". In other words, we have different levels of chance to happen.

The word "might" expresses a lower chance to happen. Thus even in the pure NL setting there is a strong relation between probability and possibility.

Respectively questions Q1–Q2 can be reformulated into *equivalent* NL forms:

(Q1.1) What is your estimate of the chance that Vera *is* 43?

(Q2.1) What is your estimate of the chance that Vera *might be* 43?

Note that question (Q3): "What is your estimate of the *grade of membership* that middle-aged Vera is 43?" contains term "*grade of membership*" which is a part of

Table 1 Webster NL definitions of words possibility and probability

Possibility	Probability
a *chance* that something *might exist, happen, or be true*	the *chance* that something *will happen*
something that might be done or might happen	something that has a *chance of happening*
abilities or qualities that could make someone or something better in the future	a measure of *how often* a particular event will happen if something (such as tossing a coin) is done repeatedly

the *professional language* not the basic NL, i.e., it requires interpretation. In contrast
Q1.1 and Q2.1 do not contain any professional words and phrases. Note that the
Webster interpretation of possibility is fully consistent with Cheeseman's modal
probability [30] with the modal word "can" that we discussed above for the modal
event "n people can be in the car", P(n people can be in the car).

2.5 Experiment: Is Possibility Easier for People than Probability in NL?

We conducted an experiment asking students to answer the question Q1.1 – Q2.1 in
the form shown below, i.e., to answer questions about probability and possibility as
it follows from the Webster dictionary:
    Please provide your personal opinion for the questions below in the following
situation. It is known that Vera is middle-aged.

  1. What is the chance that she **is 43**?

     Circle your answer: 0 0.1 0.2 0.3 0.4 0.5 0.6 0.7 0.8 0.9 1.0 X
     Circle X if it is difficult for you to assign any number.

  2. What the chance that she **might be 43**?

     Circle your answer: 0 0.1 0.2 0.3 0.4 0.5 0.6 0.7 0.8 0.9 1.0 X
     Circle X if it is difficult for you to assign any number.

    In these questions we deliberately avoided both words "probability" and "possi-
bility", but used the word "chance" in combinations with words "is" and "might be"
to provide a meanings of these concepts which is derived from the Webster dictionary.
It is important to distinguish wording from semantics of the concepts of "probability" and "possibility". Questions must preserve semantics of these concepts, but can
be worded differently. The word "chance" is more common in NL and is clearer to
the most of the people. By asking questions with the word "chance" we preserve
the semantics of concepts of "probability" and "possibility" and made the task for
the respondents easier.
    A total of 27 sophomore Computer Science university students answered these
questions. Practically all of them are English native speakers and none of them took
fuzzy logic or probability theory classes before. Answering time was not limited, but
all answers were produced in a few minutes. Two strong students took an initiative
and offered formulas and charts to compute the answers and generated answers based
on them. Table 2 shows all answers of this experiment and Table 3 shows the statistical
analysis of these answers.
    This experiment leads to the following conclusion:

  1. Most of the students (80%) do not have any problem to answer both questions
     Q1.1 (probability) and Q2.1 (possibility).

Table 2 Answers of respondents

Person	Q1.1	Q2.1
1	0.1	0.3
2	0.1	1
3	0.1	1
1	0.2	0.3
2	0.2	1
3	0.3	0.5
4	0.3	0.6
5	0.3	0.9
6	0.3	1
7	0.3	1
8	0.5	0.8
9	0.5	1
10	0.7	0.8
11	0.7	0.9
12	0.8	1
13	0.8	0.6
14	1	0
15	1	0.4
16	0.4	0.1
17	1	1
18	0.4	0.4
19	0.3	0.3
20	0.1	X
21	X	0.3
22	X	1
23	X	0.3
24	X	1
25	X	X
26	X	X
27	X	X

2. Answers produced quickly (in less than 5 min).
3. Answers are consistent with expected higher value for Q2.1 than for Q1.1. The average answer for Q2.1 is 0.67 and the average answer for Q1.1 is 0.45.
4. Out of 27 students three students found difficult to answer both questions (11.1% of all respondents).
5. Out of 27 students four students found difficult to answer only question Q1.1 (14.8% of all respondents).
6. Out of 27 students one student found difficult to answer only question Q2.1 (3.7% of all respondents).

Table 3 Statistical analysis of answers

Characteristic	Q1.1	Q2.1
Average answer (all numeric answers)	0.45	0.67
Std. dev. (all numeric answers)	0.3	0.34
Number of all numeric answers	20	24
Average answer for cases with answer Q2.1 > answer Q1.1	0.36	0.81
Std. dev. for cases with answer Q2.1 > answer Q1.1	0.23	0.26
Number of answers where answer Q2.1 > answer Q1.1	12	12
Average answer for cases with answer Q1.1 > answer Q2.1	0.63	0.49
Std. dev. for cases with answer Q1.1 > answer Q2.1	0.37	0.36
Number of answers where answer Q1.1 > answer Q2.1	4	4
Average answer for cases with answer Q1.1 = answer Q2.1	0.57	0.57
Std. dev. for cases with answer Q1.1 = answer Q2.1	0.38	0.38
Number of answers where answer Q1.1 = answer Q2.1	3	3
Number of respondents who refused to answer both questions	3	3
Number of respondents who refused to answer Q1.1 only	4	0
Number of respondents who refused to answer Q2.1 only	0	1
Total number of respondents who refused to answer.	7	4

Thus only conclusion 5 can serve as a partial support for Zadeh's claim that answering Q2.1 is impossible. However, it is applicable only to less than 15 % of participants. The other 85% of people have the same easiness/difficulty to answer both Q1.1 and Q2.1.

These results seem to indicate that the probabilistic approach to CWW is at least as good as the possibilistic approach for most of the people who participated in this experiment. The expansion of this experiment to involve more respondents is desirable to check the presented results. We strongly believe that experimental work is necessity in fuzzy logic to guide the production of meaningful formal methods and the scientific justification of existing formal methods, which is largely neglected.

2.6   Is Sum of Possibilities Greater Than 1?

Consider a probability space with two events e_{m1} and e_{m2}

$$e_{m1} = \text{"A } may \text{ happen", } e_{m2} = \text{"A } may\,not \text{ happen",}$$

Thus, for A = "rain" we will have events

$$e_{m1} = \text{"rain } may \text{ happen", } e_{m2} = \text{"rain } may\,not \text{ happen",}$$

Respectively for A = "no rain" we will have events

$$e_{m1} = \text{"no rain } may \text{ happen", } e_{m2} = \text{"no rain } may\,not \text{ happen",}$$

Note, that "dry weather" is not equal to "no rain" (negation of rain), because snow is another part of negation of rain in addition to dry weather.

Next consider another probability space with two events without "may" and "may not",

$$e_1 = \text{"A", } e_2 = \text{"notA"}$$

Denote probabilities in spaces with "may" and "may not" as P_m and without them as P_e, respectively. In both cases in accordance with the definition of probability spaces sums are equal to 1:

$$P_m(e_{m1}) + P_m(e_{m2}) = 1, P_e(e_1) + P_e(e_2) = 1,$$

e.g., $P_m(e_{m1}) = 0.4, P_e(e_{m2}) = 0.6$, and $P_e(e_1) = 0.2, P_e(e_2) = 0.8$.

Note that in example $P_m(e_{m1}) + P_e(e_2) = 0.4 + 0.8 = 1.2 > 1$. In fact, these probabilities are from different probability spaces and are not supposed to be summed up.

The confusion takes place in situations like presented below. Let

$$e_{m1} = \text{"rain } may \text{ happen" and } e_2 = \text{"no rain"}$$

instead of

$$e_{m2} = \text{"rain } may\,not \text{ happen".}$$

Probability of e_2 = "no rain" is probability of a *physical event* (absence of a physical event). In contrast probability of e_{m1} = "rain *may not* happen" is probability of a *mental event*. It depends not only on physical chances that rain will happen, but also on the *mental interpretation* of the word "might" that is quite subjective.

This important difference between events e_2 and e_{m2} can be easily missed. It actually happens with a resulting claim that the possibility theory must be completely different from the probability theory based on such mixing events from different probability spaces and getting the sum of possibilities greater than 1.

2.7   Context of Probability Spaces

Consider a sum for possibilities of events A and not A: Pos(A) + Pos(not A) and a sum of probabilities Pr(A) + Pr(not A). For the last sum it is assumed in the Probability Theory that we have the *same context* for Pr(A) and Pr(not A) when it is defined that Pr(A) + Pr(not A) = 1. Removing the requirement of the same context can make the last sum greater than 1. For instance, let Pr(A) = 0.6 for A = "rain", which is computed using data for the last spring month, but Pr(not A) = 0.8 is computed using data for the last summer month. Thus, Pr(A)+Pr(not A) = 0.6 + 0.8 = 1.4 > 1.

The same *context shift* can happen for computing Pos(A) + Pos(not A). Moreover, for mental events captured by a possibility measure checking that the context is the same and not shifted is *extremely difficult*. Asking "What is the possibility of the rain?", "What is possibility that rain will not happen?", "What is the chance that rain may happen?" and "What is the chance that no rain may happen?" without controlling the context can easily produce the sum that will be greater or less than 1. Sums of possibility values from such "context-free" questions cannot serve as a justification for rejecting the probability theory and for introducing a new possibility theory with the sum greater than 1. The experiments should specifically control that *no context shift* happen.

2.8   Natural Language Possibility Versus Modal and Non-modal Probability

Joslyn [17] provided an example of the difference between probability and possibility in the ordinary natural language for a six-sided die. Below we reformulate it to show how this difference can be interpreted as a difference between *modal probability* [30] and *non-modal probability*. A six-sided die has six *possible* outcomes (outcomes that *can occur*). It is applied to both balanced and imbalanced dies. In other words, we can say, each face is *completely possible*, possible with *possibility 1*, Pos(s) = 1 or *can occur for sure*. In contrast, different faces of the imbalanced die *occur* with different probabilities $P(s_i)$.

Thus, for all sides s_i we have $Pos(s_i) = 1$, but probability $P(s_i) < 1$. Respectively $\Sigma_{i=1:6}Pos(s_i) = 6$ and $\Sigma_{i=1:6}P(s_i) = 1$. At the first glance, it is a strong argument that possibility does not satisfy Kolmogorov's axioms of probability, and respectively possibility is not probability even from the formal mathematical viewpoint, not only as a NL concept.

In fact, $Pos(s_i)$ satisfies Kolmogorov's axioms, but considered not in a single probability space with 6 elementary events (die sides), but in 6 probability spaces with pairs of elementary events: $\{s_1, \text{not } s_1\}$, $\{s_2, \text{not } s_2\}$, ..., $\{s_6, \text{not } s_6\}$ each.

Consider $\{s_1, \text{not } s_1\}$ with $Pos(s_1)$ as an answer for the question: "What is the probability that side s_1 *may occur/happen?*" and $Pos(\text{not } s_1)$ as an answer for the question: "What is the probability that side s_1 *may not occur/happen?*". The common sense answers are $Pos(s_1) = 1$, and $Pos(\text{not } s_1) = 0$ that are fully consistent with the Kolmogorov's axioms. This approach is a core of our approach with the *set of probability spaces* and probability distributions [31–33].

Consider another version of the same situation of mixing of probabilities and possibilities from different spaces as an incorrect way to justify superadditivity for both of them. It is stated in [34]: "...probabilistic relationship between p(A) and p(not A) is fully determined. By contrast, P(A) and P(not A) are weakly dependent in real life situations like medical diagnosis. For instance, given a particular piece of *evidence*, A can be *fully possible* and not A can be *somewhat possible* at the same time. Therefore, a "superadditivity" inequality stands: Pos(A) + Pos(not A) ≥ 1."

Let us analyze this medical diagnosis example for a pair of events

$$\{A, notA\} = \{e_1, e_2\} =$$
$$= \{\text{malignant tumor occurred for the patient, malignant tumor did not occur for the patient}\}$$

Then according to the probability theory we must have $P_e(e_1) + P_e(e_2) = 1$ and it seems reasonable if these probabilities will be based on frequencies of the A and not A under the given evidence. Here in $\{e_1, e_2\}$ the physical entity (malignant tumor) is negated to get e_2 from e_1, and no modality is involved.

Next consider another pair of events (modal events with the words "might" and "might not"):

$\{e_{m1}, e_{m2}\} = \{$malignant tumor might occur for the patient, malignant tumor might *not* occur for the patient$\}$

Having $P_m(e_{m1}) + P_m(e_{m2}) = 1$ is also seems reasonable for these events if these probabilities will be based on frequencies of subjective judgments of experts about e_{m1} and e_{m2} under given evidence. The experts will estimate the same physical entity "malignant tumor" for two different (opposite) modalities "can" and "cannot". In other words here we *negate modality* ("can") not a *physical entity* (malignant tumor) as was the case with $\{e_1, e_2\}$. The same property is expected for the events:

$\{e_{m3}, e_{m4}\} = \{$benign tumor might occur for the patient, benign tumor might not occur for the patient$\}$

The pairs $\{e_{m1}, e_{m2}\}$ and $\{e_{m3}, e_{m4}\}$ differ from another pair that mix them:

$\{e_{m1}, e_{m3}\} = \{$malignant tumor might occur for the patient, *no* malignant tumor (benign tumor) *might* occur for the patient$\}$

For this pair, $P_m(e_{m1}) + P_m(e_{m3}) = 1$ seems less reasonable. Here in e_{m3} the *physical entity* (malignant tumor) is *negated*, but *modality* ("might") is the same. In essence, e_{m1} and e_{m3} are from different modal probability spaces, and should not be added, but operated with as it is done in the probability theory with probabilities from different probability spaces.

This example shows that probability spaces for modal entities must be built in a *specific way* that we identified as N1 below. Let A be a modal statement. It has two components: a physical entity E and a modal expression M, denote such A as A = (E, M). Respectively, there are three different not A:

(N1): not A = (E, not M)
(N2): not A = (not E, M)
(N3): not A = (not E, not M)

As we have seen above, it seems more reasonable to expect additivity for N1 than for N2 and N3. The next example illustrates this for N1 and N2. Consider A = (E, M) where E = "sunrise", not E = "sunset", M = "can", and not A = (E, not M) for N1, and not A = (not E, M) for N2.

For N1 the questions are:

"What is the probability that John *can* watch the sunrise tomorrow?" and

"What is the probability that John *cannot* watch the sunrise tomorrow?

For N2 these questions are:

"What is the probability that John can watch the *sunrise* tomorrow?" and

"What is the probability that John can watch the *sunset* tomorrow?

These questions in the possibilistic form can be:

"What is the possibility that John will watch the *sunrise* tomorrow?" and

"What is the possibility that John will watch the *sunset* tomorrow?"

Both probabilities/possibilities P(A) and P(not A) can reach 1 for N2, therefore building spaces with N1 and N3 should be avoided. The situation is the same as in the probability theory itself—not every set of events can be used as a set of meaningful elementary events.

Consider the next example, based on the following joke: "Can misfortune make a man a millionaire? Yes if he is a billionaire. In this example the first two pairs of questions are in N1, and the last pair is in N2:

What is the probability that misfortune *can* make a man a millionaire if he is a billionaire?

What is the probability that misfortune *cannot* make a man a millionaire if he is a billionaire?

What is the possibility that misfortune *will* make a man a millionaire if he is a billionaire?

What is the possibility that misfortune *will not* make a man a millionaire if he is a billionaire?

What is the possibility that misfortune will make a man a millionaire if he is a billionaire?

What is the possibility that negligence will make a man a millionaire if he is a billionaire?

Consider another example on possibility of n people in the car [30]. Let n = 10 and A is "10 people are in the car". What is the possibility of "10 people in the car"? Let Pos(A) = 0.8 then "What is the possibility of not A?", i.e., possibility of "not 10 people in the car". We need to define a probability space {A, notA}. Note that "not 10 people in the car" includes 9 people. Thus

$$Pos(notA) \geq Pos(9),$$

Next it is logical to assume that Pos(9) > Pos(10). Thus, Pos(not A) ≥ 0.8 with

$$Pos(A) + Pos(notA) > 1.$$

As we noted above such "superadditivity" often is considered as an argument for the separate possibility theory with superadditivity.

In fact this cannot be such argument because there is a space with normal additivity (P(A) + p(notA) = 1). In that space we have a different A. It is not an event that 10 people are in the car, but a modal statement, where A = "10 people *can be* in the car", and respectively not A = "10 people *cannot be* in the car", e.g., with P(A) = 0.8 and P(notA) = 0.2.

Thus in general superadditivity situations N2 and N3 can be modeled with a Set (pair) of probability (K-spaces) Spaces (SKS) of N1 type without any superadditivity. This is in line with our approach [32] discussed above.

Note that superadditivity can really make sense in the probability theory and be justified, but very differently, and not as an argument for the separate possibility theory. See a chapter by Resconi and Kovalerchuk in this volume.

3 Relations Between Grades of Membership of Fuzzy Set and Probabilities

3.1 What is More General Concept Fuzzy Set, Probability or Possibility?

Joslyn stated [17] that neither is fuzzy theory specially related to possibility theory, nor is possibility theory specifically related to fuzzy sets. From his viewpoint "both probability distributions and possibility distributions are *special cases of fuzzy sets*". It is based on the fact an arbitrary fuzzy set can be specialized to be probability or possibility by imposing additional properties. For probability this property is that the sum (integral) or all values must be equal to 1. For possibility this property is that the maximum must be equal to 1.

The specialization of an arbitrary fuzzy set to probability or possibility can be done as follows. Let $m(x)$ be a membership function of a fuzzy set on the interval [a, b]. We can normalize values of $m(x)$ by dividing them by the value S which is the sum (integral) of m(x) values in this interval and get probabilities, $p(x) = m(x)/S$. Alternatively, we can divide $m(x)$ by the value M which is the maximum of m(x) values in the interval [a, b], $pos(x) = m(x)/M$.

However, these transformations do not establish the actual relation between theories: Fuzzy Set Theory (FST), Fuzzy Logic (FL) and Probability Theory (PrT) where the *main role play operations* with membership functions m and probabilities p. The concept of the fuzzy set is only a part of the fuzzy set theory, we should not base the comparison of the theories based only one concept, but should analyze other critical concepts such as operations.

Probability theory operations of intersection and union ($\cap$ and $\cup$) are *contextual*, $p(x \cap y) = f(p(x), p(y|x)) = p(x)p(y|x)$, where conditional probability p(y|x) expresses contextual dependence between x and y. In contrast, operations in FST and FL, i.e., t-norms and t-conforms are *"context-free"*, $m(x \& y) = f(m(x), m(y))$, that is no conditional properties, dependencies are captured. In fact FL, FST present rather a *special case* of the Probability Theory because all t-norms are a *subset* of copulas that represent n-D probability distributions [4, 35].

Joslyn [17] also made an interesting comment that the same researchers, who object to confusion of membership grades with probability values, are not troubled with confusion of membership grades with possibility values. In fact both confusions have been resolved by:

(In1) interpreting a **fuzzy set** as a **set of probability distributions** (**SPD**) not a single probability distribution [4, 31, 32].

(In2) interpreting **possibility** as **modal probability** as proposed by Cheeseman [30].

In the next section we illustrate the first interpretation and show how fuzzy sets are interpreted/modeled by SPD. Note that the interpretation of membership functions as a *single probability distribution* is too narrow and often is really confusing. The interpretation In2 already has been discussed in the previous sections.

3.2   What Is the Relation of a Membership Function to Probability?

Piegat [25, 36] stated that a membership function of the crisp set is not pdf and not probability distribution, noting that it only informs that a set element belongs to the considered set.

While it is true that in general a membership function (MF) is not a pdf or a probability distribution it is not a necessary to compare them in a typical way, i.e., by one-to-one mapping. This relation can be expressed in different ways.

The relation can be one-to-many similarly to the situation when we compare a linear function and a piecewise linear function. While these functions differ, a set of linear functions accurately represents a piecewise linear function. In the case of MF a set of pdfs represents any MF of a crisp set or a fuzzy set [32]. This technique uses a set of probability Spaces that are *Kolmogorov's measure (K-Measure) spaces (SKM)* [4].

Let $f(x) = 1$ be a MF of the interval [a, b]. Consider a pair of events

$$\{e_1(x), e_2(x)\} = \{x \text{ belongs to } [a, b], x \text{ does not belong to } [a, b]\}$$

with pdf $P_x(e(x)) = 1$, if x belongs to [a, b] else $P_x(e(x)) = 0$. Thus, $P_x(e_1(x)) = f(x) = 1$, $P_x(e_2(x)) = 0$ and $P_x(e_1(x)) + P_x(e_2(x)) = 1$. This is a simplest probability space that satisfies all Kolmogorov's axioms of probability. A set of these pdfs $\mathbf{p} = \{P_x\}$ for all x from [a, b] is equivalent to MF of the crisp set.

Note that here we use the term probability density function (pdf) in a general sense as any function with sum(integral) of its values equal to one on its domain. It can be a traditional pdf defined on a continuous variable or on a set of any nature continuous or discrete that will be called a set of elementary events. As we will see below typically these sets will be pairs of NL sentences or phrases.

For a fuzzy set with a membership function $f(x)$ we have probabilities

$$P_x(e_1(x)) = f(x), P_x(e_2(x)) = 1 - f(x)$$

that satisfy all Kolmogorov's axioms of probability, where again a set of these pdfs $\mathbf{p} = \{P_x\}$ for all x from [a, b] is equivalent to MF of the fuzzy set $\langle [a, b], f \rangle$. Figure 1 shows these probability spaces for each x that is human's height h.

In other words MF is a *cross-section* of a set of pdfs $\mathbf{p}$. More details are in [32]. The advantage of MF over a set of pdfs is in *compactness* of the representation for many MFs such as triangular MFs. Each of these MFs "compresses" an infinite number of pdfs. A probability space S(180) is shown as a pair of circles on a vertical line at point h = 180. As we can see from this figure, just two membership functions serve as a compact representation of many simple probability spaces. This is a *fundamental representational advantage* of Zadeh's fuzzy linguistic variables [37, 38] versus multiple small probability spaces.

Fig. 1 Sets of probability spaces S(h) for elementary events {short(h), tall(h)} for each height h

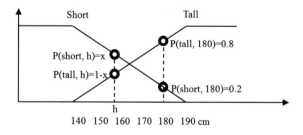

It is important to note that having two MFs $m_1(x)$ and $m_2(x)$ for two fuzzy sets the operations with them must follow the rules of probabilistic operations AND and OR with sets of respective pdfs $\mathbf{p}_1$ and $\mathbf{p}_2$ on two sets of probability spaces to be *contextual*.

An example of these operations is presented in [4] using convolution-based approach. In this context it is very important to analyze the relation of t-norms and t-conorms used in fuzzy logic with sets of probability distributions. One of such analyses is presented in [4, 35] in this volume.

The common in fuzzy logic "context-free" heuristic min and max operations and other t-norms and t-conorms for AND and OR operations on fuzzy sets are not correct probabilistic operations in general. These context-free operations can only serve as approximations of contextual probabilistic operations for AND and OR operations. For the tasks where the context is critical context-free operations will produce incorrect results.

Thus, membership functions and sets of pdfs are mutually beneficial by combining *fast* development of pdfs (coming from MFs) and rigor of *contextual operations* (coming from probability).

In this sense, fuzzy membership functions and probabilities are *complimentary not contradictory*. It would be incorrect to derive from advantages of MFs outlines above the conclusion that probabilistic interpretation only brings a complication to a simple fuzzy logic process for solving practical problems. The probabilistic interpretation of MFs removes a big chunk of heuristics from the solution and brings a rigorous mechanism to incorporate context into AND and OR operations with fuzzy sets.

3.3   Are Fuzzy Sets and Possibility Distributions Equivalent?

Zadeh [22] defined a possibility distribution as a fuzzy set claiming that possibilistic concepts are inherently more appropriate for fuzzy theory than probabilistic concepts.

This equivalence was rejected in earlier studies [17] which reference works with a *wider view* on the concept of possibility that includes:

- its generalization to the *lattice* [13],
- involvement of *qualitative relations* [16],
- the semantics of *betting* [19],
- *measurement* theory [21],
- *abstract* algebra [20].

Other alternatives include:

- *modal probabilities* [30] and the later work by
- *elaborated lattice* approach [39].

Also a general logic of uncertainty based on the lattice can be traced to Gains [18, 40].

3.4   Is Interpreting Grade of Membership as Probability Meaningful?

Zadeh [6] objects to equating a grade of membership and a truth value to probability, but equates a grade of membership and a truth value as obvious. He states that "equating grade of membership to probability, is *not meaningful*", and that "consensus-based definitions may be formulated *without* the use of probabilities".

To support this statement he provided an example of a *grade of membership* 0.7 that Vera is middle-aged if she is 43. Below we explicate and analyze his statement for this example that is reproduced as follows.

Grade of Membership that Vera is middle-aged being 43:

$$\text{GradeMembership(Vera is middle-aged|Age} = 43) = 0.7, \tag{1}$$

Truth value for Age = 43 to be middle-aged,

$$\text{TruthValue(Age} = 43 \text{ is middle-aged)} = 0.7. \tag{2}$$

Possibility that Vera is middle-aged for the Age = 43,

$$\text{Poss(Age} = 43| \text{ Vera is middle-aged)} = 0.7 \tag{3}$$

Probability that Vera is middle-aged for the Age = 43,

$$\text{Prob(Vera is middle-aged| Age} = 43) = 0.7 \tag{4}$$

In other words, Zadeh equates (1), (2) and (3) and objects to equating all of them to (4) as *not meaningful* and *not needed*.

Consider a crisp voting model that equates the grade of membership to probability *meaningfully*. In this model, the statement "Vera is middle-aged" is classified as true or false by each respondent. Respectively the grade of membership is defined as the *average* of votes. Jozo Dujmovic presented this approach in details in his chapter in [41] and in this volume [29].

In a *flexible voting model* the voter indicates a grade of membership on the scale from 0 to 1 as a subjective/declarative probability of each voter. In the case of multiple voters the consensus-based grade of membership is defined as the *average* of subjective/declarative probabilities of voters. Thus, a meaningful probabilistic interpretation of a grade of membership exists for long time and is quite simple.

Next we comment on Zadeh's statement that "consensus-based definitions may be formulated *without* the use of probabilities". If the intent of the whole theory is to get results like in (1) then it can be obtained without using the word "probability". In fact both fuzzy logic and probability theory are interested to get more complex answers than (1). This leads to the need to *justify operations* on combination such as t-norms and t-conorms used in fuzzy logic as normative or descriptive. These operations differ from probability operations. Moreover there are multiples of such operations for both conjunction and disjunction in contrast with the probability theory, where the conjunction (intersection) and disjunction (union) are unique. Thus each t-norm and t-conorm must be justified against each other and probabilistic operations. Note that experimental studies [42, 43] with humans do not justify min t-norm versus product t-norm, which is a probabilistic operation for independent events. In general the justification of operations continues to be debated [3, 44] while extensive experimental studies with humans still need to be conducted in the fuzzy logic justification research.

3.5   Unsharpness of Border And Randomness Versus Axiomatic Probability

Chang [24] differenced fuzzy logic and probability theory stating that (1) *fuzzy sets theory* deals with *unsharpness*, while the *probability theory* deals with *randomness* and (2) the problems with unsharpness *cannot be converted* into "randomness problems". This statement was made to justify that there are problems which cannot be solved by using the probability theory, but which can be solved by using the fuzzy sets and fuzzy logic. In this way, he attempted to counter my arguments that probability theory has a way to solve Zadeh's CWW test tasks by using the formal axioms of probability theory (called by Tschantz [45] axioms of *K-measure space*).

Tschantz [45] objected to Chang's statement: "As Boris has said many times, he is not attempting to convert reasoning about vague words into reasoning about randomness. He is merely attempting to *reuse the formal axioms of Kolmogorov*. While these axioms were first found useful in modeling randomness, the fact that they are useful for modeling randomness does not mean that they cannot also be useful for modeling vagueness. To put it another way, someone can use the same model M to model both X or Y even if X and Y are not the same thing. Thus, I do not buy CL Chang's argument that Boris is incorrect. If he [24] shows that vagueness has some feature that makes it behave in such a way that is so different from randomness that no model can be good at modeling both, then I would buy his argument. However, I would need to know *what that feature is*, what "good" means in this context, and how he proves that no model can be good at both. It seems far easier to point to some contradiction or unintuitive result in Boris's model."

So far no contradiction or unintuitive results have been presented by opponents to reject the approach based on the Space of K-Measures.

4 What Is the Future Rigorous Possibility Theory?

4.1 Is Possibility an Upper Estimate of Probability for Complex Events?

Below we analyze abilities to interpret possibility Pos as an upper estimate of probability Pr using the current min-max possibility theory where "&" operator is modeled as min and "or" operator is modeled as max.

Section 2.4 had shown how questions with words "possibility" and "probability" can be reworded to be in a comparable form using their Webster definitions. Thus Webster leads to the conclusion that possibility must be an upper estimate of probability in NL for every expression X,

$$Pos(X) \geq Pr(X)$$

Respectively when possibility of A&B is defined as

$$Pos(A\&B) = min(Pos(A), Pos(B))$$

it gives an *upper estimate* of Pr(A&B),

$$Pos(A\&B) = min(Pos(A), Pos(B)) \geq Pr(A\&B)$$

However in contrast when Pos(A or B) is defined as

$$Pos(A \text{ or } B) = max(Pos(A), Pos(B))$$

it gives a *lower estimate* of Pr(A or B) = P(A) + P(B) - P(A&B)

$$Pos(A \text{ or } B) = max(Pos(A), Pos(B)) \leq Pr(AorB)$$

Thus, the NL property Pos(X) ≥ Pr(X) does not hold when X = (A or B) in the min-max possibility theory advocated by Zadeh and others.

Let us compute the possibility of another composite expression X = A&(B or C) using the min-max possibility theory:

$$Pos(A\&(B \text{ or } C) = min(Pos(A), Pos(B \text{ or } C))) =$$
$$min (Pos(A), max (Pos(B), Pos(C))).$$

The value min (Pos(A), Pos(B or C))) is supposed to be an upper estimate for Pr(A&(B or C)). However, min(Pos(A), max(Pos(B), Pos(C))) is not an upper estimate for Pr(A&(B or C)) because max (Pos(B), Pos(C)) is a lower estimate for Pr(B or C). The infusion of this lower estimate to min (Pos(A), Pos(B or C))) can destroy an upper estimate.

Below we show it with a numeric example. Consider sets A, B and C such that A has a significant overlap with (B or C) with probabilities

$$\Pr(A) = 0.4, \Pr(B) = 0.3, \Pr(C) = 0.32, \Pr(B \text{ or } C) = 0.5, \Pr(A\&(B \text{ or } C)) = 0.35$$

From these numbers we see that

$$\Pr(A) = 0.4 < \Pr(B \text{ or } C) = 0.5$$

$$\Pr(B \text{ or } C) = 0.5 > \max(\Pr(B), \Pr(C)) = \max(0.3, 0.32)) = 0.32$$

$$\min(\Pr(A), \Pr(B \text{ or } C))) = \min(0.4, 0.5) = 0.4 =$$
$$\text{Pos}(A\&(B \text{ or } C)) > \Pr(A\&(B \text{ or } C)) = 0.35.$$
$$\text{Pos}(A\&(B \text{ or } C)) = \min(\Pr(A), \max(\Pr(B), \Pr(C))) =$$
$$\min(0.4, 0.32) = 0.32 < \Pr(A\&(B \text{ or } C)) = 0.35$$

Thus, upper estimate 0.4 of probability 0.35 is converted to the lower estimate 0.32 of this probability 0.35.

This confirms that for composite expressions that involve "or" operator, we cannot ensure that possibility of that expression is an upper estimate of its probability in the min-max possibility theory. Thus, a future possibility theory should differ from the current min-max possibility theory to be able to hold the property that $\text{Pos}(X) \geq \Pr(X)$ for all composite expressions X.

4.2　Should Evaluation Structure Be Limited by [0, 1] or Lattice?

In Sect. 3.3 we referenced several studies that generalized the possibility theory to a lattice (partial order of evaluations) from the full order in [0.1]. However, to the best of our knowledge, none of the previous studies went beyond a *lattice structure*. The lattice assumption means that the *absolute positive* and *negative possibility* exists with max value commonly assigned to be equal to 1 and min value to be equal to 0. It does not mean that x with $\text{Pos}(x) = 1$ always happens, but such x is one of the real alternatives, e.g., x is physically possible. Similarly $\text{Pos}(x) = 0$ can mean that x physically cannot happen.

There are NL concepts where this assumption can be too restrictive. Consider the concept of happiness and a chance that somebody might by absolutely happy. Can somebody be absolutely happy (unhappy) to the level that he/she cannot be happier in the future? The lattice assumption tells that if x is absolutely happy at time t, Happy(x) = 1, then x cannot be happier later. If x tells that he/she was absolutely happy yesterday, and even happier today then under the lattice model this person will be inconsistent or irrational. This person "exceeded" the absolute maximum. Another example is "possibility of absolute understanding". What is a chance that somebody might understand something absolutely today and cannot increase understanding later? In other words, we question the concept of absolute maximum for possibilities such as absolute happiness, Pos(Happy(x)) = 1 and absolute understanding, Pos(Understand(x)) = 1. The number of such "unlimited" NL concepts is quite large.

How to build a model to accommodate such behavior? We can simply remove limit 1 (or any constant given in advance) and allow the unlimited values of possibility of happiness, in this example and, in a general measure of possibility.

Other difficulties for constructing an evaluation medium are that often a set of possible alternatives is uncertain, and assigning the possibility values to a known alternative is highly uncertain and difficult. For instance, the NL expression "unlimited possibilities" often means that a set of alternatives is huge, and not fully known, being uncertain to a large extent. Often it is easier to say that A is more possible than B, and both alternatives are not fully possible, than to give numeric Pos(A) and Pos(B). We simply may have no stable landmark for this. The NL expression "unlimited possibilities" can mean that for a possible A, there can exist a B, which is more *possible* than A. The next difficulty is that adding more possible alternatives to the consideration changes the possibility values. Next the fixed value of max of possibility, e.g., $\max_{x \in X}$Pos(x) = 1, is not coming from the NL, but is just *imposed* by the mathematical formalisms. We advocate the need for developing the formalism with "unlimited" possibilities.

4.3 From an Exact Number to an Interval

Another common viewpoint is that possibility expresses a *less constraining form of uncertainty* than probability [46], and even that the probability theory brings in the excess of constraints [17] requiring an *exact number* instead of an interval of belief and plausibility [Bel, Pl] as in the evidence measure in the belief theory.

Again while it is true that probability is a single number, not an interval, the issue is not in the formal expansion to the interval, but in the justification of operations to produce the intervals and operations with intervals. We need well justified *empirical procedures* to get them. Without this generalization, the intervals and other alternative formalisms have little value. Somebody must bring the "life meaning" to it. Respectively the claims of successful applications need to be closely scrutinized to answer a simple question: "Is the application successful because of the method, or in spite of it?" In other words: "Does this method capture the properties of the application better than the alternative methods, in our case the pure probabilistic methods?" It short, deficiency of one method should not be substituted by deficiency of another method.

5 Conclusion

This work clarified the relationships between the probability and the possibility theories as tools for solving Computing with Words tasks that require dealing with extreme uncertainty of the Natural Language. Multiple arguments have been provided to show deficiencies of Zadeh's statement that CWW tasks are in the province of the fuzzy logic and possibility theory, and probabilistic solutions are not appropriate for these tasks. We proposed a useful constructive probabilistic approach for CWW based on sets of specialized K-Measure (Kolmogorov's measure) probability spaces. Next we clarified the relationships between probability and possibility theories on this basis, outlined a future rigorous possibility theory and open problems for its development.

References

1. Zadeh, L., Computing with Words: Principal Concepts and Ideas, Springer, 2012.
2. Zadeh, L., From Computing with Numbers to Computing with Words—From Manipulation of Measurements to Manipulation of Perceptions, IEEE Transactions on Circuits and Systems—I: Fundamental Theory and Applications, Vol. 45, No. 1, 1999, 105–119.
3. Beliakov, G., Bouchon-Meunier, B., Kacprzyk, J., Kovalerchuk, B., Kreinovich, V., Mendel J.,: Computing With Words: role of fuzzy, probability and measurement concepts, and operations, Mathware & Soft Computing Magazine. Vol.19 n.2, 2012, 27–45.
4. Kovalerchuk, B. Summation of Linguistic Numbers, Proc. of North American Fuzzy Information Processing Society (NAFIPS) and World Congress on Soft Computing, 08-17-19, 2015, Redmond, WA pp.1–6. doi:10.1109/NAFIPS-WConSC.2015.7284161.
5. Zadeh, L.: Discussion: Probability Theory and Fuzzy Logic are Complementary rather than competitive, Technometrics, vol. 37, n 3, 271–276, 1995.
6. Zadeh L., Berkeley Initiative on Soft Compting (BISC), Posts on 01/30/2014, 02/05/2014 , 04/09/2014, 05/27/2014, 10/27/2014 http://mybisc.blogspot.com.
7. Goodman, I.R., Fuzzy sets as equivalent classes of random sets, in: R. Yager (Ed.), Fuzzy Sets and Possibility Theory, 1982, pp. 327–343.
8. Nguyen, H.T., Fuzzy and random sets, Fuzzy Sets and Systems 156 (2005) 349–356.
9. Nguyen, H.T., Kreinovich, V., How to fully represent expert information about imprecise properties in a computer system: random sets, fuzzy sets, and beyond: an overview. Int. J. General Systems 43(6): 586–609 (2014).
10. Orlov A.I., Fuzzy and random sets, Prikladnoi Mnogomiernii Statisticheskii Analyz (Nauka, Moscow), 262–280, 1978 (in Russian).
11. Kovalerchuk, B., Probabilistic Solution of Zadeh's test problems, in IPMU 2014, Part II, CCIS 443, Springer, pp. 536–545, 2014.
12. Kovalerchuk, B., Berkeley Initiative on Soft Computing (BISC), Posts on 01/17/2014, 02/03/2014, 03.07.2014.03/08/2014, 03/21/2014 http://mybisc.blogspot.com.
13. Cooman, Gert de; Kerre, E; and Vanmassenhov, FR: Possibility Theory: An Integral Theoretic Approach, Fuzzy Sets and Systems, v. 46, pp. 287–299, (1992).
14. Dubois, D., Possibility Theory and Statistical Reasoning, 2006.
15. Dubois, D., Nguyen, H.T., Prade H., 2000. Possibility theory, probability and fuzzy sets: misunderstandings, bridges and gaps. In: D. Dubois H. Prade (Eds), Fundamentals of Fuzzy Sets, The Handbooks of Fuzzy Sets Series, Kluwer, Dordrecht, 343–438.
16. Dubois, D., Steps to a Theory of Qualitative Possibility, in: Proc. 6th Int. Congress on Cybernetics and Systems, pp. 10–14, 1984.

17. Joslyn, C., Possibilistic processes for complex systems modeling, Ph. D. dissertation, SUNY Binghamton, 1995. http://ftp.gunadarma.ac.id/pub/books/Cultural-HCI/Semiotic-Complex/thesis-possibilistic-process.pdf.
18. Gaines, B., Fundamentals of decision: Probabilistic, possibilistic, and other forms of uncertainty in decision analysis. Studies in Management Sciences, 20:47-65, 1984.
19. Giles, R., Foundations for a Theory of Possibility, in: Fuzzy Information and Decision Processes, pp. 183–195, North-Holland, 1982.
20. Yager, R., On the Completion of Qualitative Possibility Measures, IEEE Trans. on Fuzzy Systems, v. 1:3, 184–194, 1993.
21. Yager, R., Foundation for a Theory of Possibility, J. Cybernetics, v. 10, 177–204, 1980.
22. Zadeh, L., Fuzzy sets as a basis for a theory of possibility, Fuzzy Sets and Systems, 1, 3–28, 1978.
23. Zadeh, L., A Note on Z-numbers, Information Science 181, 2923–2932, 2011.
24. Chang, C.-L., Berkeley Initiative on Soft Computing (BISC) post, 2014 http://mybisc.blogspot.com.
25. Piegat, A., Berkeley Initiative on Soft Compting (BISC) Post on 04/10/2014. http://mybisc.blogspot.com.
26. Burdzy, K., Search for Certainty. On the Clash of Science and Philosophy of Probability. World Scientific, Hackensack, NJ, 2009.
27. Wright, G., Ayton, O.,(eds.) Subjective probability. Wiley, Chichester, NY, 1994.
28. Kolmogorov, A., Foundations of the Theory of Probability, NY, 1956.
29. Dujmovic, J., Relationships between fuzziness, partial truth and probability in the case of repetitive events. (This volume).
30. Cheeseman, Peter: Probabilistic vs. Fuzzy Reasoning. In: Kanal, Laveen N. and John F. Lemmer (eds.): Uncertainty in Artificial Intelligence, Amsterdam: Elsevier (North-Holland), 1986, pp. 85–102.
31. Kovalerchuk, B., Quest for Rigorous Combining Probabilistic and Fuzzy Logic Approaches for Computing with Words, in R.Seising, E. Trillas, C. Moraga, S. Termini (eds.): On Fuzziness. A Homage to Lotfi A. Zadeh, SFSC Vol. 216, Springer 2013. Vol. 1. pp. 333–344.
32. Kovalerchuk, B., Context Spaces as Necessary Frames for Correct Approximate Reasoning. International Journal of General Systems. vol. 25 (1) (1996) 61–80.
33. Kovalerchuk, B., Klir, G.: Linguistic context spaces and modal logic for approximate reasoni.ng and fuzzyprobability comparison. In: Proc. of Third International Symposium on Uncertainty Modeling and Analysis and NAFIPS' 95, IEEE Press, A23A28 (1995).
34. Raufaste, E., da Silva Neves, R., Mariné, C., Testing the descriptive validity of Possibility Theory in human judgments of uncertainty, Artificial Intelligence, Volume 148, Issues 1–2, 2003, 197–218.
35. Resconi G., Kovalerchuk, B., Copula as a Bridge between Probability Theory and Fuzzy Logic. (In this volume).
36. Piegat, A., A New Definition of the Fuzzy Set, Int. J. Appl. Math. Comput. Sci., Vol. 15, No. 1, 125–140, 2005
37. Zadeh, L., The concept of linguistic variable and its application to approximate reasoning—1. Information Sciences 8, 199–249, 1977.
38. Kovalerchuk, B., Vityaev, E., Data Mining in Finance: Advances in Relational and Hybrid Methods (chapter 7 on fuzzy systems), Boston: Kluwer (2000).
39. Grabisch, M., Belief Functions on Lattices, International Journal of Intelligent Systems, Vol. 24, 76–95, 2009.
40. Kovalerchuk B, Analysis of Gaines' logic of uncertainty. In: Proceedings of NAFIPS'90, Eds I.B.Turksen. v.2, Toronto, Canada, 293–295, (1990).
41. Dujmovic, J, Berkeley Initiative on Soft Compting (BISC), Post on 4/28/2014. http://mybisc.blogspot.com.
42. Kovalerchuk, B, Talianski, V., Comparison of empirical and computed values of fuzzy conjunction. Fuzzy sets and Systems 46:49–53 (1992).

43. Thole, U., Zimmermann, H-J., Zysno, P., On the Suitability of Minimum and Product Operations for the Intersection of Fuzzy Sets, Fuzzy Sets and Systems 2, 173–186 (1979).
44. Kovalerchuk B., Interpretable Fuzzy Systems: Analysis of T-norm interpretability, IEEE World Congress on Computational Intelligence, 2010. doi:10.1109/FUZZY.2010.5584837.
45. Tschantz, M., Berkeley Initiative on Soft Compting (BISC) Posts on 03/11/2014, 04/07/2014, http://mybisc.blogspot.com.
46. Dubois, D., Prade, H., Measure-Free Conditioning, Probability, and Non-Monotonic Reasoning, in: Proc. 11th Int. Joint Conf. on Artificial Intelligence, pp. 1110–1114, 1989.

Modeling Extremal Events Is Not Easy: Why the Extreme Value Theorem Cannot Be As General As the Central Limit Theorem

Vladik Kreinovich, Hung T. Nguyen, Songsak Sriboonchitta, and Olga Kosheleva

Abstract In many real-life situations, a random quantity is a joint result of several independent factors, i.e., a *sum* of many independent random variables. The description of such sums is facilitated by the Central Limit Theorem, according to which, under reasonable conditions, the distribution of such a sum tends to normal. In several other situations, a random quantity is a *maximum* of several independent random variables. For such situations, there is also a limit theorem—the Extreme Value Theorem. However, the Extreme Value Theorem is only valid under the assumption that all the components are identically distributed—while no such assumption is needed for the Central Limit Theorem. Since in practice, the component distributions may be different, a natural question is: can we generalize the Extreme Value Theorem to a similar general case of possible different component distributions? In this paper, we use simple symmetries to prove that such a generalization is not possible. In other words, the task of modeling extremal events is provably more difficult than the task of modeling of joint effects of many factors.

V. Kreinovich (✉)
Department of Computer Science, University of Texas at El Paso 500 W. University,
El Paso, TX 79968, USA
e-mail: vladik@utep.edu

H.T. Nguyen
Department of Mathematical Sciences, New Mexico State University Las Cruces,
New Mexico 88003, USA
e-mail: hunguyen@nmsu.edu

H.T. Nguyen · S. Sriboonchitta
Department of Economics, Chiang Mai University, Chiang Mai, Thailand
e-mail: songsakecon@gmail.com

O. Kosheleva
University of Texas at El Paso 500 W. University, El Paso, TX 79968, USA
e-mail: olgak@utep.edu

© Springer International Publishing AG 2017
V. Kreinovich (ed.), *Uncertainty Modeling*, Studies in Computational
Intelligence 683, DOI 10.1007/978-3-319-51052-1_8

1 Sums and Maxima of Independent Factors: Formulation of the Problem

Why normal distributions are ubiquitous: sums of many independent factors. In many practical situations, we have a joint effects of many independent small factors. This happens, e.g., in measurements, when:

- after eliminating all major sources of possible measurement error,
- we end up with a measurement error which results from a joint effect of multiple difficult-to-eliminate independent sources of measurement uncertainty.

In this case, the measurement error—i.e., the difference $\Delta X = \widetilde{X} - X$ between the measurement result $\widetilde{X}$ and the actual value X of the corresponding quantity—can be represented as a sum $\Delta X = \sum_{i=1}^{n} X_i$ of a large number n of small independent random variables X_i.

The description of the resulting probability distribution for this sum ΔX is facilitated by the well-known *Central Limit Theorem*, according to which, under some reasonable conditions, the distribution of such a sum tends to Gaussian as n tends to infinity; see, e.g., [13]. This limit result means that for large n, the distribution of the sum ΔX is close to Gaussian.

This is indeed the case for most measuring instruments: experimental analysis shows that for about 60% of them, the measurement error is normally distributed [10, 11]. The Central Limit Theorem also explains why normal distribution is ubiquitous in many other areas as well: the familiar bell-shaped curve indeed describes many phenomena, from distribution of people by height or by weight to distribution of molecules by velocity.

Extremal events: maxima of many independent factors. In many other practical situations, we are interested in describing the *maxima* of several independent factors. For example, in structural engineering, to estimate the structure's stability under catastrophic events such as hurricanes and earthquakes, it is important to estimate the probability that this structure collapses—i.e., that in one of its components, the tension exceeds the stability threshold. The condition that one of the tension values X_i exceeds the threshold x is equivalent to $X \geq x$, where $X \stackrel{\text{def}}{=} \max_i X_i$ is the maximum of several independent components. Thus, to study such extremal events, it is important to know the probability distribution of such maxima.

Similar arguments show the need to study similar maxima in earth sciences, in finances, in hydrology, and in many other areas where rare disastrous events are possible; see, e.g., [1–7, 9, 12].

Limit theorems for extreme events: what is known. Similarly to the Central Limit Theorem that describes the limit of sums, there are the limit theorems that describe the limits of maxima of many independent random variables. The most well-known of these limit theorems is the *Extreme Value Theorem* (also known as the *Fisher-Tippett-Gnedenko Theorem*), according to which, if we have a sequence of independent identically distributed random variables X_i and the distributions of their maxima $M_n \overset{\text{def}}{=} \max(X_1, \ldots, X_n)$ has a limit, then this limit has one of the following forms [5]:

- *Weibull law*, with cumulative distribution function (cdf)

$$F(x) = \exp\left(-\left|\frac{x-b}{a}\right|^{\alpha}\right) \text{ for } x \leq b \text{ and } F(x) = 1 \text{ for } x \geq b;$$

- *Gumbel law* $F(x) = \exp\left(-\exp\left(\frac{b-x}{a}\right)\right)$; and

- *Fréchet law* $F(x) = \exp\left(-\left(\frac{x-b}{a}\right)^{-\alpha}\right)$ for $x > b$ and $F(x) = 0$ for $x \leq b$.

Formulation of the problem: what is available for the central limit theorems but missing for extreme value theorems. In many formulations of the Central Limit Theorem, it is not necessary to require that all components random variables X_i are identically distributed. These theorems are applicable to many situations in which different variables X_i have different distributions.

In contrast, the Extreme Value Theorem is only known for the case when all the component random variables X_i are identically distributed. In practical applications, the distributions of the corresponding random variables X_i—e.g., variables describing stability of different part of the construction—are, in general, different. A natural question arises: can we generalize the Extreme Value Theorem so that it can be applied to the case when we have different distributions X_i?

What we prove in this paper. In this paper, we prove that such a generalization is not possible. In this sense, the task of modeling extremal events is provably harder than the task of modeling a joint effect of several factors.

2 Analysis of the Problem

What do we mean by the desired generalization? Both in case of the Central Limit Theorem and in the case of the Extreme Value Theorem, we have a finite-parametric family of limit distributions such that, under certain reasonable conditions, the distribution of the corresponding sum or maxima tends to one of the distributions from this class.

From this viewpoint, when we say that we are looking for a generalization of the Extreme Value Theorem—which would be similar to the current (generalized) versions of the Central Limit Theorem—we mean that:

- we are looking for a finite-dimensional family $\mathcal{F}$ of probability distributions,
- such that, that, under certain reasonable conditions, the distribution of the corresponding maxima tends to one of the distributions from the class $\mathcal{F}$.

Let us describe which properties of the class $\mathcal{F}$ follow from this desired description.

First desired property: the class $\mathcal{F}$ should be invariant under shifts and scalings.
We are interested in the distributions of physical quantities X_i. Of course, in all
data processing, we deal with the numerical values of the corresponding physical
quantities. The numerical value of a quantity depends on the choice of a measuring
unit and on the choice of a starting point. For example, we can measure time in years
or days or seconds, we can start measuring time with year 0 or with year 1950 (as is
sometimes done in astronomical computations), etc.

When we change a starting point for measuring X by a new starting point which
is smaller by b units, then the numerical value X changes to $X' = X + b$. This shift
changes the numerical expression for the cdf: instead of the original probability
$F(x) = \text{Prob}(X \leq x)$ that $X \leq x$, for $X' = X + b$, we have

$$F'(x) = \text{Prob}(X' \leq x) = \text{Prob}(X + b \leq x) = \text{Prob}(X \leq x - b) = F(x - b).$$

When we change a measuring unit to a new one which is $a > 0$ times smaller,
then the numerical value X changes to $X' = a \cdot X$. For example, if instead of meters,
we use centimeters, a unit which is $a = 100$ times smaller, then all numerical values
of length are multiplied by 100: e.g., $X = 2$ m becomes $X' = 200$ cm. This scaling
changes the numerical expression for the cdf: instead of the original probability
$F(x) = \text{Prob}(X \leq x)$ that $X \leq x$, for $X' = a \cdot X$, we have

$$F'(x) = \text{Prob}(X' \leq x) = \text{Prob}(a \cdot X \leq x) = \text{Prob}\left(X \leq \frac{x}{a}\right) = F\left(\frac{x}{a}\right).$$

In general, if we change both the starting point and the measuring unit, we get a
new cdf

$$F'(x) = F\left(\frac{x - b}{a}\right).$$

If we perform this transformation, then all the values X_i are replaced by the new
values $X'_i = a \cdot X_i + b$. For the maximum, we similarly have

$$M'_n = \max(X'_1, \ldots, X'_n) = \max(a \cdot X_1 + b, \ldots, a \cdot X_n + b) =$$

$$a \cdot \max(X_1, \ldots, X_n) + b = a \cdot M_n + b.$$

Thus, if in the original units, we had a limit distribution $F(x)$, in the new units, we
will have a limit distribution $F'(x) = F\left(\frac{x - b}{a}\right)$.

The desired limit theorem should not depend on the choice of the starting point
or on the choice of a measuring unit. Thus, it is reasonable to require that if the class
$\mathcal{F}$ of limit distributions contains a cdf $F(x)$, then it should also contain a re-scaled
and shifted distribution $F'(x) = F\left(\frac{x - b}{a}\right)$.

Second desired property: the class $\mathcal{F}$ should be closed under multiplication of cdfs. Let us assume that $F(x)$ and $F'(x)$ are two cdfs from the desired class $\mathcal{F}$. By definition of the class $\mathcal{F}$, all distributions from this class are limits of distributions of the maxima. In particular:

- the cdf $F(x)$ is the limit of the distributions $F_n(x)$ of $M_n = \max(X_1, \ldots, X_n)$ for some sequence of independent random variables X_i, and
- the cdf $F'(x)$ is the limit of the distributions $F'_n(x)$ of $M'_n = \max(X'_1, \ldots, X'_n)$ for some sequence of independent random variables X'_i.

Then, for a combined sequence $X''_i \stackrel{\text{def}}{=} X_1, X'_1, \ldots, X_n, X'_n, \ldots$, the corresponding maxima will have the form

$$M''_{2n} = \max(X_1, X'_1, \ldots, X_n, X'_n) =$$

$$\max(\max(X_1, \ldots, X_n), \max(X'_1, \ldots, X'_n)) = \max(M_n, M'_n).$$

The distribution of M_n is close to $F(x)$, the distribution of M'_n is close to $F'(x)$. The cdf $F''_{2n}(x)$
for the maximum M''_{2n} can be thus described as follows:

$$F''_{2n}(x) = \text{Prob}(M''_n \leq x) = \text{Prob}(\max(M_n, M'_n) \leq x) =$$

$$\text{Prob}(M_n \leq x \;\&\; M'_n \leq x).$$

Since the variables X_i and X'_i are independent, their maxima M_n and M'_n are also independent, so

$$F''_{2n}(x) = \text{Prob}(M_n \leq x \;\&\; M_n \leq x) = \text{Prob}(M_n \leq x) \cdot \text{Prob}(M'_n \leq x),$$

i.e., $F''_{2n}(x) = F_n(x) \cdot F'_n(x)$. In the limit, the distribution of M_n tends to $F(x)$ and the distribution of M'_n tends to $F'(x)$, so the distribution for the new sequence tends to the product $F(x) \cdot F'(x)$.

Thus, with every two cdfs $F(x)$ and $F'(x)$, the class $\mathcal{F}$ should also contain their product $F(x) \cdot F'(x)$.

The class $\mathcal{F}$ should be finite-dimensional. The previous property is easier to describe if we consider logarithms $\ln(F(x))$: the logarithm of product of the cdfs is the sum of their logarithms, so the class $\mathcal{L}$ of such logarithms should be closed under addition.

One can easily check that if the original class $\mathcal{F}$ is closed under shifts and scalings, then the class $\mathcal{L}$ of logarithms of functions $F(x) \in \mathcal{F}$ should also be similarly closed.

Each such class can be naturally extended to a linear space. We can show that this space should also be closed under shift and scaling.

The fact that the original set is finite-dimensional (= finite-parametric) implies that this space should also be finite-dimensional, i.e., all its functions should have the form

$$\ell(x) = C_1 \cdot e_1(x) + \cdots + C_m \cdot e_m(x),$$

where m is the dimension of this space, $e_i(x)$ are given functions, and C_i are arbitrary real values.

Now, we are ready to formulate and prove our main result.

3 Definitions and the Main Result

Definition We say that a finite-dimensional linear space $\mathcal{L}$ of differentiable functions is *shift-* and *scale-invariant* if with every function $\ell(x)$ and for every two real numbers $a > 0$ nd b, this class also contains the function

$$\ell'(x) = \ell\left(\frac{x - b}{a}\right).$$

Proposition *For every shift- and scale-invariant finite-dimensional linear space $\mathcal{L}$ of differentiable functions, all its elements are polynomials.*

Comment. All the proofs are given in the following section.

Corollary *If $F(x)$ is a cdf, then its logarithm $\ln(F(x))$ cannot be an element of a shift- and scale-invariant finite-dimensional linear space.*

Discussion. This result shows that a finite-dimensional limit class $\mathcal{F}$ is not possible. Thus, the Extreme Value Theorem indeed cannot be extended to the general case when variables X_i are not necessarily identically distributed.

4 Proofs

Proof of the Proposition. The main ideas of this proof can be found in [8].

$1°$. The fact that the linear space $\mathcal{L}$ is shift-invariant means, in particular, that for every basis function $e_i(x)$ and for every real value b, the shifted function $e_i(x + b)$ also belongs to this linear space. Since all the function from the linear space are linear combinations of the basis functions $e_1(x), \ldots, e_m(x)$, this means that for every b, there exist values $C_{i,j}(b)$ for which

$$e_i(x + b) = C_{i,1}(b) \cdot e_1(x) + \cdots + C_{i,m}(b) \cdot e_m(x). \tag{1}$$

For each b, we can take m different values $x_1, \ldots, x_m$, and get m resulting equalities:

$$e_i(x_1 + b) = C_{i,1}(b) \cdot e_1(x_1) + \cdots + C_{i,m}(b) \cdot e_m(x_1);$$

$$\cdots$$

$$e_i(x_j + b) = C_{i,1}(b) \cdot e_1(x_j) + \cdots + C_{i,m}(b) \cdot e_m(x_j); \tag{2}$$

$$\cdots$$

$$e_i(x_m + b) = C_{i,1}(b) \cdot e_1(x_m) + \cdots + C_{i,m}(b) \cdot e_m(x_m).$$

We thus get a system of m linear equations for m unknowns $C_{i,1}(b)$, ..., $C_{i,m}(b)$. By using Cramer's rule, we can describe the values $C_{i,j}(b)$ as ratios of polynomials in terms of the coefficients $e_j(x_k)$ and the right-hand sides $e_i(x_j + b)$. Since the functions $e_i(x)$ are differentiable, we can conclude that the dependence $C_{i,j}(b)$ on b is differentiable as well.

$2°$. We can now combine the Eq. (1) corresponding to different functions $e_i(x)$. As a result, we get the following system of m equations:

$$e_1(x + b) = C_{1,1}(b) \cdot e_1(x) + \cdots + C_{1,m}(b) \cdot e_m(x);$$

$$\cdots$$

$$e_i(x + b) = C_{i,1}(b) \cdot e_1(x) + \cdots + C_{i,m}(b) \cdot e_m(x); \tag{3}$$

$$\cdots$$

$$e_m(x + b) = C_{m,1}(b) \cdot e_1(x) + \cdots + C_{m,m}(b) \cdot e_m(x).$$

Differentiating both sides of these equations by b and taking $b = 0$, we get the following system of differential equations:

$$\frac{de_1(x)}{dx} = c_{1,1} \cdot e_1(x) + \cdots + c_{1,m} \cdot e_m(x);$$

$$\cdots$$

$$\frac{de_i(x)}{dx} = c_{i,1} \cdot e_1(x) + \cdots + c_{i,m} \cdot e_m(x); \tag{4}$$

$$\cdots$$

$$\frac{de_m(x)}{dx} = c_{m,1} \cdot e_1(x) + \cdots + c_{m,m} \cdot e_m(x),$$

where we denoted $c_{i,j} \overset{\text{def}}{=} \dfrac{dC_{i,j}(b)}{db}\Big|_{b=0}$.

We have a system (4) of linear differential equations with constant coefficients. It is known that a general solution to this system is a linear combination of expressions of the type $x^k \cdot \exp(\lambda \cdot x)$, where:

- the value λ is a (possible complex) eigenvalue of the matrix $c_{i,j}$, and
- the value k is a natural number; this number should be smaller than the multiplicity of the corresponding eigenvalue.

Thus, each function $e_i(x)$ is such a linear combination.

$3°$. Let us now use scale-invariance. The fact that the linear space $\mathcal{L}$ is scale-invariant means, in particular, that for every basis function $e_i(x)$ and for every real value a, the shifted function $e_i(a \cdot x)$ also belongs to this linear space. Since all the function from the linear space are linear combinations of the basis functions $e_1(x), \ldots, e_m(x)$, this means that for every a, there exist values $A_{i,j}(a)$ for which

$$e_i(a \cdot x) = A_{i,1}(a) \cdot e_1(x) + \cdots + A_{i,m}(a) \cdot e_m(x). \tag{5}$$

For each a, we can take m different values $x_1, \ldots, x_m$, and get m resulting equalities:

$$e_i(a \cdot x_1) = A_{i,1}(a) \cdot e_1(x_1) + \cdots + A_{i,m}(a) \cdot e_m(x_1);$$

$$\cdots$$

$$e_i(a \cdot x_j) = A_{i,1}(a) \cdot e_1(x_j) + \cdots + A_{i,m}(a) \cdot e_m(x_j); \tag{6}$$

$$\cdots$$

$$e_i(a \cdot x_m) = A_{i,1}(a) \cdot e_1(x_m) + \cdots + A_{i,m}(a) \cdot e_m(x_m).$$

We thus get a system of m linear equations for m unknowns $A_{i,1}(a), \ldots, A_{i,m}(a)$. By using Cramer's rule, we can describe the values $A_{i,j}(a)$ as ratios of polynomials in terms of the coefficients $e_j(x_k)$ and the right-hand sides $e_i(a \cdot x_j)$. Since the functions $e_i(x)$ are differentiable, we can conclude that the dependence $A_{i,j}(a)$ on a is differentiable as well.

$4°$. We can now combine the Eq. (5) corresponding to different functions $e_i(x)$. As a result, we get the following system of m equations:

$$e_1(a \cdot x) = A_{1,1}(a) \cdot e_1(x) + \cdots + A_{1,m}(a) \cdot e_m(x);$$

$$\cdots$$

$$e_i(a \cdot x) = A_{i,1}(a) \cdot e_1(x) + \cdots + A_{i,m}(a) \cdot e_m(x); \tag{7}$$

$$\cdots$$

$$e_m(a \cdot x) = C_{m,1}(a) \cdot e_1(x) + \cdots + A_{m,m}(a) \cdot e_m(x).$$

Differentiating both sides of these equations by a and taking $a = 1$, we get the following system of differential equations:

$$x \cdot \frac{de_1(x)}{dx} = a_{1,1} \cdot e_1(x) + \cdots + a_{1,m} \cdot e_m(x);$$

$$\cdots$$

$$x \cdot \frac{de_i(x)}{dx} = a_{i,1} \cdot e_1(x) + \cdots + a_{i,m} \cdot e_m(x); \qquad (8)$$

$$\cdots$$

$$x \cdot \frac{de_m(x)}{dx} = a_{m,1} \cdot e_1(x) + \cdots + a_{m,m} \cdot e_m(x),$$

where we denoted $a_{i,j} \overset{\text{def}}{=} \dfrac{dA_{i,j}(a)}{da}\Big|_{a=1}$.

$5°$. To solve this new system of equations, we can introduce a new variable $t \overset{\text{def}}{=} \ln(x)$, for which $\dfrac{dx}{x} = dt$. Here, $x = \exp(t)$, so for the new functions $E_i(t) \overset{\text{def}}{=} e_i(\exp(t))$, the system (8) takes the following form:

$$\frac{dE_1(t)}{dt} = a_{1,1} \cdot E_1(t) + \cdots + a_{1,m} \cdot E_m(t);$$

$$\cdots$$

$$\frac{dE_i(t)}{dt} = a_{i,1} \cdot E_1(t) + \cdots + a_{i,m} \cdot E_m(t); \qquad (9)$$

$$\cdots$$

$$\frac{dE_m(x)}{dx} = a_{m,1} \cdot E_1(t) + \cdots + a_{m,m} \cdot E_m(t).$$

This is a system of linear differential equations with constant coefficients; so, each function $E_i(t)$ is linear combination of the expressions of the type $t^k \cdot \exp(\lambda \cdot t)$, Thus, for $e_i(x) = E_i(\ln(x))$, we conclude that each function $e_i(x)$ is a linear combination of the expressions

$$(\ln(x))^k \cdot \exp(\lambda \cdot \ln(x)) = (\ln(x))^k \cdot x^\lambda.$$

6°. We have proven that:

- on the one hand, each function $e_i(x)$ is a linear combination of the expressions $x^k \cdot \exp(\lambda \cdot x)$, where k is a natural number;
- on the other hand, each function $e_i(x)$ is a linear combination of the expressions $(\ln(x))^k \cdot x^\lambda$.

One can easily see that the need to be represented in the second form excludes the possibility of $\lambda \neq 0$. Thus, each function $e_i(x)$ is a linear combination of the expressions of the type x^k with natural k—i.e., a polynomial. Every function from the linear space $\mathcal{L}$ is a linear combination of the basis functions $e_i(x)$ and is, thus, also a polynomial.

The proposition is proven.

Proof of the Corollary. Let us prove this result by contradiction. Let us assume that for some cdf $F(x)$, its logarithm $\ln(F(x))$ belongs to a shift-and scale-invariant linear space $\mathcal{L}$. Due to Proposition, this implies that this logarithm is a polynomial $P(x)$: $\ln(F(x)) = P(x)$ and thus, $F(x) = \exp(P(x))$.

When $x \to -\infty$, we have $F(x) \to 0$, so we should have

$$P(x) = \ln(F(x)) \to -\infty.$$

For the corresponding polynomial $P(x) = a_0 \cdot x^k + a_1 \cdot x^{k-1} + \cdots$, this means that:

- either k is even and $a_0 < 0$,
- or k is odd and $a_0 > 0$.

When $x \to +\infty$, then:

- in the first case, we have $P(x) \to -\infty$, while
- in the second case, we have $P(x) \to +\infty$.

However, we should have $F(x) \to 1$ and thus, $P(x) = \ln(F(x)) \to \ln(1) = 0$.

This contradiction shows that our assumption was wrong, and logarithms for cdfs cannot belong to shift-and scale-invariant linear spaces.

Acknowledgements We acknowledge the partial support of the Center of Excellence in Econometrics, Faculty of Economics, Chiang Mai University, Thailand. This work was also supported in part by the National Science Foundation grants HRD-0734825 and HRD-1242122 (Cyber-ShARE Center of Excellence) and DUE-0926721.

References

1. J. Beirlant, Y. Goegebeur, J. Segers, and J. Teugels, *Statistics of Extremes: Theory and Applications*, Wiley, New York, 2004.
2. E. Castillo, A. S. Hadi, N. Balakrishnan, and J. M. Sarabia, *Extreme Value and Related Models with Applications in Engineering and Science*, Wiley, New York, 2004.

3. S. Coles, *An Introduction to Statistical Modeling of Extreme Values*, Springer Verlag, London, 2001.
4. L. de Haan and A. Ferreira, *Extreme Value Theory: An Introduction*, Springer Verlag, New York, 2006.
5. P. Embrechts, C. Klüppelberg, and T. Mikosch, *Modelling Extremal Events for Insurance and Finance*, Spring Verlag, Berlin 1997.
6. E. J. Gumbel, *Statistics of Extremes*, Dover, New York, 2013.
7. S. Kotz and S. Nadarajah, *Extreme Value Distributions: Theory and Applications*, Imperial College Press, 2000.
8. H. T. Nguyen and V. Kreinovich, *Applications of Continuous Mathematics to Computer Science*, Kluwer, Dordrecht, 1997.
9. S. Y. Novak, *Extreme Value Methods with Applications to Finance*, Chapman & Hall/CRC Press, London, 2011.
10. P. V. Novitskii and I. A. Zograph, *Estimating the Measurement Errors*, Energoatomizdat, Leningrad, 1991 (in Russian).
11. A. I. Orlov, "How often are the observations normal?", *Industrial Laboratory*, 1991, Vol. 57. No. 7, pp. 770–772.
12. S. I. Resnick, "Extreme Values, Regular Variation and Point Processes", Springer Verlag, New York, 2008.
13. D. J. Sheskin, *Handbook of Parametric and Nonparametric Statistical Procedures*, Chapman & Hall/CRC, Boca Raton, Florida, 2011.

Information Quality and Uncertainty

Marie-Jeanne Lesot and Adrien Revault d'Allonnes

Abstract The quality of a piece of information depends, among others, on the certainty that can be attached to it, which relates to the degree of confidence that can be put in it. This paper discusses various components to be considered when assessing this certainty level. It shows that they cover a wide range of different types of uncertainty and provide a highly relevant application domain for theoretical questioning about uncertainty modelling. It also describes several frameworks that have been considered for this task.

Keywords Information scoring · Information processing · Uncertainty type · Competence · Reliability · Plausibility · Credibility · Linguistic uncertainty

1 Introduction

Information quality (see e.g. [1]) and its implementation in the domain of information evaluation (see e.g. [2]) aim at providing guidance and help to users in the drowning quantity of information they are nowadays overwhelmed with, in particular due to the dramatic increase of Web usage, e.g. through blogs and social networks, such as Facebook and Twitter. One specificity of these new media is that everyone can participate in the information spread and be a source of information, making the question of a relevance measure of the available information crucial. As a consequence, it is necessary to dispose of tools for automatically assessing their quality: there is an acute need for automatic methods to identify the "best", e.g. understood as the most useful, pieces of information.

M.-J. Lesot (✉)
Sorbonne Universités, UPMC Univ Paris 06, CNRS, LIP6 UMR 7606,
4 Place Jussieu, 75005 Paris, France
e-mail: Marie-Jeanne.Lesot@lip6.fr

A. Revault d'Allonnes
Université Paris 8, EA 4383, LIASD, 93526 Saint-Denis, France
e-mail: Allonnes@ai.univ-paris8.fr

© Springer International Publishing AG 2017
V. Kreinovich (ed.), *Uncertainty Modeling*, Studies in Computational
Intelligence 683, DOI 10.1007/978-3-319-51052-1_9

Numerous criteria and properties have been proposed and considered to that aim [1, 2]. This paper[1] focuses on the certainty dimension, numerically evaluated as a degree of certainty that can be attached to any piece of information. In a schematic view, it exploits the argument according to which a certain piece of information is worthier than a doubtful one. Insofar, it is related to the task that aims at assessing the trust that can be put in a piece of information. It can be underlined that such a degree of trust can mean either evaluating the reality of the fact the piece of information reports [3–5] or the extent to which the rater is convinced, based on the process with which he forms an opinion about this piece of information [6–8].

Even if uncertainty is only one of its components, information quality appears as a highly relevant application framework for the theoretical domain of uncertainty modelling. Indeed, it turns out to be a very challenging one, raising critical requirements that lead to question existing models and possibly to develop new ones. As discussed in this paper, information processing involves several types of uncertainty that must be distinguished, appropriately modelled and possibly combined: information-related uncertainty covers a wide spread spanning over several dimensions. As detailed in the following, one can mention distinctions between objective and subjective uncertainty, as well as between general versus contextual uncertainty.

This paper first discusses various kinds of uncertainty that can be attached to a piece of information in Sect. 2, organising them according to their cause, i.e. the characteristic of the considered piece of information that triggers them. Section 3 discusses the two axes objective-subjective and general-contextual. Section 4 briefly describes some theoretical frameworks that have been proposed to model uncertainty for information evaluation.

2 Sources of Uncertainty in the Information Processing Framework

This section discusses 5 sources of uncertainty that can be considered in the framework of information processing, structuring them according to their cause: it distinguishes the uncertainties respectively triggered by the content of a piece of information, its source, its context, its formulation and its automatic extraction.

In order to illustrate these types, it considers the following fictitious piece of information together with two basic meta-data, namely author and publication date:

On February 15th 2015, the International Olympic Committee declared
"In 2048, the Summer Olympic Games will probably take place in November"

---

[1]This paper is based on part of the panel which has been organised by Prof. Kovalerchuk at IPMU2012 on the general topic "Uncertainty Modelling".

2.1 Content-Related Uncertainty: What is Said?

The degree of uncertainty attached to a piece of information obviously depends on its content, i.e. the answer to the question "what does it say?": for the running example, it for instance relates to the assertion that can be schematically written as "2048 Summer Olympic Games dates = November".

More precisely, the piece of knowledge provided by the considered information can trigger a surprise effect that influences its uncertainty level: an unexpected fact can, at least at first, appear as more uncertain than a known one. The surprise effect can be measured with respect to two types of background, leading to distinguish between the notions of plausibility and credibility.

Knowledge context: plausibility Surprise can be defined as compared to the personal background of the information rater, i.e. as the compatibility of the considered piece of information with his/her knowledge, which is defined as *plausibility* [8].

For instance for the running example, the asserted date may appear highly atypical, in particular to people living in the North hemisphere, who usually do not associate November with summer. As a consequence, they may receive the information with more caution and consider it as more uncertain than people living in the South hemisphere. Along the same lines, for someone with knowledge about the history of the Olympic Games, for instance knowing that the situation where the summer games take place in November already occurred (in 1956, for the Melbourne Games), the fact may appear as less uncertain.

Plausibility can be considered as the first component in the conviction establishing process [8], that determines an a priori confidence level attached to a considered piece of information.

Other information context: credibility Surprise can also be defined with respect to other available pieces of information, e.g. other assertions provided in the same period regarding the location and dates of the Olympic Games: in this case, the considered piece of information is compared to other statements, building the *credibility* component [3, 8].

More precisely, the assessment of credibility relies on the identification of corroboration or invalidation of the considered piece of information, defining another type of background for the evaluation of its attached uncertainty. This dimension both depends on the content of the information and the context of its assertion, it is more detailed in the section discussing the latter (Sect. 2.3).

2.2 Source-Related Uncertainty: Who Says it?

The uncertainty attached to an assertion also depends on its source, i.e. the answer to the question "who says it?": for the running example, it for instance relates to the fact that the International Olympic Committee provides it, who can be considered as a qualified source. The question is then to define the characteristics that make a source "qualified", this section discusses some of them, a more complete discussion can be found in [9] for instance.

It must be underlined that, altogether, the qualification of a source is contextual: it may not be the same for all pieces of information and may for instance depend on their topics, i.e. on their contents. However, some of its components remain topic-independent and general.

Source trustworthiness: reliability The reliability of the source corresponds to an a priori assessment of its quality, independently of the considered piece of information: it indicates whether, in general, the assertions it provides can be trusted or should be considered with caution.

In the seminal model for information evaluation [3], reliability plays a major role: this model represents the information score as a bi-gram defined as the concatenation of two symbols measured on two discrete graded scales associated to linguistic labels. The first one is called reliability, although it may depend on several distinct dimensions [10]: its explicit and direct presence in the final score underlines its crucial role.

This subjective dimension, that may take different values for different raters, is difficult to define formally and thus to measure. It can be related to the concept of source reputation although the latter may be as difficult to model and quantify. In the case of Twitter sources, it has for instance been proposed to establish it from measurable quantities such as the number of followers or the source social status [11].

Reliability can also be assessed by comparing previous source assertions with the ground truth when the occurrence of events has make it possible to establish whether the source was right or wrong [12, 13]. This approach highlights the fact that reliability is a dynamic concept whose measure should evolve with time. It also relates this dimension to validity [4], according to which if the source produces a piece of information, then the latter is true.[2]

Another component of reliability can be derived from the formulation used by the source: the number of citations that are contained in its publications allows to evaluate the extent to which it cites its own sources [9, 14, 15]. Now, offering a possibility to track back the origin of the provided information contributes to its reliability. Another indication can be derived from the amount of grammatical and spelling errors [9, 14, 15]: it is argued that a grammatically mistake-free text advocates for analysis capacity and critical way of thinking, which are desirable qualities for reliable sources. Although these quantities are related to the question "how is it said", discussed in Sect. 2.4, they capture an uncertainty originated from the source, allowing to infer some of the source characteristics, whereas the components described in Sect. 2.4 measure uncertainty originated from the expression itself.

Source expertise level: competence A distinct component of source-related uncertainty comes from its competence, that measures the extent to which it is entitled to provide the information, i.e. whether it is legitimated to give it [7, 9].

---

[2]Conversely, a source is said to be *complete* if, when a piece of information is true, the source provides it [4]. This useful characterisation, related both to the source omniscience and "sharing communication type", is however less central for the assessment of the information uncertainty.

In the considered example, it can for instance be considered that the IOC is much more competent than a taxi driver would be, leading to a lower uncertainty regarding the date of the 2048 Olympic Games than would occur if the latter provided the information.

Competence relates to the source expertise and appears to be a topic-dependent component: the IOC would be significantly less competent to provide information about the World Football Champions' cup; likewise, the taxi driver would be a legitimate source about efficient routes or traffic jams for instance, leading to less uncertain pieces of information regarding these topics.

It is worth noticing that two types of competence can be distinguished, an essential one and a more accidental one, that respectively apply to experts and witnesses [9]. Indeed, an essential competence can be established from the source fields of study and possibly diplomas, or from official roles: they provide a theoretical expertise and indeed entitle a source to make assertions about a given topic. On the other hand, a geographical or temporal proximity to an event provides an empirical competence, granting witnesses a local expertise level.

Source intention The assessment of the certainty degree attached to a piece of information, or the degree of trust put in it, can also depend on source characteristics even more difficult to establish, related to its *intention*: indeed, a source may for instance pursue an aim of desinformation, with the intention to lure the information rater. The certainty degree should obviously be reduced in such a communication paradigm, if it can be recognised as such.

This dimension is related to a *sincerity* feature, which captures the tendency of the source to tell the truth or not (see also [4]): it can be considered that sincerity is a general characteristic of the source, describing its global tendency, whereas its intention is more contextual and varies for each piece of information. Sincerity can be considered as being related to the source reliability, as they both depend on the truth of the source assertions. The notion of sincerity may be seen as integrating a judgment component, that takes into account the source intention when interpreting the reason why it is wrong.

Source implication Another source characteristic is captured by its *commitment degree*, i.e. the extent to which it is involved in the propagation of the information it produces. Commitment depends on what the source may loose if it produces erroneous information, and, insofar, can be seen as related to its reputation.

It has for instance been proposed, in the case of Twitter sources, to measure the commitment degree as a function of the energy they put in their accounts [9, 15], in turn quantified by the richness of their profile, e.g. the number of filled fields, the presence of a picture or the number of publications.

The source commitment also influences the uncertainty that can be attached to its assertions, under the interpretation that a highly committed source should be less prone to produce erroneous content and may be trusted.

Successive sources: hearsay A specific case for the evaluation of the source of an information occurs when the piece of information is not directly obtained, i.e. when it results from a series of successive sources, following a scheme of the form "S_1 says that S_2 says that ... S_n says that F" where F is the fact and $S_i, i = 1, \ldots, n$

the sources. Dedicated models have been proposed to process such cases, see e.g. [16, 17].

Indeed, for such pieces of information, all previous source-uncertainty related components are measured not with respect to F (except for S_n) but, for S_i, with respect to "S_{i+1} says that... S_n says that F": competence then for instance measures whether S_i is entitled to report the assertions of S_{i+1}.

2.3 Context-Related Uncertainty: When is it Said?

Another meta-data that influences the certainty attached to a considered piece of information relates to the context of its assertion, understood as the global answer to the question "when is it said?". Different components can be considered, a purely temporal one as well as a more global one that depends on other available assertions.

Temporal context The date associated to an assertion contributes to the certainty level that can be attached to it, both in comparison with the date of the reported event and with the current date.

Indeed, the gap between the reported event and the assertion influences the uncertainty: information provided too much in advance may be considered with caution, decreasing their certainty level. For instance for the running example, if the assertion is about the Olympic Games in 2084, it may be interpreted as less certain.

On the other hand, a comparison with the current date can influence the importance that should be granted to a considered piece of information: when faced with an information stream, it can be useful to take into account older, and possibly out-of-date, pieces of information to a lesser degree than the more recent ones. It has for instance been proposed to associate each assertion with a currentness score [5], so as to weight down the pieces of information according to their possible obsolescence. It can be underlined that such a model makes the evaluation sensitive to the information order, possibly leading to different results if a piece of information I_1 is published before I_2 or reciprocally. Such a behaviour can be considered as a realistic approach to model the uncertainty evolution when faced with an information stream.

It can be noted that beside these relative date comparisons, with respect to the information content and the current date, an absolute effect can be considered: some dates do bear meaning and influence the evaluation of their content. This component depends on a cultural dimension that makes difficult its general implementation. For instance, one can consider that information produced on April 1st is less certain than others; announcements contained in election campaigns may also require a specific processing.

Other assertion context: credibility The evaluation of the uncertainty attached to a piece of information classically includes a cross-checking step, aiming at identifying complementary information backing up or undermining it: confirmations and invalidations respectively increase and decrease its certainty level. The *credibility* dimension can be understood as a degree of confirmation resulting from comparison of the piece of information to be rated with the available information [3, 5, 7, 18].

In the seminal model [3], the second symbol of the bigram measures this confirmation degree, as indicated by the description it is accompanied by. It can be underlined that its linguistic labels mainly describe the information certainty, across the scale *improbable, doubtful, possibly true, probably true, confirmed by other sources*, showing the relation with this underlying essential component.

The principle of credibility evaluation [5, 7] consists in aggregating several assertions, said to be homologous, that refer to the same content. It thus depends on the choice of a similarity measure that measures the degree of confirmation by assessing the extent to which an homologous piece of information corroborates the information to be rated (see e.g. [5] for a discussion on such eligible measures and their components).

The aggregation step can take into account various dimensions, among which the previous degree of confirmation, the individual uncertainty attached to the homologous information [5, 7, 18], but also the relations between the sources [5]: one can consider a refined notion of confirmation and invalidation, weighting them according to affinity or hostility relations between sources. Indeed, a confirmation provided by sources known to be in affinity relation should have a lower influence than a confirmation by independent, not to say hostile, sources: friendly sources are expected to be in agreement and to produce somehow redundant information.

As the temporal component, the credibility dimension makes uncertainty evaluation sensitive to the order of the pieces of information in a stream, taking into account more subtle relations than their publication dates only. This dynamical behaviour, source of many a theory of argumentation, considers that two confirmations followed by an invalidation may lead to a different level of uncertainty than a confirmation followed by a contradiction and another confirmation might [18].

2.4   Formulation-Related Uncertainty: How is it Said?

The words used in a piece of information play a major role on the attached uncertainty level, both because of the imprecision they convey and the uncertainty they intrinsically convey. The additional role of linguistic quality, that influences the assessment of the source reliability, has been discussed in Sect. 2.2.

Natural language is often imprecise (see e.g. [19]), allowing for fuzziness of the conveyed message, which can lead to uncertainty: if, for instance, the IOC asserts that the 2048 Games will take place "around the end of the year", some uncertainty is attached to the fact that the games will take place in November. In this case, uncertainty arises from the approximate compatibility between the rated piece of information and the query (e.g. regarding the Games date): only a partial answer is available. Such imprecision also plays a role in the identification of homologous information involved in the cross-checking step of credibility assessment discussed in Sect. 2.3.

Beside imprecision, the used words also convey uncertainty: they give indication regarding the source own level of uncertainty and influence the overall evaluation of the uncertainty [20]. In the case of the considered example for instance, the linguistic expression contains the adverb "probably" whose presence increases the uncertainty of the final evaluation.

Linguistic works (see e.g. [21, 22]) propose classification of uncertainty bearing terms, making it possible to assess the global expressed uncertainty. Such terms include adjectives (such as certain, likely or improbable), modal verbs (e.g. may, might, could, should), adverbs (such as certainly, possibly or undeniably) or complex idiomatic structures. Modifiers such as "very" can be used to reinforce or weaken the previous linguistic tags.

2.5 Automatic Processing-Related Uncertainty: How is it Extracted?

A fifth level of uncertainty comes from the fact that the available pieces of information are automatically processed, which can introduce errors in the content identification and thus for many of the components mentioned in the previous sections.

Indeed, the evaluation of the uncertainty attached to a piece of information according to the previously cited dimensions for instance include the use of tools for named entity detection, event and relationship identification and date extraction [22]. They also require to solve difficult linguistic tasks, as negation handling and anaphora resolution, that still are challenges for automatic text processing systems. These uncertainties can be measured automatically, for instance through performance rates of the corresponding methods, i.e. using recognition rate, recall or precision.

Among the examples of the encountered difficulties, one can for instance mention possible errors in the text topic identification, possibly leading to erroneous assessment of the source competence (see Sect. 2.2). Similarly, the identification of the date in the processed document may result in mistakes in the evaluation of the temporal content (see Sect. 2.3). The most impacted dimension is probably credibility (Sect. 2.3), that relies on the extraction of homologous pieces of information, and therefore both on all the documents processing and the computation of their similarities. It can be noticed that this task is sometimes performed semi-automatically, in order to guarantee its quality, crucial for the whole system [5].

3 Uncertainty Types for Information

Form a formal point of view, the various uncertainty types discussed in the previous section can be classified according to two axes, opposing objective versus subjective uncertainties as well as general versus contextual ones.

It can be underlined that the considered uncertainties also differ in their very nature: for instance, some express structural doubts about the phenomena, as content plausibility or recognition rate for instance, whereas the linguistically triggered uncertainty on the other hand captures an imprecision level.

Objective versus subjective uncertainty A first axis discriminating the listed uncertainty types refers to the position of the rater and his/her implication in the evaluation: some of them actually do not depend on the rater and constitutes objective dimensions, whereas others are subjective.

Indeed, the evaluation of the uncertainty triggerend by the automatic processing step for instance is objective and can be automatically measured. Similarly, the evaluation of the degree of confirmation between two pieces of information, i.e. the credibility dimension, does not depend on the rater and is identical for all users.

On the other hand, the plausibility dimension is subjective: it is measured by comparison to the rater's background knowledge and therefore varies from one rater to another. Likewise, most source evaluation criteria can be considered as subjective: for instance, not all users may agree on the competence fields of a given source, nor on its intention.

General versus contextual uncertainty Another discriminating axis refers to the dependence of the dimension to the rated piece of information: some criteria are evaluated generally, a priori, i.e. independently of any information, whereas others characterise the considered one.

As an example, the source reliability does not depend on the rated piece of information and similarly applies to all the source assertions. The category of general criteria also involve the evaluation of the uncertainty triggered by automatic processing step, which is measured globally, for all types of information. Similarly, the measure of the formulation-related uncertainty relies on a linguistic modelling of uncertainty expression: the latter is built generally, not for a specific piece of information.

On the other hand, the source competence for instance is topic-dependent and thus varies from one piece of information to the other. In that sense, it is considered to be contextual. Obviously, the content credibility, as well as the temporal dimension, are contextual too.

4 Formal Frameworks for Information Scoring

As discussed in the previous sections, the uncertainty to be considered in the domain of information quality covers different types. As a consequence, distinct formal frameworks have been considered to represent it or some of its components. A central issue is to dispose of aggregation operators to combine the individual uncertainty scores obtained for each considered component. It can be observed that some propositions focus on this aggregation issue, in a multi-criteria aggregation approach, using for instance Choquet integrals [9].

This section briefly discusses the main existing uncertainty modelling frameworks applied to the case of information evaluation, distinguishing them depending on whether they model symbolic, ordered or numerical uncertainties.

Symbolic framework Symbolic approaches in the domain of information evaluation include logical representation, in particular in the framework of modal logics

[4, 23–25], that allow to perform logical inferences to characterise the sources and the pieces of information. However, they usually do not model the attached uncertainty.

The first formal framework for information evaluation considering uncertainty has been proposed in the seminal model [3]: it represents the information score as a bi-gram defined as the concatenation of two symbols measured on two discrete graded scales associated to linguistic labels: according to the descriptions they are accompanied by, the first one captures the source reliability and the second one the information credibility. However it has been shown [6, 10, 26] that this symbolic approach raises some difficulties, among others regarding the manipulation and comparison of the obtained scores.

Ordered framework: extended multivalued logic In order to ease the manipulation of uncertainty scores, it has been proposed to exploit an extended multivalued logic framework [8, 27] to model the process of trust building: trust can be defined on a single discrete graded scale, clarified with linguistic labels, improving the legibility of a unique degree with a semantic interpretation. Moreover, this framework is equipped with formal tools to combine the truth degrees through logical operations that generalise conjunction, disjunction or implication, as well as arithmetical ones [28].

The extension [8, 27] of classical multivalued logic consists in introducing an additional degree that allows to distinguish between facts that are 'neither true nor false', i.e. that have a neutral truth value, and facts whose truth values cannot be evaluated: it makes it possible to distinguish between ignorance and neutral knowledge, which is for instance required to distinguish between a source whose reliability is unknown from a source with intermediate reliability.

Numerical frameworks: probability, possibility and evidence Probability theory is one of the most frequent framework used to model uncertainties. In the case of information evaluation, it can for instance naturally be used to quantify the uncertainty related to the extraction process, e.g. to measure error recognition rates of the applied automatic tools. However, many components of information evaluation uncertainty cannot be considered as having a probabilistic nature. Moreover, they need to distinguish between ignorance and uniform distribution, as sketched above, which cannot be implemented in the probabilistic framework. Furthermore, probabilities impose strong axiomatic constraints, restricting the choice of aggregation operators. Finally, probability theory often requires to set a priori distributions, which may be a difficult task in the case of information evaluation.

Possibility theory [29] allows to represent the ignorance case separately from the neutral one and offers a wide variety of aggregation operators allowing to model many different behaviours for the combination of the considered uncertainty dimensions. It has for instance be applied to assess the uncertainty that can be attached to an event e, to answer the question "did e take place?", based on a set of pieces of information, enriching the binary answer yes/no with a confidence level [5].

The theory of belief functions [30] generalises the probability and the possibility theories, offering a very rich expression power. It has been applied to information evaluation in particular to the issues of reported information [16, 26] and source reliability measures [25, 31].

5   Conclusion

This chapter considered the issue of uncertainty in the domain of information evaluation, discussing the various types of uncertainty that can be attached to a piece of information, describing either the event it reports or its intrinsically attached trust. Many components can be distinguished, whose combination builds to a complex notion for which several theoretical frameworks have been considered, so as to capture its diverse facets.

Among other topics related to uncertainty in the context of information evaluation, dynamics and validation offer challenging issues opening the way to research directions. The need for modelling the temporal evolution of uncertainty comes from the availability of information streams, beyond the individual pieces of information, as briefly mentioned previously. It also comes from the possible evolution of the general components of the source characteristics: if, for instance, the reliability of a source proves to change over time, it may require to re-evaluate the uncertainty attached to previously assessed pieces of information this source had provided, and, consequently, also to the information they are analogous to.

The issue of validation aims at assessing the quality of the proposed uncertainty models, both regarding the considered components and the chosen formal framework. Now its difficulty comes from the lack of data allowing to perform empirical studies: in the case of real data, it is difficult to dispose of expected scores to which the computed ones can be compared. The use of artificial data raises the challenge of their realistic generation controlling their relevance.

References

1. Berti-Equille, L., Comyn-Wattiau, I., Cosquer, M., Kedad, Z., Nugier, S., Peralta, V., Si-Saïd Cherfi, S., Thion-Goasdoué, V.: Assessment and analysis of information quality: a multidimensional model and case studies. Int. Journal of Information Quality **2**(4) (2011) 300–323
2. Capet, P., Delavallade, T., eds.: Information Evaluation. Wiley (2014)
3. OTAN: Annex to STANAG2022. Information handling system (1997)
4. Demolombe, R.: Reasoning about trust: a formal logical framework. In Jensen, C., Poslad, S., Dimitrakos, T., eds.: Proc. of the Int. Conf. on Trust Management, iTrust. Number 2995 in LCNS, Springer (2004) 291–303
5. Lesot, M.-J., Delavallade, T., Pichon, F., Akdag, H., Bouchon-Meunier, B.: Proposition of a semi-automatic possibilistic information scoring process. In: Proc. of the 7th Conf. of the European Society for Fuzzy Logic And Technology EUSFLAT'11 and LFA'11. (2011) 949–956
6. Besombes, J., Revault d'Allonnes, A.: An extension of STANAG2022 for information scoring. In: Proc. of the Int. Conf. on Information Fusion. (2008) 1635–1641
7. Revault d'Allonnes, A.: An architecture for the evolution of trust: Definition and impact of the necessary dimensions of opinion making. In Capet, P., Delavallade, T., eds.: Information Evaluation. Wiley (2014) 261–294
8. Revault d'Allonnes, A., Lesot, M.J.: Formalising information scoring in a multivalued logic framework. In: Proc. ofthe Int. Conf. on Information Processing and Management of Uncertainty in Knowledge-Based Systems, IPMU'14. Volume CCIS 442., Springer (2014) 314–323
9. Pichon, F., Labreuche, C., Duqueroie, B., Delavallade, T.: Multidimensional approach to reliability evaluation of information sources. In Capet, P., Delavallade, T., eds.: Information Evaluation. Wiley (2014) 129–160

10. Capet, P., Revault d'Allonnes, A.: Information evaluation in the military domain: Doctrines, practices and shortcomings. In Capet, P., Delavallade, T., eds.: Information Evaluation. Wiley (2014) 103–128
11. Ulicny, B., Kokar, M.: Toward formal reasoning with epistemic policies about information quality in the twittersphere. In: Proc. of the 14th Int. Conf. on Information Fusion. (2011)
12. Ellouedi, Z., Mellouli, K., Smets, P.: Assessing sensor reliability for mutisensor data fusion with the transferable belief model. IEEE Trans. on System, Man and Computing **34**(1) (2004) 782–787
13. Destercke, S., Chojnacki, E.: Methods for the evaluation and synthesis of multiple sources of information applied to nuclear computer codes. Nuclear Engineering and Design **238**(2) (2008) 2482–2493
14. Castillo, C., Mendoza, M., Poblete, B.: Information credibility on twitter. In: Proc. of the 20th Int. Conf. on World Wide Web, WWW'11. 675–684
15. Morris, M., Counts, S., Roseway, A., Hoff, A., Schwarz, J.: Tweeting is believing? understanding microblog credibility perceptions. In: ACM Conf. on Computer Supported Cooperative Work, CSCW'12. (2012) 441–450
16. Cholvy, L.: Evaluation of information reported: a model in the theory of evidence. In: Proc. of the Int. Conf. on Information Processing and Management of Uncertainty in Knowledge-Based Systems, IPMU'10. (2010) 258–267
17. Cholvy, L.: Collecting information reported by imperfect agents. In: Proc. of the Int. Conf. on Information Processing and Management of Uncertainty in Knowledge-Based Systems, IPMU'12. Volume CCIS 299., Springer (2012) 501–510
18. Revault d'Allonnes, A., Lesot, M.-J.: Dynamics of trust bulding: models of information cross-checking in a multivalued logic framework. In: Proc. of the IEEE Int. Conf. on Fuzzy Systems, fuzzIEEE'15. (2015)
19. Zadeh, L.: Precisiated natural language (PNL). AI Magazine **25**(3) (2004) 74–91
20. Auger, A., Roy, J.: Expression of uncertainty in linguistic data. In: Proc. of the Int. Conf. on Information Fusion. (2008)
21. Rubin, V., Liddy, E., Kando, N.: Certainty identification in texts: categorization model and manual tagging results. In Shanahan, J., Qu, Y., Wiebe, J., eds.: Computing attitudes and affect in text: theory and applications. Information retrieval. Springer (2005) 61–74
22. Ba, H., Brizard, S., Dulong, T., Goujon, B.: Uncertainty of an event and its markers in natural language processing. In Capet, P., Delavallade, T., eds.: Information Evaluation. Wiley (2014) 161–186
23. Demolombe, R., Lorini, E.: A logical account of trust in information sources. In: Proc. of the Workshop on trust in agent societies. (2008)
24. Herzig, A., Lorini, E., Hübner, J.F., Vercouter, L.: A logic of trust and reputation. Logic Journal of the IGPL **18**(1) (2010) 214–244
25. Cholvy, L.: When reported information is second hand. In Capet, P., Delavallade, T., eds.: Information Evaluation. Wiley (2014) 231–260
26. Cholvy, L.: Information evaluation in fusion: a case study. In: Proc. of the Int. Conf. on Information Processing and Management of Uncertainty in Knowledge-Based Systems, IPMU'04. (2004) 993–1000
27. Revault d'Allonnes, A., Akdag, H., Poirel, O.: Trust-moderated information-likelihood. a multivalued logics approach. In: Proc. of the 3rd Conf. on Computability in Europe, CiE 2007. (2007) 1–6
28. Seridi, H., Akdag, H.: Approximate reasoning for processing uncertainty. J. of Advanced Comp. Intell. and Intell. Informatics **5**(2) (2001) 110–118
29. Zadeh, L.: Fuzzy sets as the basis for a theory of possibility. Fuzzy Sets and Systems **1** (1978) 3–28
30. Shafer, G.: A mathematical theory of evidence. Princeton University Press (1976)
31. Pichon, F., Dubois, D., Denoeux, T.: Relevance and truthfulness in information correction and fusion. Int. Journal of Approximate Reasoning, IJAR **53** (2012) 159–175

Applying Anomalous Cluster Approach to Spatial Clustering

Susana Nascimento and Boris Mirkin

Abstract The concept of anomalous clustering applies to finding individual clusters on a digital geography map supplied with a single feature such as brightness or temperature. An algorithm derived within the individual anomalous cluster framework extends the so-called region growing algorithms. Yet our approach differs in that the algorithm parameter values are not expert-driven but rather derived from the anomalous clustering model. This novel framework successfully applies to the issue of automatically delineating coastal upwelling from Sea Surface Temperature (SST) maps, a natural phenomenon seasonally occurring in coastal waters.

1 Introduction

In our previous work [1, 2], we automated the process of delineation of upwelling regions and boundaries using a fuzzy clustering method supplemented with the anomalous cluster initialization process [3]. Yet that method operates over temperature data only, without any relation to the spatial arrangement of the pixels involved. Therefore, we decided to modify the anomalous cluster framework in such a way that it applies to the pixels spatially located on geographical maps. We apply the view that an upwelling region grows step-by-step by involving nearest cold water pixels. The process is controlled by a function expressing the similarity of pixel temperatures to those already in the region. In a self-tuning version of the algorithm the homogeneity

S. Nascimento (✉)
Department of Computer Science and NOVA Laboratory for Computer Science
and Informatics (NOVA LINCS), Faculdade de Ciências e Tecnologia, Universidade
Nova de Lisboa, 2829-516 Caparica, Portugal
e-mail: snt@fct.unl.pt

B. Mirkin
Department of Data Analysis and Machine Intelligence, National Research
University Higher School of Economics, Moscow, Russian Federation
e-mail: bmirkin@hse.ru; mirkin@dcs.bbk.ac.uk

B. Mirkin
Department of Computer Science, Birkbeck University of London, London, UK

© Springer International Publishing AG 2017
V. Kreinovich (ed.), *Uncertainty Modeling*, Studies in Computational
Intelligence 683, DOI 10.1007/978-3-319-51052-1_10

threshold is locally derived from the approximation criterion over a window around the pixels under consideration. This window serves as a boundary regularize.

The paper is organized in three main sections. Section 2 describes a version of the anomalous cluster model and method relevant to the task. Section 3 describes a modification of this method applicable to clustering pixels according to the sea surface temperature maps. Section 4 gives a glimpse on application of the method to real temperature map data. Conclusion outlines the contents and sketches future work.

2 Anomalous Cluster Model and Alternating Method to Optimize It

Consider a set of objects P characterized by just one feature x so that, for every $p \in P$, $x(p)$ is a real number representing the value of the feature at p. A subset $C \subseteq P$ will be characterized by a binary vector $z = (z_p)$ such that $z_p = 1$ if $p \in C$ and $z_p = 0$ otherwise. We find such a C that approximates the feature x as closely as possible. To adjust the quantitative expression of C with z to the measurement scale of x, vector z should be supplied with an adjustable scaling coefficient λ. Also, a preliminary transformation of the x scale can be assumed by shifting the zero point of x into a point of interest, e.g. the mean value of x. Therefore, following Mirkin [3, 4], an approximation model can be stated as

$$y_p = \lambda z_p + e_p \tag{1}$$

where y_p are the preprocessed feature values, $z = (z_p)$ is the unknown cluster membership vector and λ is the scaling factor value, also referred to as the cluster intensity value [3, 4]. The items e_p represent errors of the model; they emerge because the vector λz_p may have only two different values, 0 and λ, whereas rescaled feature y may have different values. Anyway, the model requires that the errors should be made as small as possible.

Consider the least squares criterion $\Phi^2 = \sum_{p \in P} e_p^2 = \sum_{p \in P} (y_p - \lambda z_p)^2$ for fitting the model (1).

This is a more or less conventional statistics criterion. Yet in the clustering context, Φ^2 bears a somewhat unconventional meaning. Indeed, any $z_p = 0$ contributes y_p^2 to Φ^2 independently of the λ value. Therefore, to minimize the criterion, $z_p = 0$ should correspond to those objects at which pre-processed feature values $y(p)$ are zero or near zero. In contrast, those objects $p \in P$ at which maximum or almost maximum absolute values of the feature hold, should be assigned with $z_p = 1$. Moreover, these must be either maximum positive values or minimum negative values but not both. Indeed, as it is well known, the optimum λ at any given C must be the average of $y(p)$ over all $p \in C$, $\lambda(C) = \sum_{p \in C} y_p / |C|$. Substituting this value into criterion Φ^2, one can easily derive the following decomposition

$$\Phi^2 = \sum_{p \in P} y_p^2 - \lambda^2(C)|C| \tag{2}$$

or, equivalently,

$$\sum_{p \in P} y_p^2 = |C|\lambda^2(C) + \Phi^2. \tag{3}$$

The latter expression is a Pythagorean decomposition of the data scatter (on the left) in explained and unexplained parts. The smaller the unexplained part, Φ^2, the greater the explained part,

$$g^2(C) = |C|\lambda^2(C). \tag{4}$$

Equation (4) gives an equivalent reformulation to the least-squares criterion: an optimal C must maximize it. This cannot be achieved by mixing in C objects with both high positive and high negative y values because this would make the average $\lambda(C)$ smaller.

Therefore, an optimal C must correspond to either highest positive values of y or lowest negative values of y. Assume, for convenience, the latter case and consider a local search algorithm for finding a suboptimal C by adding objects one by one starting from a singleton $C = \{p\}$. What singleton? Of course that one corresponding to the lowest negative value of y, to make the value of criterion (4) as high as possible.

In publications [3, 4] only local search algorithms were considered. In these algorithms, entities are added (or removed) one by one to warrant a maximum possible local increment of the criterion until that becomes negative. Here, we develop a method which is more suitable for temperature map data. In this new method iterations are carried on in a manner similar to that of the well known clustering k-means method which is described in every text on data mining or clustering (see, for example [3]). Specifically, given a central value $c = \lambda(C)$, we add to cluster C all the relevant objects at once, after which the central value is recomputed and another iteration is applied. The computations converge to a stable solution that cannot be improved with further iterations.

To arrive at this "batch" clustering method, let us derive a different expression for the criterion Φ^2 by "opening" parentheses in it. Specifically, since $z_p^2 = z_p$ because z_p accepts only 0 and 1 values, we may have

$$\Phi^2(C, \lambda) = \sum_{p \in P}(y_p - \lambda z_p)^2 = \sum_{p \in P} y_p^2 - 2\lambda \sum_{p \in P}(y_p - \lambda/2)z_p \tag{5}$$

As the data scatter $\sum_{p \in P} y_p^2$ is constant, minimizing (5) is equivalent to maximizing the scoring function

$$f(C, \lambda) = \sum_{p \in P} \lambda(y_p - \lambda/2)z_p = \sum_{p \in C} \lambda(y_p - \lambda/2). \tag{6}$$

This can be rewritten as

$$f(C, \lambda, \pi) = \sum_{p \in C} \left(\lambda y_p - \pi\right) \tag{7}$$

where π is a parameter that is, optimally, equals $\pi = \lambda^2/2$, and yet can be considered a user-defined threshold in criterion (7). This criterion may be considered as depending on two variables, that are to be determined: λ and C. Therefore, the method of alternating optimization can be applied to maximize it. Each iteration of this method would consist of two steps. First, given λ, find all $p \in P$ such that $\lambda y_p > \pi$ and put them all as C. Of course, these y-values must be negative since $\lambda < 0$ in our setting. Second, given a C, find λ as the within-C average of y_p. Since both steps are optimal with respect to the corresponding variable, this method increases the value of g^2 at each step and, therefore, must converge because there are a finite number of different subsets C. In the follow-up the alternating anomalous clustering algorithm will be referred to as AA-clustering.

3 Adapting AA-clustering to the Issue of Delineating Upwelling Areas on Sea Surface Temperature Maps

Consider the set of pixels of a Sea Surface Temperature (SST) map of an ocean part as the set P, the feature x being the surface temperature. Such maps are used in many applications of which we consider the problem of automatic delineation of coastal upwelling. This is a phenomenon that occurs when the combined effect of wind stress over the coastal oceanic waters and the Coriolis force cause these surface waters to move away from the coast. Therefore, deep, cold and nutrient-rich waters move to the surface to compensate for the mass deficiency due to this surface water circulation. As such, it has important implications in ocean dynamics and for the understanding of climate models. The identification and continuing monitoring of upwelling is an important part of oceanography.

Unfortunately the current state is far from satisfactory. Although a number of approaches for segmentation of upwelling have been proposed, they suffer from too complex computational processes needed to get more or less satisfactory results (see, for instance, [5–9]).

Therefore, we decided to apply the self-tuning AA-clustering to pixels of the temperature map starting from the coldest pixel which, in fact, corresponds to the nature of upwelling.

Let $P = R \times L$ be a map under consideration where R is the set of rows and L, the set of columns, so that a pixel p can be presented as $p = (i, j)$ where $i \in L$ is its row coordinate and $j \in L$ its column coordinate. Then a corresponding sea surface temperature map can be denoted as $x = (x(i, j))$, for all $i \in R$ and $j \in L$. First of all, let us center the temperature, that is, subtract the average temperature

$t^* = mean(x)$ of the temperature map x from the temperature values at all pixels in $R \times L$. Let the centered values be denoted as $t(i, j), (i, j) \in R \times L$. The algorithm finds a cluster $C \subseteq R \times L$ in the format of a binary map $Z(R, L)$ with elements z_{ij} defined as $z_{ij} = 1$ if $(i, j) \in C$ and $z_{ij} = 0$, otherwise. Since the pixels are not just arbitrary objects but rather elements of a spatial grid, we need to introduce this property into the AA clustering approach.

For this purpose, let us consider each pixel $p = (i, j)$ as an element of a square window of a pre-specified size $W(i, j)$ centered at p. Based on preliminary experimentation we define the window size as 7×7 (pixels). The usage of a window system appears to be useful not only as a device for maintaining continuity of the cluster C being built, but also that its boundary is of more or less smooth shape. We refer to the pixels in window $W(i, j)$ as the neighborhood of $p = (i, j)$.

The algorithm starts by selecting a seed pixel, $o = (i_o, j_o)$, as a pixel with the lowest temperature value. The cluster C is initialized as the seed $o = (i_o, j_o)$ together with pixels within the window $W(i_o, j_o)$ satisfying the similarity condition

$$c \times t(i, j) \geq \pi, \tag{8}$$

where c is the reference temperature taken at the start as the temperature of the seed pixel o, and π, a similarity threshold, as described in the previous section. For convenience, let us refer to pixels in cluster C as labeled.

Once cluster C is initialized, its boundary set F is defined as the set of such unlabeled pixels, that their neighborhood intersects the cluster. Therefore,

$$F = \{(i', j') \notin C | W(i', j') \cap C \neq \varnothing\} \tag{9}$$

Then the algorithm proceeds iteratively expanding the cluster C step by step by dilating its boundary F until it is empty. For each boundary pixel (i', j') in F we define the boundary expansion region as the subset of pixels (i, j) of C that intersect the exploring window centered at pixel (i', j'), that is, $(i, j) \in W(i', j') \cap C$, and we define c^* as the average temperature of those pixels.

The homogeneity criterion of the algorithm is defined by the following similarity condition (10). This condition involves the reference temperature $c^* = mean\left(T\left(W(i', j') \cap C\right)\right)$, the mean temperature of the window pixels within the expanding region:

$$c^* \times t(i', j') \geq \pi \tag{10}$$

Therefore, in the following text we take the self-tuned value for the similarity threshold as half the squared average temperature over the cluster C.

A more or less smooth shape of the growing region is warranted by the averaging nature of the similarity criterion and by involving windows around all pixels under consideration in the frontline.

This method, called Seed Expanding Cluster (SEC) [25], is a specific case of Seeded Region Growing (SRG) approach introduced by Adams and Bischof [10] for region based image segmentation (see also [11–14]). The SRG approach tries to find a region which is homogeneous according to a certain *feature of interest* such as intensity, color, or texture. The algorithm follows a strategy based on the growth of a region, starting from one or several 'seeds' and by adding to them similar neighboring pixels. The growth is controlled by using a homogeneity criterion, so that the adding decision is generally taken based only on the contrast between the evaluated pixel and the region. However, it is not always easy to decide when this difference is small (or large) enough to make a reasonable decision.

The Seeded Region Growing image segmentation approach has been widely used in various medical image applications like magnetic resonance image analysis and unsupervised image retrieval in clinical databases [15–18]. The approach has been also successfully applied in color image segmentation with applications in medical imaging, content-based image retrieval, and video [14, 19, 20], as well as in remote sensing image analysis [21, 22].

Main challenging issues that arise with SRG methods are:

(i) selection of the initial seed(s) in practical computations to find a good segmentation;
(ii) choosing the homogeneity criterion and specifying its threshold;
(iii) efficiently ordering pixels for testing whether they should be added to the region.

Most approaches of SRG involve homogeneity criteria in the format of the difference in the feature of interest between its value at the pixel to be labeled and the mean value at the region of interest [10, 11, 13, 14]. A weak point of these algorithms is the definition of the non-homogeneity threshold at which the pixels under consideration are considered as failing the homogeneity test and, therefore, cannot be added to the region. Such a definition is either expert driven or supervised in most of the currently available algorithms [11, 14].

Many SRG algorithms grow the regions using a sequential list which is sorted according to the dissimilarity of unlabeled pixels to the growth region [10, 14, 15]. The disadvantage is that the segmentation results are very much sensitive to this order.

As can be readily seen, our approach avoids these issues altogether. It utilizes a mathematically derived, though somewhat unusual, homogeneity criterion, in the format of a product rather than the conventional difference between the pixel and the mean of the region of interest. To this end, we first subtract the average temperature value from all the temperature values. This process is implemented by using the concept of window of a pre-specified size around the pixels under consideration: only those within the window are involved in the comparison processes. This provides for both the spatial homogeneity and smoothness of the growing region. Indeed, only borderline pixels are subject to joining in, because the windows around remote pixels just do not overlap the growing region. Therefore, there is no need in specifying the order of testing for labeling among pixels: all those borderline pixels can be considered and decided upon simultaneously. The process starts from a cluster consisting of just one pixel, the coldest one, according to the approximation clustering criterion. The preprocessed temperature of this pixel is negative with a relatively large absolute value. Our region growing process initializes with a fragment of the coldest pixels, which is rather robust. Moreover, the simultaneous borderline labeling considerably speeds up the SRG procedure.

In our experiments, we used also the similarity threshold π derived according to Otsu's thresholding method [23]. This method fine-tunes the similarity threshold by finding the maximum inter-class variance that splits between warm and cold waters, and is considered one of the most popular threshold method in the literature [24].

4 Experimental Testing

The newly developed method and its Otsu's competitor have been applied to a group of 28 images of size 500×500 showing the upwelling phenomenon in Portugal coastal waters; a detailed description can be found in [25]. The selected images cover different upwelling situations. Specifically:

(i) SST images with a well characterized upwelling situation in terms of fairly sharp boundaries between cold and warm surface waters measured by relatively contrasting thermal gradients and continuity along the coast (two topmost images);
(ii) SST images showing distinct upwelling situations related to thermal transition zones offshore from the North toward the South and with smooth transition zones between upwelling regions;
(iii) noisy SST images with clouds, so that information for defining the upwelling front is lacking (fourth-line image).

Figure 1 (left column) illustrate these types of situations. These images have been manually annotated by expert oceanographers regarding the upwelling regions (binary ground truth maps), which are shown in the right column of Fig. 1.

Here we report of experiments on the SEC method at which the value of parameter π has been determined by either as the optimal $\lambda^2/2$ (SelfT-SEC) or by using the Otsu method (Otsu-SEC) applied to ground truth maps.

To compare the performance of seed region growing algorithms, we use the popular precision and recall characteristics, as well as their harmonic mean, the F-measure. Precision corresponds to the probability that the detection is valid, and recall to the probability that the ground truth is detected.

Overall, the segmentations are rather good, with 82% of F-scores ranging between 0.7 and 0.98. On analyzing segmentations obtained by the self-tuning threshold version of the algorithm we obtained good results in 75% of the cases. The majority of the lower value scores occur for the images with weak gradients. Figure 2 (left column) illustrates the segmentation results obtained by the self-tuning SEC algorithm for three SST images presented in Fig. 1.

By comparing the relative performances of the two unsupervised thresholding versions of SEC algorithm (Otsu-SEC and SelfT-SEC), we came up with the following conclusions. The Otsu-SEC wins in 53.6% of the cases whereas the self-tuning version wins in 46.4% of images.

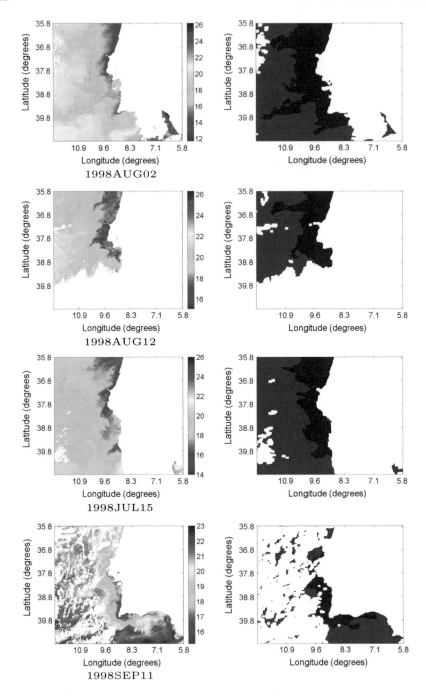

Fig. 1 Four SST images of Portugal showing different upwelling situations (*left column*); corresponding binary ground-truth maps (*right column*)

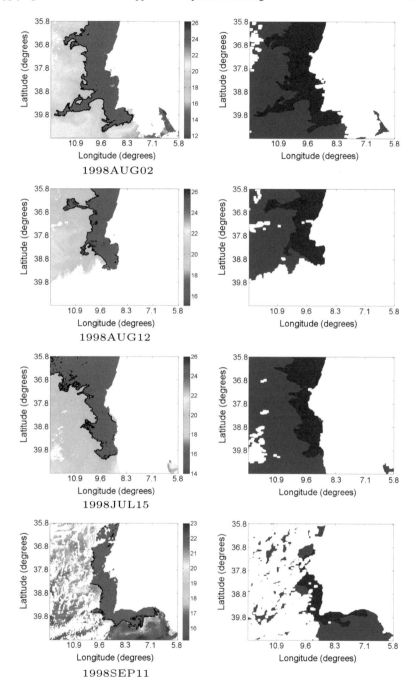

Fig. 2 Upwelling areas found by the self-tuning version of SEC algorithm on SST images of Portugal and her coastal waters (*left column*) versus the binary ground-truth maps (*right column*)

The two versions of the algorithm are implemented in MatLab R2013a. The experiments have been run on a computer with a 2.67 GHz Intel(R) core(TM) i5 processor and 6 Gbytes of RAM. The operating system is Windows 8.1 Pro, 64-bit. The elapsed time of segmentation of an SST image with the Otsu's thresholding version takes 25 s, whereas the self-tuning version takes 22 s for the task.

5 Conclusion

We have proposed a new method for image segmentation combining ideas of AA clustering and Seed Region Growing. This algorithm involves a novel homogeneity criterion (10), no order dependence of the pixel testing, and a version with self-tuning threshold derived from the approximation criterion.

The Otsu's version of the algorithm leads to high F-measure values at segmenting SST images showing different upwelling situations. The self-tuning version of the algorithm succeeds at all images presenting contrasting gradients between the coastal cold waters and the warming offshore waters of the upwelling region, and at some images with weak gradients for upwelling.

Further research should be directed toward both extending the SEC algorithm to situations with many clusters and applying it to other image segmentation problems.

Acknowledgements The authors thank the colleagues of Centro de Oceanografia and Department de Engenharia Geográfica, Geofísica e Energia (DEGGE), Faculdade de Ciências, Universidade de Lisboa for providing the SST images examined in this study. The authors are thankful to the anonymous reviewers for their insightful and constructive comments that allowed us to improve our paper.

References

1. Nascimento, S., Franco, P.: Segmentation of upwelling regions in sea surface temperature images via unsupervised fuzzy clustering. In: Corchado, E. and Yin, H. (Eds.), Procs. Intelligent Data Engineering and Automated Learning (IDEAL 2009), LNCS 5788, Springer-Verlag, 543–553 (2009)
2. Nascimento, S., Franco, P., Sousa, F., Dias, J., Neves, F.: Automated computational delimitation of SST upwelling areas using fuzzy clustering, Computers & Geosciences **43**, 207–216 (2012)
3. Mirkin, B.: Clustering: A Data Recovery Approach, 2nd Edition, Chapman and Hall, Boca Raton (2012)
4. Mirkin, B.: A sequential fitting procedure for linear data analysis models. Journal of Classification **7**, 167–195 (1990)
5. Arriaza, J., Rojas, F., Lopez, M., Canton, M.: Competitive neural-net-based system for the automatic detection of oceanic mesoscalar structures on AVHRR scenes. IEEE Transactions on Geoscience and Remote Sensing **41**(4), 845–885 (2003)
6. Chaudhari, S., Balasubramanian, R., Gangopadhyay, A.: Upwelling Detection in AVHRR Sea Surface Temperature (SST) Images using Neural-Network Framework. 2008 IEEE International Geoscience & Remote Sensing Symposium II, 926–929 (2008)

7. Kriebel, S. T., Brauer, W., Eifler, W.: Coastal upwelling prediction with a mixture of neural networks. IEEE Transactions on Geoscience and Remote Sensing **36**(5), 1508–1518 (1998)
8. Marcello, J., Marques, F., Eugenio, F.: Automatic tool for the precise detection of upwelling and filaments in remote sensing imagery. IEEE Transactions on Geoscience and Remote Sensing **43**(7), 1605–1616 (2005)
9. Nieto, K., Demarcq, H., McClatchie, S.: Mesoscale frontal structures in the Canary Upwelling System: New front and filament detection algorithms applied to spatial and temporal patterns. Remote Sensing of Environment **123**, 339–346 (2012)
10. Adams, R., Bischof, L.: Seeded region growing. IEEE Transactions on Pattern Analasys and Machine Intelligence **16**, 641–647 (1994)
11. Fan, J., Zeng, G., Body, M., Hacid, M.-S.: Seeded region growing: an extensive and comparative study. Pattern Recognition Letters **26**(8), 1139–1156 (2005)
12. Mehnert, A., Jackway, P.: An improved seeded region growing algorithm. Pattern Recognition Letters **18**(10), 1065–1071 (1997)
13. Shih, F., Cheng, S.: Automatic seeded region growing for color image segmentation. Image and Vision Computing, **23**, 877–886 (2005)
14. Verma, O., Hanmandlu, M., Seba, S., Kulkarni, M., Jain, P.: A Simple Single Seeded Region Growing Algorithm for Color Image Segmentation using Adaptive Thresholding, Procs. of the 2011 International Conference on Communication Systems and Network Technologies, IEEE Computer Society, Washington, DC, USA, pp. 500–503 (2011)
15. Harikrishna-Rai, G. N., Gopalakrishnan-Nair, T.R.: Gradient Based Seeded Region Grow method for CT Angiographic Image Segmentation. International Journal of Computer Science and Networking **1**(1), 1–6 (2011)
16. Mancas, M., Gosselin, B., Macq, B.: Segmentation Using a Region Growing Thresholding. Proc. SPIE 5672, Image Processing: Algorithms and Systems IV 388 (2005)
17. Wu, J., Poehlman, S., Noseworthy, M. D., Kamath, M.: Texture feature based automated seeded region growing in abdominal MRI segmentation. Journal of Biomedical Science and Engineering **2**, 1–8 (2009)
18. Zanaty, E.A.: Improved region growing method for magnetic resonance images (MRIs) segmentation. American Journal of Remote Sensing **1**(2), 53–60 (2013)
19. Fan, J., Yau, D.K.Y., Elmagarmid, A. K., Aref, W. G.: Automatic image segmentation by integrating color-based extraction and seeded region growing. IEEE Transactions on Image Processing **10**(10), 1454–1466 (2001)
20. Ugarriza, L. G., Saber, E., Vantaram, S.R., Amuso, V., Shaw, M., Bhaskar, R.: Automatic Image Segmentation by Dynamic Region Growth and Multiresolution Merging. IEEE Transactions on Image Processing **18**(10), 2275–2288 (2009)
21. Byun, Y., Kim, D., Lee, J., Kim, Y.: A framework for the segmentation of high-resolution satellite imagery using modified seeded-region growing and region merging. International Journal of Remote Sensing **32**(16), 4589–4609 (2011)
22. Zhang, T., Yang, X., Hu, S., Su, F.: Extraction of Coastline in Aquaculture Coast from Multispectral Remote Sensing Images: Object-Based Region Growing Integrating Edge Detection. Remote Sensing **5**(9), 4470–4487 (2013)
23. Otsu, N.: A Threshold Selection Method from Gray-Level Histograms. IEEE Transactions on System, Man, and Cybernetics SMC-**9**(1), 62–66 (1979)
24. Sezgin, M., Sankur, B.: Survey over image thresholding techniques and quantitative performance evaluation. Journal of Electronic Imaging **13**(1), 146–168 (2004)
25. Nascimento, S., Casca, S., Mirkin, B.: A Seed Expanding Cluster Algorithm for Deriving Upwelling Areas on Sea Surface Temperature Images, Computers & Geosciences Special issue on "Statistical learning in geoscience modelling: novel algorithms and challenging case studies", Elsevier 85(Part B), 74–85 (2015) doi:10.1016/j.cageo.2015.06.002

Why Is Linear Quantile Regression Empirically Successful: A Possible Explanation

Hung T. Nguyen, Vladik Kreinovich, Olga Kosheleva, and Songsak Sriboonchitta

Abstract Many quantities describing the physical world are related to each other. As a result, often, when we know the values of certain quantities $x_1, \ldots, x_n$, we can reasonably well predict the value of some other quantity y. In many application, in addition to the resulting estimate for y, it is also desirable to predict how accurate is this approximate estimate, i.e., what is the probability distribution of different possible values y. It turns out that in many cases, the quantiles of this distribution linearly depend on the values $x_1, \ldots, x_n$. In this paper, we provide a possible theoretical explanation for this somewhat surprising empirical success of such linear quantile regression.

1 Formulation of the Problem

What is regression: a brief reminder. Many things in the real world are related to each other. As a result, if we know the values of some quantities $x_1, \ldots, x_n$, then we can often reasonable well estimate the value of some other quantity y.

In some cases, the dependence of y on $x_1, \ldots, x_n$ is known. In many other situations, we do not know this dependence, so we need to find this dependence based on

H.T. Nguyen
Department of Mathematical Sciences, New Mexico State University,
Las Cruces, NM 88003, USA
e-mail: hunguyen@nmsu.edu

H.T. Nguyen · S. Sriboonchitta
Department of Economics, Chiang Mai University, Chiang Mai, Thailand
e-mail: songsakecon@gmail.com

V. Kreinovich (✉)
Department of Computer Science, University of Texas at El Paso 500 W. University,
El Paso, TX 79968, USA
e-mail: vladik@utep.edu

O. Kosheleva
University of Texas at El Paso, 500 W. University, El Paso, TX 79968, USA
e-mail: olgak@utep.edu

© Springer International Publishing AG 2017   159
V. Kreinovich (ed.), *Uncertainty Modeling*, Studies in Computational
Intelligence 683, DOI 10.1007/978-3-319-51052-1_11

the empirical data. The desired dependence of y on $x_1, \ldots, x_n$ is known as a *regression function* $y \approx f(x_1, \ldots, x_n)$, and the methodology of determining the regression function from the empirical data is known as *regression analysis*.

In many practical situations, the dependence of y on $x_1, \ldots, x_n$ is well described by a linear function

$$y \approx \beta_0 + \sum_{i=1}^{n} \beta_i \cdot x_i. \tag{1}$$

Such linear dependence is known as *linear regression*.

What is quantile regression. Traditionally, the emphasis of regression analysis has been on finding the actual dependence $y \approx f(x_1, \ldots, x_n)$. However, finding this dependence is not enough. As we have mentioned earlier, the value $f(x_1, \ldots, x_n)$ is only an *approximation* to y. It is good to know this approximation, but it is also important to know *how accurate* is this approximation. In other words, we want to know not only the estimate of y for given x_i, we also want to know how the conditional probability distribution of y depends on the inputs $x_1, \ldots, x_n$.

One of the empirically efficient ways for finding this dependence is the method of *quantile regression*. One of the possible ways to describe the conditional probability distribution $P(y \mid x_1, \ldots, x_n)$ is to describe, for each probability p, the p-th quantile y_p of this distribution, i.e., that value for which the conditional probability $\mathrm{Prob}(y \leq y_p \mid x_1, \ldots, x_n)$ is equal to p. In particular:

- for $p = 0.5$, we get the median,
- for $q = 0.25$ and $q = 0.75$, we get the quartiles, etc.

One of the most empirically successful methods of describing the dependence of the conditional probability distribution on x_i is a method of *quantile regression*, when, for each p, we find a regression function $y_p = f_p(x_1, \ldots, x_n)$ that describes the dependence of the corresponding quantile y_p on the inputs x_i.

In particular, somewhat surprisingly, in many practical situations, this dependence turns out to be linear:

$$y_p \approx \beta_{0,p} + \sum_{i=1}^{n} \beta_{i,p} \cdot x_i \tag{2}$$

for appropriate coefficients $\beta_{i,p}$; see, e.g., [2–6].

Why is linear quantile regression empirically successful? Why is this linear quantile regression empirically successful in many practical applications? In this paper, we provide a possible explanation for this empirical success.

The structure of this paper is as follows. First, in Sect. 2, we provide fundamental reasons why linear regression is often empirically successful. In Sect. 3, we expand this result to the case of *interval uncertainty*, when instead of predicting the exact value of the quantity y, we predict the interval of its possible values. Finally, in Sect. 4, we show how this result can be expanded from interval to probabilistic uncertainty—thus explaining the empirical success of linear quantile regression.

2   Why Linear Regression Is Often Empirically Successful: A General Explanation

What we do in this section. Our goal is to explain why linear *quantile* regression is empirically successful. To explain this empirical phenomenon, let us first provide a possible explanation of why linear regression *in general* is empirically successful.

Empirical fact. Linear regression is empirically successful in many real-life situations, often in situations when the known empirical dependence is non-linear.

In this section, we will provide a possible explanation for this empirical fact.

Basis for our explanation: possibility of different starting points for measuring the corresponding quantities. We are interested in the dependence between the *quantities* x_i and y. To describe this dependence between *quantities*, we describe the dependence between the *numerical values* of these quantities.

The difference between the quantity itself and its numerical value may be perceived as subtle but, as will show, this difference is important—and it provides the basis for our explanation. The reason why there is a difference in the first place is that the numerical value of a quantity depends on the starting point for measuring this quantity. If we change this starting point to the one which is a units earlier, then all the numerical values of this quantity change from the previous value x to the new value $x + a$.

For example, we can start measuring temperature with the absolute zero (as in the Kelvin scale) or with the temperature at which the ice melts (as in the Celsius scale), the corresponding numerical values differ by $a \approx 273°$. Similarly, we can start measuring time with the birth year of Jesus Christ or, as the French Revolution decreed, with the year of the French Revolution.

It may be not so clear, but when we gauge many economic and financial quantities, there is also some arbitrariness in the selection of the starting point. For example, at first glance, unemployment is a well-defined quantity, with a clear starting point of 0%. However, economists who seriously study unemployment argue that starting it from 0 is somewhat misleading, since this may lead to an unrealistic expectation of having 0 unemployment. There is a natural minimal unemployment level of approximately 3%, and a more natural way of measuring unemployment is:

- not by its absolute value,
- but by the amount by which the current unemployment level exceeds its natural minimum.

Similarly, a person's (or a family's) income seems, at first glance, like a well-defined quantity with a natural starting point of 0. However, this does not take into account that 0 is not a possible number, a person needs to eat, to get clothed. Thus, a more reasonable way to gauge the income is:

- not by the absolute amount,
- but by how much the actual income exceeds the bare minimum needed for the person's survival.

The changes in the starting points should not affect the actual form of the dependence. In general, we can have different starting points for measuring each of the input quantities x_i. As a result, for each i, instead of the original numerical values x_i, we can have new values $x_i' = x_i + a_i$, for some constants a_i.

The change in the starting point:

- *changes* the numerical value, but
- *does not change* the actual quantity.

Thus, it is reasonable to require that the exact form of the dependence between x_i and y should not change if we simply change the starting points for all the inputs.

Of course, even for the simplest dependence $y = x_1$, if we change the starting point for x_1, then the numerical value of y will change as well, by the same shift—and thus, while the numerical value of y changes, the quantity y does not change—because the change in the starting point for x_1 simply implies that we correspondingly change the starting point for y.

In general, it is therefore reasonable to require that for each combination of shifts $a_1, \ldots, a_n$:

- once we shift the inputs to $x_i' = x_i + a_i$ and apply the function f to these shifted values,
- the resulting value $y' = f(x_1', \ldots, x_n')$ should simply be obtained from the original pre-shifted value $y = f(x_1, \ldots, x_n)$ by an appropriate shift:

$$y' = y + s(a_1, \ldots, a_n).$$

Thus, we arrive at the following definition.

Definition *We say that a function $f(x_1, \ldots, x_n)$ is shift-invariant if for every tuple $(a_1, \ldots, a_n)$ there exists a value $s(a_1, \ldots, a_n)$ such that for all tuples $(x_1, \ldots, x_n)$, we have*

$$f(x_1 + a_1, \ldots, x_n + a_n) = f(x_1, \ldots, x_n) + s(a_1, \ldots, a_n). \qquad (3)$$

The desired dependence should be continuous. The values x_i are usually only approximately known—they usually come from measurements, and measurement are always approximate. The actual values x_i^{act} of these quantities are, in general, slightly different from the measurement results x_i that we use to predict y. It is therefore reasonable to require that when we apply the regression function $f(x_1, \ldots, x_n)$ to the (approximate) measurement results, then the predicted value $f(x_1, \ldots, x_n)$ should be close to the prediction $f(x_1^{\text{act}}, \ldots, x_n^{\text{act}})$ based on the actual values x_i^{act}.

In other words, if the inputs to the function $f(x_1, \ldots, x_n)$ change slightly, the output should also change slightly. In precise terms, this means that the function $f(x_1, \ldots, x_n)$ should be *continuous*.

Now that we have argued that the regression function be shift-invariant and continuous, we can explain why linear regression is empirically successful.

Proposition *Every shift-invariant continuous function $f(x_1, \ldots, x_n)$ is linear, i.e., has the form*

$$f(x_1, \ldots, x_n) = \beta_0 + \sum_{i=1}^{n} \beta_i \cdot x_i \tag{4}$$

for appropriate coefficients β_i.

Proof Substituting the values $x_i = 0$ into the equality (3), we conclude that

$$f(a_1, \ldots, a_n) = f(0, \ldots, 0) + s(a_1, \ldots, a_n) \tag{5}$$

for all possible tuples $(a_1, \ldots, a_n)$. In particular, this is true for the tuples $(x_1, \ldots, x_n)$ and $(x_1 + a_1, \ldots, x_n + a_n)$, i.e., we have:

$$f(x_1, \ldots, x_n) = f(0, \ldots, 0) + s(x_1, \ldots, x_n) \tag{6}$$

and

$$f(x_1 + a_1, \ldots, x_n + a_n) = f(0, \ldots, 0) + s(x_1 + a_1, \ldots, x_n + a_n). \tag{7}$$

Substituting the expressions (6) and (7) into the equality (3) and cancelling the common term $f(0, \ldots, 0)$ in both sides of the resulting equality, we conclude that

$$s(x_1 + a_1, \ldots, x_n + a_n) = s(x_1, \ldots, x_n) + s(a_1, \ldots, a_n) \tag{8}$$

Such functions are known as *additive*.

From the equality (5), we conclude that

$$s(a_1, \ldots, a_n) = f(a_1, \ldots, a_n) - f(0, \ldots, 0). \tag{9}$$

Since the function $f(a_1, \ldots, a_n)$ is continuous, we can conclude that the function $s(a_1, \ldots, a_n)$ is continuous as well. So, the function $s(x_1, \ldots, x_n)$ is continuous and additive.

It is known (see, e.g., [1]) that every continuous additive function is a homogeneous linear function, i.e., has the form

$$s(x_1, \ldots, x_n) = \sum_{i=1}^{n} \beta_i \cdot x_i \tag{10}$$

for some real numbers β_i. Thus, from the formula (5), we can conclude that

$$f(x_1, \ldots, x_n) = \beta_0 + s(x_1, \ldots, x_n) = \beta_0 + \sum_{i=1}^{n} \beta_i \cdot x_i, \tag{11}$$

where we denoted $\beta_0 \stackrel{\text{def}}{=} f(0, \ldots, 0)$.

The proposition is proven.

Comment. It is easy to see that, vice versa, every linear function (4) is continuous and shift-invariant: namely, for each such function, we have:

$$f(x_1 + a_1, \ldots, x_n + a_n) = \beta_0 + \sum_{i=1}^{n} \beta_i \cdot (x_i + a_i) =$$

$$\beta_0 + \sum_{i=1}^{n} \beta_i \cdot x_i + \sum_{i=1}^{n} \beta_i \cdot a_i = f(x_1, \ldots, x_n) + s(a_1, \ldots, a_n),$$

where we denoted $s(a_1, \ldots, a_n) \stackrel{\text{def}}{=} \sum_{i=1}^{n} \beta_i \cdot a_i$.

3   Case of Interval Uncertainty

Description of the case. In the previous section, we have shown that when we try to predict a numerical value y, then it is often beneficial to use linear regression. As we have mentioned, predicting a single value y is often not enough:

- in addition to the approximate value y,
- it is also necessary to know how accurate is this approximate value, i.e., which values y are possible.

Because of this necessity, in this section, we consider a situation, in which, for each inputs $x_1, \ldots, x_n$:

- instead of predicting a *single* value y,
- we would like to predict the *interval* $\left[\underline{y}(x_1, \ldots, x_n), \overline{y}(x_1, \ldots, x_n) \right]$ of all the values of y which are possible for given inputs $x_1, \ldots, x_n$.

Why linear regression. In the case of interval uncertainty, instead of a *single* regression function $y = f(x_1, \ldots, x_n)$, we have *two* regression functions:

- a regression function $\underline{y} = \underline{f}(x_1, \ldots, x_n)$ that describes the lower endpoint of the desired interval, and
- a regression function $\overline{y} = \overline{f}(x_1, \ldots, x_n)$ that describes the upper endpoint of the desired interval.

It is reasonable to require that each of these two functions is

- continuous, and
- does not change if we change the starting points for measuring the inputs—i.e., is *shift-invariant* (in the sense of the above Definition).

Thus, due to our proposition, each of these functions is linear, i.e., we have

$$\underline{f}(x_1, \ldots, x_n) = \underline{\beta}_0 + \sum_{i=1}^{n} \underline{\beta}_i \cdot x_i \qquad (12)$$

and

$$\overline{f}(x_1, \ldots, x_n) = \overline{\beta}_0 + \sum_{i=1}^{n} \overline{\beta}_i \cdot x_i. \qquad (13)$$

for appropriate values $\underline{\beta}_i$ and $\overline{\beta}_i$.

4   Case of Probabilistic Uncertainty

Description of the case. We consider the situation in which for each combination of inputs $x_1, \ldots, x_n$, in addition to the set of possible values of y, we also know the probability of different possible values of y. In other words, for each tuple of inputs $x_1, \ldots, x_n$, we know the corresponding (conditional) probability distribution on the set of all possible values y.

What is the relation between a probability distribution and the set of possible values? From the previous section, we know how to describe regression in the case of interval uncertainty. We would like to extend this description to the case of probabilistic uncertainty. To be able to do that, let us recall the usual relation between the probability distribution and the set of possible values.

This relation can be best illustrated on the example of the most frequently used probability distribution—the normal (Gaussian) distribution, with the probability density

$$\rho(y) = \frac{1}{\sqrt{2\pi} \cdot \sigma} \cdot \exp\left(-\frac{(y - \mu)^2}{2\sigma^2}\right). \qquad (14)$$

The ubiquity of this distribution comes from the Central Limit Theorem, according to which the probability distribution caused by the joint effect of many small independent random factors is close to Gaussian; see, e.g., [7].

From the purely mathematical viewpoint, a normally distributed random variable can attain any real value. Indeed, the corresponding probability density is always positive, and thus, there is always a non-zero probability that we will have a value far away from the mean μ.

However, values which are too far from the mean have such a low probability that from the practical viewpoint, they are usually considered to be impossible. It is well know that:

- with probability 95%, the normally distributed random variable y is inside the two-sigma interval $[\mu - 2\sigma, \mu + 2\sigma]$;

- with probability 99.9%, y is inside the three-sigma interval $[\mu - 3\sigma, \mu + 3\sigma]$, and
- with probability $1-10^{-8}$, y is inside the six-sigma interval $[\mu - 6\sigma, \mu + 6\sigma]$.

In general, a usual way to transform a probability distribution into an interval of practically possible values is to use a *confidence interval*, i.e., an interval for which the probability to be outside this interval is equal to some pre-defined small value p_0. A usual way to select such an interval is to select the bounds $\underline{y}$ and $\overline{y}$ for which:

- the probability to have y smaller than $\underline{y}$ is equal to $\dfrac{p_0}{2}$, and
- the probability to have y larger than $\overline{y}$ is equal to $\dfrac{p_0}{2}$.

One can easily see that:

- the lower endpoint $\underline{y}$ of this confidence interval is the quantile $y_{p_0/2}$, and
- the upper endpoint $\underline{y}$ of this confidence interval is the quantile $y_{1-p_0/2}$.

Depending on the problem, we can have different probabilities p_0, so we can have all possible quantiles.

Conclusion: why linear quantile regression is empirically successful. For each combination of inputs $x_1, \ldots, x_n$, based on the related (conditional) probability distribution of y, we can form the interval of practically possible values, in which both endpoints are quantiles y_p corresponding to some values p.

In the previous section, we have shown that reasonable requirements imply that each of these endpoints is a linear function of the inputs. Thus, we conclude that for each p, we have

$$y_p \approx \beta_{0,p} + \sum_{i=1}^{n} \beta_{i,p} \cdot x_i, \tag{2}$$

for appropriate values $\beta_{i,p}$.

This is exactly the formula for linear quantile regression. Thus, we have provided the desired first-principles for linear quantile regression formulas. The existence of such a justification can explain why linear quantile regression is empirically successful.

Acknowledgements We acknowledge the partial support of the Center of Excellence in Econometrics, Faculty of Economics, Chiang Mai University, Thailand. This work was also supported in part by the National Science Foundation grants HRD-0734825 and HRD-1242122 (Cyber-ShARE Center of Excellence) and DUE-0926721.

References

1. J. Aczél and J. Dhombres, *Functional Equations in Several Variables*, Cambridge University Press, 2008.
2. J. D. Angrist and J.-S. Pischke, *Mostly Harmless Econometrics: An Empiricist's Companion*, Princeton University Press, Princeton, New Jersey, 2009.

3. C. Davino, M. Furno, and D. Vistocco, *Quantile Regression: Theory and Applications*, Wiley, New York, 2013.
4. B. Fitzenberger, R. Koenker, and J. A. F. Machado (editors), *Economic Applications of Quantile Regression*, Physika-Verlag, Heidelberg, 2002.
5. L. Hao and D. Q. Naiman, *Quantile Regression*, SAGE Publications, Thousands Oaks, California, 2007.
6. R. Koenker, *Quantile Regression*, Cambridge University Press, New York, 2005.
7. D. J. Sheskin, *Handbook of Parametric and Nonparametric Statistical Procedures*, Chapman & Hall/CRC, Boca Raton, Florida, 2011.

A Rapid Soft Computing Approach to Dimensionality Reduction in Model Construction

Vesa A. Niskanen

Abstract A rapid soft computing method for dimensionality reduction of data sets is presented. Traditional approaches usually base on factor or principal component analysis. Our method applies fuzzy cluster analysis and approximate reasoning instead, and thus it is also viable to nonparametric and nonlinear models. Comparisons are drawn between the methods with two empiric data sets.

Keywords Dimension reduction · Factor analysis · Principal component analysis · Fuzzy cluster analysis · Fuzzy reasoning

1 Introduction

In model construction large observation or input variable sets arouse various problems and thus we usually attempt to simplify our examinations by applying data compression or dimensionality reduction. Then we may operate with fewer observations or variables. In statistical multivariate analysis this means that in the former case we may compress our data matrices by applying cluster analysis (CA), whereas the number of variables is reduced by combining similar original variables for variable groups with such methods as the principal component analysis (PCA) and factor analysis (FA) [1].

Today soft computing (SC) systems have also proven to be useful in statistical modeling and model construction in general, and its CA and regression models provide good examples of these [2–4]. On the other hand, we still face certain challenges when applying SC techniques to dimensionality reduction.

One open problem is how to reduce the dimensionality in our data matrix if the traditional PCA and FA approaches are insufficient. Typical limitations of PCA and FA are that they are only appropriate for linear models, their data sets should be sufficiently large and their variables are expected to be normally distributed [1, 5].

V.A. Niskanen (✉)
Department of Economics & Management, University of Helsinki,
PO Box 27, 00014 Helsinki, Finland
e-mail: vesa.a.niskanen@helsinki.fi

© Springer International Publishing AG 2017
V. Kreinovich (ed.), *Uncertainty Modeling*, Studies in Computational
Intelligence 683, DOI 10.1007/978-3-319-51052-1_12

In many practical cases, however, actual relationships are nonlinear and/or sample sizes are small. In such situations, SC techniques have often been very successful in data processing. It is therefore reasonable to believe that SC techniques will be also efficient in reducing dimensionality.

Below we apply SC models to dimensionality reduction and we aim at a simple and an easily understandable method that is also robust and good in practice. This approach thus provides us with a "quick-and-dirty" method in the practice of model construction. We also attempt to draw an analogy between our approach and the PCA and FA. Some SC approaches to dimensionality reduction are available already, but they seem to be fuzzified versions of PCA or FA [6–18].

On the other hand, some papers have applied fuzzy similarity measures to this problem area, but they do not seem to correspond with the theoretical background or the goodness criteria of PCA and FA [13, 14]. Our approach, in turn, applies fuzzy similarity measures but we also yield and assess our outcomes according to PCA and FA and their goodness criteria. We also maintain Lotfi Zadeh's original idea on fuzzy systems, viz. instead of only using fuzzy mathematics or set theory, we also apply fuzzy reasoning in an understandable manner [19–21]. Thanks for the good available fuzzy methods in CA and approximate reasoning, that are well-known in the fuzzy community already, we adopt a general, a meta-level, approach, and thus detailed calculations are precluded.

Section 2 presents basic ideas on PCA and FA. Section 3 introduces our method. Section 4 provides two real-world examples. Section 5 concludes our examination.

2 Dimensionality Reduction in Statistics

In the explorative studies in human sciences [22] we aim at reducing the number of the original variables by grouping first the similar variables and then specifying such new variables that constitute these variable groups. These new variables are often referred to as sum variables because they are usually the (possibly weighted) sums the original variables. In this manner we may understand better the nature of our data and can perform simpler calculations. We may even attempt to find new "latent" variables behind the original variables in which case we can also label these new variables according to our interpretations, if necessary. For example, if we notice that a certain group of variables in our data matrix actually measures the same general feature, such as person's mathematical skills, from various standpoints, we may specify the sum variable "mathematical skills" that captures the meanings of the corresponding original variables. In the confirmatory studies, in turn, we may apply the available background theories to our dimensionality reduction.

In the traditional statistics we may specify our sum variables directly on our intuitive basis by calculating the sums of the selected original variables. We may also apply PCA or FA, in which case the obtained sum variables are referred to as principal components or factors, respectively. In fact, these methods operate with the standard scores of the original variables, i.e., for each variable its mean is subtracted from the observations, and then these differences are divided by its standard deviation, standard score = (observation − mean)/standard deviation. This transformation indicates us how the observations are distributed around the mean in the units of the standard deviation. Hence, when in PCA and FA we calculate the sum variable values for the original observations, viz. the principal component or factor scores, we also obtain the standard scores, and this may complicate the interpretation of our outcomes. Thus simpler sum variable specifications are also available, and these are discussed below.

We usually proceed as follows with the traditional PCA and FA [1]:

1. We assume that there are sufficiently high linear inter-correlations between the original variables, because the similarities between the variables are usually based on these correlations.
2. We also assume that the sample size is sufficiently large (e.g., at least five observations per variable), outliers are excluded, and the variables are measured at least at the level of good ordinal scales. In FA the variables should also be normally distributed and multicollinearity is not accepted.
3. We calculate the so-called principal component or factor loadings that are the correlations between the original variables and the principal components or factors.
4. We apply rotation to these components or factors in order to better interpret our loadings.
5. We select the appropriate principal components or factors, and these will be used in our sum variable specifications.

One distinction between the PCA and FA is that in the former case we always obtain the same unique components, whereas in FA the factors may vary according to the established number of factors.

In a sense, PCA and FA aim to find variable clusters according to the linear inter-correlations between the variables. The variable groups having intercorrelations will constitute variable clusters, and thus the obtained components or factors are the corresponding cluster centers. The principal component or factor loadings may now be regarded as being the "degrees of membership" of the variables to these clusters. Another approach is to consider that the principal components or factors span such vector spaces in which the loadings of the variables denote their coordinates.

On the other hand, we may also assume that in dimensionality reduction we aim to find clusters of variables according to their distances, and thus we can apply CA or multidimensional scaling. Hence, instead of the so-called Q-techniques of clustering, we are now applying the R-techniques [23]. In practice, when our data set is not too large, we may then operate with the transposed version of the original data matrix and then apply CA, and this approach is adopted below. We also apply fuzzy rule-based reasoning in our analyses. Thanks for the good fuzzy clustering techniques and usable approximate reasoning methods, the SC approach is more robust, applicable and user-friendly than the traditional methods.

Table 1 Example of a data matrix and its principal components

Case nr.	Variables and principal components					
	X_1	$X_2 \ldots$	X_n	C_1	$C_2 \ldots$	$C_{q \leq n}$
1	X_{11}	X_{12}	X_{1n}	C_{11}	C_{12}	C_{1q}
2	X_{21}	X_{22}	X_{2n}	C_{21}	C_{22}	C_{2q}
$\ldots$	$\ldots$			$\ldots$		
m	X_{m1}	$X_{m2} \ldots$	X_{mn}	C_{m1}	C_{m2}	C_{mq}

3   Soft Computing and Dimensionality Reduction

Our SC approach below applies both fuzzy clustering and approximate reasoning directly to dimensionality reduction. In this manner we may also apply dimensionality reduction in a nonparametric manner to nonlinear data sets. Since we adopted a meta-level approach, i.e., only methods of general nature are considered and detailed mathematical analyses are precluded, we apply the prevailing fuzzy clustering methods and fuzzy rule-based models, and these have also proved to be useful in practice. However, for the sake of consistency, our goodness criteria for the outputs are those of PCA and FA. In this respect fuzzy mountain clustering [24, 25] and fuzzy c-means clustering methods [10, 11, 26–34] are analogous to PCA and FA, respectively.

For example, in mountain clustering and PCA we specify the unique outputs iteratively one at a time starting from the densest or largest group of observations or variables, whereas within the fuzzy c-means method and FA our outcomes vary according to the established number of clusters or factors. We focus on the mountain clustering method and PCA below, because these techniques may bring better understanding to our approach.

Given now the original data matrix with m cases or observations (rows) and n variables (columns), if we apply fuzzy clustering method to dimensionality reduction, we may proceed as follows (Table 1).

1. We focus on such groups of variables that are close to each other. In other words, the distances between these variables, X_i, are small. In practice, we may operate, for example, with their standard scores, ZX_i, and our task stems from the calculation of the norms, $||ZX_i - ZX_j||$ ($i \neq j$). Alternative transformations may also be used, but in any case the original variables should be transformed into similar scales, because otherwise our variables have unequal weights.
2. Our method uses the transpose of the original data matrix (columns become rows), and then we apply fuzzy cluster analysis to the variables.
3. The obtained cluster centers of the variables, C_k, will be our "principal components" or "factors", and our loadings are now the correlations between the variables and these centers. We may also use linguistic values in this context.
4. We assess the goodness of our outcomes by applying such prevailing criteria as the communalities and eigenvalues of the variables.

When we specify the sum variables according to our loadings, we use functions that express sufficiently well the relationships between the variables and the cluster centers. Traditionally we may apply linear regression analysis in order to obtain such corresponding sum variables, $S_k \approx C_k$,

$$S_k = \sum_i w_{ik} \cdot ZX_i, (i = 1, 2, \ldots, n), \tag{1}$$

in which C_k is the dependent variable, the weights, w, are the regression coefficients and ZX are the standard scores of the original variables [1, 5, 35–37]. This method is used for calculating the principal component or factor scores.

However, (1) also includes the irrelevant variables in the sum variables, i.e., the variables with low loadings, and thus it may yield more or less misleading outcomes. Hence, in practice, we quite often simply calculate the sum of the relevant variables in each component, i.e., we only select the variables with the high loadings. In the case of the similar original scales we prefer the averages of these variables. This idea is widely used in PCA and FA in the human sciences. We may then justify our decisions by applying item analysis with Cronbach's alpha reliability coeffiecints, which are based on correlation coefficients, even though this method is not foolproof for this task [1, 5]. Another, sometimes more reliable method is based on the factor score covariance matrix [1]. The former usually minimizes and the latter maximizes the reliability coefficients.

Within our SC framework, we may also apply the fuzzy rule-based models, F, for all relevant variables in a component, i.e., the cluster center, C_k, is the dependent variable in the model

$$S_k = F_k(X_q, \ldots, X_r), q \geq 1, r \leq n, \tag{2}$$

Method (2) seems better in practice because it is also appropriate for nonlinear relationships.

Our SC approach may nevertheless arouse problems if variable clusters are unavailable or we have large observation sets. In the former case it may be difficult to find plausible cluster centers with any method, and the latter case may lead to quite heavy computations due to the large number of parameters.

Below we apply the idea on mountain clustering that is analogous to PCA [24, 25]. Hence, our method specifies the first cluster center according to the greatest or densest variable cluster. The second center is assigned to the second densest cluster with the restriction that it is not in the neighborhood of the first cluster. The next center, in turn, belongs to the third densest cluster, but it should also locate far from the previous clusters, and so forth. The number of cluster centers is determined by the user in the manner of PCA.

From the mathematical standpoint, the general idea for our clustering is that our first cluster center, C_1, is obtained when we minimize this type of penalty function,

$$\Sigma_i \mu_{C1}(ZX_i) \cdot ||ZX_i - C_1||, i = 1, 2, \ldots, n, \tag{3}$$

in which μ is an appropriate fuzzy triangular or bell-shaped membership function with its maximum value at C_1. This method finds the vector C_1 to be the center of the densest variable cluster.

The second cluster center, C_2, should represent the second densest cluster, and thus the neighborhood of C_1 should be excluded from our analysis. Hence, our minimizing penalty function should now also contain the exclusion function, Ex, that excludes the first cluster,

$$\Sigma_i Ex_{C2}(ZX_i) \cdot \mu_{C2}(ZX_i) \cdot ||ZX_i - C_2||, i = 1, 2, \ldots, n, \tag{4}$$

in which, for example, $Ex_{C2}(ZX_i) = (1 - \mu_{C1}(ZX_i))^s (s > 1)$. This means that in the second round the variables close to C_1 are irrelevant.

In the third round, the variables close to C_1 and C_2 are irrelevant, i.e.,

$$Ex_{C3}(ZX_i) = \min(Ex_{C2}(ZX_i), (1 - \mu_{C2}(ZX_i))^s) \tag{5}$$

and the penalty function for C_3 is

$$\Sigma_i(Ex_{C3}(ZX_i) \cdot \mu_{C3}(ZX_i) \cdot ||ZX_i - C_3||, i = 1, 2, \ldots, n. \tag{6}$$

We will continue till all the values of $Ex_{Ci}(ZX_i)$ are small, this meaning that we have examined all variable clusters.

These operations may be carried out conveniently with such methods as the genetic algorithms if custom-made models are preferred. We may also apply the original mountain clustering method directly, if the number of variables is sufficiently large. Below we will provide examples with the empiric data sets.

4   Real-World Examples

We examine below two real-world data sets with both PCA and our SC method. We aim to demonstrate that our method, that is simpler and more robust than PCA and FA, is also plausible for dimensionality reduction. We use Matlab^{TM} version 2014b and IBM SPSS^{TM} version 22 in our calculations.

4.1   The Iris Data

Fisher's Iris data is the widely-used benchmark data in cluster analysis. Fuzzy clustering methods have already proved their usability in this context of grouping the objects, but only some indirect methods have been suggested for dimensionality reduction of variables. This data set is challenging to us because it contains problematic clusters.

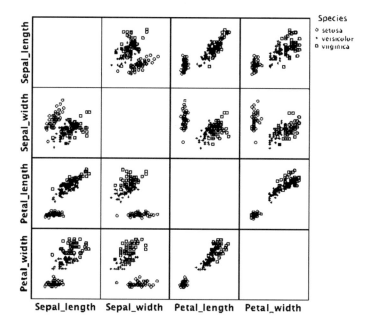

Fig. 1 Scatter plots of flowers in the Iris data

The Iris data contains 150 of these flowers and four feature variables that measure in millimeters their *Sepal lengths* (Sl), *Sepal widths* (Sw), *Petal lengths* (Pl) and *Petal widths* (Pw; Fig. 1). In the cluster analysis for these flowers we should find three clusters, and good fuzzy CA methods are able to perform this. Since sufficiently high inter-correlations prevail between the variables, we may also attempt to use PCA and our SC method for dimensionality reduction.

4.1.1 The PCA Approach

In PCA the values of the inter-correlations between our variables may first be analyzed with such rules of thumb as the Kaiser-Mayer-Olkin measure and Bartlett's test, and if the former yields values greater than 0.6 and the latter rejects its null hypothesis, our correlations seem to be sufficiently high [1, 5]. In our data the former yields 0.54 and the latter rejects the null hypothesis (at the level of significance <0.05). Hence, the former value is not fully satisfactory, but the latter fulfills the conditions.

On the other hand, the communalities of the variables are higher than 0.9, and hence we assume that PCA is justified with all our feature variables in this context. The communalities are the *rsquares* in those regression models in which the feature variable is the dependent variable and the principal components are the independent variables. Hence, the communalities indicate how well the selected principal components can explain or predict the variances of the variables.

PCA yields first at the extraction stage the initial component loadings (i.e., the correlations between the variables and the components) for the variables in each component by starting from the largest variable group. In the first component we thus obtain the highest absolute values of the loadings, the second component has the second highest values, and so forth. There is also the restriction that the components must be orthogonal, i.e., they have no intercorrelations.

We also calculate the sums of squares of these loadings in each component, and these sums are referred to as the eigenvalues. We are usually interested in those principal components that yield eigenvalues greater than or equal with unity. The sums of the eigenvalues of our components divided by the number of the original variables, in turn, reveals us how much our components explain of the total variance of our variables.

In order to better understand our outputs, rotation is also carried out, and it modifies our original principal component loadings. The rotation aims to yield either high or low loadings, and in addition to the orthogonal methods, we may now apply oblique methods. The latter methods allow intercorrelations between the principal components, and this situation is usual in the human sciences. However, unlike in orthogonal rotation, in oblique rotation the loadings are not the correlations between the components and variables, but rather the weights that show us the importance of the variables in each component. Typical examples of orthogonal and oblique methods are Varimax and Promax, respectively.

Summing up the foregoing measures, given a table of original or orthogonally rotated principal component loadings, the row-wise sums of squares of the loadings yield the communalities, whereas the corresponding column-wise values yield the eigenvalues.

We prefer the oblique "Procrustean" Promax method in rotation below, and these principal component loadings are presented in Table 2 and Fig. 2 (the loadings less than the absolute value of 0.3 are omitted below because they are irrelevant). We select two principal components, because they already explain approximately 96% of the total variance of the variables (i.e., the sum of these two eigenvalues/

Table 2 Rotated component matrix of Iris data

	Component	
	1	2
Sepal_length	1.000	
Petal_length	0.933	
Petal_width	0.929	
Sepal_width		1.000

Extraction Method: Principal Component Analysis, Rotation Method: Promax with Kaiser Normalization, Rotation converged in 3 iterations

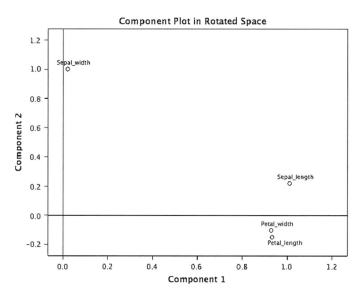

Fig. 2 Component plot in the rotated space with the Iris data

Table 3 The intercorrelations between the variables and the components in Iris data

	Component	
	1	2
Sepal_length	0.982	−0.455
Petal_length	0.962	−0.407
Petal_width	0.938	
Sepal_width	−0.307	0.995

Extraction Method: Principal Component Analysis, Rotation Method: Promax with Kaiser Normalization

$4 \times 100\% = 96\%$) even though the eigenvalue of the second component was slightly less than unity.

We notice that, according to the first component in our rotated table, we may generate a sum variable that includes the variables *Sepal length*, *Petal length* and *Petal width*. The second component only includes one high loading, viz. for *Sepal width*. Hence, instead of the original variables, we may use two principal variables within the Iris data, if necessary. Table 3 presents the loadings that are also the correlations between the variables and the components because in oblique rotation the loadings in Table 2 are not correlations (as in the orthogonal rotation). The latter loadings are better comparable with our SC analyses below. We notice that these loadings are slightly more blurred with respect to sum variable specification. Both of these loading tables are nevertheless used in the conduct of inquiry.

According to Table 4, that presents the regression coefficients for the component scores, our first sum variable would now be

Table 4 Component score coefficient matrix for Iris data

	Component	
	1	2
Sepal_length	0.366	−0.200
Petal_length	0.005	−0.923
Petal_width	0.339	−0.140
Sepal_width	−0.338	−0.097

Extraction Method: Principal Component Analysis, Rotation Method: Promax with Kaiser Normalization

$$S_1 = 0.366 \cdot ZSl + 0.005 \cdot ZSw + 0.339 \cdot ZPl + 0.338 \cdot ZPw \qquad (7)$$

if this prevailing method is applied to the standardized feature variables. Since *Sepal width* is irrelevant to S_1 and the rest of the loadings are quite similar, in practice we may use for the original variables their nonweighted sum instead,

$$S_1 = Sl + Pl + Pw, \qquad (8)$$

or their average, if their standard scores are used. In item analysis Cronbach's alpha is greater than 0.9 for S_1, and this result also corresponds to this sum variable construction.

Hence, PCA provided us with one plausible sum variable, and this was due to the high linear intercorrelations between the feature variables.

4.1.2   The Soft Computing Approach

If we apply our SC method, we principally utilize the distances between the variables and, in the manner of the PCA, we operate with the standard scores of the original variables. Then, within the Iris data, we notice in the dendrogram in Fig. 3 that Sepal width is clearly distinct from the others and the rest of the feature variables seem to belong to same cluster. The multidimensional scaling analysis (SPSS Proxcal), that allocates the variables into a 2-D space according to their distances, also seems to support quite well this resolution (Fig. 4). Hence, it seems that we may specify one sum variable as above.

According to our cluster analysis approach, we will proceed as follows:

1. We specify two cluster centers, and these are our principal components, C_i. The correlations between the variables and principal components will be our component loadings.
2. The communalities are the *rsquares* of the fuzzy models F_i: $(C_1,C_2) \rightarrow ZX_i$, $i = 1, 2, \ldots, n$, i.e., we consider how well our components explain or predict the variables.
3. The eigenvalues are the squared column-wise sums of the loadings as above.

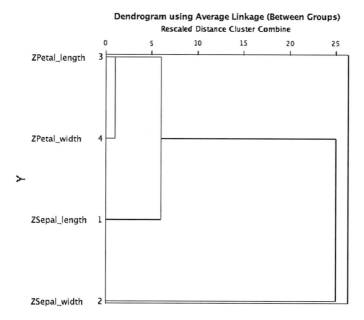

Fig. 3 Dendrogram based on the average linkage method and distances between four standard score variables in the Iris data

Fig. 4 The locations of the standardized variables according to the multidimensional scaling

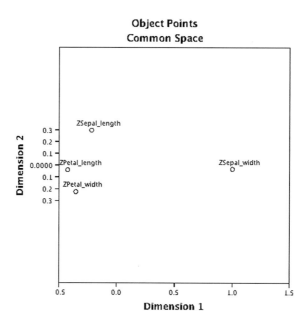

Table 5 Intercorrelations between the standardized variables and principal components in the Iris data

Variable	C_1	C_2	Communalities
ZSepal length	0.999		1.000
ZSepal width		1.000	1.000
ZPetal length	0.888		0.944
ZPetal width	0.838		0.873
Eigenvalues	2.49	1.000	

Table 6 Rmse values of the fuzzy models $F_{ij}: C_j \rightarrow ZX_i$

	C1	C2
ZX1	0.042	0.342
ZX2	0.282	0.001
ZX3	0.099	0.503
ZX4	0.134	0.522

4. The final sum variable, S_1, only constitutes the relevant variables of the first principal component, and its specification is similar to that of the PCA method. We may apply the fuzzy model $S_1 = F_1(ZX_1, ZX_3, ZX_4)$ by using C_1 as the dependent variable instead, if necessary.

Our mountain clustering method for the feature variables seems to yield two plausible cluster centers, and the corresponding principal components contain sufficiently high loadings. We used Matlab's Fuzzy Logic Toolbox and Takagi-Sugeno reasoning for these tasks [38]. Table 5 presents these correlations, or loadings (the absolute values less than 0.3 are omitted as above). This Table is analogous to Table 3 within the PCA. For the illustrative purposes, we also calculated the corresponding *rmse* values, and naturally they were consistent with our loadings (Table 6).

Table 5 also presents the communalities and eigenvalues, and fuzzy rule-based systems with seven rules and Takagi-Sugeno reasoning were used in an above-mentioned manner in this context. Our eigenvalues and communalities indicate that two components yield high loadings and all the variables are relevant in this context. These values also correspond to the PCA outcomes above. There is a slight negative correlation between C_1 and C_2, and in this respect we have an oblique resolution.

Figure 5 depicts the locations of our variables in the principal component space and it also corresponds quite well to the PCA approach. Figure 6 depicts the locations of the variables as well as the principal components based on the PCA and our method when the multidimensional scaling is applied. We notice that our outcomes are slightly dissimilar to those of the PCA. In fact, our components are closer to singular variables and thus we should possibly fine-tune our model. This procedure is nevertheless precluded here because we have adopted the meta-level approach and our outcomes are already sufficiently plausible.

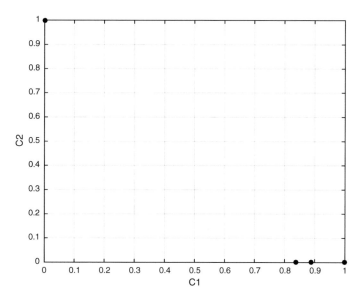

Fig. 5 The loadings of the standardized variables in the principal component space with Iris data

Fig. 6 The locations of the standardized variables as well as the PCA components (Pca) and SC components (Fc) according to the multidimensional scaling

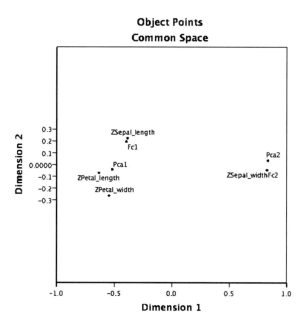

Fig. 7 Scatterplots of the
PCA (Pca) and SC (Fc)
components in the Iris data

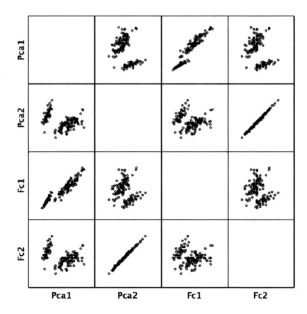

If we will generate the linear sum variable, $S_1 \approx C_1$, it would be similar to that of
(7). The corresponding fuzzy-model approach, in turn, will base on the rule-based
system, F_1,

$$S_1 = F_1(\text{Sl, Pl, Pw}) \tag{9}$$

when C_1 is used as the dependent variable in the model construction.

The intercorrelations between the PCA and our components are depicted in Fig. 7,
and, as expected, they indicate high positive correlations.

Since fuzzy systems are now applied, we could also establish that the closer the
variables are to the components, the higher their degrees of membership, and vice
versa. We could even replace our loadings with these memberships, if necessary.
However, then the comparison between the distinct component extractions would be
more difficult than in the case of correlations.

4.2 The World95 Data

Our second example deals with the benchmark data collected within the international
world survey from 109 countries in 1995 (World95 data), and this is included in the
SPSS example data sets, inter alia [39]. We focus on seven variables, *Average female
life expectancy*, *Average male life expectancy*, *People who read (%)*, *Population
increase (% per year)*, *Daily calorie intake*, *Log (base 10) of GDP per capita* and
Birth to death ratio. Figure 8 depicts the inter-correlations between our variables.

Fig. 8 Scatter plots of variables in the World95 data

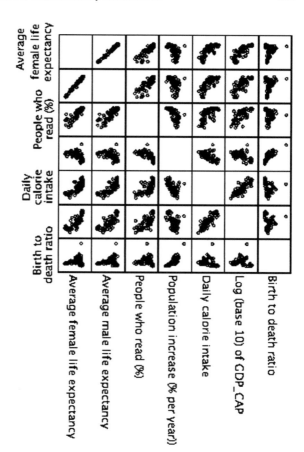

4.2.1 The PCA Approach

We used again PCA with Promax oblique rotation. Both the Kaiser-Mayer-Olkin measure and Bartlett's test now fulfilled the conditions on the satisfactory inter-correlations. The communalities were at least 0.91 and thus all the original variables seemed relevant in our analysis. We selected three components, even though only two components had eigenvalues higher than unity, because our decision seemed to reflect better the variable groups. These components explained approximately 95% of the total variance of the variables. The first two components have a quite high correlation (.753), and thus oblique rotation is justified.

Table 7 and Fig. 9 indicate that three variables have high loadings in the first principal component (the loadings less than 0.3 are omitted). The second and the third components seem to include two variables with high loadings. In our outcome *Population increase* is not having a clear membership to any component. Since the foregoing table will not yield the correlations in oblique rotation, Table 8 presents the corresponding loading matrix based on the correlations between the variables and

Table 7 Rotated component matrix of Word95 data

	Component		
	1	2	3
Average female life expectancy	0.849		
Average male life expectancy	0.830		
People who read (%)	1.000		
Population increase (% per year)	−0.432		0.759
Daily calorie intake		0.951	
Log (base 10) of GDP/CAP		0.767	
Birth to death ratio			1.000

Extraction Method: Principal Component Analysis, Rotation Method: Promax with Kaiser Normalization

Fig. 9 Component plot in the rotated space with the World95 data

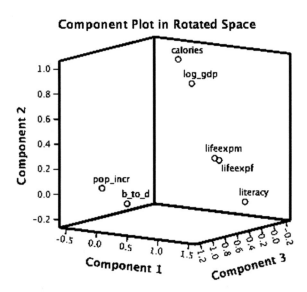

components. We notice that if we relied on this, the sum variable specification would be more problematic, but on the other hand, these values are better comparable to our SC-method outputs below.

Table 8 The intercorrelations between the variables and the components in the World95 data

	Component		
	1	2	3
Average female life expectancy	0.973	0.815	
Average male life expectancy	0.955	0.807	
People who read (%)	0.955	0.672	−0.396
Population increase (% per year)	−0.668	−0.569	0.892
Daily calorie intake	0.733	−0.968	−0.349
Log (base 10) of GDP/CAP	0.800	0.941	−0.373
Birth to death ratio			970

Extraction Method: Principal Component Analysis, Rotation Method: Promax with Kaiser Normalization

If we specify now three sum variables, we may proceed as above by merely calculating the sums of those variables that have high loadings in the principal components. For example,

$$S_2 = \text{Daily calorie intake} + \text{Log (base 10) of GDP per CAP}, \qquad (10)$$

or their averages, if the standard scores are used.

Hence, it seems plausible to specify sum variables among this data set even though now this task is more challenging than with the Iris data. Next we apply our SC method to this task.

4.2.2 The Soft Computing Approach

When our SC method is applied to three components, the correlation between the first two components is −0.565, and thus in this respect we also apply an "oblique" method. According to multidimensional scaling, three cluster centers also seem plausible, even though clear clusters are now unavailable (Fig. 10).

Our intercorrelations between the variables and the components seem somewhat distinct from the PCA outcomes with oblique loadings (Table 9, Figs. 11 and 12). Now the first component seems to comprise three variables common to both the PCA and SC approaches. In the SC model *ZLog (base 10) of GDP per CAP* is also having a high loading in the first component, and the same outcome is found in the correlation Table 8 above, and we must bear in mind that this table presents the correlations between the variables and the components in the manner of our SC method. The third component in PCA and the second component in the SC approach, in turn, provide quite similar outcomes.

Fig. 10 The locations of the standardized variables according to the multidimensional scaling

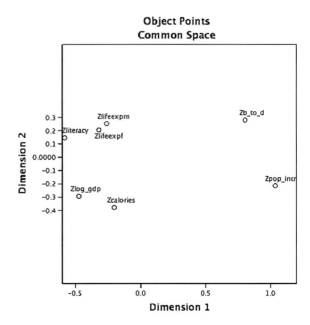

Table 9 Intercorrelations between the standardized variables and principal components in the World95 data

Variable	C_1	C_2	C3	Communalities
ZAverage femal life expectancy	0.999			0.999
ZAverage male life expectancy	0.982			0.982
ZPeople who read (%)	0.771			0.795
ZPopulation increase (% per year)		0.999		1.000
ZDaily calorie intake			1.000	1.000
ZLog (base 10) of GDP per CAP	0.837			0.876
ZBirth to death ratio		0.816		0.939
Eigenvalues	3.261	1.665	1.000	

Hence, the variables *ZLog (base 10) of GDP per CAP* and *ZDaily calorie intake* seem to yield distinct outcomes, but even in this case the correlation Table 8 corresponds quite well to our results. Figure 13 depicts the scatter plots of our components.

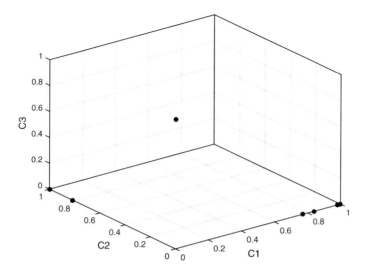

Fig. 11 The loadings of the standardized variables in the principal component space with World95 data

Fig. 12 The locations of the standardized variables as well as the PCA components (Pca) and SC components (Fc) according to the multidimensional scaling

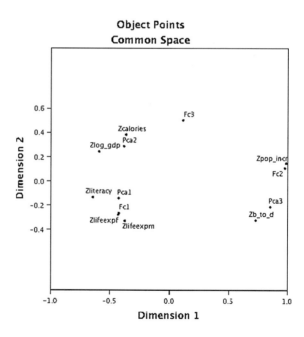

As regards our communalities (that were calculated according to the fuzzy models) and eigenvalues, they seem to fulfill the given conditions.

Our sum variables, again, are the sums of the relevant variables in each component, for example,

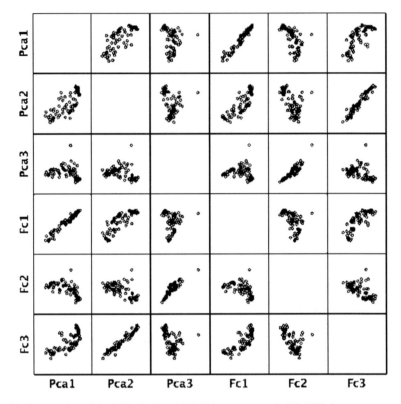

Fig. 13 Scatter plots of the PCA (Pca) and SC (Fc) components in World95 data

$$S_2 = \text{Population increase (\% per year)} + \text{Birth to death ratio,} \qquad (11)$$

or, by applying the corresponding fuzzy model,

$$S_2 = F_2(\text{Population increase(\% per year), Birth to death ratio),} \qquad (12)$$

with the dependent variable C_2 (Fig. 14).

Hence, in this context the SC method yields somewhat distinct outcomes, and this is due to our clustering approach and merely tentative calculations based on our general approach. In addition, the three-component approach seemed not fully justified in this context. On the other hand, in the human sciences the variables in the real world data often contain quite much noise and borderline cases.

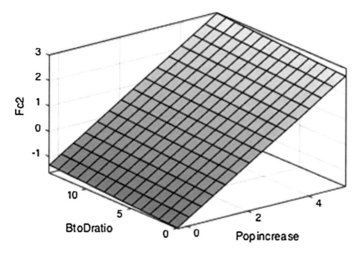

Fig. 14 A fuzzy model fitting to sum variable $S_2 = F$ (Population increase (% per year), Birth to death ratio)

5 Conclusions

We have considered above how the dimensionality reduction analogous to PCA may be carried out with fuzzy clustering method and fuzzy reasoning in an understandable manner, and two real-world data sets were also used as examples. Such prevailing traditional methods as PCA and FA are only appropriate to fairly limited usage because they presuppose linear correlations between the variables and normally distributed data sets, inter alia. Our SC approach, in turn, also seems usable to nonlinear and nonparametric data sets. The central idea in our approach is that we use fuzzy clustering method for finding the appropriate cluster centers to our variables, and these centers provide a basis for our sum variable construction.

In order to draw comparisons to the traditional approaches, our component loadings and goodness criteria based on various intercorrelations and *rsquares* between the variables and the principal components, and in this context we also applied fuzzy reasoning. However, in the long term we could replace the loadings with the degrees membership, as well as use even more fuzzy reasoning models. The loadings could also be linguistic values, if necessary. In this manner we could even better attain Lotfi Zadeh's recent idea on the fuzzy extended logic.

Since the SC community can already provide us with good model construction methods, we did not formulate any novel calculation technics but rather we focused at meta-level on constructing a tentative and an analogous system to PCA. Our contribution was to apply the fuzzy R-technique to data matrix and then construct fuzzy models for assessing the goodness of our outcomes with the loadings, communalities and eigenvalues. We also used fuzzy models for sum variable specifications because they are also appropriate to nonlinear cases.

We still have some open questions. First, due to the clustering approach, we encounter such their prevailing problems as the nonspherical clusters, selection of the correct metrics or the appropriate number of clusters. The lack of variable clusters or the great number of observations and original variables may also arouse problems.

Second, we still expect such standard goodness criteria for our dimensionality reduction within SC that may replace those of the PCA and FA. Examples of these are communalities and eigenvalues. Finally, if the degrees of membership are used for loadings, we still have various alternatives for specifying them.

One new frontier is to apply the fuzzy c-means clustering to this problem area, and this method would be analogous to FA. This is an interesting topic for the future studies and now it was mainly precluded due to the lack of space.

Despite the foregoing open problems, our SC approach seems nevertheless promising in practice as a "quick-and-dirty" method for the dimensionality reduction. However, further studies are still expected in this problem area.

References

1. H. Harman, "Modern Factor Analysis", Univ. of Chigago Press, Chigago, 1976.
2. H. Bandemer, and W. Näther, "Fuzzy Data Analysis", Kluwer, Dordrecht, 1992.
3. P. Grzegorzewski, et al. (Eds.), "Soft Methods in Probability, Statistics and Data Analysis", Physica Verlag, Heidelberg, 2002.
4. M. Smithson, "Fuzzy Set Analysis for Behavioral and Social Sciences", Springer, New York, 1986.
5. J. Zar, "Biostatistical Analysis", Prentice-Hall, Englewood Cliffs, New Jersey, 1984.
6. H.-M. Lee et al., "An Efficient Fuzzy Classifier with Feature Selection Based on Fuzzy Entropy", IEEE Transactions on Systems, Man and Cybernetics, Part B, Vol. 31/3, 2001, pp. 426–432.
7. T. Denoeux and M. Masson, "Principal Component Analysis of Fuzzy Data Using Autoassociative Neural Networks", IEEE Transactions on Fuzzy Systems, vol. 12/3, pp. 336–349, 2004.
8. P. Giordani and H. Giers, "A Comparison of Three Methods for Principal Component Analysis of Fuzzy Interval Data", Computational Statistics & Data Analysis, vol. 51, 2006, pp. 379–397.
9. A. Gonzales and R. Perez, "Selection of Relevant Features in a Fuzzy Genetic Learning Algorithm", IEEE Transactions on Systems, Man and Cybernetics, Part B, Vol. 31/3, 2001, pp. 417–425.
10. T.-P. Hong and J.-P. Chen, "Finding Relevant Attributes and Membership Values", Fuzzy Sets and Systems 103/3, 1999, pp. 389–404.
11. E. Hruschka et al., "A Survey of Evolutionary Algorithms for Clustering", IEEE Transactions on Systems, Man and Cybernetics, Part C, 39/2, 2009, pp. 133–155.
12. Y. Jin, "Fuzzy Modelling of High-Dimensional Systems: Complexity Reduction and Interpretability Improvement, IEEE Transactions on Fuzzy Systems, vol. 8/2, 2000, pp. 212–221.
13. F. Klawonn and J. Castro, J, "Similarity in Fuzzy Reasoning" Mathware & Soft Computing, 2/3, 1995, pp. 197–228.
14. P. Luukka, "A New Nonlinear Fuzzy Robust PCA Algorithm and Similarity Classifier in Classification of Medical Data Sets", Fuzzy Systems, vol. 13/3, 2011, pp. 153–162.
15. C.-H. Oh et al., "Fuzzy Clustering Algorithm Extracting Principal Components Independent of Subsidiary Variables", Proceedings of the IEEE-INNS-ENNS, vol. 3, 2000, pp. 377–380.

16. M. Rezaee and A. Moini, "Reduction Method Based on Fuzzy Principal Component Analysis in Multi-Objective Possibilistic Programming", The International Journal of Advanced Manufacturing Technology vol. 67, 1–4, 2013, pp. 823–831.
17. Y. Yaeibuuchi and J. Watada, "Fuzzy Principal Component Analysis for Fuzzy Data", Proceedings of the Sixth IEEE International Conference, 1997, pp. 1127–1132.
18. T. Yang and S. Wang, "Robust algorithms for Principal Component Analysis", Pattern Recognition Letters, vol. 20, 1999, pp. 927–933.
19. L. Zadeh, "Fuzzy Logic = Computing with Words", IEEE Transactions on Fuzzy Systems, vol. 2, 1996, pp. 103–111.
20. L. Zadeh, "Similarity Relations and Fuzzy Orderings", Information Sciences, vol. 3, 1971, pp. 177–200.
21. L. Zadeh, "Toward Extended Fuzzy Logic – A First Step", Fuzzy Sets and Systems, vol. 160, 2009, pp. 3175–3181.
22. V. A. Niskanen, "Soft Computing Methods in Human Sciences", Studies in Fuzziness and Soft Computing, 134, Springer Verlag, Berlin, 2004.
23. P. Sneath, and R. Sokal, "Numerical Taxonomy", Freeman, San Francisco, 1973.
24. S. Chiu, "Fuzzy Model Identification Based on Cluster Estimation", Journal of Intelligent and Fuzzy Systems vol. 2, 1994, 267–278.
25. R. Yager, and D. Filev, "Generation of Fuzzy Rules by Mountain Clustering", Journal of Intelligent and Fuzzy Systems, vol. 2, 1994, pp. 209–219.
26. J. Bezdek, J. and S. Pal, "Fuzzy Models for Pattern Recognition", IEEE Press, New York, 1992.
27. J. Bezdek et al., "Visual Assessment of Clustering Tendency for Rectangular Dissimilarity Matrices", IEEE Transactions on Fuzzy Systems, vol. 15/5, 2007, pp. 890–903.
28. A. Celikyilmaz and B. Turksen, "Enhanced Fuzzy System Models with Improved Fuzzy Clustering Algorithm", IEEE Transactions on Fuzzy Systems 16/3, 2008, pp. 779–794.
29. H. Frigui and C. Hwang, "Fuzzy Clustering and Aggregation of Relational Data with Instance-Level Constraints", IEEE Transactions on Fuzzy Systems, vol. 16/6, 2008, pp. 1565–1581.
30. J. Handl and J. Knowles, "An Evolutionary Approach to Multiobjective Clustering", IEEE Transactions on Evolutionary Computation, vol. 11/1, 2007, pp. 56–76.
31. R. Hathaway and Y. Hu, "Density-Weighted Fuzzy C-Means Clustering", IEEE Transactions on Fuzzy Systems, vol. 17/1, 2009, pp. 243–253.
32. J. Kacprzyk, J. Owsinski, and D. Viattchenin, "A New Heuristic Possibilistic Clustering Algorithm for Feature Selection", Journal of Automation, Mobile Robotics & Intelligent Systems vol. 8, 2, 2014, pp. 40–46.
33. N. R. Pal et al., "A Possibilistic Fuzzy C-Means Clustering Algorithm", IEEE Transactions on Fuzzy Systems, vol. 13/4, 2005, pp. 517–530.
34. W. Pedrycz et al., "P-FCM: a Proximity-Based Fuzzy Clustering", Fuzzy Sets and Systems 146/1, 2004, pp. 21–42.
35. R. Hathaway and J. Bezdek, "Switching Regression Models and Fuzzy Clustering", IEEE Transactions on Fuzzy Systems, vol. 1/3, 1993, pp. 195–204.
36. J. Kacprzyk and M. Fedrizzi, "Fuzzy regression analysis", Physica Verlag, Heidelberg, 1992.
37. M.-S. Yang et al., "Alpha-Cut Implemented Fuzzy Algorithms and Switching Regression", IEEE Transactions on Systems, Man and Cybernetics, Part B, vol. 38/3, 2008, pp. 588–603.
38. T. Takagi and M. Sugeno, "Fuzzy Identification of Systems and Its Applications to Modeling and Control", IEEE Transactions on Systems, Man and Cybernetics, 15/1, 1986, pp. 116–132.
39. World95 data: http://19-577-spring-2012.wiki.uml.edu/file/detail/World95.sav

Physics of the Mind, Dynamic Logic, and Monotone Boolean functions

Leonid I. Perlovsky

Abstract The chapter discusses physics of the mind, a mathematical theory of higher cognition developed from the first principles, including concepts, emotions, instincts, the knowledge instinct, and aesthetic emotions leading to understanding of the emotions of the beautiful. The chapter briefly discusses neurobiological grounds as well as difficulties encountered by previous attempts at mathematical modeling of the mind encountered since the 1950s. The mathematical descriptions are complemented with detailed conceptual discussions so the content of the chapter can be understood without necessarily following mathematical details. Formulation of dynamic logic in terms of monotone Boolean functions outlines a possible future direction of research.

Keywords Physics of the mind · Concepts · Cognition · The instinct for knowledge · Aesthetic emotions · Beautiful · Cognitive science · Psychology · Dynamic logic · Monotone boolean functions

1 Mechanisms of the Mind

How the mind works has been a subject of discussions for millennia, from Ancient Greek philosophers to mathematicians, to modern cognitive scientists [1]. This chapter describes physics of the mind, a mathematical theory built from the first principles including dynamic logic, higher cognitive functions, and its further development using monotone Boolean functions. This technique could serve two purposes. First, it would lead to the development of smart computers and intelligent robots. Second, it would help to unify and clarify complex issues in philosophy, psychology, neurobiology, and cognitive science. This chapter is a step toward developing "physics of the mind," a theory of the mind concentrating on developing a mathematical model of the mind from a limited number of the first principles.

A broad range of opinions exists about the mathematical methods suitable for the description of the mind. Founders of artificial intelligence, including

L.I. Perlovsky (✉)
Psychology Department, Northeastern University, Boston, USA
e-mail: lperl@rcn.com

© Springer International Publishing AG 2017   193
V. Kreinovich (ed.), *Uncertainty Modeling*, Studies in Computational
Intelligence 683, DOI 10.1007/978-3-319-51052-1_13

Newell and Minsky [64] and Marvin [62], thought that formal logic was sufficient and no specific mathematical techniques would be needed to describe the mind. An opposite view was advocated by [37, 67], suggesting that the mind cannot be understood within the current knowledge of physics; new unknown yet physical phenomena will have to be accounted for explaining the working of the mind; quantum computational processes are necessary for understanding the mind [27, 37, 67]. This chapter develops a point of view that the main difficulty toward a mathematical theory of the mind since the 1950s has been using logic, and the new mathematical theory of dynamic logic enables explaining the mind from the "the first principles" [68, 69, 71, 91, 95].

This chapter presents a mathematical theory of dynamic logic proposed to be intrinsic to operations of the mind, a suggestion that have been experimentally proven [6, 56]. It discusses difficulties encountered by previous attempts at mathematical modeling of the mind and how the new theory overcomes these difficulties. I show an example of solving a problem related to perception that was unsolvable in the past, argue that the theory is related to an important mechanism of "the knowledge instinct", KI, as well as to other cognitive functions, including interactions of language and cognition. I discuss neurobiological foundations, cognitive, psychological, and philosophical connections, experimental verifications, and further mathematical developments using monotone Boolean functions originally developed jointly with Prof. Boris Kovalerchuk [50–53, 109, 110].

2 Logic and the Mind

For a long time logic was considered the best way to deduce scientific truths. In the 1930s [23] proved that logic is inconsistent and cannot serve this foundational purpose, nevertheless artificial intelligence, mathematical and psychological models of the mind until today are logical, misleading intuitions of psychologists and mathematicians modeling the mind.

The beginning of this story is usually attributed to Aristotle, the inventor of logic. He was proud of this invention and emphasized, "nothing in this area existed before us" (Aristotle, IV BCE). However, Aristotle did not think that the mind works logically; he invented logic as a supreme way to argument already discovered truths, not as a theory of the mind. To explain the mind, Aristotle developed a theory of forms, the fundamental mechanism of the mind as a process, in which an illogical "form-as-potentiality" "meets matter" and becomes a logical "form-as-actuality". Today this process is called an interaction between top-down and bottom-up neural signals (BU, TD). A mathematical model of this process, dynamic logic is described later.

During centuries following Aristotle not all subtleties of his thoughts were understood. With the advent of science, the idea that intelligence is equivalent to logic was gaining grounds. In the 19th century mathematicians turned their attention to logic. George Boole thought that Aristotle did not complete a theory of the mind, and it should be improved by making logic more exact. The foundation of logic, since Aristotle (Aristotle, IV BCE, c), was the law of excluded middle (or excluded third): every statement is either true or false, any middle alternative is excluded.

Boole thought that the contradiction between exactness of the law of excluded third and vagueness of language should be corrected, and a new branch of mathematics, formal logic was born. Prominent mathematicians contributed to the development of formal logic including, in addition to Boole, Gottlob Frege, Georg Cantor, Bertrand Russell, David Hilbert, and Kurt Gödel. Logicians 'threw away' uncertainty of language and founded formal mathematical logic based on the law of excluded middle. Hilbert developed an approach named formalism which rejected the intuition as a part of scientific investigation and thought to define scientific objects formally in terms of axioms or rules. Hilbert was sure that his logical theory also described mechanisms of the mind, "The fundamental idea of my proof theory is none other than to describe the activity of our understanding, to make a protocol of the rules according to which our thinking actually proceeds" (see [31]). In the 1900 he formulated famous Entscheidungsproblem: to define a set of logical rules sufficient to prove all past and future mathematical theorems. This entailed formalization of scientific creativity and the entire human thinking. This illustrates the difference between mathematics and physics, whereas mathematics concentrates on internal structure of the theory, physics concentrates on the fundamental laws of nature and their mathematical description.

Almost as soon as Hilbert formulated his formalization program the first hole appeared. In 1902 Russell exposed an inconsistency of formal procedures by introducing a set R as follows: *R is a set of all sets which are not members of themselves.* Is R a member of R? If it is not, then it should belong to R according to the definition, but if R is a member of R, this contradicts the definition [103]. Thus either way we get a contradiction. This became known as the Russell's paradox. Its jovial formulation is as follows: A barber shaves everybody who does not shave himself. Does the barber shave himself? Either answer to this question (yes or no) leads to a contradiction. This barber, like Russell's set can be logically defined, but cannot exist. For the next 25 years mathematicians where trying to develop a self-consistent mathematical logic, free from paradoxes of this type. But in 1931, Gödel proved that it is not possible, formal logic was inexorably inconsistent and self-contradictory [23].

Belief in logic has deep psychological roots related to functioning of the human mind. A major part of any perception and cognition, illogical Aristotelian process involving forms-as-potentialities is not accessible to consciousness. We are conscious about the 'final states' of these processes, crisp forms-as-actualities which are perceived by our minds as 'concepts' approximately obeying formal logic. For this reason prominent mathematicians believed in logic. Even after Gödelian proof, founders of artificial intelligence still insisted that logic is sufficient to explain how the mind works.

3   Cognition, Logic, and Complexity

Object perception involves signals from sensory organs and mental representations of objects. During perception, the mind associates subsets of signals corresponding to objects with object representations in the memory. This recognition activates brain signals leading to mental and behavioral responses.

Mathematical models of this *recognition* step in this seemingly simple association-recognition-understanding process has not been easy, a number of difficulties have been encountered over the last sixty years. These difficulties were summarized under the notion of combinatorial complexity, CC, [70]. CC refers to multiple combinations of various elements in a complex system; for example, recognition of an object usually requires concurrent recognition of the multiple elements of the scene that could be encountered in various combinations. CC is prohibitive because the number of combinations is very large: for example, the number of combinations of 100 elements (not too large a number) is 100^{100}, exceeding the number of all elementary particle events in the life of the Universe; no computer would ever be able to compute that many combinations.

It has been proven that CC is mathematically similar to the Gödelian incompleteness. Therefore using logic leads to CC. But logic is used by all popular mathematical approaches to developing "cognitive algorithms" and modeling the mind, even fuzzy logic and neural networks, approaches specifically designed to overcome limitations of logic, still use logic in their training or learning procedures, every training example is a separate logical statement (e.g. "this is food"). CC of engineering algorithms and mathematical approaches to theories of the mind is related to the fundamental inconsistency of logic.

4   Fundamental Mechanisms of the Mind

Concepts are representations of the world events in the mind. According to [40], they are the contents of pure reason. They model events or simulate them. For this reason [7] calls them "simulators", [35] prefers to use the word *model*.

Instincts are sensory-like neural mechanism measuring vital bodily parameters. When these parameters are outside safe bounds, the instinctual mechanism sends evaluative neural signals to decision-making parts of the brain [26].

Emotions refer to many different mechanisms in the mind and body. According to [26] emotional signals evaluate concepts for the purpose of instinct satisfaction. Recognition of objects or situations that can potentially satisfy vital needs of the organism (instincts) receive preferential attention. A number of neural and physiological realizations of this mechanism have been identified [22].

Aesthetic emotions [41] emphasized the role of emotions in learning: aesthetic emotions are related to learning; they are judgments about correspondence between an object-event and its concept, which today we relate to interactions between BU and TD neural signals. Grossberg and Levine [26] theory of instinctual drives and emotions has been developed along Kantian ideas by introducing KI [71, 74, 75, 77, 98]. KI is an instinct measuring similarities between object-events and their concepts. Satisfactions or dissatisfactions of KI are indicated by aesthetic emotions [92, 93] that drive improvement of concepts in correspondence with experience. KI and aesthetic emotions are the foundations of all human higher cognitive abilities [58, 72, 82, 92], including emotions of the beautiful.

Emotions evaluating satisfaction or dissatisfaction of the knowledge instinct are not directly related to bodily needs. Therefore, they are 'spiritual' or aesthetic emotions. I would like to emphasize that aesthetic emotions are not peculiar to perception of art; they are inseparable from every act of perception and cognition. In the next sections we describe a mathematical theory of conceptual-emotional recognition and understanding. As we discuss, in addition to concepts and emotions, it involves mechanisms of intuition, imagination, conscious, and unconscious. This process is intimately connected to an ability of the mind to think, to operate with symbols and signs. The mind involves a hierarchy of multiple layers of concept-models, from simple perceptual elements to concept-models of objects, to relationships among objects, to complex scenes, and up the hierarchy... toward the concept-models of the meaning of life and purpose of our existence. Hence the tremendous complexity of the mind, yet relatively few fundamental principles of the mind organization go a long way explaining this system. The mind is not a kludge [84].

5   Physics of the Mind

Physics of the mind is different from psychology in that it searches for the fundamental principles of the mind. Among these principles are mechanisms of dynamic logic, concept-models, emotions, the knowledge instinct, and aesthetic emotions discussed above. Few more fundamental principles will be gradually introduced and discussed. These fundamental mechanisms are organized in a multi-layer, hierarchical system [25, 71]. The mind is not a strict hierarchy; there are multiple feedback connections among several adjacent layers, this approximately-hierarchical organization is another fundamental principle. At each layer there are concept-models encapsulating the mind's knowledge; they generate TD signals interacting with BU signals; the interaction between TD and BU neural signals is a fundamental principle constituting the essence of thinking and cognition. These interactions are governed by the knowledge instinct, which drives concept-model learning, adaptation, and formation of new concept-models for better correspondence to experience.

I begin with describing a basic mechanism of interaction between TD and BU signals between two adjacent hierarchical layers. At each layer, output signals are concepts recognized in (or formed from) BU signals. BU signals are associated with (or recognized, or grouped into) concepts according to the models and the knowledge instinct at this layer. This general structure of the theory of the mind corresponds to our knowledge of neural structures in the brain; although it is a physical intuition of the general principle, not necessarily one-to-one mapping to actual neural connections. How actual brain neuronal connection "implement" models and the knowledge instinct is an area of active research. The knowledge instinct is modeled mathematically as maximization of a similarity measure between bottom-up and top-down signals. In the process of learning and understanding input signals, models are adapted for better representation of the input signals so that similarity between the models and signals increases. This increase in similarity satisfies the knowledge instinct and is felt as aesthetic emotions.

5.1   The Knowledge Instinct

At a particular hierarchical layer, we enumerate neurons by index $n = 1, ... N$. These neurons receive BU signals, $\mathbf{X}(n)$, from lower layers in the processing hierarchy. $\mathbf{X}(n)$ is a field of BU neuronal synapse activations, coming from neurons at a lower layer. Each neuron has a number of synapses; for generality, we describe each neuron activation as a set of numbers, $\mathbf{X}(n) = \{X_d(n), d = 1, ... D\}$. TD, or priming signals to these neurons are sent by concept-models, $\mathbf{M}_m(\mathbf{S}_m, n)$; we enumerate models by index $m = 1, ... H$. Each model is characterized by its parameters, $\mathbf{S}_m$; in the neuron structure of the brain they are encoded by strength of synaptic connections, mathematically, we describe them as a set of numbers, $\mathbf{S}_m = \{S^a_m, a = 1, ... A\}$. Models *represent* signals in the following way. Say, signal $\mathbf{X}(n)$, is coming from sensory neurons activated by object m, characterized by parameters $\mathbf{S}_m$. These parameters may include position, orientation, or lighting of an object m. Model $\mathbf{M}_m(\mathbf{S}_m, n)$ predicts a value $\mathbf{X}(n)$ of a signal at neuron n. For example, during visual perception, a neuron n in the visual cortex receives a signal $\mathbf{X}(n)$ from retina and a priming signal $\mathbf{M}_m(\mathbf{S}_m, n)$ from an object-concept-model m. A neuron n is activated if both BU signal from lower-layer-input and TD priming signal are strong. Various models compete for evidence in the BU signals, while adapting their parameters for better match as described below. This is a simplified description of perception. The most benign everyday visual perception uses many layers from retina to object perception. Perception of minute features, or everyday objects, or cognition of complex abstract concepts is due to the same mechanism described below. Perception and cognition involve models and learning. In perception, models correspond to objects; in cognition models correspond to relationships, situations, and more abstract entities.

Learning is an essential part of perception and cognition, and it is driven by the knowledge instinct. It increases a similarity measure between the sets of models and signals, $L(\{\mathbf{X}\},\{\mathbf{M}\})$. The similarity measure is a function of model parameters and associations between the BU and TD signals. For concreteness I refer here to an object perception using a simplified terminology, as if perception of objects in retinal signals occurs in a single layer.

In constructing a mathematical description of the similarity measure, it is important to acknowledge two principles. First, the exact content of the visual field is unknown before perception occurred. Important information could be contained in any BU signal; therefore, the similarity measure is constructed so that it accounts for all input information, $\mathbf{X}(n)$,

$$L(\{\mathbf{X}\}, \{\mathbf{M}\}) = \prod_{n \in N} l(\mathbf{X}(n)). \tag{1}$$

This expression contains a product of partial similarities, $l(\mathbf{X}(n))$, over all BU signals; therefore it forces the mind to account for every signal (even if one term in the product is zero, the product is zero, the similarity is low and the knowledge instinct is not satisfied); this is a reflection of the first principle. Second, before perception occurs, the mind does not know which retinal neuron corresponds to which object. Therefore a partial similarity measure is constructed so that it treats each model as an alternative (a sum over models) for each input neuron signal. Its constituent elements are conditional partial similarities between signal $\mathbf{X}(n)$ and model $\mathbf{M}_m$, $l(\mathbf{X}(n)|m)$. This measure is "conditional" on object m being present,[1] therefore, when combining these quantities into the overall similarity measure, L, they are multiplied by r(m), which represent the measure of object m actually being present. Combining these elements with the two principles noted above, a similarity measure is constructed as follows:

$$L(\{\mathbf{X}\}, \{\mathbf{M}\}) = \prod_{n \in N} \sum_{m \in M} r(m)l(\mathbf{X}(n)|m). \tag{2}$$

The structure of (2) follows standard principles of the probability theory: a summation is taken over alternatives, m, and various pieces of evidence, n, are multiplied. This expression is not necessarily a probability, but it has a probabilistic structure. The name "conditional partial similarity" for $l(\mathbf{X}(n)|m)$ (or simply $l(n|m)$) follows the probabilistic terminology. If learning is successful, $l(n|m)$ becomes a conditional probability density function, a probabilistic measure that signal in neuron n originated from object m. Coefficients r(m), called priors in probability theory, contain preliminary biases or expectations, expected objects m have relatively high r(m) values; their true values are usually unknown and should be learned, like other parameters

---

[1] Mathematically, the condition that the object m is present with 100 % certainty, is expressed by normalization condition: $\int l(X|m)dX = 1$. We should also mention another normalization condition: $\int l(X(n))dX(n) = 1$, which expresses the fact that, if a signal is received, some object or objects are present with 100 % certainty.

S_h. If learning is successful, L approximates a total likelihood of observing signals $\{X(n)\}$ coming from objects described by models $\{M_m\}$ and leads to near-optimal Bayesian decisions.

Note. In the probability theory, a product of probabilities usually assumes that evidence is independent. Expression (2) contains a product over n, but it does not assume independence among various signals $X(n)$. There is a dependence among signals due to models: each model $M_m(S_m, n)$ predicts expected signal values in many neurons n.

During the learning process, concept-models are constantly modified. From time to time a system forms a new concept, while retaining an old one as well; alternatively, old concepts are sometimes merged or eliminated. Even so functional forms of models, $M_m(S_m, n)$, are fixed and learning-adaptation involves only model parameters, S_m, still structural variation of models can be achieved as we discuss below [73, 74]. Formation of new concepts and merging or elimination-forgetting of old ones require a modification of the similarity measure (2); the reason is that more models always result in a better fit between the models and data. This is a well known phenomenon, a similarity measure fit to the data is biased toward a larger value. To obtain an unbiased estimation the similarity (2) should be reduced by using a "penalty function," $p(N,M)$ that grows with the number of models M, and this growth is steeper for a smaller amount of data N. For example, an asymptotically unbiased maximum likelihood estimation leads to multiplicative $p(N, M) = \exp(-N_{par}/2)$, where N_{par} is a total number of adaptive parameters in all models (using this penalty function is known as Akaike Information Criterion, see [74] for further discussion and references).

5.2   Dynamic Logic

The learning process consists in estimating model parameters S and associating signals with concepts by maximizing the similarity (2). Note, all possible combinations of signals and models are accounted for in expression (2). This can be seen by expanding a sum in (2), and multiplying all the terms; it would result in M^N items, a huge number. This is the number of combinations between all signals (N) and all models (M). Here is the source of CC of many popular algorithms. For example, a popular multiple hypothesis testing algorithm [106] attempts to maximize similarity L over model parameters and associations between signals and models, in two steps. First it takes one of the M^N items, that is one particular association between signals and models; and maximizes it over model parameters. Second, the largest item is selected (that is the best association for the best set of parameters). Such a program inevitably faces a wall of CC, the number of computations on the order of M^N.

Our theory solves this problem by using dynamic logic [68, 71, 74]. An important aspect of dynamic logic is matching vagueness or fuzziness of similarity measures to the uncertainty of models. Initially, parameter values are not known, and uncertainty of models is high; so is the fuzziness of the similarity measures. In the process of learning, models become more accurate and the similarity measure more crisp, the value of the similarity increases. This is the mechanism of dynamic logic in which vague models-representations evolve into crisp models-representations. This vague-to-crisp process is a mathematical description of the Aristotelian theory of the mind, in which illogical "forms-as-potentialities" "meet matter" and turn into logical "forms-as-actualities" [1].

Mathematically it is described as follows. First, assign any values to unknown parameters, $\{S_m\}$. Then, compute association variables $f(m|n)$,

$$f(m|n) = r(m)\, l\,(\mathbf{X}(n)|m)/ \sum_{m' \in M} r(m')l(\mathbf{X}(n)|m'). \tag{3}$$

Equation (3) looks like the Bayes formula for a posteriori probabilities; if $l\,(n|m)$ in the result of learning become conditional likelihoods, $f(m|n)$ become Bayesian probabilities for signal n originating from object m. The next step defines a joint dynamics of the association variables and model parameters,

$$df(m|n)/dt = f(m|n) \sum_{m' \in M} [\delta_{mm'} - f(m'|n)] \cdot [\partial \ln l(n|m')/\partial \mathbf{M}_{m'}]\partial \mathbf{M}_{m'}/\partial \mathbf{S}_{m'} \cdot d\mathbf{S}_{m'}/dt, \tag{4}$$

$$d\mathbf{S}_m/dt = \sum_{n \in N} f(m|n)[\partial \ln l(n|m)/\partial \mathbf{M}_m]\partial \mathbf{M}_m/\partial \mathbf{S}_m, \tag{5}$$

here

$$\delta_{mm'} \text{ is } 1 \text{ if } m = m',\ 0 \text{ otherwise.} \tag{6}$$

The following theorem was proven [71].

Theorem *Equations (3) through (6) define a convergent dynamic system with stationary states defined by* $\max_{\{Sm\}} L$.

It follows that the stationary states of the system are the maximum similarity states satisfying the knowledge instinct. When partial similarities are specified as probability density functions (pdf), or likelihoods, the stationary values of parameters $\{S_m\}$ are asymptotically unbiased and efficient estimates of these parameters [15]. A computational complexity of the MF method is linear in N.

In plain English, this means that dynamic logic is a convergent process. It converges to the maximum of similarity, and therefore satisfies the knowledge instinct. Several aspects of this convergence are discussed in the next section. If likelihood is used as similarity, parameter values are estimated efficiently (that is, in most cases, parameters cannot be better learned using any other procedure). Moreover, as a part of the above theorem, it is proven that the similarity measure increases at each iteration (until the system receives new data). The psychological interpretation is that the knowledge instinct is satisfied at each step: a modeling field system with dynamic logic *enjoys* learning.

5.3 Example of Dynamic Logic Operations

Finding patterns below noise could be an exceedingly complex problem. If an exact pattern shape is not known and depends on unknown parameters, these parameters should be found by fitting the pattern model to the data. However, when location and orientation of patterns are not known, it is not clear which subset of the data points should be selected for fitting. A standard approach for solving this kind of problems, as discussed, multiple hypothesis testing [106] tries all combinations of subsets and models, it faces combinatorial complexity. In this example, we are looking for 'smile' and 'frown' patterns in noise shown in Fig. 1a without noise, and in Fig. 1b with noise, as actually measured. The image size in this example is 100×100 points (N = 10,000). The true number of patterns is 3 plus 1 for noise. The number of patterns is not known, therefore, at least 4 patterns $+1$ noise (= 5) should be fit to the data, to decide that 3 patterns fit best. This yields an incomputable combinatorial complexity $M^N = 5^{10,000} = 10^{3,000}$.

Nevertheless, this problem unsolvable due to CC becomes solvable using dynamic logic, as illustrated in Fig. 1: (c) illustrates initial vague dynamic logic state, corresponding to randomly selected parameter values; on the first iteration, (d), vagueness is somewhat reduced; (d) through (h) show improved models at various iteration stages (total of 22 iterations). Between iterations (d) and (e) the algorithm decided, that it needs three Gaussian models for the 'best' fit. There are several types of models: one uniform model describing noise (it is not shown) and a variable number of blob models and parabolic models, which number, location and curvature are estimated from the data. Until about stage (g) the algorithm used simple blob models, at (g) and beyond, the algorithm decided that it needs more complex parabolic models to describe the data. Iterations stopped at (h), when similarity stopped increasing.

5.4 The Mind Hierarchical Organization

The previous sub-sections described a single processing layer of a cognitive mind system. At each layer of a hierarchy there are BU signals from lower layers, models, similarity measures (2), aesthetic emotions, which psychologically measure satisfaction of KI, and mathematically are changes in similarity (2), and actions; actions include adaptation, behavior satisfying the knowledge instinct—maximization of similarity, equations (3) through (6). An input to each layer, the BU signals are a set of signals $X(n)$, or in neural terminology, an input field of neuronal activations. The result of learning at a given layer are activated models, or concepts m recognized in the BU signals n; these models along with the corresponding instinctual signals and emotions may activate behavioral models and generate behavior at this layer.

The activated models initiate other actions. They serve as input signals to the next, higher processing layer, where more general concept-models are recognized

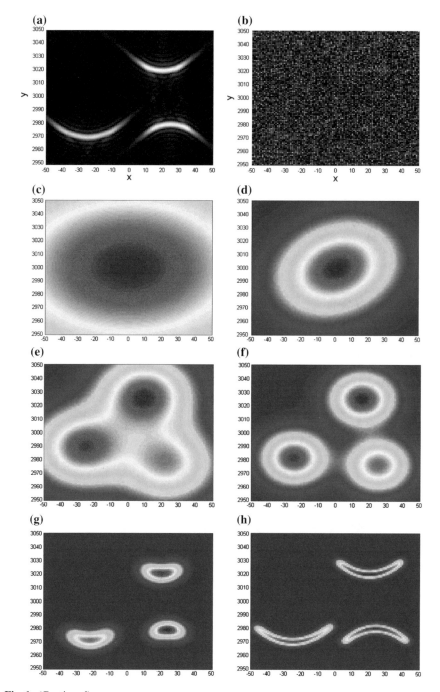

Fig. 1 (Continued)

◀**Fig. 1** Finding 'smile' and 'frown' patterns in noise, an example of dynamic logic operation: **a** true 'smile' and 'frown' patterns are shown without noise; **b** actual image available for recognition (signal is below noise, signal-to-noise ratio is between –2 dB and –0.7 dB); **c** an initial fuzzy blob-model, the fuzziness corresponds to uncertainty of knowledge; **d** through **h** show improved models at various iteration stages (total of 22 iterations). Between stages **d** and **e** the algorithm tried to fit the data with more than one model and decided, that it needs three blob-models to 'understand' the content of the data. There are several types of models: one uniform model describing noise (it is not shown) and a variable number of blob-models and parabolic models, which number, location and curvature are estimated from the data. Until about stage **g** the algorithm 'thought' in terms of simple blob models, at (**g**) and beyond, the algorithm decided that it needs more complex parabolic models to describe the data. Iterations stopped at (**h**), when similarity (2) stopped increasing. This example is discussed in more details in [93]

Fig. 2 The hierarchical cognitive system. At each layer of a hierarchy there are models, similarity measures, and actions (including adaptation, maximizing the knowledge instinct—similarity). High levels of partial similarity measures correspond to concepts recognized at a given layer. Concept activations are output signals at this layer and they become BU input signals to the next layer, propagating knowledge up the hierarchy

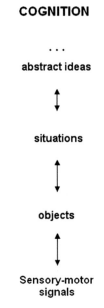

or created. Output signals from a given layer, serving as input, BU to the next layer, could be model activation signals, a_m, defined as

$$a_m = \sum_{n \in N} f(m|n). \tag{7}$$

In addition, output signals may include model parameters. The hierarchical cognitive system is illustrated in Fig. 2. Within the hierarchy of cognition, each concept-model finds its "mental" meaning and purpose at a higher layer (in addition to other purposes). For example, consider a concept-model "chair." It has a "behavioral" purpose of initiating sitting behavior (if sitting is required by the body), this is the "bodily" purpose at the same hierarchical layer. In addition, it has a "purely mental" purpose at a higher layer in the hierarchy, a purpose of helping to recognize a more general concept, say of a "concert hall," which model contains rows of chairs.

A mathematical formulation of the hierarchical similarity is a product of similarity measures at every layer. We denote similarity (1) at the layer h as L(h). Then, the total similarity is

$$L = \prod_{h \in H} L(h). \tag{8}$$

In this total hierarchical similarity, concepts formed at the level h become BU signals for the layer h+1, whereas concepts formed at the level h+1 become TD signals for the layer h. Whereas conditional similarities l (n|m) in Eq. (2), used in the example of Fig. 1 were designed for this example using specific functional shapes corresponding to image patterns, now a uniform conditional similarity is needed, suitable for every hierarchical level and capable to be learned for all current and future concepts. This new conditional similarity is based on the idea that higher-level concepts are formed from subsets of the BU signals. For simplicity of notations for now we leave out the index h. At every level in the hierarchy, each higher-level concept-model m is characterized by parameters-probabilities, $\mathbf{p}_m = (p_{m1}, \,.. \, p_{mi}, \,...)$, where p_{mi} is the probability of BU signal x_{ni} being part of the higher model m. The model-conditional similarity measures, l (n|m), are defined as follows,

$$l(n|m) = \prod_{i=1} p_{mi}{}^{x_{ni}} (1 - p_{mi})^{(1-x_{ni})}. \tag{9}$$

Applying the dynamic logic learning equation we obtain a surprisingly simple learning-estimation equation for the model parameters, p_{mi},

$$p_{mi} = \sum_{n \in N} f(m|n) x_{ni} / \sum_{n' \in N} f(m|n'). \tag{10}$$

The dynamic logic learning converges in few iterations [97]. Equations (8, 9, 10) complete the definition of the hierarchical system of the mind.

Models at higher layers in the hierarchy are defined as subsets of lower-level models and therefore more general than models at lower layers. For example, at the very bottom of the hierarchy, if we consider vision system, models correspond (roughly speaking) to retinal ganglion cells and perform similar functions; they detect simple features in the visual field; at higher layers, models correspond to functions performed at V1 and higher up in the visual cortex, that is detection of more complex features, such as contrast edges, their directions, elementary moves, etc. Visual hierarchical structure and models are studied in details ([24, 111]). At still higher cognitive layers, models correspond to objects, to relationships among objects, to situations, and relationships among situations, etc. [71, 97]. Still higher up are even more general models of complex cultural notions and relationships, like family, love, friendship, and abstract concepts, like law, rationality, etc. Contents of these models correspond to cultural wealth of knowledge, including writings of Shakespeare and Tolstoy.

Let me repeat that concept-models at every layer evolved in evolution with the purpose to unify lower-level concepts. This is the fundamental organizational principle of the hierarchy that helps us to understand contents and meanings of the concepts near the top of the hierarchy. These "top" concepts evolve with the purpose to unify the entire life experience. We perceive these concepts as the meaning of life. This clarifies Kantian conclusion that at the top of the hierarchy of the mind, there are concept-models of the meaning and purpose of our existence, unifying our knowledge, and the corresponding behavioral models aimed at achieving this meaning.

We discussed that satisfaction of the knowledge instinct is accompanied with aesthetic emotions. At lower levels of the hierarchy, say below objects, these aesthetic emotions usually are below the threshold of consciousness. Near the top of the hierarchy, where concepts address meanings important for life, aesthetic emotions.

6   Higher Cognition, Beautiful, and the Dual Hierarchy

We discussed that satisfaction of the knowledge instinct is accompanied by aesthetic emotions. At lower levels of the hierarchy, say below objects, these aesthetic emotions usually are below the threshold of consciousness. We are not getting emotionally excited when we understand an everyday object, say a refrigerator. Near the top of the hierarchy, where concepts address meanings important for life, aesthetic emotions acquire similarly important meanings. Aesthetic emotions felt when understanding the meaning and purpose of life are emotions of the beautiful [83]. The essential aspects of this theory of the beautiful follows the Kantian aesthetics (1790), the following mathematical theory helps understanding the details that have escaped Kantian analysis.

Can this mathematical analysis be used for exactly elucidating what is the meaning of life and emotions of the beautiful? The answer is unexpectedly negative: there cannot be a clear-cut prescription for what the meaning of life is and what the beautiful is. This negative answer is not entirely unexpected, it corresponds to our intuitions about these highest concepts and emotions. And yet, it calls for clarification of how it is possible that mathematical descriptions leads to denial of clear-cut definitions. To say it more exactly, the mathematical analysis clarifies the meanings of these highest concepts and emotions [92], as well as the corresponding meaning of "exactness."

Let us examine more closely exactness of perception of everyday objects. Look at an object in front of you, then close eyes and imagine this object with closed eyes. Imaginations are not as exact as perception of objects with opened eyes, it is impossible to recollect all the details with closed eyes. We know the mechanism of visual imagination: imagination, say of an object, is experiencing a neural projections of the object concept-model-representation onto the visual cortex. Vagueness of imaginations testifies to the vagueness of concept-representations. This fact has been proven in [6, 80]. Moreover this publication demonstrated that vague states and processes in the mind are also less accessible to consciousness.

If representations of the everyday objects are vague and not completely conscious, what does it say about representations near the top of the hierarchy. These "top" representations built on multiple layers of representations throughout the hierarchy must be much vaguer and much less conscious. But does not it contradict our ability to discuss in details the meaning of life and what is beautiful? To answer this question we have to analyze the difference between cognition and language, and the function of language in cognition.

6.1 Language and Cognition in Thinking

Do we think with language, or is language just a communication device used for expression of completed thoughts? What is a difference between language and cognition? Chomsky [12] suggested that these two abilities are separate and independent. Cognitive linguistics emphasizes a single mechanism for both [16]. Evolutionary linguistics considers the process of transferring language from one generation to the next one [11, 13, 34]. This process is a "bottleneck" that forms the language. Brighton, Smith, and Kirby [10] demonstrated emergence of compositional language due to this bottleneck. Still, none of these approaches resulted in a computational theory explaining how humans acquire language and cognition. Below I discuss a computational model overcoming previous difficulties and unifying language and cognition as two separate and closely integrated abilities. I identify their functions and discuss why human thinking ability requires both language and cognition. The fundamental difference between language and cognition is that language does not directly interacts with the world, language is learned from culturally evolved surrounding language. Cognition interacts with the world and is learned from experience in real world under the guidance of language.

We have discussed the difficulty of mathematical models of cognition, it is related to a need to consider combinations of sensor signals, objects, and events. The number of combinations is very large and even a limited number of signals or objects form a very large number of combinations, exceeding all interactions of all elementary particles in a lifetime of the Universe, a combinatorial complexity, CC. This difficulty in modeling the mind has been overcome by dynamic logic. Whereas classical logic considers static statements such as "this is a chair," dynamic logic models processes from vague to crisp representations. These processes do not need to consider combinations, an initial vague state of a "chair" matches any object in the field of view, and at the end of the process it matches the chair actually present, without CC.

Yet, there remains another difficulty, similar still even more complex. It is related to the fact that "events" and "situations" in the world do not exist "ready for cognition." There are many combinations of percepts and objects, a near infinity, events and situations important for understanding and learning have to be separated from those that are just random collections of meaningless percepts or random objects [97].

Events and situations recognized by non-human animals are very limited compared to human abilities to differentiate events in the world. Human cognitive abilities acquire their power due to language. Language is "easier" to learn than cognitive representations for the following reason, language representations: words, phrases exist in the surrounding language "ready-made," created during millennia of cultural evolution. Therefore language could be learned without much real-life experience; only interactions with language speakers are required. Every child learns language early in life before acquiring full cognitive understanding of events and their cognitive meanings. Thus language is learned early in life with only limited cognitive understanding of the world [78, 87, 91]. Cognitive representations of situations and abstract concepts are vague. Throughout the rest of life, language guides acquisition of cognitive representations from experience. Vague cognitive representations become more crisp and concrete. Thinking involves both language and cognition, and as we discuss later thinking about abstract ideas usually involves language more than cognition, not too different from thinking by children.

6.2   The Dual Hierarchy

A mathematical description of this interacting language-cognition system, connecting language, world and cognition, requires the dual hierarchy illustrated in Fig. 3. Cognitive hierarchy from sensor-motor percepts near "bottom," to objects "higher up," to situations, and to still more abstract cognitive representations have been illustrated in Fig. 2 and mathematically modeled in the previous sections. Language representations are organized in a parallel hierarchy from sounds, and words for objects and situations, to phrases, and to more abstract language representations. Our previous discussion is mathematically modeled as a dual hierarchy [78, 87, 91, 97] illustrated in Fig. 3.

Language is learned from the surrounding language, where it exists "ready-made" for cognition, and therefore after 5 or 7 language representations are crisp and conscious throughout the hierarchy. But cognitive representations, as discussed are vague and less conscious. For this reason abstract concepts are mostly known to us through language. This explains why concepts at the top of the hierarchy, such as meaning of life, and the corresponding emotions of the beautiful can be discussed in great details, still their exact contents are not known.

Hierarchical organization of cognition and related brain structures are reviewed in [4]. In particular, anterior-posterior axis corresponds to a gradient of abstract-concrete cortex functions. Hierarchical organization of language functions is also well established. However, hierarchical organization of language does not correspond to a particular spatial axis in the brain, it is distributed [100]. Therefore, the dual hierarchy in Fig. 3 is a functional hierarchy not organized along a spatial axis in the brain as in this figure. A fundamental aspect of acquiring language is interaction with the surrounding language as well as between BU and TD representations. Language learning is grounded in experience with the surrounding language. Learning

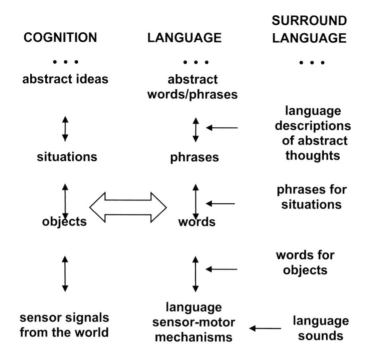

Fig. 3 The dual hierarchy. Language and cognition are organized into approximate dual hierarchy. Learning language is grounded in the surrounding language throughout the hierarchy. Cognitive hierarchy is grounded in experience only at the very *bottom*

of cognitive mental representations is directly grounded in experience only at the very bottom of the hierarchy; the rest of the hierarchy is located inside the brain. But in addition to interactions between BU and TD signals (indicated in Fig. 3 by vertical arrows), cognition interacts with language representations, and learning cognition is grounded in experience guided by language. In this interaction a lower layer representations are organized in more abstract and general concept-representations at a higher layer.

The dual model makes a number of experimentally testable predictions, and some of these have been confirmed [78, 87, 91, 95, 97]. It explains functions of language and cognition in thinking: cognitive representations model surrounding world, relations between objects, events, and abstract concepts. Language stores culturally accumulated knowledge about the world, yet language is not directly connected to objects, events, and situations in the world. Language guides acquisition of cognitive representations from random percepts and experiences, according to what is considered worth learning and understanding in culture. Events that are not described in language are likely not even noticed or perceived in cognition. (2) Whereas language is acquired early in life, acquiring cognition takes a lifetime. The reason is that language representations exist in surrounding language "ready-made," acquisition of language requires only interaction with language speakers, but does not require much life experience. Cognition on the opposite requires life experience. (3) This is the reason why abstract words excite only language regions of brain, whereas concrete words excite also cognitive regions [8]. The dual model predicts that abstract concepts are often understood as word descriptions, but not in terms of objects, events, and relations among them. (4) This model explains why language is acquired early in life, whereas cognition takes a lifetime. It also explains why children can acquire the entire hierarchy of language including abstract words without experience necessary for understanding them. (5) Since dynamic logic is the basic mechanism for learning language and cognitive representations, the dual model suggests that language representations become crisp after language is learned (5–7 years of age), however cognitive representations may remain vague for much longer; the vagueness is exactly the meaning of "continuing learning", this takes longer for more abstract and less used concepts. (6) The dual model gives mathematical description of the recursion mechanism [97]. Whereas Hauser et al. [30] postulate that recursion is a fundamental mechanism in cognition and language, the dual model suggests that recursion is not fundamental, hierarchy is a mechanism of recursion.

(7) Another mystery of human-cognition, not addressed by cognitive or language theories, is basic human irrationality. This has been widely discussed and experimentally demonstrated following discoveries of Tversky and Kahneman [108], leading to the 2002 Nobel Prize. According to the dual hierarchy model, the "irrationality" originates from the dichotomy between cognition and language. Language is crisp and conscious while cognition might be vague and ignored when making decisions. Yet, collective wisdom accumulated in language may not be properly adapted to one's personal circumstances, and therefore be irrational in a concrete situation. In the 12th c. Maimonides wrote that Adam was expelled from paradise because he refused original thinking using his own cognitive models, but ate from the tree of knowledge and acquired collective wisdom of language [57].

6.3 Emotional Prosody and Its Cognitive Function

The language instinct drives the development of the language part of the dual hierarchy [99]. The knowledge instinct drives the development of the cognitive hierarchy and the language guidance of the cognition, indicated by a wide horizontal arrow in Fig. 3. These neural connections have to be developed and maintained. This requires motivation, in other words, emotions. These aesthetic emotions corresponding to the KI must be in addition to utilitarian meanings of words, otherwise only practically useful words would be connected to their cognitive meanings. Also these emotions must "flow" from language to cognition and they must act fast, so that language is able to perform its cognitive function of guiding acquisition of cognitive representations, organizing experience according to cultural contents of language. These emotions therefore must be contained in language sounds, before cognitive contents are acquired [90].

This requirement of emotionality of language sounds is surprising and contradictory to assumed direction of evolution of language. Evolution of the language ability required rewiring of human brain in the direction of freeing vocalization from uncontrollable emotions [19, 79]. Still, the dual model requires that language sounds be emotional. This emotionality is necessary to motivate connecting sounds and cognitive meanings. Emotionality of human voice is most pronounced in songs [81, 85, 88, 90]. Emotions of everyday speech are low, unless affectivity is specifically intended. We may not notice emotions in everyday "non-affective" speech. Nevertheless, this emotionality is important for developing the cognitive part of the dual model. If language is highly emotional, speakers are passionate about what they say, however evolving new meanings might be slow, emotional ties of sounds to old meanings might be "too strong." If language is low-emotional, new words are easy to create, however motivation to develop the cognitive part of the dual model might be low, the real-world meaning of language sound might be lost. Cultural values might be lost as well. Indeed languages differ in how strong are emotional connections between sounds and meanings. This leads to cultural differences. Thus the dual model leads to Emotional Sapir-Whorf Hypothesis [76, 79, 86]. Strength of emotional connections between sound and meaning depends on language inflections. In particular, after English lost most of its inflections, it became a low emotional language, powerful for science and engineering. At the same time English is losing autonomous connections to cultural values that used to be partially inherent in language sounds. Fast change of cultural values during recent past in English-speaking cultures is usually attributed to progress in thinking, whereas effects of change in emotionality of language sounds have not been noticed.

6.4 Musical Emotions

At lower levels KI acts automatically: sensory-motor experiences are directly embodied. But at higher levels abstract knowledge is called abstract exactly because it does not exist pre-formed in the world, it is created through the interaction of the world and the mind. But cognitive dissonance (CD), a mechanism opposite to KI, might interfere at higher levels. CD is a discomfort caused by holding conflicting cognitions [14, 21, 29]. This discomfort is usually resolved by devaluing or discarding the conflicting cognition. This discarding often occurs below the level of consciousness; it is fast and momentary [36]. It is also known that the majority of new knowledge originates through the differentiation of previous knowledge, which is the mechanism for several broad empirical laws: Zipf's law, the power law, Pareto law emerge when new entities (or usage) evolve from pre-existing ones [65, 66, 105]. Therefore, almost all knowledge contradicts other knowledge to some extent and according to CD theory, any knowledge should be discarded before its usefulness becomes established.

As language began emerging, every word brought new knowledge. However, CD should have interfered with this process. Before the usefulness of new knowledge could be established, it should have been discarded along with language [89]. To overcome CD and enable the evolution of language and knowledge, a word should "grab attention" preconsciously. Emotion of prosody accomplishes this [91]. This same mechanism is alive and well today. Over millions of years many musical and prosodial emotions have evolved and have been culturally inherited by everyone. Yet the diversity of culture is "too much" for some people, leading to new knowledge often being discarded in school and later. Recall that Einstein never received a Nobel Prize for Theory of Relativity. The emotions of language prosody embody the meanings of language and connect sounds to cognitive meanings in everyday conversations as well as in science (experimental evidence is discussed below). While language evolved toward more semantic and less emotional sounds, emotionality of voice, inherited from our animal past, evolved toward stronger emotions, to songs and music. And today song lyrics may affect us stronger than the same text without music.

Did music evolve to connect abstract thoughts and KI? This hypothesis has been confirmed experimentally. In [3] children devalued a toy if they could not play with it. The desire 'to have' contradicts the inability 'to attain'; this creates CD, which is resolved by discarding the contradiction. This experiment repeated many times [14] was first described in Aesop's fable 2500 years ago: the fox unable to attain the grape devalues a contradictory cognition by deciding: "the grape is sour."

Does music help in overcoming CD? [60, 85, 86] have reproduced the above experiment with music in the background and observed that the toy is *not* devalued. Another experiment demonstrated that academic test performance may improve while listening to music. Perlovsky and Cabanac [94] demonstrated (1) that students allocate *less* time to more difficult and stressful tests (as expected from CD theory), and (2) with music in the background students *can* tolerate stress, allocate *more* time to stressful tests, and improve grades. These experiments confirmed the hypothesis that music helps in overcoming CD. It is likely that music emerged and evolved for a fundamental cognitive function: music makes the accumulation of knowledge and human evolution possible.

For thousands of years philosophers and psychologists wondered about the origin of dissonances and consonances. Masataka and Perlovsky [61] demonstrated that consonant music helps "everyday" decision-making in the presence of cognitive interfering evidence, whereas dissonant music increases interference effects. Is music limited to a few emotions, or does every musical phrase evoke a different shade of emotion? Researchers take opposite sides of this issue [17, 38, 39, 44, 107, 112].

How can multiplicity of emotions be explained and justified from a cognitive and evolutionary standpoint, and why has this ability emerged? The proposed hypothesis relating music to CD suggests the following explanation. CD produces a variety of emotional discomforts, different emotions for every combination of knowledge—in other words, a huge number of emotions. Most of these emotions are barely noticed because they lie below the level of consciousness, and in these unconscious states they produce disincentives for knowledge. Music helps to overcome these emotional discomforts by developing a huge number of conscious musical emotions. The mind being conscious of the multiplicity of emotions can bring into consciousness emotions of CD, and thus be prepared to tolerate them. We enjoy even sad and difficult musical emotions for their positive effect of overcoming difficult CD. Possibly this explains the mysterious enjoyment of sad music: it helps us to overcome CD of life's unavoidable disappointments, including the ultimate one, the knowledge of our finiteness in the material world.

Melody, harmony, and other musical devices produce complex, uniquely human musical emotions—they are related to knowledge and therefore are aesthetic emotions ([9, 96]). They expand KI toward a differentiated instinct sensitive not only to unifying knowledge and the world, but also to unifying multiplicity of contradictions among various aspects of knowledge. While CD split our psyche into differentiated knowledge, KI and music unify our psyche. Musical emotions embody abstract knowledge and unify our mental life, language, and body.

These are the reasons why music affects us so strongly. Music connects thinking and intuition to the world. Our spiritual life is embodied through music. Uniquely human refined musical emotions embody our abstract thoughts from the everyday to the most exalted experience. Our highest mental representations near the top of the mental hierarchy attempt to unify our entire life experience. As discussed we perceive them as the meaning of life; their cognitive representations are vague but their feelings are strong, we feel them as emotions of the beautiful [82, 83]. These representations cannot be matched to anything "objectively existing" in the world outside of our brain-mind. Their deep meanings have been created in cultural evolution. Every individual human being receives this cultural knowledge through language. However, this cultural wisdom is not received in an embodied form. It might remain as meaningless disembodied text in books. It is up to everyone's personal effort to create an embodied meaning of life from one's own life experience. Music helps us to embody the meaning of life. The beautiful and sublime, art and religious experience, emotions that embody the meaning of life, as well as the highest spiritual experiences are all embodied through music.

For thousands of years music has been an unexplainable mystery. Aristotle [2] listed the power of music among the great unsolved problems. Darwin wrote (1871) that musical ability "must be ranked amongst the most mysterious with which (man) is endowed." Nature published a series of essays on music [20]. The authors of these essays agreed that "none... has yet been able to answer the fundamental question: why does music have such power over us?" [5]. Today with the help of the physical theory of the mind we have an answer to this question.

7 Future Mathematical Development. Monotone Boolean Functions

This section uses mathematical formalization of dynamic logic for further development of the foundation of the mathematical theory of the mind formulated above; here we name the above theory modeling field theory, MFT. This section is based on previous joint development with Prof. Boris Kovalerchuk [50–53, 109].

Empirical data, E is any data that can be used to identify a model. In logical terms it is defined as a pair

$$E = <A, \Omega>,$$

where A is a set of objects, $\Omega = \{P_i\}$ is a signature, that is a set of predicates P_i of arity n_i, e.g., predicate $P_1(x,y)$ with $n_i = 2$ can mean that length of x is no less than the length of y, $l(x) \geq l(y)$.

Definition The pair $<A, \Omega>$ is called an *empirical system* in Krantz et al. [55].

Definition A pair $<A, \Omega>$ is knows in logic as a *model* (**of the system of axioms T**) [59].

Alfred Tarski proposed the name 'model theory' in 1954 [32]. A variety of other names are also used that include a *relational system* Krantz et al. [55], a *protocol of experiment* ([47, 104]). To distinguish this model from model in MFT we will also call this model a **logic model** or first-order model [33]. Application of algebraic methodology to System Software is presented in [102].

The next important concept in MFT is a concept of **a priori model** (of reality), M. In logic formalization it can be matched with a **system of axioms** T.

Definition A **system of axioms** T is a set of closed first order logic (FOL) formulas (sentences) in the signature Ω.

This means that every variable x_i is presented in the formula with the existential ($\exists$) or the universal ($\forall$) quantifier, e.g., $\forall x_i \exists x_j P_1(x_i, x_j)$.

The quick comparison of MFT and logic approaches in these two concepts reveals a fundamental difference between them. Logic is going from a *very formal* (syntactical) axiomatic system T to something more real called a model $A_T = <A_T, \Omega>$ of that formal system T. MFT is going in the opposite way—from *very informal* reality to more formal models. As a result the concepts of a model are quite different in two theories. Empirical data in MFT is a model $E = <A, \Omega>$ in logic, if we interpret empirical data as an empirical system E [55]. On the other hand MFT model is not a model in logic it leans more to a set of axioms about the class of the logic models.

This difference was well described in Hodges [32]: "To *model* a phenomenon is to construct a formal theory that describes and explains it. In a closely related sense, you *model* a system or structure that you plan to build, by writing a description of it. These are very different senses of 'model' from that in model theory: the 'model' of the phenomenon or the system is not a structure but a theory, often in a formal language".

The next MFT concept is a **similarity measure** L(M, E) between empirical data E and a priory model M that is assigned individually to each specific problem and E:

$$L : \{(M, E)\} \rightarrow R,$$

where r is a set or real numbers.

The closest concept in the logic for this concept is a statement that $m = <a, \Omega>$ is a model of the system of the axioms t.

Definition Pair $e = <a, \Omega>$ is a model of the system of the axioms t if every formula from t is true on e.

Definition Boolean similarity measure $b(t, e)$ is defined to be equal to 1, $l(t, e) = 1$, if m is a model of t, else $l(t, e) = 0$.

7.1   Concept of Uncertainty, Generality and Simplicity

7.1.1   Uncertainty, Generality and Simplicity Relations Between Models

Below we introduce concepts of uncertainty, generality and simplicity relations. These concepts can be applied to both logic and MFT models.

Notation An **uncertainty relation between models** is denoted as "$\geq_{Mu}$", relation $M_i \geq_{Mu} M_j$ is read: "Model M_i is equal in uncertainty or *more uncertain* than model M_j" or "Model M_j is no less certain than model M_i" This relation is a partial order.

Notation A **generality relation between models** is denoted as "$\geq_{Mg}$" and relation $M_i \geq_{Mg} M_j$ is read: "Model M_j is a *specialization* of the measure M_i" or "Model M_i is a *generalization* of the measure M_j". This relation is a partial order too.

Notation A **simplicity relation between model** is denoted as "$\geq_{Ms}$" and relation $M_i \geq_{Ms} M_j$ is read: "Model M_i is equal in simplicity of *simpler* than Model M_j". This relation also is a partial order.

For the MFT models that are represented as a system of axioms the generality relation can be defined as follows.

Definition $T_i \geq_{gen} T_j$ if and only if $T_i \subset T_j$, i.e., system of axioms T_i is equal to or more general than the system of axioms T_j if and only if T_i contains less axioms than T_j, $T_i \subset T_j$.

7.1.2   Uncertainty, Generality and Simplicity Relations Between Similarity Measures

Below we introduce concepts of uncertainty, generality and simplicity relations for similarity measures. These concepts can be applied to both logic and MFT similarity measures.

Notation An **uncertainty relation** between similarity measures is denoted as "$\geq_{Lu}$" and relation $L_i \geq_{Lu} L_j$ is read: "Measure L_i is equal to in uncertainty or more uncertain than measure L_j". This is a partial order relation.

Notation A **generality relation between similarity measures** is denoted as "$\geq_{Lg}$" and relation $L_i \geq_{Lg} L_j$ is read: "Measure L_j is a *specialization* of measure L_i and measure L_i is a *generalization* of the measure L_j" This relation also is a partial order.

Notation A **simplicity relation between similarity measures** is denoted as "$\geq_{Ls}$" and relation $L_i \geq_{Ls} L_j$ is read: "Measure L_j is equal in simplicity or *simpler* than measure L_i". This relation also is a partial order.

Definition Mapping F between a set of models {M} and a set of similarity measures {L}

$$F : \{M\} \rightarrow \{L\}$$

is called a **match mapping** if F preserves uncertainty, generality and simplicity relations between models and measures in the form of homomorphism from relational system $<\{M\}, \geq_{Mg}, \geq_{Mu}>$ to relational system $<\{L\}, \geq_{Lg}, \geq_{Lu}>$, i.e.,

$$\forall M_a, M_b(M_a \geq_{Mg} M_{b\Rightarrow}F(M_a) \geq_{Lg} F(M_b)),$$

$$\forall M_a, M_b(M_a \geq_{Mu} M_{b\Rightarrow}F(M_a) \geq_{Lu} F(M_b)).$$

A homomorphism in contrast with an isomorphism allows several models to be mapped to the same similarity measure L.

Specific match mappings may need additional properties such as simplicity relation between models, that is if two models that are equal in generality and uncertainty the preference is given to a simpler one.

7.2 Partial Order of Models

Two different models can be at the same level of uncertainty ($M_1 =_u M_2$), one model can be more uncertain than another one ($M_1 >_u M_2$), or these models can be incomparable for uncertainty. Symbol "$\geq_u$" is also can be viewed as a disjunction of relations $>_u$ and $=_u$.

We may define *model uncertainty* in such way that two different quadratic models $M_1: 2x^2 + 3y$ and model $M_2: 5x + 4y^2$ will have the same level of uncertainty $M_1 =_u M_2$. The *number of unknown coefficients* is a possible ways to do this. For M_1 and M_2 these numbers m_1 and m_2 are equal to zero. All coefficients are known and thus both models are certain.

Definition *NUC measure of polynomial model uncertainty* is defined as the Number of Unknown Coefficients (NUC) in the model.

The generality relation between models M_1 and M_2 can also be defined. For instance, it can be the *highest power n* of the polynomial model. Both models M_1 and M_2 are quadratic with $n_1 = n_2 = 2$ and, thus, have the same generality.

Definition *HP measure of polynomial model generality* is defined as the *Highest Power n* of the polynomial model.

Alternatively we may look deeper and notice that M_1 contains x^2 and M_2 contains y^2. We may define the *generality* of a polynomial model as its *highest polynomial variable,* which are x^2 for M_1 and y^2 for M_2. We cannot say that one of them is more general and can call them incomparable in generality.

Definition *HPV measure of polynomial model generality* is defined as the *Highest Power Variable (HPV)* of the polynomial model.

Consider model M_3: $5x + by^2$. Using NUC measure this model is *more uncertain* than model M_2: $5x+4y^2$, $M_3 >_{Mu} M_2$, because coefficient b in M_3 is not known, that is NUC for M_2 is $n_2 = 0$ and NUC for M_3 is $n_3 = 1$ and $n_3 > n_2$.

We can also consider M_3 as *more general* than M_2: $5x+4y^2$, $M_3 >_{Mg} M_2$, because M_2 is a specialization of M_3 with $b = 4$. Similarly model M_4: $ax+cx^2+by^2$ is more general and uncertain than models M_1: $2x^2 +3y$, M_2: $5x+4y^2$ and M_3: M_3: $5x + by^2$, because all coefficients in M_4 are uncertain, but none of the coefficients is uncertain in M_1, M_2 and M_3. In these examples uncertainty and generality are consistent and it is hard to distinguish them. In the next section we provide an example with clear difference between them. To formalize this idea we need to introduce some concepts that also will be described in the next section.

7.3 Examples

7.3.1 Uncertainty and Generality of Polynomial Models

In this section we discuss uncertainty and generality of polynomial models based on a parameterization idea. At first we consider an example of models with increasing levels of uncertainty:

Level 0: $3x + 4y + 5y^2$. All coefficients are known at level 0 (no uncertainty).
Level 1: $ax + 4y + 5y^2$. One coefficient is unknown at level 1.
Level 2: $ax + by + 7y^2$. Two coefficients are unknown at level 2.

We may notice that models $3x + 4y + 5y^2$, $ax + 4y + 5y^2$and $ax+by+7y^2$ form a chain from a more specific model (level 0) to a less specific model (level 2) as well as from a more certain model to a more uncertain model.

In contrast models $3x + 4y + 5y^2$, $ax + 9y + 5y^2$and $ax + by + 7y^2$ form an *increasing uncertainty chain* by UNC measure, but they *do not form an increasing generality*. We cannot get $3x + 4y + 5y^2$ by specializing $ax+9y+5y^2$and $ax + by + 7y^2$, because y coefficients 4 and 9 are different. Similarly we cannot get $ax + 9y + 5y^2$ by specializing $ax + by + 7y^2$, because y^2 coefficients 5 and 7 are different. This example illustrates the difference between uncertainty and generality relations.

Another example provides us five models at five uncertainty levels:

Uncertainty level n = 4: $M_4 = ax^2 + by + cx + d$
Uncertainty level n = 3: $M_3 = ax^2 + by + 7x + 10$
Uncertainty level n = 2: $M_2 = ax^2 + 3y + 7x + 10$
Uncertainty level n = 1: $M_1 = x^2 + 3y + 7x + 10$
Uncertainty level n = 0: $M_0 = 9x^2 + 3y + 7x + 10$

Models M_4, M_3, M_2, M_1, M_0 form a *uncertainty decreasing chain* with UNC uncertainty relation defined above:

$$M_4 >_{Mu} M_3 >_{Mu} M_2 >_{Mu} M_1 >_{Mu} M_0$$

They also form a *generality decreasing chain*

$$M_4 >_{Mg} M_3 >_{Mg} M_2 >_{Mg} M_1 >_{Mg} M_0$$

Here model M_3 can be obtained by specialization of parameters of model M_4 and so on, but we did not define the generality concept for them formally yet. Below it is done by using parameterization approach.

Each considered model has 4 parameters, p_1, p_2, p_3, and p_4. For instance, for model $M_2 = ax^2 + 3y + 7x + 10$ parameter $p_1 = 1$ represents uncertainty of ax^2, where coefficient a is unknown. Similarly, $p_2 = p_3 = p_4 = 0$, because further coefficients 3, 7 and 10 are known. In this notation we can represent each model as a Boolean vector, $\mathbf{v}_i = (v_{i1}, v_{i2}, .., v_{ik}, ..., v_{in})$:

$M4 : v_4 = 1111; M3 : v_3 = 1110; M2 : v_2 = 1100; M1 : v_1 = 1000; M0 : v_0 = 0000.$

Definition Parametric model M_i is *no less general* than model M_j if

$$\mathbf{v}_i \geq \mathbf{v}_j, \text{ i.e., } \forall k \, v_{ik} \geq v_{jk}.$$

In accordance with this definition we have

$$1111 \geq 1110 \geq 1100 \geq 1000 \geq 0000$$

that is isomorphic to $M_4 >_{Mg} M_3 >_{Mg} M_2 >_{Mg} M_1 >_{Mg} M_0$.

7.4   Uncertainty and Generality of Kernel Models

In this section we discuss uncertainty and generality of parametric kernel models. Consider a model that consists of n Gaussian kernels. Each kernel K_i has two parameters, p_{i1} and p_{i2} that are mean and standard deviation respectively (or covariation matrix in a multidimensional case). We define the following levels of uncertainty:

Level 0: All 2n coefficients p_{i1} and p_{i2} are known at level 0 (no uncertainty).
Level 1: Only one coefficient is unknown at level 1.
Level 2: Two coefficients are unknown at level 2.
Level 2n: All 2n coefficients are unknown at level 2n.

Consider level n with all p_{i1} are known and all p_{i2} are not known. Assume also that the maximum possible value p_{2max} of all p_{i2} is known. This value is considered as a priory value of p_{i2} for all i in the initial a priory model M_0. Assume that a learning operator $C(M_0, E)$ produced a model M_1 that shrinks the standard deviation max value p_{2max}, to smaller numbers $p_{i2max}(M_1)$ for i = 1.

The **similarity measure** $S(M_1, E)$ can be defined by kernel overlap. If kernels do not overlap in the 2 standard deviations, $\pm 2p_{i2max}(M_1)$ then $S(M_1, E) = 1$, else $S(M_1, E) < 1$.

Now we can apply learning operator $C(M_j, E)$ and produce a chain of models, where each model M_{j+1} is more specific then model M_j with decreasing parameters p_{i2}.

$$M_n >_{Mg} M_{Mn-1}....M_{j+1} >_{Mg} M_j....M_2 >_{Mg} M_1 >_{Mg} M_0.$$

Each considered model has 2n parameters, p_{11}, p_{12}, p_{21}, p_{22},..., p_{n1}, p_{n2}. We already assumed that n parameters p_{i1} are known. Now we encode known parameters as 1 and unknown as 0, thus we have for model M_0 a 2n-dimensional Boolean vector

$$\mathbf{v}_0 = (v_{01}, v_{02}, .., v_{0k}, ..., v_{0n}) = (1010....10)$$

In this notation we can represent each model for n = 3

$$M_3 : v_2 = 111111; M_2 : v_2 = 011111; M_1 : v_1 = 011110; M_0 : v_0 = 101010.$$

and

$$111111 \geq 011111 \geq 011110 \geq 101010.$$

that is consistent with $M_3 >_{Mg} M_2 >_{Mg} M_1 >_{Mg} M_0$.

A more detailed uncertainty parameterization can be developed if Boolean vectors are substituted by k-valued vectors $\mathbf{u}_i = (u_{i1}, u_{i2}, .., u_{ik}, ..., u_{in})$ with

$$u_{ij} \in U = \{0, 1/(k-1), 2/(k-1), ...k-2/(k-1), 1\}.$$

Definition Parametric model M_i is *no less general* than model M_j if

$$\mathbf{u}_i \geq \mathbf{u}_j, \text{ i.e., } \forall k \ v_{im} \geq v_{jm}.$$

Above we encoded known parameters as 1 and unknown as 0. Now we can assign a level of parameter uncertainty u_{i2} by computing $p_{i2}(Mj)/p_{i2max}$ and assigning u_{i2} as a nearest number from $\{0, 1/(k\text{-}1), 2/(k\text{-}1), \ldots k\text{-}2/(k\text{-}1), 1\}$, e.g., let $p_{i2}(Mj)/p_{i2max} = 0.75$, but the nearest k-value if 0.8, then $u_{i2} = 0.8$.

7.5   Similarity Maximization

Now we can define **a similarity maximization** problem in MFT using the definition of the similarity measure provided above.

Definition A similarity L_{fin} measure is called a **final similarity measure** if

$$\forall\, M, E, L_i\ \ L_i(M, E)_{\geq Lu} L_{fin}(M, E)$$

The goal of setting up the final similarity measure is to set up the level of certainty of model similarity of the data that we want to reach.

Definition The **static model optimization problem (SMOP)** is to find a model M_a such that

$$L_{fin}(M_a, E) = \text{Max}_{i\in I} L_{fin}(M_i, E) \tag{11}$$

subject to conditions (12) and (13):

$$\forall\, M_j\ L_{fin}(M_a, E) = L_{fin}(M_j, E) \Rightarrow M_a \geq_{Mu} M_{j,} \tag{12}$$

$$\forall\, M_j((L_{fin}(M_a, E) = L_{fin}(M_j, E)\ \&\ ((M_j \geq_{Mg} M_a)\vee((M_a \geq_{Mg} M_j))) \Rightarrow$$
$$M_a \geq_{Mg} M_j)) \tag{13}$$

The goal of conditions (12) and (13) is prevent model overfitting. In addition conditions (12) and (13) can be beneficial computationally if further specification of the model requires more computations.

Condition (12) means that if M_a and M_j have the same similarity measure with E, $L_{fin}(M_a, E) = L_j(M_j, E)$, then uncertainty of M_a should be no less than uncertainty of M_j. Say if model M_a has three uncertain coefficients and model M_j has only one uncertainty coefficient then we prefer model M_a.

Condition (13) means that if M_a and M_j have the same similarity measure with E, $L_a(M_a, E) = L_j(M_j, E)$, and M_j and M_a are comparable relative to generality relation "$\geq_{Mg}$" then M_a should be no less general than M_j, $M_j \geq_{Mu} M_a$. This means that model M_j can be obtained by specification of model M_a. Say if models M_a and M_j have all the same coefficients, but coefficient c, and c can be any number from [1, 3] in M_a then we would prefer this model to a model M_j with a more specific c = 2.1.

Definition The dynamic logic model optimization (DLPO) problem is to find a model M_a such that

$$L_a(M_a, E) = Max_{i \in I} L_i(M_i, E) \tag{14}$$

subject to conditions (14) and (15):

$$\forall M_j \, L_a(M_a, E) = L_j(M_j, E) \Rightarrow M_a \geq_{Mu} M_j, \tag{15}$$

$$\forall M_j ((L_{fin}(M_a, E) = L_j(M_j, E) \, \& \, ((M_j \geq_{Mg} M_a) \vee ((M_a \geq_{Mg} M_j))) \Rightarrow$$
$$M_a \geq_{Mg} M_j)). \tag{16}$$

This is a non-standard optimization problem. In standard optimization problems the optimization criterion L is static, which is given at the beginning of the optimization process and is not changed in the course of the optimization and does not depend on the model M_i to be optimized. Only models M_i are changed in the standard (static) optimization process:

$$Max_{i \in I} L(M_i, E). \tag{17}$$

In the dynamic logic model optimization problem the criterion L is changing dynamically with models M_i. MFT shows that this is an effective way to cut down computational (combinatorial) complexity of finding an optimal model. Since the focus of MFT approach is in cutting **computational complexity (CC)** of model optimization a **dual optimization problem** can be formulated.

Definition Mapping $\{M\} \rightarrow \{M\}$ is called a **learning (adaptation) operator** C,

$$C(M_i, E) = M_{i+1},$$

where E are data and $M_i \geq_{Mu} M_{i+1}$, $M_i \geq_{Mg} M_{i+1}$, this operation represents a *cognitive learning process* c of a new model m_{i+1} from a given model m_i and data e. in other words it is an adaptation of model m_i to data e that produce model m_{i+1}.

Definition An optimization problem of finding a **shortest sequence** of matched pairs (M_i, L_i) of models M_i and optimization criteria (similarity measures) L_i that **solves** the optimization problem (4)–(6) for a given data E is called a **dual dynamic logic model optimization (DDLMO)** problem, that is finding a sequence of n matching pairs

$$(M_1, L_1), (M_2, L_2), \ldots, (M_n, L_n)$$

such that

$$L(M_n, E) = \text{Max}_{i \in I} L(M_i, E).$$

$$\forall M_i L_i = F(M_i), C(M_i, E) = M_{i+1}, M_i \geq_{Mu} M_{i+1}, M_i \geq_{Mg} M_{i+1}, M_n = M_a, L_n = L_a$$

This means finding a sequence of more specific and certain models for given data E, matching operator F and learning operator C that maximizes $L(M_i, E)$.

7.6 Monotonicity, Monotone Boolean and K—valued Functions

We consider a Boolean function f: $\{0,1\}^n \rightarrow \{0,1\}$.

Definition A Boolean function f is a **monotone Boolean function** if:

$$\forall v_i \geq v_j \ \& \ \Rightarrow f(v_j) \geq f(v_i).$$

This means that

$$\forall (v_i \geq v_j \ \& \ f(v_i) = 0) \Rightarrow f(v_j) = 0$$

$$\forall (v_i \geq v_j \ \& \ f(v_j) = 1) \Rightarrow f(v_i) = 1.$$

Function f is a non-decreasing decreasing function.

Now we consider fixed E, M_i parameterized by v_i and explore interpretation of $L(M_i, E)$ as $f(v_i)$, i.e., $L(M_i, E) = f(v_i)$. Let us assume for now that $L(M_i, E)$ has only two values (unacceptable-0 and acceptable-1). Later on it can be generalized to a k-value case. If $L(M_i, E)$ is monotone then

$$\forall v_i \geq v_j \Rightarrow L(M_i, E)) \geq L(M_j, E),$$

e.g., if $L(M_{3-1110}, E) = L(M_{2-1100}, E) = 0$ then

$$\forall\, (\mathbf{v}_i \geq \mathbf{v}_j \,\&\, L(M_{3-1110}, E) = 0) \Rightarrow L(M_{2-1100}, E) = 0 \qquad (18)$$

$$\forall\, (\mathbf{v}_i \geq \mathbf{v}_j \,\&\, L(M_{2-1100}, E) = 1) \Rightarrow L(M_{3-1110}, E) = 1 \qquad (19)$$

This means that if a model with more unknown parameters $\mathbf{v}_i$ failed then a model with less unknown parameters $\mathbf{v}_j$ will also fail. In other words, if at higher level of uncertainty the model in not acceptable $L(M_i, E) = 0$, then it can not be acceptable on the lower level of uncertainty, $L_j(M_j, E) = 0$. If we conclude that a quadratic polynomial model (M_2) is not acceptable, $L(M_2, E) = 0$, then a more specific quadratic model M_3 also cannot be acceptable, $L(M_3, E) = 0$. Thus, we do not need to test model M_3. This is an idea how monotonicity can help to decrease computational complexity. To be able to use this principle of monotonicity in a task we need to check that it takes place for that task.

In essence, we can use

$$\forall(\mathbf{v}_i \geq \mathbf{v}_j \,\&\, f(\mathbf{v}_i) = 0) \Rightarrow f(\mathbf{v}_j) = 0$$

for rejection models and we can use

$$\forall\, (\mathbf{v}_i \geq \mathbf{v}_j \,\&\, f(\mathbf{v}_j) = 1) \Rightarrow f(\mathbf{v}_i) = 1$$

for confirming models.

In the case of **model rejection test** for data E the main focus is not quick building a model but quick rejecting a model M (Popper's principle). In essence the test $L_3(M_3, E) = 0$ means that the whole class of the models M_3 with 3 unknown parameters fails. For testing M_3 positively for data E we need to find 4 correct parameters. This may mean searching in a large 4-D parameter space $[-100, +100]^4$ for single vector, say $(p_1, p_2, p_3, p_4) = (9, 3, 7, 10)$, if each parameter varies in the interval $[-100, 100]$. For rejection we may need only, 4 training vectors (x,y,u) from data E and 3 test vectors. The first four vectors will allow us to build a quadratic surface in 3-D as a model. We would just need to test that three test vectors from E do not fit this quadratic surface.

Definition K-valued function f of n variable f: $U^n \to U$ is called a **monotone k-valued function** if

$$\mathbf{u}_i \geq \mathbf{u}_j \Rightarrow f(\mathbf{u}_i) \geq f(\mathbf{u}_j)$$

This function can be applied similarly in the case when we have more uncertainty levels between 0 and 1.

7.7   Search Process

7.7.1   Search Process in Monotone Functions Terms

In the optimization process we want to keep a track of model rejections and be able to guide dynamically what model will be tested next to minimize the number of tests. This is a part of dynamic logic in MFT that we formalize below using the theory of Monotone Boolean and k-valued Functions. Formulas (a) and (b) are key formulas to minimize tests, but we need the whole strategy how to minimize the number of tests and formalize it. One of the ways of formalization is to minimize **Shannon function** φ [28, 45].

$$\min_{A \in \mathbf{A}} \max_{f \in \mathbf{F}} \; \varphi(f, A),$$

where $\mathbf{A}$ is a set of algorithms, F is a set of monotone functions and $\varphi(f,A)$ is a number of tests that algorithm A does to fully restore function f. Each test means computing a value $f(\mathbf{v})$ for a particular vector $\mathbf{v}$. In the theory of monotone Boolean functions it is a assumed that there is an **oracle** that is able to produce the value $f(\mathbf{v})$, thus each test is equivalent to a request to the oracle [28, 45, 48, 49]. Minimization of Shannon function means that we search for the algorithm that needs smallest number of tests for its worst case (function f that needs maximum number of tests relative to other functions). This is a classic min-max criterion.

It was proven in [28, 45] that

$$\min_{A \in \mathbf{A}} \max_{f \in \mathbf{F}} \; \varphi(f, A) = \binom{n}{\lfloor n/2 \rfloor} + \binom{n}{\lfloor (n/2) \rfloor + 1},$$

$\lfloor x \rfloor$ is a floor of x (an integer that smaller than x and closest to x).

The proof of this theorem allows us to derive an algorithm based on the structure called Hansel chains. These chains designed by Hansel cover the whole n-dimensional binary cube $\{1,0\}^n$. The steps of the algorithm are presented in detail in [48]. The main idea of these steps is building Hansel chains, starting from testing his smallest chains, expanding each tested value using (a) and (b) formulas presented above and test values that are left not expanded on the same chains then move to larger chains until no chains left.

In mathematical terms the goal of the search is to find a **smallest lower unit v**, i.e., a Boolean vector such that $f(\mathbf{v}) = 1$, and for every $w < v$ $f(w) = 0$, and for every $u > v$ $|u| > |v|$. A less challenging problem could be to find any lower unit of f.

The difference of the approach based on the Hansel chains from traditional one when individual parameters are added sequentially to the list of certain parameters is that the simple sequence does not optimize Shannon function, that is it may require more steps than Hansel chains. Mathematical results are also known for k-valued monotone functions [42, 43].

7.7.2 Search in Logic and Probabilistic Terms

The search problem in logic terms can be formulated as a satisfiability problem:
Find a system of axioms T_a such that

$$L_a(T_a, E) = 1$$

subject to the condition

$$\forall T_j L_a(T_a, E) = L_j(T_j, E) \Rightarrow T_j \geq_{Mu} T_a,$$

i.e., if T_a and T_j have the same similarity with E, $L_a(T_a, E) = L_j(T_j, E)$, then M_a should have a lower uncertainty than T_j, e.g., $T_j \geq_{Mu} M_a$.

If a similarity measure is defined as a **probabilistic measure** with values in [0, 1] then the probabilistic version of the task of finding system of axioms T for model A that maximizes a probabilistic similarity measure is:

$$Max_{i \in I} L(M_i, E),$$

where $L = F(M_i)$ and I is a set of models. This task can be further developed by using Occam principle—to select the simplest model out of two equally appropriate.

7.8 *Summary of Formalization*

In this section formalization of the concept of the dynamic logic in the terms of the first order logic, logic model theory and theory of Monotone Boolean functions has been introduced. It concentrates on the main idea of dynamic logic of matching levels of uncertainty of the problem/model and levels of uncertainty of the evaluation criterion used dramatically minimize search for model identification. When a model becomes more certain then the evaluation criterion is also adjusted dynamically to match an adjusted model. This dynamic logic process of model construction is likely mimics a process of a natural evolution.

This section introduced the concepts of partial order on the models with respect to their uncertainty, generality and simplicity. These concepts are also introduced for similarity measures and examples provided for models and measures. Next these partial orders are represented using a set of Boolean parameters. The theory of monotone Boolean functions is used for guiding and visualizing search in the parameter space in the dynamic logic setting.

The proposed formalization creates a framework for developing specific applications, that will consist of a sets of models, matching similarity measures, processed for testing them and model learning processed for specific problems in pattern recognition, data mining, optimization, cognitive process modeling and decision making.

Further theoretical studies may reveal a dipper links with classical optimization search processes and significantly advance then by adding an extra layer of actual constructing optimization criteria not only using them.

In the area of logic further theoretical studies may also reveal a dipper links with classical logic problems such as decidability, completeness and consistency.

In the areas of machine learning further theoretical studies may also reveal a dipper links with analytical machine learning [63], inductive and probabilistic logic programming [18, 101], relational machine learning [54], where a priory models play a critical role in the learning process.

8 Conclusion

This chapter summarizes previous development of physics of the mind, a theory of cognition based on fundamental principles of the mind operation, and dynamic logic a mathematical foundation of the physics of the mind that enables to overcome combinatorial complexity, which has prevented previous developments of cognitive theories since the 1950s. Mathematical formulations of the fundamental principles of the mind have been presented including dynamic logic, the knowledge instinct, mechanisms of concepts, emotions, aesthetic emotions, emotions of the beautiful, language, the dual model of the interactions between language and cognition, including the fundamental role of prosodial emotions in this interaction. Physics of the mind predicted a number of psychological phenomena, many of which have been confirmed in experiments.

The chapter concludes with the development of mathematical formalization of dynamic logic using first order logic, logic model theory, and monotone Boolean functions that could be used for further mathematical development.

Acknowledgements It is my pleasure to thank people whose thoughts helped to develop ideas in this chapter, Moshe Bar, Boris Kovalerchuk, Evgeny Vityaev.

References

1. Aristotle. (1995a). The complete works. The revised Oxford translation, ed. J. Barnes, Princeton, NJ: Princeton Univ. Press. Original work VI BCE.
2. Aristotle. (1995b). Organon. The complete works. The revised Oxford translation, ed. J. Barnes, Princeton, NJ: Princeton Univ. Press. Original work VI BCE, 18a28-19b4; 1011b24-1012a28.
3. Aronson, E. and Carlsmith, J. M. (1963). Effect of the severity of threat on the devaluation of forbidden behavior. J Abnor Soc Psych 66, 584–588.
4. Badre, D. (2008). Cognitive control, hierarchy, and the rostro–caudal organization of the frontal lobes. Trends in Cognitive Sciences, 12(5), 193–200.
5. Ball, P. (2008). Facing the music. Nature, 453, 160–162.

6. Bar, M.; Kassam, K.S.; Ghuman, A.S.; Boshyan, J.; Schmid, A.M.; Dale, A.M.; Hämäläinen, M.S.; Marinkovic, K.; Schacter, D.L.; Rosen, B.R.; et al. (2006). Top-down facilitation of visual recognition. Proc. Natl. Acad. Sci. USA, 103, 449–454.

7. Barsalou L. W. (1999). Perceptual symbol systems. Behav. Brain Sci. 22:577–660.

8. Binder, J.R., Westbury, C.F., McKiernan, K.A., Possing, E.T., & Medler, D.A. (2005).Distinct Brain Systems for Processing Concrete and Abstract Concepts. Journal of Cognitive Neuroscience 17(6), 1–13.

9. Bonniot-Cabanac, M.-C., Cabanac, M., Fontanari, F., and Perlovsky, L.I. (2012). Instrumentalizing cognitive dissonance emotions. Psychology 3, 1018–1026.

10. Brighton, H., Smith, K., & Kirby, S. (2005). Language as an evolutionary system. Phys. Life Rev., 2005, 2(3), 177–226.

11. Cangelosi A. & Parisi D., Eds. (2002). Simulating the Evolution of Language. London: Springer.

12. Chomsky, N. (1995). The minimalist program. Cambridge: MIT Press.

13. Christiansen, M. H., & Kirby, S. (2003). Language evolution. New York: Oxford Univ. Press.

14. Cooper, J. (2007). Cognitive dissonance: 50 years of a classic theory. Los Angeles, CA: Sage.

15. Cramer, H. (1946). *Mathematical Methods of Statistics*, Princeton University Press, Princeton NJ.

16. Croft, W. & Cruse, D.A. (2004). Cognitive Linguistics. Cambridge: Cambridge University Press.

17. Cross, I., & Morley, I. (2008). The evolution of music: theories, definitions and the nature of the evidence. In S. Malloch, & C. Trevarthen (Eds.), Communicative musicality (pp. 61-82). Oxford: Oxford University Press.

18. Cussens, J., Frisch, A. (2000). Inductive Logic Programming, Springer.

19. Deacon, T.W. (1997). The symbolic species: the co-evolution of language and the brain. New York: Norton.

20. Editorial. (2008). Bountiful noise. Nature, 453, 134.

21. Festinger, L. (1957). A Theory of Cognitive Dissonance. Stanford, CA: Stanford University Press.

22. Gnadt, W. & Grossberg, S. (2008). SOVEREIGN: An autonomous neural system for incrementally learning planned action sequences to navigate towards a rewarded goal. Neural Networks, 21(5), 699–758.

23. Gödel, K. (2001). Collected Works, Volume I, Publications 1929–1936. Feferman, S., Dawson, J.W., Jr., Kleene, S.C., Eds.; Oxford University Press: New York, NY.

24. Grossberg, S. (1988) Nonlinear neural networks: Principles, mechanisms, and architectures. Neural Networks, 1, 17–61.

25. Grossberg, S. (2000). Linking Mind to Brain: the mathematics of biological intelligence. *Notices of the American Mathematical Society*, 471361–1372.

26. Grossberg, S. & Levine, D.S. (1987). Neural dynamics of attentionally modulated Pavlovian conditioning: blocking, inter-stimulus interval, and secondary reinforcement. Psychobiology, 15(3), 195–240.

27. Hameroff, S. & Penrose, R. (2014). Consciousness in the universe. A review of the 'Orch OR' theory. Physics of Life Reviews, 11, 39–78.

28. Hansel, G., Sur le nombre des fonctions Boolenes monotones den variables. C.R. Acad. Sci. Paris, v. 262, n. 20, 1088–1090, 1966.

29. Harmon-Jones, E., Amodio, D. M., and HarmonJones,C. (2009). "Action-based model of dissonance: a review, integration, and expansion ofconceptions of cognitive conflict," in Advances in Experimental Social Psychology, M. P. Zanna (Burlington, MA: Academic Press), 119–166.

30. Hauser, M.D., Chomsky, N., & Fitch, W. T. (2002). The faculty of language: what is it, who has it, and how did it evolve?" Science, 298(5988), 1569–1579. doi:10.1126/science.298.5598.1569.

31. Hilbert, D. (1928). The Foundations of Mathematics. In J. van Heijenoort, Ed., *From Frege to Gödel*. Cambridge, MA: Harvard University Press, 1967, p. 475.

32. Hodges, W. (2005a). Model Theory, Stanford Encyclopedia of Philosophy. http://plato.stanford.edu/entries/model-theory/#Modelling.
33. Hodges, W. (2005b). First-order Model Theory, Stanford Encyclopedia of Philosophy. http://plato.stanford.edu/entries/modeltheory-fo/.
34. Hurford, J. (2008). The evolution of human communication and language. In P. D'Ettorre & D. Hughes, Eds. Sociobiology of communication: an interdisciplinary perspective. New York: Oxford University Press, pp. 249–264.
35. Jackendoff, R. (2002). Foundations of Language: Brain, Meaning, Grammar, Evolution, Oxford Univ Pr., New York, NY.
36. Jarcho, J. M., Berkman, E. T., & Lieberman, M. D. (2011). The neural basis of rationalization: cognitive dissonance reduction during decision-making. Soc Cogn Affect Neurosci, 6(4), 460–467.
37. Josephson, B. 1997. An Integrated Theory of Nervous System Functioning embracing Nativism and Constructivism. International Complex Systems Conference. Nashua, NH.
38. Juslin, P.N. (2013). From everyday emotions to aesthetic emotions: Towards a unified theory of musical emotions. Physics of Life Reviews, 10(3), 235–266.
39. Juslin, P.N. & Västfjäll, D. (2008) Emotional responses to music: The Need to consider underlying mechanisms. Behavioral and Brain Sciences, 31(05), 559–575.
40. Kant, I. (1781). Critique of Pure Reason.Tr. J.M.D. Meiklejohn, 1943. Willey Book, New York, NY.
41. Kant, I. (1790). Critique of Judgment, Tr. J.H.Bernard, Macmillan & Co., London, 1914.
42. Katerinochkina, N. N. (1981). Search for the maximal upper zero for a class of monotone functions of k-valued logic. (Russian) Zh. Vychisl. Mat. i Mat. Fiz. 21(2), 470–481, 527.
43. Katerinochkina, N. N. (1989). Efficient realization of algorithms for searching for a maximal zero of discrete monotone functions, Reports in Applied Mathematics, Akad. Nauk SSSR, Vychisl. Tsentr, Moscow, 16(2), 178–206 (in Russian).
44. Koelsch, S. (2011). Towards a neural basis of processing musical semantics. Physics of Life Reviews, 8(2), 89–105.
45. Korobkov V.K. (1965). On monotone Boolean functions of algebra logic, In Problemy Cybernetiki, v.13, "Nauka" Publ., Moscow, 5–28, (in Russian).
46. Kosslyn, S. M. (1994). Image and Brain. MIT Press, Cambridge.
47. Kovalerchuk, B. (1973). Classification invariant to coding of objects. Computational Systems (Novosibirsk), 55, 90–97 (in Russian).
48. Kovalerchuk, B., Triantaphyllou, E., Aniruddha, S. Deshpande, S. Vityaev, E. (1996). Interactive learning of monotone Boolean functions. 94 (1–4), 87–118.
49. Kovalerchuk, B., Lavkov, V. (1984). Retrieval of the maximum upper zero for minimizing the number of attributes in regression analysis. USSR Computational Mathematics and Mathematical Physics, 24 (4), 170–175.
50. Kovalerchuk, B. & Perlovsky, L.I. (2008). Dynamic Logic of Phenomena and Cognition. IJCNN 2008, Hong Kong, pp. 3530–3537.
51. Kovalerchuk, B. & Perlovsky, L.I. (2009). Dynamic Logic of Phenomena and Cognition. IJCNN 2009, Atlanta, GA.
52. Kovalerchuk, B. & Perlovsky, L.I.. (2011). Integration of Geometric and Topological Uncertainties for Geospatial Data Fusion and Mining. Applied Imagery Pattern Recognition Workshop (AIPR), IEEE. doi:10.1109/AIPR.2011.6176346.
53. Kovalerchuk, B., Perlovsky, L., & Wheeler, G. (2012). Modeling of Phenomena and Dynamic Logic of Phenomena. Journal of Applied Non-classical Logics, 22(1), 51–82. arXiv:abs/1012.5415.
54. Kovalerchuk, B., Vityaev E. (2000). Data Mining in Finance: Advances in Relational and Hybrid Methods, Kluwer.
55. Krantz, D. H., Luce, R. D., Suppes, P., Tversky, A. (1971–1990). Foundations of Measurement. New York, London: Academic Press.
56. Kveraga, K., Boshyan, J., & Bar, M. (2007) Magnocellular projections as the trigger of top-down facilitation in recognition. Journal of Neuroscience, 27, 13232–13240.

57. Levine, D.S., Perlovsky, L.I. (2008). Neuroscientific Insights on Biblical Myths: Simplifying Heuristics versus Careful Thinking: Scientific Analysis of Millennial Spiritual Issues. Zygon, Journal of Science and Religion, 43(4), 797–821.
58. Levine, D.S. & Perlovsky, L.I. (2010). Emotion in the pursuit of understanding. International Journal of Synthetic Emotions, 1(2), 1–11.
59. Malcev, A.I. (1973). Algebraic Systems. Springer-Verlag.
60. Masataka, N. & Perlovsky, L.I. (2012). The efficacy of musical emotions provoked by Mozart's music for the reconciliation of cognitive dissonance. Scientific Reports 2, Article number: 694. doi:10.1038/srep00694; http://www.nature.com/srep/2013/130619/srep02028/full/srep02028.html.
61. Masataka, N. & Perlovsky, L.I. (2013). Cognitive interference can be mitigated by consonant music and facilitated by dissonant music. Scientific Reports 3, Article number: 2028 (2013). doi:10.1038/srep02028; http://www.nature.com/srep/2013/130619/srep02028/full/srep02028.html.
62. Minsky, M. (1988). *The Society of Mind*. MIT Press, Cambridge, MA.
63. Mitchell, T. (1997). Machine Learning, McGraw Hill.
64. Newell, A. (1983). Intellectual Issues in the History of Artificial Intelligence. In the *Study of Information*, ed. F.Machlup & U.Mansfield, J.Wiley, New York, NY.
65. Newman, M. E. J. (2005). Power laws, Pareto distributions and Zipf's law. Contemporary physics, 46(5), 2005, 323–351.
66. Novak. J. D. (2010). Learning, Creating, and Using Knowledge: Concept maps as facilitative tools in schools and corporations. Journal of e-Learning and Knowledge Society, 6(3), 21–30.
67. Penrose, R. (1994). *Shadows of the Mind*. Oxford University Press, Oxford, England.
68. Perlovsky, L.I. (1996). *Gödel Theorem and Semiotics*. Proceedings of the Conference on Intelligent Systems and Semiotics '96. Gaithersburg, MD, v.2, pp. 14–18.
69. Perlovsky, L.I.(1997). Physical Concepts of Intellect. *Proc. Russian Academy of Sciences*, 354(3), pp. 320–323.
70. Perlovsky, L.I. (1998). Conundrum of Combinatorial Complexity. *IEEE Trans. PAMI*, 20(6) p. 666–70.
71. Perlovsky, L.I. (2001). Neural Networks and Intellect: using model-based concepts. Oxford University Press, New York, NY (3rd printing).
72. Perlovsky, L.I. (2002). Aesthetics and mathematical theories of intellect. Iskusstvoznanie, 2/02, 558–594 (Russian).
73. Perlovsky, L.I. (2004). Integrating Language and Cognition. *IEEE Connections*, Feature Article, 2(2), pp. 8–12.
74. Perlovsky, L.I. (2006). Toward Physics of the Mind: Concepts, Emotions, Consciousness, and Symbols. Phys. Life Rev. 3(1), pp. 22–55.
75. Perlovsky, L.I. (2007a). Neural Dynamic Logic of Consciousness: the Knowledge Instinct. Chapter in Neurodynamics of Higher-Level Cognition and Consciousness, Eds.Perlovsky, L.I., Kozma, R. ISBN 978-3-540-73266-2, Springer Verlag, Heidelberg, Germany, pp. 73–108.
76. Perlovsky, L.I. (2007b). Evolution of Languages, Consciousness, and Cultures. IEEE Computational Intelligence Magazine, 2(3), pp. 25–39.
77. Perlovsky, L.I. (2008a). Sapience, Consciousness, and the Knowledge Instinct. (Prolegomena to a Physical Theory). In Sapient Systems, Eds. Mayorga, R. Perlovsky, L.I., Springer, London, pp. 33–60.
78. Perlovsky, L.I. (2009a). Language and Cognition.Neural Networks, 22(3), 247–257. doi:10.1016/j.neunet.2009.03.007.
79. Perlovsky, L.I. (2009b). Language and Emotions: Emotional Sapir-Whorf Hypothesis. Neural Networks, 22(5–6); 518–526. doi:10.1016/j.neunet.2009.06.034.
80. Perlovsky, L.I. (2009c). 'Vague-to-Crisp' Neural Mechanism of Perception. IEEE Trans. Neural Networks, 20(8), 1363–1367.
81. Perlovsky, L.I. (2010a). Musical emotions: Functions, origin, evolution. Physics of Life Reviews, 7(1), 2–27. doi:10.1016/j.plrev.2009.11.001.

82. Perlovsky, L.I. (2010b). Neural Mechanisms of the Mind, Aristotle, Zadeh, & fMRI, IEEE Trans. Neural Networks, 21(5), 718–33.
83. Perlovsky, L.I. (2010c). Intersections of Mathematical, Cognitive, and Aesthetic Theories of Mind, Psychology of Aesthetics, Creativity, and the Arts, 4(1), 11–17. doi:10.1037/a0018147.
84. Perlovsky, L.I. (2010d). The Mind is not a Kludge, Skeptic, 15(3), 51–55.
85. Perlovsky, L.I. (2012a). Cognitive function, origin, and evolution of musical emotions. Musicae Scientiae, 16(2), 185–199; doi:10.1177/1029864912448327.
86. Perlovsky, L.I. (2012b). Cognitive Function of Music, Part I. Interdisciplinary Science Reviews, 37(2), 129–42.
87. Perlovsky, L.I. (2012c). Brain: conscious and unconscious mechanisms of cognition, emotions, and language. Brain Sciences, Special Issue "The Brain Knows More than It Admits", 2(4):790–834. http://www.mdpi.com/2076-3425/2/4/790.
88. Perlovsky L. I. (2012d). Cognitive Function of Music Part I. Interdisc. Science Rev, 7(2),129–42.
89. Perlovsky, L.I. (2013a). A challenge to human evolution—cognitive dissonance. Front. Psychol. 4:179. doi:10.3389/fpsyg.2013.00179; http://www.frontiersin.org/cognitive_science/10.3389/fpsyg.2013.00179/full.
90. Perlovsky, L.I. (2013b). Language and cognition—joint acquisition, dual hierarchy, and emotional prosody. Frontiers in Behavioral Neuroscience, 7:123; doi:10.3389/fnbeh.2013.00123; http://www.frontiersin.org/Behavioral_Neuroscience/10.3389/fnbeh.2013.00123/full.
91. Perlovsky, L.I. (2013c). Learning in brain and machine—complexity, Gödel, Aristotle. Frontiers in Neurorobotics. doi:10.3389/fnbot.2013.00023; http://www.frontiersin.org/Neurorobotics/10.3389/fnbot.2013.00023/full.
92. Perlovsky, L.I. (2014a). Aesthetic emotions, what are their cognitive functions? Front. Psychol. 5:98. http://www.frontiersin.org/Journal/10.3389/fpsyg.2014.00098/full; doi:10.3389/fpsyg.2014.0009.
93. Perlovsky, L. I., Bonniot-Cabanac, M.-C., Cabanac, M. (2010). Curiosity and Pleasure. WebmedCentral PSYCHOLOGY 2010;1(12):WMC001275. http://www.webmedcentral.com/article_view/1275; http://arxiv.org/ftp/arxiv/papers/1010/1010.3009.pdf.
94. Perlovsky, L.I., Cabanac, A., Bonniot-Cabanac, M-C., Cabanac, M. (2013). Mozart Effect, Cognitive Dissonance, and the Pleasure of Music; Behavioural Brain Research, 244, 9–14. arXiv:1209.4017.
95. Perlovsky, L.I., Deming R.W., & Ilin, R. (2011). Emotional Cognitive Neural Algorithms with Engineering Applications. Dynamic Logic: from vague to crisp. Springer, Heidelberg, Germany.
96. Perlovsky, L.I. & Ilin R. (2010). Grounded Symbols in The Brain, Computational Foundations for Perceptual Symbol System. WebmedCentral PSYCHOLOGY 2010;1(12):WMC001357
97. Perlovsky, L.I. & Ilin, R. (2012). Mathematical Model of Grounded Symbols: Perceptual Symbol System. Journal of Behavioral and Brain Science, 2, 195–220. doi:10.4236/jbbs.2012.22024; http://www.scirp.org/journal/jbbs/.
98. Perlovsky, L.I. & McManus, M.M. (1991). Maximum Likelihood Neural Networks for Sensor Fusion and Adaptive Classification. *Neural Networks*, 4(1), pp. 89–102.
99. Pinker, S. (1994). The language instinct: How the mind creates language. New York: William Morrow.
100. Price, C.J. (2012). A review and synthesis of the first 20 years of PET and fMRI studies of heard speech, spoken language and reading. NeuroImage, 62, 816–847.
101. Raedt, Luc De. (2006). From Inductive Logic Programming to Multi-Relational Data Mining. Springer.
102. Rus, T., Rus, D. L. (1990). System Software and Software Systems: Concepts And Methodology—V.1: Systems Methodology for Software, http://citeseer.ist.psu.edu/351353.html.
103. Russell, B. (1967). The History of Western Philosophy. Simon & Schuster/Touchstone, New York, NY.

104. Samokhvalov, K. (1973). On theory of empirical prediction, Computational Systems (Novosibirsk), 55, 3–35 (in Russian).
105. Simonton, D. K. (2000). Creativity. Cognitive, personal, developmental, and social aspects American Psychologist, 55(1), 151–158.
106. Singer, R.A., Sea, R.G. and Housewright, R.B. (1974). Derivation and Evaluation of Improved Tracking Filters for Use in Dense Multitarget Environments, *IEEE Transactions on Information Theory*, IT-20, pp. 423–432.
107. Scherer, K.R. (2004). Which emotions can be induced by music? what are the underlying mechanisms? and how can we measure them? Journal of New Music Research, 33(3), 239–251; doi:10.1080/0929821042000317822.
108. Tversky, A., Kahneman, D. (1974). Judgment under Uncertainty: Heuristics and Biases. Science, 185, 1124–1131.
109. Vityaev, E.E., Perlovsky, L.I., Kovalerchuk, B.Y., Speransky, S.O. (2011). Probabilistic dynamic logic of the mind and cognition, Neuroinformatics, 5(1), 1–20.
110. Vityaev, E.E., Perlovsky, L.I., Kovalerchuk, B. Y., & Speransky, S.O. (2013). Probabilistic dynamic logic of cognition. Invited Article. Biologically Inspired Cognitive Architectures 6, 159–168.
111. Zeki, S. (1993). A Vision of the Brain. Blackwell, Oxford, UK.
112. Zentner, M., Grandjean, D., Scherer, K. R. (2008). Emotions evoked by the sound of music: Characterization, classification, and measurement. Emotion, 8(4), 494–521.

Fuzzy Arithmetic Type 1 with Horizontal Membership Functions

Andrzej Piegat and Marek Landowski

Abstract The chapter shortly (because of the volume limitation) presents multidimensional fuzzy arithmetic based on relative-distance-measure (RDM) and horizontal membership functions which considerably facilitate calculations. This arithmetic will be denoted as MD-RDM-F one. It delivers full, multidimensional problem solutions that further enable determining, in an accurate and unique way, various representations of the solutions such as span (maximal uncertainty of the solution), cardinality distribution of possible solution values, center of gravity of the solution granule, etc. It also allows for taking into account relations and dependencies existing between variables, what is absolutely necessary e.g. in calculations with fuzzy probabilities that always should sum up to 1 or in equation system solving.

Keywords Fuzzy arithmetic · Fuzzy mathematics · Uncertainty theory · Granular computing · Soft computing

1 Introduction

Fuzzy arithmetic [10–12, 15, 17–19, 25–27, 34] is extension of interval arithmetic [20, 22–24, 28–32] from calculation on standard intervals to calculation on fuzzy intervals and fuzzy numbers. This arithmetic is connected with uncertainty theory [6], soft computing [14], granular computing [25], grey systems [21], etc. It is necessary for solving problems of Computing with Words [37, 38], for solving linear and nonlinear equations and fuzzy equation systems [1, 2, 5, 7, 8, 13] which occur in many real problems of economy, engineering, medicine, environmental protection, etc. [3, 4, 9, 16, 17, 36, 39]. Fuzzy arithmetic has been developed for many

A. Piegat (✉)
Faculty of Computer Science and Information Systems, West Pomeranian University
of Technology, Zolnierska 49, 71–210 Szczecin, Poland
e-mail: apiegat@wi.zut.edu.pl

M. Landowski
Maritime University of Szczecin, Waly Chrobrego 1–2, 70–500 Szczecin, Poland
e-mail: m.landowski@am.szczecin.pl

© Springer International Publishing AG 2017
V. Kreinovich (ed.), *Uncertainty Modeling*, Studies in Computational
Intelligence 683, DOI 10.1007/978-3-319-51052-1_14

Fig. 1 Trapezoidal
membership function or a
fuzzy interval

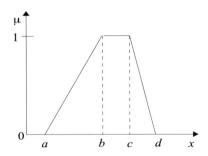

years. There exist many methods of fuzzy arithmetic such as L-R arithmetic [10, 17, 18, 26], fuzzy arithmetic based on discretized fuzzy numbers and Zadeh's extension principle, fuzzy arithmetic based on decomposed fuzzy numbers (α-cuts and Moore interval arithmetic), standard fuzzy arithmetic based on decomposed fuzzy numbers, advanced fuzzy arithmetic based on transformation method or on extended transformation method [17], constrained fuzzy arithmetic [19] and other. Overview of many methods can be found in [12, 17, 25]. Simpler types of fuzzy arithmetic (FA) allow for analytical determining (by hand) of solutions. More complicated methods determine solutions numerically and require computer application. The fact that new types of FA have been developed all the time means that existing methods of FA are not ideal and can be improved. Such improvement proposal in form of MD-RDM-F arithmetic will be presented in this chapter. Concept of this arithmetic has been elaborated by A. Piegat. This arithmetic uses horizontal membership functions (horizontal MFs). They were already, introductory presented in [35, 36]. Figure 1 shows a trapezoidal MF (fuzzy interval).

A vertical model (1) of the fuzzy interval from Fig. 1 expresses the dependence $\mu = f(x)$ in the usual way known from literature of fuzzy sets.

$$\mu(x) = \begin{cases} (x-a)/(b-a) & \text{for } x \in [a,b] \\ 1 & \text{for } x \in (b,c] \\ (d-x)/(d-c) & \text{for } x \in (c,d] \end{cases} \qquad (1)$$

Formula (1) expresses unique dependence of the "vertical" variable μ from the "horizontal" variable x. However, a question can be asked: can a "horizontal" model of MF $x = f(\mu)$ be determined? It seems impossible because such dependence would be not unique and thus it would not be function. However, let us consider horizontal cut of the MF on level μ, which further on will be called not α-cut but μ-cut, Fig. 2a.

Variable α_x, $\alpha_x \in [0, 1]$, called RDM-variable [20, 30, 31] determines relative distance of a point $x^* \in [x_L(\mu), x_R(\mu)]$ from the origin of the local coordinate-system positioned on the left side of MF-cut made on level μ, Fig. 2. Thus, this variable introduces Cartesian-coordinate-system inside of interval. The left border $x_L(\mu)$ and the right border $x_R(\mu)$ of MF is determined by formula (2).

(a)

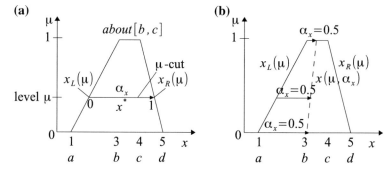

(b)

Fig. 2 Visualization of μ-cut (**a**) and of the horizontal approach to description of membership functions (**b**)

Fig. 3 Horizontal membership function $x = (1 + 2\mu) + (4 - 3\mu)\alpha_x, \alpha_x \in [0, 1]$, corresponding to the function shown in Fig. 2, as unique function in 3D-space

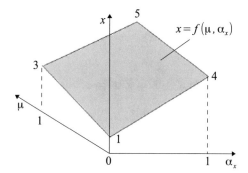

$$x_L = a + (b - a)\mu, \ x_R = d - (d - c)\mu \tag{2}$$

The left border $x_L(\mu)$ is transformed into the right border $x_R(\mu)$ by the RDM-variable α_x. Contour line $x(\mu, \alpha_x)$ of constant values of α_x is determined by formula (3).

$$x(\mu, \alpha_x) = x_L + (x_R - x_L)\alpha_x, \ \alpha_x \in [0, 1] \tag{3}$$

This line is set of points lying at equal relative distance α_x from the left border $x_L(\mu)$ of MF. Precise form of formula (3) given by formula (4) can be called horizontal membership function.

$$x = [a + (b - a)\mu] + [(d - a) - \mu(d - a + b - c)]\alpha_x, \ \alpha_x \in [0, 1] \tag{4}$$

Formula (4) describes function $x = f(\mu, \alpha_x)$, which is function of two variables defined in 3D-space. As Fig. 3 shows, this function is unique.

Because the horizontal function $x = f(\mu, \alpha_x)$ defines not one value of variable x but a set of possible values of this variable corresponding to a given μ-cut level, it defines an information granule and will be denoted as x^{gr}. In this chapter the horizontal model (4) for the trapezoidal MF was shown. However, if $b = c$ then model (4) describes triangular MF and if $a = b$ and $c = d$ then it describes rectangular MF. Figure 2 presents MF with line borders $x_L(\mu)$ and $x_R(\mu)$. However, the borders can also be nonlinear, e.g. of Gauss-type. To derive formulas for horizontal MF in this case, nonlinear formulas for the left and right border of the MF should be determined and inserted in the general formula (2).

2 Basic Operations of MD-RDM-F Arithmetic

Let $x^{gr}(\mu, \alpha_x)$ be horizontal MF representing fuzzy interval X and $y^{gr}(\mu, \alpha_y)$ be horizontal MF representing fuzzy interval Y, formula (5) and (6).

$$X : x^{gr} = [a_x + (b_x - a_x)\mu] + [(d_x - a_x) - \mu(d_x - c_x + b_x - a_x)]\alpha_x, \ \mu, \alpha_x \in [0, 1] \tag{5}$$

$$Y : y^{gr} = [a_y + (b_y - a_y)\mu] + [(d_y - a_y) - \mu(d_y - c_y + b_y - a_y)]\alpha_y, \ \mu, \alpha_y \in [0, 1] \tag{6}$$

Addition $X + Y = Z$ of two independent intervals

$$X + Y = Z : x^{gr}(\mu, \alpha_x) + y^{gr}(\mu, \alpha_y) = z^{gr}(\mu, \alpha_x, \alpha_y), \ \mu, \alpha_x, \alpha_y \in [0, 1] \tag{7}$$

Example: for $x^{gr}(\mu, \alpha_x) = (1 + 2\mu) + (4 - 3\mu)\alpha_x$ representing trapezoidal MF $(1, 3, 4, 5)$ and $y^{gr}(\mu, \alpha_y) = (1 + \mu) + (3 - 2\mu)\alpha_y$ representing MF $(1, 2, 3, 4)$ the sum is given by (8).

$$z^{gr}(\mu, \alpha_x, \alpha_y) = x^{gr}(\mu, \alpha_x) + y^{gr}(\mu, \alpha_y) = (2 + 3\mu) + (4 - 3\mu)\alpha_x + (3 - 2\mu)\alpha_y,$$
$$\mu, \alpha_x, \alpha_y \in [0, 1] \tag{8}$$

If we are interested in the span $s(z^{gr})$ of the 4-D result granule $z^{gr}(\mu, \alpha_x, \alpha_y)$ then it can be determined with known methods of function examination, formula (9).

$$s(z^{gr}) = \left[\min_{\alpha_x, \alpha_y} z^{gr}\left(\mu, \alpha_x, \alpha_y\right), \max_{\alpha_x, \alpha_y} z^{gr}\left(\mu, \alpha_x, \alpha_y\right) \right] \tag{9}$$

In frame of the function examination its extremes should be found. The extremes can lie on borders of the function domain or inside of it, in zeroing points of derivatives in respect of RDM-variables. Because the addition result function is monotonic, its extrema lie on the domain borders. Examination of function (8) shows that its minimum corresponds to $\alpha_x = \alpha_y = 0$ and maximum to $\alpha_x = \alpha_y = 1$. Thus, the span of the result granule is finally expressed by formula (10).

$$s(z^{gr}) = [2 + 3\mu, 9 - 2\mu], \ \mu \in [0, 1] \tag{10}$$

One should noticed that span of the 4D-result $z^{gr}(\mu, \alpha_x, \alpha_y)$ is not the full result but only 2D-information about the maximal uncertainty of the result. On the basis of formula (9) determining the full, multidimensional addition result, apart from the span also other features of the result can be determined, such as cardinality distribution $card(z)$ of all possible result values, position of center of gravity (x_{CofG}, y_{CofG}), core position $core(z)$ etc. However, it is not discussed in this chapter because of its volume limitation.

Subtraction $X - Y = Z$ of two independent intervals

$$X - Y = Z : x^{gr}(\mu, \alpha_x) - y^{gr}(\mu, \alpha_y) = z^{gr}(\mu, \alpha_x, \alpha_y), \ \mu, \alpha_x, \alpha_y \in [0, 1] \quad (11)$$

x^{gr} and y^{gr} are determined by general formulas (5) and (6). If $x^{gr}(\mu, \alpha_x) = (1 + 2\mu) + (4 - 3\mu)\alpha_x$ and $y^{gr}(\mu, \alpha_y) = (1 + \mu) + (3 - 2\mu)\alpha_y$ then the subtraction result is given by (12).

$$z^{gr} = x^{gr} - y^{gr} = \mu + (4 - 3\mu)\alpha_x - (3 - 2\mu)\alpha_y, \ \mu, \alpha_x, \alpha_y \in [0, 1] \quad (12)$$

If we interested in span $s(z^{gr})$ of the 4D-result granule, then it can be determined from (13). $\text{Min}(z^{gr})$ corresponds in this formula to $\alpha_x = 0, \alpha_y = 1$ and $\max(z^{gr})$ to $\alpha_x = 1, \alpha_y = 0$.

$$s(z^{gr}) = \left[\min_{\alpha_x, \alpha_y} z^{gr}, \max_{\alpha_x, \alpha_y} z^{gr} \right] = [-3 + 3\mu, 4 - 2\mu], \ \mu \in [0, 1] \quad (13)$$

Multiplication $XY = Z$ of two independent intervals

$$XY = Z : x^{gr}(\mu, \alpha_x)y^{gr}(\mu, \alpha_y) = z^{gr}(\mu, \alpha_x, \alpha_y), \ \mu, \alpha_x, \alpha_y \in [0, 1] \quad (14)$$

x^{gr} and y^{gr} are determined by general formulas (5) and (6). If $x^{gr}(\mu, \alpha_x) = (1 + 2\mu) + (4 - 3\mu)\alpha_x$ and $y^{gr}(\mu, \alpha_y) = (1 + \mu) + (3 - 2\mu)\alpha_y$ then the multiplication result z^{gr} is given by (15).

$$z^{gr} = x^{gr} \cdot y^{gr} = [(1 + 2\mu) + (4 - 3\mu)\alpha_x][(1 + \mu) + (3 - 2\mu)\alpha_y], \ \mu, \alpha_x, \alpha_y \in [0, 1] \quad (15)$$

If we are interested in span $s(z^{gr})$ of the result, then it is expressed by formula (16), where $\min(z^{gr})$ corresponds to $\alpha_x = \alpha_y = 0$ and $\max(z^{gr})$ to $\alpha_x = \alpha_y = 1$.

$$s(z^{gr}) = \left[\min_{\alpha_x, \alpha_y} z^{gr}, \max_{\alpha_x, \alpha_y} z^{gr} \right] = [(1 + 2\mu)(1 + \mu), (5 - \mu)(4 - \mu)] \quad (16)$$

Division X/Y of two independent intervals

$$X/Y = Z : x^{gr}(\mu, \alpha_x)/y^{gr}(\mu, \alpha_y) = z^{gr}(\mu, \alpha_x, \alpha_y), \ \mu, \alpha_x, \alpha_y \in [0, 1], \ 0 \notin Y \quad (17)$$

x^{gr} and y^{gr} are determined by general formulas (5) and (6). If $x^{gr}(\mu, \alpha_x) = (1 + 2\mu) + (4 - 3\mu)\alpha_x$ and $y^{gr}(\mu, \alpha_y) = (1 + \mu) + (3 - 2\mu)\alpha_y$ then the division result z^{gr} is given by (18).

$$z^{gr} = x^{gr}/y^{gr}$$
$$= [(1 + 2\mu) + (4 - 3\mu)\alpha_x]/[(1 + \mu) + (3 - 2\mu)\alpha_y], \quad \mu, \alpha_x, \alpha_y \in [0, 1] \quad (18)$$

If we are interested in span $s(z^{gr})$ of the multidimensional result then it is expressed by (19).

$$s(z^{gr}) = \left[\min_{\alpha_x, \alpha_y} z^{gr}, \max_{\alpha_x, \alpha_y} z^{gr} \right] = [(1 + 2\mu)/(4 - \mu), (5 - \mu)/(1 + \mu)] \quad (19)$$

3 Some Mathematical Properties of MD-RDM-F Arithmetic

Commutativity
MD-RDM-F arithmetic is commutative. For any fuzzy intervals X and Y Eqs. (20) and (21) are true.

$$X + Y = Y + X \quad (20)$$

$$XY + YX \quad (21)$$

Associativity
MD-RDM-F arithmetic is associative. For any fuzzy intervals X, Y, Z Eqs. (22) and (23) are true.

$$X + (Y + Z) = (X + Y) + Z \quad (22)$$

$$X(YZ) = (XY)Z \quad (23)$$

Neutral Elements of Addition and Multiplication
In MD-RDM-F arithmetic there exist additive and multiplicative neutral elements such as degenerate fuzzy interval 0 and 1 for any interval X, as shown in Eqs. (24) and (25).

$$X + 0 = 0 + X = X \quad (24)$$

$$X \cdot 1 = 1 \cdot X = X \quad (25)$$

Inverse Elements in MD-RDM-F Arithmetic
In MD-RDM-F arithmetic an element (fuzzy interval) $-X : -x^{gr} = -[a + (b - a)\mu] - [(d - a) - \mu(d - a + b - c)]\alpha_x, \alpha_x \in [0, 1]$ is an additive inverse element

of fuzzy interval $X : x^{gr} = [a + (b - a)\mu] + [(d - a) - \mu(d - a + b - c)]\alpha_x$, $\alpha_x \in [0, 1]$. It is explained by formula (26).

$$X - X : x^{gr} - x^{gr} = \{[a + (b - a)\mu] + [(d - a) - \mu(d - a + b - c)]\alpha_x\}$$
$$- \{[a + (b - a)\mu] + [(d - a) - \mu(d - a + b - c)]\alpha_x\} = 0, \qquad (26)$$
$$\mu, \alpha_x \in [0, 1]$$

If parameters of two fuzzy intervals are equal: $a_x = a_y, b_x = b_y, c_x = c_y, d_x = d_y$, then interval $-Y$ is only then the additive inverse interval of X when also the inner RDM-variables are equal, $\alpha_x = \alpha_y$, which means full coupling (correlation) of both uncertain values modeled by the intervals. A multiplicative inverse element of fuzzy interval $X : x^{gr} = [a + (b - a)\mu] + [(d - a) - \mu(d - a) - \mu(d - a + b - c)]\alpha_x$, $\alpha_x \in [0, 1]$, if $0 \notin X$, in MD-RDM-F arithmetic is $1/X : x^{gr} = 1/ = [a + (b - a)\mu] + [(d - a) - \mu(d - a) - \mu(d - a + b - c)\alpha_x$, $\alpha_x \in [0, 1]$. The above is explained by formula (27).

$$X/X : x^{gr}/x^{gr} = \{[a + (b - a)\mu] + [(d - a) - \mu(d - a + b - c)]\alpha_x\}$$
$$/ \{[a + (b - a)\mu] + [(d - a) - \mu(d - a + b - c)]\alpha_x\} = 1, \qquad (27)$$
$$\mu, \alpha_x \in [0, 1]$$

It should be noted that if parameters of two fuzzy intervals X and Y are equal: $a_x = a_y, b_x = b_y, c_x = c_y, d_x = d_y$, then interval $1/Y$ is the multiplicative inverse interval of X only when also the inner RDM-variables are equal, $\alpha_x = \alpha_y$, which means full coupling (correlation) of both uncertain values modeled by the intervals. Full or partial couplings between uncertain variables occur in many real problems.

Sub-distributive Law
The sub-distributive law is given by (28).

$$X(Y + Z) = XY + XZ \qquad (28)$$

In MD-RDM-F arithmetic this law holds. Proof: for any three fuzzy intervals described in terms of the RDM horizontal notation,

$$X : x^{gr} = [a_x + (b_x - a_x)\mu] + [(d_x - a_x) - \mu(d_x - a_x + b_x - c_x)]\alpha_x, \ \alpha_x \in [0, 1]$$

$$Y : y^{gr} = [a_y + (b_y - a_y)\mu] + [(d_y - a_y) - \mu(d_y - a_y + b_y - c_y)]\alpha_y, \ \alpha_y \in [0, 1]$$

$$Z : z^{gr} = [a_z + (b_z - a_z)\mu] + [(d_z - a_z) - \mu(d_z - a_z + b_z - c_z)]\alpha_z, \ \alpha_z \in [0, 1]$$

analysis of the sub-distributive law results in conclusions expressed by (29).

$$X(Y+Z): x^{gr}(y^{gr} + z^{gr}) =$$
$$= \{[a_x + (b_x - a_x)\mu] + [(d_x - a_x) - \mu(d_x - c_x + b_x - a_x)]\alpha_x\}$$
$$\cdot \{[a_y + (b_y - a_y)\mu] + [(d_y - a_y) - \mu(d_y - c_y + b_y - a_y)]\alpha_y$$
$$+ [a_z + (b_z - a_z)\mu] + [(d_z - a_z) - \mu(d_z - c_z + b_z - a_z)]\alpha_z\}$$
$$= \{[a_x + (b_x - a_x)\mu] + [(d_x - a_x) - \mu(d_x - c_x + b_x - a_x)]\alpha_x\} \qquad (29)$$
$$\cdot \{[a_y + (b_y - a_y)\mu] + [(d_y - a_y) - \mu(d_y - c_y + b_y - a_y)]\alpha_y\}$$
$$+ \{[a_x + (b_x - a_x)\mu] + [(d_x - a_x) - \mu(d_x - c_x + b_x - a_x)]\alpha_x\}$$
$$\cdot \{[a_z + (b_z - a_z)\mu] + [(d_z - a_z) - \mu(d_z - c_z + b_z - a_z)]\alpha_z\}$$
$$= x^{gr}y^{gr} + x^{gr}z^{gr} : XY + XZ, \quad \mu, \alpha_x, \alpha_y, \alpha_z \in [0, 1]$$

Because in MD-RDM-F arithmetic the sub-distributive law holds transformations of equations are admissible because they do not change the result.

Cancellation Law for Addition
Cancellation law (30) for addition of fuzzy intervals holds both for the standard C-αC-F arithmetic and for MD-RDM-F one.

$$X + Z = Y + Z \Rightarrow X = Y \qquad (30)$$

Cancellation Law for Multiplication
This law has form of (31).
$$XZ = YZ \Rightarrow X = Y \qquad (31)$$

Let us assume, for simplicity, three triangle fuzzy numbers: $X = (1, 2, 3)$, $Y = (2, 2.5, 3)$ and $Z = (-1, 0, 1)$.

The fuzzy numbers are expressed in terms of horizontal MFs (32).

$$X = \{x : x = [a_x + (b_x - a_x)\mu] + (d_x - a_x)(1 - \mu)\alpha_x, \quad \mu, \alpha_x \in [0, 1]\} \qquad (32)$$

Fuzzy numbers $X = (1, 2, 3)$, $Y = (2, 2.5, 3)$, $Z = (-1, 0, 1)$ are expressed by (33).

$$X = \{x : x = (1 + \mu) + 2(1 - \mu)\alpha_x, \quad \mu, \alpha_x \in [0, 1]\}$$
$$Y = \{y : y = (2 + 0.5\mu) + (1 - \mu)\alpha_y, \quad \mu, \alpha_y \in [0, 1]\} \qquad (33)$$
$$Z = \{z : z = (-1 + \mu) + 2(1 - \mu)\alpha_z, \quad \mu, \alpha_z \in [0, 1]\}$$

Now, particular products can be calculated, (34).

$$XZ = \{xz : xz = [(1 + \mu) + 2(1 - \mu)\alpha_x][(-1 + \mu) + 2(1 - \mu)\alpha_z]\}$$
$$YZ = \{yz : yz = [(2 + 0.5\mu) + (1 - \mu)\alpha_y][(-1 + \mu) + 2(1 - \mu)\alpha_z]\} \qquad (34)$$
$$\text{where } \mu, \alpha_x, \alpha_y, \alpha_z \in [0, 1]$$

Analysis of products XZ and YZ given by (34) shows that these products are different because $Y \neq Z$. If necessary, in frame of MD-RDM-F arithmetic, on the basis of (34) spans of XZ and YZ can be calculated, formulas (35) and (36).

$$s(XZ) = \left[\min_{\alpha_x, \alpha_z} XZ, \max_{\alpha_x, \alpha_z} XZ \right] = [(3 - \mu)(-1 + \mu), (3 - \mu)(1 - \mu)] \tag{35}$$
$$\mu, \alpha_x, \alpha_z \in [0, 1]$$

Because XZ and YZ are monotonic functions, $\min(XZ)$ occurs on its domain border for $\alpha_x = 1, \alpha_y = 0$. Max$(XZ)$ occurs for $\alpha_x = 1, \alpha_z = 1$. Min$(YZ)$ occurs for $\alpha_y = 1$ and $\alpha_z = 0$, and max(YZ) for $\alpha_y = 1$ and $\alpha_z = 1, \alpha_x, \alpha_y, \alpha_z \in [0, 1]$.

$$s(YZ) = \left[\min_{\alpha_y, \alpha_z} YZ, \max_{\alpha_y, \alpha_z} YZ \right] = [(3 - \mu)(-1 + \mu), (3 - \mu)(1 - \mu)] \tag{36}$$
$$\mu, \alpha_y, \alpha_z \in [0, 1]$$

As formula (33) shows precise, multidimensional products XZ and YZ are different. Only their spans (35) and (36), in this special case, are identical. However, the spans are not precise multiplication results of the intervals, they only are information about

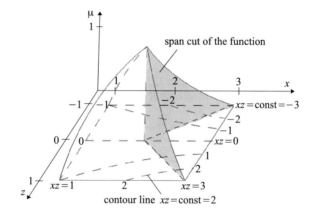

Fig. 4 3D-projection of the 4D result granule of multiplication $XZ = (1, 2, 3)(-1, 0, 1)$ of two fuzzy numbers from the full space $XZ \times \mu \times X \times Z$ on the space $\mu \times X \times Z$

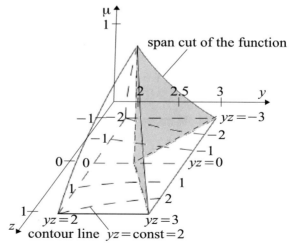

Fig. 5 3D-projection of the 4D result-granule of multiplication $YZ = (2, 2.5, 3)(-1, 0, 1)$ of two fuzzy numbers from the full space $YZ \times \mu \times Y \times Z$ on the space $\mu \times Y \times Z$

Fig. 6 Span function
min $s = f_L(\mu)$ and
max $s = f_R(\mu)$ identical for
two different 4D
multiplication results
$XZ = (1, 2, 3)(-1, 0, 1)$
and $YZ = (2, 2.5, 3)$
$(-1, 0, 1)$

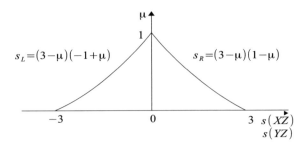

the results. To better understand this on Figs. 4 and 5 were presented 3D-projections of 4D-products XZ and YZ. Values of the fourth variable z were shown on this projections by contour lines of constant z-values. On these figures also cuts were shown which generate the span function shown in Fig. 6. The span functions $s(XZ)$ and $s(YZ)$ are identical in this case. They inform about the span width on particular μ-levels.

4   Paradox of Hukuhara Difference

Let us assume triangle fuzzy numbers $X = (4, 6, 9)$ and $Y = (0, 2, 3)$. The standard difference Z_s of these numbers has form (37).

$$Z_s = X - Y = (4, 6, 9) - (0, 2, 3) = (4 - 3, 6 - 2, 9 - 0) = (1, 4, 9) \quad (37)$$

However, equation $Z_s = X - Y$ is not equivalent to $X = Y + Z_s$ (38).

$$Y + Z_s = (0, 2, 3) + (1, 4, 9) = (1, 6, 12) \neq X = (4, 6, 9) \quad (38)$$

This was the reason of introducing by Hukuhara the second way of interval difference calculation Z_H that is achieved on the basis of equation $X = Y + Z_H$ [4, 33], later called Hukuhara difference (H-difference). Before Hukuhara, this difference had been developed by Kaucher. Hence, it also can be called Kaucher-difference.

Definition 1 Given $X, Y \in F$, the H-difference Z_H of X and Y is defined by $X \ominus Y \Leftrightarrow X = Y + Z_H$: if $X \ominus Y$ exists, it is unique and its μ-cuts are $[X - Y]_\mu$.

Example of H-difference Z_H of X and Y is given by (39).

$$X = Y + Z_H = (4, 6, 9) = (0, 2, 3) + (z_{H1}, z_{H2}, z_{H3})$$
$$= (z_{H1}, 2 + z_{H2}, 3 + z_{H3}) : Z_H = (4, 4, 6) \quad (39)$$

H-difference is frequently used in solving e.g. fuzzy differential equations [25, 33], but not only. Introducing H-difference in fuzzy arithmetic created a very strange and paradoxical situation. Now, there exist two different ways of the difference calculation: the standard way from equation $Z_s = X - Y$ and Hukuhara-way from equation $X = Y + Z_H$. However, to be on this trail, one could also add two next ways of the difference calculation: difference Z_3 calculated from equation $Y = X - Z_3$ and difference Z_4 calculated from equation $X - Y - Z_4 = (0, 0, 0)$! It is a paradoxical situation. The paradox of many various differences $Z = X - Y$ does not exist in MD-RDM-F arithmetic, where only one difference of fuzzy numbers and intervals exist, and equations $Z = X - Y$, $X = Y + Z$, $Y = X - Z$, $X - Y - Z = (0, 0, 0)$ are equivalent! What was said above will now be explained. Formulas (40) and (41) present horizontal models of fuzzy numbers $X = (4, 5, 6)$ and $Y = (0, 2, 3)$.

$$X = \{x : x = 4 + 2\mu + 5(1 - \mu)\alpha_x, \ \mu, \alpha_x \in [0, 1]\} \tag{40}$$

$$Y = \{y : y = 2\mu + 3(1 - \mu)\alpha_y, \ \mu, \alpha_y \in [0, 1]\} \tag{41}$$

Formula (42) shows their difference $Z_s = X - Y$ calculated in terms of MD-RDM-F arithmetic.

$$\begin{aligned} Z_s &= X - Y \\ &= \{z_s = x - y : x - y = [4 + 2\mu + 5(1 - \mu)\alpha_x] - [2\mu + 3(1 - \mu)\alpha_y], \\ &\quad \mu, \alpha_x, \alpha_y \in [0, 1]\} \end{aligned} \tag{42}$$

Formula (43) shows H-difference Z_H calculated from equation $X = Y + Z_H$.

$$\begin{aligned} Z_H &: X = Y + Z_H \\ &= \{z_H : x = y + z_H = [4 + 2\mu + 5(1 - \mu)\alpha_x] = [2\mu + 3(1 - \mu)\alpha_y] + z_H : \\ &\quad z_H = [4 + 2\mu + 5(1 - \mu)\alpha_x] - [2\mu + 3(1 - \mu)\alpha_y], \mu, \alpha_x, \alpha_y \in [0, 1]\} \end{aligned} \tag{43}$$

As (43) shows H-difference Z_H calculated from equation $X = Y + Z_H$ is identical as the difference Zs calculated in the standard way from equation $Z_s = X - Y$. Similar results are achieved if the difference Z is calculated from other possible formulas $Y = X - Z$ or $X - Y - Z = (0, 0, 0)$. It means that in MD-RDM-F arithmetic all possible equations forms $Z = X - Y$, $X = Y + Z$, $Y = X - Z$, $X - Y - Z = (0, 0, 0)$ are equivalent. If we are interested in span $s(Z)$ of the difference then it can be calculated from (44).

$$\begin{aligned} s(Z) &= \left[\min_{\alpha_x, \alpha_y} Z, \ \max_{\alpha_x, \alpha_y} Z \right] \\ &= \left[\min_{\alpha_x, \alpha_y} (4 + (1 - \mu)(5\alpha_x - 3\alpha_y)), \ \max_{\alpha_x, \alpha_y} (4 + (1 - \mu)(5\alpha_x - 3\alpha_y)) \right] \end{aligned} \tag{44}$$

Because $Z = f(\mu, \alpha_x, \alpha_y)$ is monotonic function, its extrema occur on its domain borders. Examination of this function shows that its minimum occurs for $\alpha_x = 0$, $\alpha_y = 1$ and the maximum for $\alpha_x = 1$ and $\alpha_y = 0$. Thus, span $s(Z)$ is determined by formula (45).

$$s(Z) = [1 + 3\mu, 9 - 5\mu] \tag{45}$$

It should be once more reminded that span $s(Z)$ of the difference $Z = f(\mu, \alpha_x, \alpha_y)$ is not the difference itself but only a 2D-representation of it.

5 Application Example of Fuzzy RDM-Arithmetic with Horizontal Membership Functions

Experts have formulated following prognoses of economic growth in a country: *"Strong economic growth (SG) will take place with a medium (M) probability, moderate growth with a less than medium (LM) probability, stabilization of the growth (ST) will occur with a small probability (S) and recession (R) with a very small (VS) probability."* Experts are also of the opinion that $M \geq LM \geq S \geq VS$ independently of how large are precise values of particular probabilities. Fuzzy definitions of understanding linguistic values of growth are given in Fig. 7 and of probability in Fig. 8.

The task consists in determining expected value of the economic growth x_{exp}^{gr} in multidimensional granular form and its 2D-representation in form of span $s(x_{exp}^{gr}) = f(\mu)$.

Horizontal MFs of particular linguistic values of economic growth are given by (46) and of probability by (47), α_i and β_j are RDM-variables.

Fig. 7 Membership functions of linguistic values of the economic growth

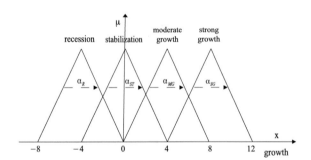

Fig. 8 Membership values of linguistic values of probability

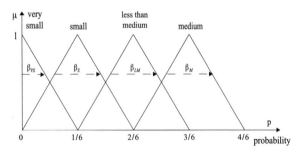

$$x_R = (-8 + 4\mu) + \alpha_R(8 - 8\mu)$$
$$x_{ST} = (-4 + 4\mu) + \alpha_{ST}(8 - 8\mu)$$
$$x_{MG} = 4\mu + \alpha_{MG}(8 - 8\mu) \tag{46}$$
$$x_{SG} = (4 + 4\mu) + \alpha_{SG}(8 - 8\mu)$$
$$\mu, \alpha_R, \alpha_{ST}, \alpha_{MG}, \alpha_{SG} \in [0, 1]$$

$$p_{VS} = [\beta_{VS}(1 - \mu)]/6$$
$$p_S = [\mu + \beta_S(2 - 2\mu)]/6$$
$$p_{LM} = [(1 + \mu) + \beta_{LM}(2 - 2\mu)]/6 \tag{47}$$
$$p_M = [(2 + \mu) + \beta_M(2 - 2\mu)]/6$$
$$\mu, \beta_{VS}, \beta_S, \beta_{LM}, \beta_M \in [0, 1]$$

Precise numerical values p_{VS}, p_S, p_{LM}, p_M are not known. However, it is known that their values have to sum to 1, condition (48).

$$p_{VS} + p_S + p_{LM} + p_M = 1 \tag{48}$$

Also, experts are of the opinion that between probability values order relations (49) exist.

$$p_S \geq p_{VS}, p_{LM} \geq p_S, p_M \geq p_{LM} \tag{49}$$

After inserting in condition (48) horizontal MFs given by formulas (47) a new RDM-form (50) of this condition is achieved.

$$\beta_{VS} + 2(\beta_S + \beta_{LM} + \beta_M) = 3$$
$$\beta_{VS}, \beta_S, \beta_{LM}, \beta_M \in [0, 1] \tag{50}$$

After inserting in conditions (49) horizontal MFs given by (47) a new RDM-form (51) of these conditions is achieved.

$$\beta_{VS} - 2\beta_S \leq \mu/(1 - \mu)$$
$$\beta_S - \beta_{LM} \leq 0.5/(1 - \mu)$$
$$\beta_{LM} - \beta_M \leq 0.5/(1 - \mu) \tag{51}$$
$$\mu, \beta_{VS}, \beta_S, \beta_{LM}, \beta_M \in [0, 1]$$

Expected value of the economic growth expresses formula (52).

$$x_{exp}^{gr} = x_R p_{VS} + x_{ST} p_S + x_{MG} p_{LM} + x_{SG} p_M \tag{52}$$

After inserting in (52) horizontal MFs given by (46) and (47) formula (53) has been achieved. It expresses the granular form of the expected growth value x_{exp}^{gr} with use of RDM-variables.

$$
\begin{aligned}
x_{\exp}^{gr} = 1/6\{&[(-8 + 4\mu) + \alpha_R(8 - 8\mu)][\beta_{VS}(1 - \mu)] \\
&+ [(-4 + 4\mu) + \alpha_{ST}(8 - 8\mu)][\mu + \beta_S(2 - 2\mu)] \\
&+ [4\mu + \alpha_{MG}(8 - 8\mu)][(1 + \mu) + \beta_{LM}(2 - 2\mu)] \\
&+ [(4 + 4\mu) + \alpha_{SG}(8 - 8\mu)][(2 + \mu) + \beta_M(2 - 2\mu)]\} \\
&\mu, \alpha_R, \alpha_{ST}, \alpha_{MG}, \alpha_{SG}, \beta_{VS}, \beta_S, \beta_{LM}, \beta_M \in [0, 1]
\end{aligned}
\tag{53}
$$

Formula (53) shows how simply calculations with horizontal MFs can be made. They are inserted in mathematical formulas as usual numbers in formulas of the classic mathematics. Formula (53) defines a multidimensional information granule $x_{\exp}^{gr} = f(\mu, \alpha_R, \alpha_{ST}, \alpha_{MG}, \alpha_{SG}, \beta_{VS}, \beta_S, \beta_{LM}, \beta_M)$ or with other words, the full, multidimensional set of possible values of the expected growth x. They can be generated only by values of RDM-variables which satisfy condition $\mu, \alpha_R, \alpha_{ST}, \alpha_{MG}, \alpha_{SG}$, $\beta_{VS}, \beta_S, \beta_{LM}, \beta_M \in [0, 1]$ and conditions (50), (51). For example, set of RDM-values $\alpha_R = \alpha_{ST} = \alpha_{MG} = \alpha_{SG} = 1/8$ and $\beta_{VS} = \beta_S = \beta_M = 1/3$ and $\beta_{LM} = 2/3$ satisfies the required conditions and generates for the level $\mu = 0$ one of possible growth value $x = 1.9$. Another value set of $\mu, \alpha_R, \alpha_{ST}, \alpha_{MG}, \alpha_{SG}, \beta_{VS}, \beta_S, \beta_{LM}, \beta_M$ will generate, in the general case, another value of possible growth x. The set of possible point-solutions $x_{\exp}^{gr} = f(\mu, \alpha_R, \alpha_{ST}, \alpha_{MG}, \alpha_{SG}, \beta_{VS}, \beta_S, \beta_{LM}, \beta_M)$ determined by (53) and conditions (50), (51) creates in the space information granule of irregular form which is not a hyper-cubicoid. This multidimensional granule is difficult to imagine. Hence, scientists usually try to determine its simplified 2D-representation $s(x_{\exp}^{gr})$ called "span" or "spread" being the widest 2D-cut through the solution granule or its widest 2D-shadow. It is expressed by (54).

$$
\begin{aligned}
s\left(x_{\exp}^{gr}\right) = [&\min x_{\exp}^{gr}(\mu, \alpha_R, \alpha_{ST}, \alpha_{MG}, \alpha_{SG}, \beta_{VS}, \beta_S, \beta_{LM}, \beta_M), \\
&\max x_{\exp}^{gr}(\mu, \alpha_R, \alpha_{ST}, \alpha_{MG}, \alpha_{SG}, \beta_{VS}, \beta_S, \beta_{LM}, \beta_M)] \\
&\mu, \alpha_R, \alpha_{ST}, \alpha_{MG}, \alpha_{SG}, \beta_{VS}, \beta_S, \beta_{LM}, \beta_M \in [0, 1]
\end{aligned}
\tag{54}
$$

Values $\min x_{\exp}^{gr}$ and $\max x_{\exp}^{gr}$ have to be determined with taking into account conditions (50) and (51). Analysis of (53) allows for detection that $\min x_{\exp}^{gr}$ occurs only for $\alpha_R = \alpha_{ST} = \alpha_{MG} = \alpha_{SG} = 0$ and $\max x_{\exp}^{gr}$ for $\alpha_R = \alpha_{ST} = \alpha_{MG} = \alpha_{SG} = 1$. Hence, formula (54) can be simplified to (55).

$$
\begin{aligned}
s\left(x_{\exp}^{gr}\right) = [&\min x_{\exp}^{gr}(\mu, 0, 0, 0, 0, \beta_{VS}, \beta_S, \beta_{LM}, \beta_M), \\
&\max x_{\exp}^{gr}(\mu, 1, 1, 1, 1, \beta_{VS}, \beta_S, \beta_{LM}, \beta_M)] \\
&\mu, \beta_{VS}, \beta_S, \beta_{LM}, \beta_M \in [0, 1]
\end{aligned}
\tag{55}
$$

Determining optimal values of RDM-variables $\beta_{VS}, \beta_S, \beta_{LM}, \beta_M$ is not always easy because they are coupled (correlated) by conditions (50) and (51). Determining of their optimal values can be realized analytically (in this case extremes occur on borders of the solution domain). However, analytical domain examination for various levels of membership μ would take too much work. Therefore, an easier way can be used: numerical examination of the domain with MATLAB for particular variables $\mu, \beta_{VS}, \beta_S, \beta_{LM}, \beta_M$ changed with appropriately small descretization step $\Delta \le 0.01$.

In the result of such a numerical searching the span-MF has been found as shown in Fig. 9.

For example, on the membership level $\mu = 0$ the minimal value of $x = -1$ and it occurs for $\alpha_R = \alpha_{ST} = \alpha_{MG} = \alpha_{SG} = 0$ and $\beta_{VS} = 1$, $\beta_S = 0.75$, $\beta_{LM} = 0.25$, $\beta_M = 0$ and the maximal value of $x = 10.67$ and it occurs for $\alpha_R = \alpha_{ST} = \alpha_{MG} = \alpha_{SG} = 1$ and $\beta_{VS} = \beta_S = 0$, $\beta_{LM} = 0.5$, $\beta_M = 1$. Another simple 1D-representation of the solution span can be position of its center of gravity $x_{CoG} = 4.96$ or position $x_{\mu=1} = 5.33$ of the highest membership of the span (Fig. 9) and of the solution granule x_{exp}^{gr}. However, one should always realize that a simplified representation of the solution is not the solution itself and that without the full-dimensional solution set x_{exp}^{gr} it is not possible (in the general case) to precisely determine solution representations. Figure 10 shows span $s(x_{exp}^{gr})$ of the solution granule x_{exp}^{gr} determined without constraint conditions (50) and (51).

One can see (compare supports in Figs. 9 and 10) that uncertainty of the span $s(x_{exp}^{gr})$ determined without taking into account constraint conditions (50) and (51) is greater than when these conditions are taken into account (for $\mu = 0$ supports' widths are correspondingly equal to 14.66 and 11.67). Thus, application of RDM-variables allow for decreasing of solutions uncertainty because they allow for modeling and taking into account dependences existing between variables occurring in problems.

Fig. 9 Membership function of the span $s(x_{exp}^{gr})$ of the multidimensional solution granule x_{exp}^{gr} of the expected economic growth $x[\%]$ achieved with taking into account constraint conditions (50) and (51), CoG—center of gravity

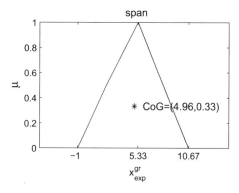

Fig. 10 Membership function of the span $s(x_{exp}^{gr})$ of the multidimensional solution granule x_{exp}^{gr} of the expected economic growth $x[\%]$ achieved without taking into account constraint conditions (50) and (51)

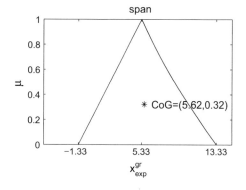

6 Conclusions

The paper shows how in terms of MD-RDM-F arithmetic basic operations as addition subtraction, multiplication and division should be carried out. It also shows certain significant properties of this arithmetic. Arithmetic operations are carried out with use of horizontal MFs, in an easy way, without the extension principle. New MD-RDM-F arithmetic has many advantages. Not all of them could be presented in this chapter because of its volume limitation. MD-RDM-F arithmetic delivers full-dimensional, precise problems' solutions that are multidimensional information granules. Without possessing this full-dimensional solution one cannot precisely and uniquely determine the solution span that is measure of the solution uncertainty. Instead, the span is what standard methods of fuzzy arithmetic try directly to determine. Unfortunately, in the general case it is not possible. Therefore the present fuzzy arithmetic is characterized by many paradoxes, as e.g. paradox of Hukuhara difference. On the basis of the full-dimensional solution achieved with MD-RDM-F arithmetic not only precise and unique solution span can be determined but also other solution representations as 2D cardinality distribution of all possible solutions, the core of the solution, center of gravity (*CofG*) of the solution, etc. MD-RDM-F arithmetic has such important mathematical properties as inverse additive element, inverse multiplicative element. In this arithmetic hold such very important laws as sub-distributive law $X(Y + Z) = XY + XZ$ and cancellation law $IF(XZ = YZ)THEN(X = Y)$. With use of MD-RDM-F arithmetic both linear and nonlinear multidimensional equations and equation systems can relatively easily be solved. In solving them very helpful are RDM-variables that allow for taking into account couplings existing between particular intervals. These possibilities will be shown in next publications of authors.

References

1. Abbasbandy, S., Asady, B.: Newton's method for solving fuzzy nonlinear equations. Applied Mathematics and Computation 159, 349–356 (2004).
2. Abbasbandy, S., Jafarian, A.: Steepest descent method for system of fuzzy linear equations. Applied Mathematics and Computation 175, 823–833 (2006).
3. Aliev, R.A., Pedrycz, W., Fazlollahi, B., Huseynow, O., Alizadeh, A., Gurimov, B.: Fuzzy logic-based generalized decision theory with imperfect information. Information Sciences 189, 18–42 (2012).
4. Aliev, R.A.: Fundamentals of the fuzzy logic-based generalized theory of decisions. Springer-Verlag, Berlin, Heidelberg, 2013.
5. Bhiwani, R.J., Patre, B.M.: Solving first order fuzzy equations: a modal interval approach. Proceedings of Second International Conference on Emerging Trends and Technology ICETET-09, IEEE Computer Society, 953–956 (2009).
6. Boading, L.: Uncertainty theory. 2nd ed., Springer-Verlag, 2007.
7. Buckley, J.J., Qu, Y.: Solving linear and Quadratic Fuzzy Equations. Fuzzy Sets and Systems, Vol. 38, 43–59 (1990).
8. Buckley, J.J., Feuring, T., Hayashi, Y.: Solving Fuzzy Equations using Evolutionary Algorithms and Neural Nets. Soft Computing - A Fusion of Foundations, Methodologies and Applications, Vol. 6, No. 2, 116–123 (2002).

9. Chen, M.-C., Wang W.-Y., Su, S.-F., Chien, Y.-H.: Robust T-S fuzzy-neural control of a uncertain active suspension systems. International Journal of Fuzzy Systems, Vol. 12, No. 4, 321–329 (December 2010).
10. Dubois, D., Prade, H.: Operations on fuzzy numbers. International Journal of Systems Science 9(6), 613–626 (1978).
11. Dubois, D., Fargier, H., Fortin, J.: A generalized vertex method for computing with fuzzy intervals. In: Proceedings of the IEEE International Conference on Fuzzy Systems, 541–546 (2004).
12. Dutta, P., Boruah, H., Al, T.: Fuzzy arithmetic with and without using α-cut method: A comparative study. International Journal of Latest Trends in Computing. Vol. 2, Issue 1, 99–107 (March 2011).
13. Dymova L.: Fuzzy solution of interval nonlinear equations. Parallel Processing and Applied Mathematics, LCNS 6068, Part II, Springer, 418–426 (2010).
14. Dymova, L.: Soft computing in economics and finance. Springer-Verlag, Berlin, Heidelberg, 2011.
15. Fodor, J., Bede, B.: Arithmetics with fuzzy numbers: a comparative overview. Proceedings of 4th Hungarian Joint Symposium on Applied Machine Intelligence, Herlany, Slovakia, 54–68 (2006).
16. Hanss, M.: On the implementation of fuzzy arithmetical operations for engineering problems. Proceedings of the 18th International Conference of the North American Fuzzy Information Processing Society - NAFIPS'99, New York, 462–466 (1999).
17. Hanss, M.: Applied fuzzy arithmetic. Springer-Verlag, Berlin, Heidelberg, 2005.
18. Kaufmann, A., Gupta, M.M.: Introduction to fuzzy arithmetic. Van Nostrand Reinhold, New York, 1991.
19. Klir, G.J.: Fuzzy arithmetic with requisite constraints. Fuzzy Sets and Systems, 165–175 (1997).
20. Landowski, M.: Differences between Moore and RDM Interval Arithmetic. Proceedings of the 7th International Conference Intelligent Systems IEEE IS'2014, September 24–26, 2014, Warsaw, Poland, Volume 1: Mathematical Foundations, Theory, Analyses, Intelligent Systems'2014, Advances in Intelligent Systems and Computing, Springer International Publishing Switzerland, Vol. 322, 331–340 (2015).
21. Li, Q.X., Liu, Si.F.: The foundation of the grey matrix and the grey input-output analysis. Applied Mathematical Modelling 32, 267–291 (2008).
22. Moore, R.E., Lodwick, W.: Interval analysis and fuzzy set theory. Fuzzy Sets and Systems 135, 5–9 (2003).
23. Moore, R.E.: Interval analysis. Prentice-Hall, Englewood Cliffs, NJ, 1966.
24. Moore, R.E., Kearfott, R.B., Cloud, J.M..: Introduction to interval analysis. SIAM, Philadelphia, 2009.
25. Pedrycz, W., Skowron, A., Kreinovich, V. (eds.): Handbook of granular computing. John Wiley & Sons, Chichester, 2008.
26. Piegat, A.: Fuzzy modeling and control. Phisica-Verlag, Heidelberg, New York, 2001.
27. Piegat, A.: Cardinality Approach to Fuzzy Number Arithmetic. IEEE Transactions on Fuzzy Systems, Vol. 13, No. 2, 204–215 (2005).
28. Piegat, A., Landowski, M.: Is the conventional interval-arithmetic correct? Journal of Theoretical and Applied Computer Science, Vol. 6, No. 2, 27–44 (2012).
29. Piegat, A., Tomaszewska, K.: Decision-Making under uncertainty using Info-Gap Theory and a New Multi-dimensional RDM interval arithmetic. Przeglaad Elektrotechniczny (Electrotechnic Review), R.89, No. 8, 71–76 (2013).
30. Piegat, A., Landowski, M.: Multidimensional approach to interval uncertainty calculations. In: New Trends in Fuzzy Sets, Intuitionistic: Fuzzy Sets, Generalized Nets and Related Topics, Volume II: Applications, ed. K.T. Atanassov et al., IBS PAN - SRI PAS, Warsaw, Poland, 137–151 (2013).
31. Piegat, A., Landowski, M.: Two Interpretations of Multidimensional RDM Interval Arithmetic—Multiplication and Division. International Journal of Fuzzy Systems, vol. 15 no. 4, Dec. 2013, 488–496 (2013).

32. Piegat, A., Landowski, M.: Correctness-checking of uncertain-equation solutions on example of the interval-modal method. In: Atanassov T. et al., Modern Approaches in Fuzzy Sets, Intuitionistic Fuzzy Sets, Generalized Nets and Related Topics, Volume I: Foundations, SRI PAS - IBS PAN, Warsaw, Poland, 159–170 (2014).
33. Stefanini, L.: A generalization of Hukuhara difference and division for interval and fuzzy arithmetic. Fuzzy Sets and Systems, Vol. 161, Issue 11, 1564–1584 (2010).
34. Stefanini, L.: New tools in fuzzy arithmetic with fuzzy numbers. In: Hullermeier E., Kruse R., Hoffmann F. (eds), IPMU 2010, Part II, CCIS 81, Springer-Verlag, Berlin, Heidelberg, 471–480 (2010).
35. Tomaszewska, K.: The application of horizontal membership function to fuzzy arithmetic operations. Journal of Theoretical and Applied Computer Science, Vol. 8, No. 2, 3-10 (2014).
36. Tomaszewska, K., Piegat, A.: Application of the horizontal membership function to the uncertain displacement calculation of a composite massless rod under a tensile load. Proceedings of International Conference Advanced Computer Systems (ACS 2014), Miedzyzdroje, Poland, October 22–24, 2014.
37. Zadeh, L.A.: Fuzzy logic = computing with words. IEEE Transactions on Fuzzy Systems 4(2), 103–111 (1996).
38. Zadeh, L.A.: From computing with numbers to computing with words - from manipulation of measurements to manipulation of perceptions. International Journal of Applied Mathematics and Computer Science 12(3), 307–324 (2002).
39. Zhow, Ch.: Fuzzy-arithmetic-based Lyapunov synthesis in the design of stable fuzzy controllers: a Computing with Words approach. International Journal of Applied Mathematics and Computer Science (AMCS), Vol. 12, No. 3, 411–421 (2002).

Copula as a Bridge Between Probability Theory and Fuzzy Logic

Germano Resconi and Boris Kovalerchuk

Abstract This work shows how dependence in many-valued logic and probability theory can be fused into one concept by using copulas and marginal probabilities. It also shows that the t-norm concept used in fuzzy logic is covered by this approach. This leads to a more general statement that axiomatic probability theory covers logic structure of fuzzy logic. This paper shows the benefits of using structures that go beyond the simple concepts of classical logic and set theory for the modeling of dependences.

1 Introduction

Modern approaches for modeling uncertainty include Probability Theory (PT), Many-Valued Logic (MVL), Fuzzy Logic (FL), and others. PT is more connected with the classical set theory, classical propositional logic, and FL is more connected with fuzzy sets and MVL. The link of the concept of probability with the classical set theory and classical propositional logic is relatively straightforward for the case of a *single variable* that is based on the use of set operations and logic propositions. In contrast, modeling joint probability for *multiple variables* is more challenging because it involves complex *relations* and *dependencies* between variables. One of the goals of this work is to clarify the *logical structure* of a joint probability for multiple variables. Another goal is to *clarify links* between different logics and joint probability for multiple variables.

Differences between concepts of probability, fuzzy sets, possibility and other uncertainty measures have been studied for a long time [1]. Recently several new attempts have been made to connect the concepts of probability, possibility, and

G. Resconi (✉)
Department of Mathematics and Physics, Catholic University, 25121 Brescia, Italy
e-mail: resconi@speedyposta.it

B. Kovalerchuk
Department of Computer Science, Central Washington University,
Ellensburg, WA 98926-7520, USA
e-mail: borisk@cwu.edu

© Springer International Publishing AG 2017
V. Kreinovich (ed.), *Uncertainty Modeling*, Studies in Computational
Intelligence 683, DOI 10.1007/978-3-319-51052-1_15

fuzzy logic into a single theory. Dubois, Prade and others [2–7] discuss conditional probability and subjective probability as a mixed concept that includes both fuzzy and probability concepts.

In [8–11] we proposed the Agent-based Uncertainty Theory (AUT) with a model for fuzzy logic and many-valued logic that involves agents and conflicts among agents. Later we extended AUT, providing a more formal description of uncertainty as the conflict process among agents with the introduction of a: new concept denoted as an *active set* [12]. This agent-based approach to uncertainty has a connection with the recent works in the theory of team semantics [13], and dependence/independence logic [13, 14].

The *dependence logic* [13–21] assumes that the logic truth value of a particular proposition *depends* on the truth value of other propositions. Dependence logic is based on *team semantics*, in which the truth of formulas is evaluated in sets of assignments provided by a team, instead of single assignments, which is common in the classical propositional logic, which is *truth functional*. The basic idea of dependence logic is that certain properties of sets *cannot be expressed* merely in terms of properties satisfied by each set individually. This situation is more complex than that which is considered in the classical logic approach. It is closer to the agent-based uncertainty theory [8], and the active sets [12] that deal with the sets of assignments, generated by teams of agents, which can conflict and produce uncertainty of the truth value, leading to a many-valued logic. Dependence logic is a special type of the first order predicate logic. While the goal to go beyond truth-functionality is quite old in fuzzy logic [22] and the dependence logic is not the first attempt outside of the fuzzy logic, the major issue is that to reach this goal in practice we need more data than truth-functional algorithms use. The current survey of attempts to go beyond truth-functionality is presented in [23].

In PT and statistics, a *copula* is a multivariate probability distribution function for "normalized" variables [24]. In other words, copula is a multivariate composition of marginal probabilities. The copula values represent degrees of dependence among values of these variables when the variables are "cleared" from differences in scales and frequencies of their occurrence. This "clearing" is done by "normalizing" arguments of copula to uniform distributions in [0, 1] interval.

Quite often the probabilistic approach is applied to study the frequency of independent phenomena. In the case of dependent variables, we cannot derive a joint probability $p(x_1, x_2)$ as a product of independent probabilities, $p(x_1)p(x_2)$ and must use the multidimensional probability distribution with dependent valuables. The common technique for modeling it is a Bayesian network. In the Bayesian approach, the evidence about the true state of the world is expressed in terms of degrees of belief in the form of Bayesian conditional probabilities. These probabilities can be causal or just correlational.

In the probability theory, the conditional probability is the main element to express the dependence or inseparability of the two states x_1 and x_2. The joint probability $p(x_1, x_2, \ldots, x_n)$ is represented via multiple conditional probabilities to express the dependence between variables. The *copula approach* introduces a *single* function $c(u_1, u_2)$ denoted as *density of copula* [25] as a way to model the *dependence* or *inseparability* of the variables with the following property in the case of two variables. The copula allows representing the joint probability $p(x_1, x_2)$ as a combination (product) of single dependent part $c(u_1, u_2)$ and independent parts: probabilities $p_1(x_1)$ and $p_2(x_2)$.

It is important to note that probabilities $p_1(x_1)$ and $p_2(x_2)$ belong to two different pdfs which is expressed by their indexes in p_1 and p_2. Below to simplify notation we will omit these indexes in similar formulas.

The investigation of copulas and their applications is a rather recent subject of mathematics. From one point of view, copulas are functions that join or 'couple' one-dimensional distribution functions u_1 and u_2 and the corresponding joint distribution function.

Copulas are of interest not only in probability theory and statistics, but also in many other fields requiring the aggregation of incoming data, such as multi-criteria decision making [26], and probabilistic metric spaces [27, 28]. Associative copulas are special continuous triangular norms (t-norms for short, [29, 30]). They are studied in several domains such as many-valued logic [31], fuzzy logic and sets [32], agent-based uncertainty theory and active sets, along with t-conorms to model the uncertainty.

2 Copula: Notation, Definitions, Properties, and Examples

2.1 Notation, Definitions and Properties

Definitions and properties below are based on [24, 33–36].
   A *joint probability distribution* is

$$p(x_1 x_2, \ldots x_n) = $$
$$p(x_1)p(x_2|x_1)p(x_3|x_1, x_2)\ldots p(x_n|x_1, x_2, \ldots, x_{n-1}).$$

A function $c(u_1, u_2)$ is a *density of copula* for $p(x_1, x_2)$ if

$$p(x_1, x_2) = c(u_1 u_2)p(x_1)p(x_2) = p(x_1)p(x_2|x_1)$$

where u_1 and u_2 are *inverse functions*,

$$\frac{du_1(x_1)}{dx_1} = p(x_1), u_1(x_1) = \int_{-\infty}^{x_1} p(s)ds,$$

$$\frac{du_2(x_2)}{dx_2} = p(x_2), u_2(x_2) = \int_{-\infty}^{x_2} p(r)dr$$

To simplify notation of $u_1(x_1)$, $u_2(x_2)$, ..., $u_n(x_n)$ sometimes we will write u_1, u_2, ..., u_n omitting arguments.

A *cumulative function* $C(u_1(x_1), u_2(x_2))$ with inverse functions u_i as arguments is called a *copula* and its derivative $c(u_1, u_2)$ is called a *density of copula*:

$$C(u_1(x_1), u_2(x_2)) = \int_{-\infty}^{x_1} \int_{-\infty}^{x_2} c(u_1(s), u_2(r)) p(s) p(r) ds dr =$$

$$= \int_{-\infty}^{x_1} \int_{-\infty}^{x_2} c(u_1(s), u_2(r)) du_1(s) du_2(r)$$

$$c(u_1, u_2) = \frac{\partial^2 C(u_1, u_2)}{\partial u_1 \partial u_2}$$

Copula properties for 2-D case:

$$p(x_1) p(x_2|x_1) = \frac{\partial^2 C(u_1, u_2)}{\partial u_1 \partial u_2} p(x_1) p(x_2)$$

$$p(x_2|x_1) = \frac{\partial^2 C(u_1, u_2)}{\partial u_1 \partial u_2} p(x_2)$$

Copula properties for 3-D case:
Joint density function:

$$p(x_1, x_2, x_n) = p(x_1) p(x_2|x_1) p(x_3|x_1, x_2) =$$
$$c(u_1, u_2, u_3) p(x_1) p(x_2) p(x_3)$$

Copula, copula density, joint and conditional probabilities.
Below in $c(u_1, u_2, u_3)$ we omitted arguments s_1, s_2, s_3 of u_1, u_2, u_3 for simplicity.

$$C(u_1(x_1), u_2(x_2), u_3(x_3)) = \int_{-\infty}^{x_1} \int_{-\infty}^{x_2} \int_{-\infty}^{x_3} p(s_1, s_2, s_3) ds_1 ds_2 ds_3 =$$

$$\int_{-\infty}^{x_1} \int_{-\infty}^{x_2} \int_{-\infty}^{x_3} c(u_1, u_2, u_3) p(s_1) p(s_2) p(s_3) ds_1 ds_2 ds_3 =$$

$$\int_{-\infty}^{x_1} \int_{-\infty}^{x_2} \int_{-\infty}^{x_3} p(s_1) p(s_2|s_1) p(s_3|s_1, s_2) ds_1 ds_2 ds_3$$

$$p(x_1) = \frac{du_1}{dx_1}, \ p(x_2) = \frac{du_2}{dx_2}, \ p(x_3) = \frac{du_3}{dx_2}$$

$$C(u_1, u_2, u_3) = \int\limits_{-\infty}^{x_1} \int\limits_{-\infty}^{x_2} \int\limits_{-\infty}^{x_3} c(u_1, u_2, u_3) p(s_1) p(s_2) p(s_3) ds_1 ds_2 ds_3$$

$$c(u_1, u_2, u_3) = \frac{\partial^3 C(u_1, u_2, u_3)}{\partial u_1 \partial u_2 \partial u_3}$$

$$p(x_1, x_2, x_n) = p(x_1) p(x_2|x_1) p(x_3|x_1, x_2) =$$

$$\frac{\partial^3 C(u_1, u_2, u_3)}{\partial u_1 \partial u_2 \partial u_3} p(x_1) p(x_2) p(x_3)$$

$$p(x_2|x_1) p(x_3|x_1, x_2) = \frac{\partial^2 C(u_1, u_2, u_3)}{\partial u_1 \partial u_2} p(x_2) p(x_3|x_1, x_2) =$$

$$\frac{\partial^3 C(u_1, u_2, u_3)}{\partial u_1 \partial u_2 \partial u_3} p(x_2) p(x_3)$$

$$\frac{\partial^2 C(u_1, u_2, u_3)}{\partial u_1 \partial u_2} p(x_3|x_1, x_2) = \frac{\partial^3 C(u_1, u_2, u_3)}{\partial u_1 \partial u_2 \partial u_3} p(x_3)$$

$$p(x_3|x_1, x_2) = \frac{\partial^3 C(u_1, u_2, u_3)}{\partial u_1 \partial u_2 \partial u_3} \frac{1}{\frac{\partial^2 C(u_1, u_2, u_3)}{\partial u_1 \partial u_2}} p(x_3)$$

Copula properties and definitions for general n-D case:

$$p(x_n|x_1, x_2, \ldots, x_{n-1}) =$$

$$= \frac{\partial^n C(u_1, u_2, \ldots, u_n)}{\partial u_1, \ldots, \partial u_n} \frac{1}{\frac{\partial^{n-1} C(u_1, u_2, \ldots, u_n)}{\partial u_1, \ldots, \partial u_{n-1}}} p(x_n)$$

Conditional copula:

$$c(x_n|x_1, x_2, \ldots, x_{n-1}) =$$

$$= \frac{\partial^n C(u_1, u_2, \ldots, u_n)}{\partial u_1, \ldots, \partial u_n} \frac{1}{\frac{\partial^{n-1} C(u_1, u_2, \ldots, u_n)}{\partial u_1, \ldots, \partial u_{n-1}}}$$

Density of copula:

$$\frac{\partial^n C(u_1, u_2, \ldots, u_n)}{\partial u_1, \ldots, \partial u_n} = c(u_1, u_2, \ldots, u_n)$$

2.2 Examples

When $u(x)$ is a marginal probability $F_i(x)$, $u(x) = F_i(x)$ and u is uniformly distributed, then the inverse function $x(u)$ is not uniformly distributed, but has values concentrated in the central part as in the Gaussian distribution. The inverse process is represented graphically in Figs. 1 and 2.

Consider another example where a joint probability density function p is defined in the two dimensional interval $[0, 2] \times [0, 1]$ as follows,

$$p(x, y) = \frac{x + y}{3}$$

Then the cumulative function in this interval is

$$C(x, y) = \int\limits_{-\infty}^{x} \int\limits_{-\infty}^{y} p(s, r)dsdr = \int\limits_{-\infty}^{x} \int\limits_{-\infty}^{y} \frac{s + r}{3} dsdr = \frac{xy(x + y)}{6} \tag{1}$$

Next we change the reference (x, y) into the reference (u_1, u_2). For

$$p(x_1) = \frac{du_1}{dx_1}, \quad p(x_2) = \frac{du_2}{dx_2}$$

Fig. 1 Relation between marginal probability $F_i(x)$ and the random variable x

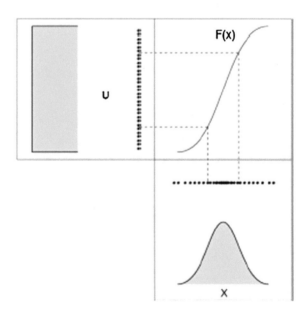

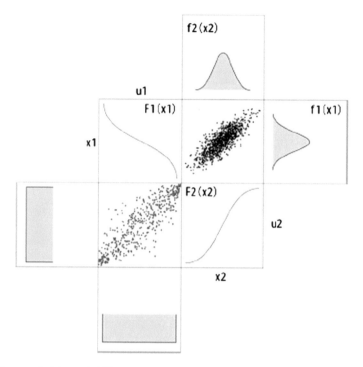

Fig. 2 Symmetric joint probability and copula

and use the marginal probabilities

$$u_1(x) = \int_{-\infty}^{x} p(s)ds, u_2(y) = \int_{-\infty}^{y} p(r)dr$$

to get

$$u_1(x) = C(x, 1) = \frac{x(x + 1)}{6}, \ x \in [0, 2]$$

$$u_2 = C(2, y) = \frac{y(y + 2)}{6} \ y \in [0, 1]$$

This allows us to compute the inverse function to identify the variables x and y as functions of the marginal functions u_1 and u_2:

$$x(u_1) = \frac{\sqrt{1 + 24u_1} - 1}{2} \quad y(u_2) = \sqrt{1 + 3u_2} - 1$$

Then these values are used to compute the copula C in function (1),

$$C(u_1, u_2) =$$

$$\frac{\sqrt{1+24u_1}-1}{2}(\sqrt{1+3u_2}-1)\frac{\dfrac{\sqrt{1+24u_1}-1}{2}+(\sqrt{1+3u_2}-1)}{6} \qquad (2)$$

$$= \frac{\sqrt{1+24u_1}-1}{2}(\sqrt{1+3u_2}-1)\frac{\dfrac{\sqrt{1+24u_1}-1}{2}+(\sqrt{1+3u_2}-1)}{6}$$

3 Approach

3.1 Meaning of Copulas

The main idea of the copula approach is *transforming a multivariate probability distribution* $F(x_1, x_2, \ldots, x_n)$ into a "normalized" multivariate distribution called a *copula*. The normalization includes transforming each of its arguments x_i into the [0, 1] interval and transforming the marginal distributions $F_i(x_i)$ into uniform distributions. The main result of the copula approach (Sklar's Theorem [24, 35, 36]) is the basis for this. It splits modeling the marginal distributions $F_i(x)$ from the modeling of the *dependence structure* that is presented by a normalized multivariate distribution (copula).

What is benefit of splitting? The copula is invariant under a strictly increasing transformation. Thus variability of copulas is less than the variability of n-D probability distributions. Therefore the selection of the copula can be simpler than the selection of a generic n-D probability distribution.

Next building a multivariate probability distribution, using the copula, is relatively simple in two steps [37, 38]:

(1) Selecting the univariate margins by transforming each of the one-factor distributions to be a uniform by setting $u_i = F_i(x_i)$ where the random variable u_i is from [0, 1].
(2) Selecting a copula connecting them by using the formula (3).

$$C(u_1, u_2, \ldots, u_n) = F^{-1}(F_1^{-1}(u_1), F_2^{-1}(u_2), \ldots, F_n^{-1}(u_n)) \qquad (3)$$

These selections are done by using available data and a collection of parametric models for the margins $F_i(x_i)$ and the copula to fit the margins and the copula parameters with that data.

Next the property $\dfrac{\partial^n C(u_1, u_2, \ldots, u_n)}{\partial u_1, \ldots, \partial u_n} = c(u_1, u_2, \ldots, u_n)$ is used to produce pdf

$$p(x_1, x_2, \ldots, x_n) = c(u_1, u_2, \ldots, u_n)p(x_1)p(x_2)\ldots p(x_n)$$

where c is a copula density. Another positive aspect of using copulas is the abilities to generate a significant number of samples of n-D vectors that satisfy the copula probability distribution. After a copula in a "good" analytical form is approximated from the limited given data, a copula distribution can be simulated to generate more samples of data distributed according to it.

This is a known practical value of copulas. Does it ensure the quality/value of the produced multivariate distribution? It is noted in [24] that copulas have limited capabilities for constructing *useful* and *well–understood* multivariate pdfs, and much less for multivariate stochastic processes, quoting [39]: "Religious Copularians have *unshakable faith* in the value of transforming a multivariate distribution to its copula". These authors expect future techniques that will go beyond copulas. Arguments of heated discussions on copulas can be found in [40]. The question "which copula to use?" has no obvious answer [24]. This discussion reminds us a similar discussion in the fuzzy logic community on justification of t-norms in fuzzy logic.

3.2   Concept of NTF-Logical Dependence

Dependences traditionally are modeled by two distinct types of techniques based on the probability theory or logic (classical and/or many-valued). It is commonly assumed that dependencies are either stochastic or non-stochastic. Respectively it is assumed that if the dependence is not stochastic then it must be modeled by methods based on the classical and many-valued logic not by methods based on the probability theory.

This is an unfortunate and *too narrow* interpretation of probabilistic methods. In fact Kolmogorov's measures (K-measures) from the probability theory allow modeling both stochastic and non-stochastic "logical" dependencies (both classical and multi-valued). The basis for this is simply considering these measures as general measures of dependence, not as the frequency-based valuations of dependence. For instance, we can measure the dependence between two circles by computing a ratio S/T, where S is the area of their overlap and T is the total area of two circles. This ratio satisfies all of Kolmogorov's axioms of the probability theory, which is a "deterministic" method to measure the dependence in this example without any frequency.

Moreover, many dependencies are a mixture of stochastic and non-stochastic components. These dependencies also can be modeled by K-measures. Below we show this. Consider a dependency that is a mixture of stochastic and non-stochastic components. We start from a frequency-based joint probability distribution $F(x_1, \ldots, x_n)$ and *extract* from it a *dependence* that is not a *stochastic* dependence, but is a "*logical*", "deterministic" dependence. How do we define a logical dependence?

We will call dependence between variables $x_1, \ldots, x_n$ a Non-Truth-Functional *Logical Dependence* (**NTF-logical dependence**) if it is derived from F, but *does not* depend on the frequencies of occurrence of $x_1, \ldots, x_n$.

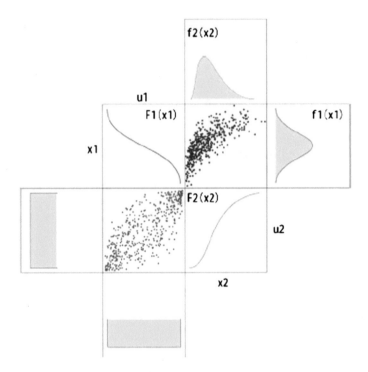

Fig. 3 Asymmetric joint probability and symmetric copula

How do we capture this logical dependence? We use a copula approach outlined above. First we convert (*normalize*) all marginal distributions $F_i(x_i)$ of F to uniform distributions $u_1, \ldots, u_n$ in [0, 1]. This will "eliminate" the dependence from frequencies of variables and differences in scales of variables. Next we build a copula: dependence C between $u_1, \ldots, u_n$, C: $\{(u_1, \ldots, u_n)\} \to [0, 1]$.

Figure 3 illustrates the difference between two dependencies. The first one is captured by F and the second one is captured by C. Copula C tells us that the logical dependence between x_1 and x_2 is symmetric, but differences in frequencies make the dependence asymmetric that is captured by F.

A similar idea is used in *machine learning* and *data mining* for discovering dependencies in data with *imbalanced classes*. For instance, in cancer diagnostics we can get many thousands of benign training cases and fortunately a much less number of malignant training cases. The actual discovery of the dependency (diagnostic rule) is conducted by selecting a similar number of training cases from both classes [41] which can be done by "normalizing" the marginal distributions of training data to the uniform distributions in each class in a coordinated way.

If F is not based on the frequencies then the interpretation of the marginal distributions is not that obvious. We still can formally convert marginal distributions $F_i(x_i)$ to uniform distributions and analyze two distributions F and C. In this case both distributions can be considered as logical dependencies. Here the copula will indicate dependence where some properties F are eliminated (lost). The usefulness of such elimination must be justified in each particular task. In essence we should show that eliminated properties are not about the mutual dependence of the variables, but are about other aspects of the variables, or even completely irrelevant, insignificant or "noise".

Note that the defined logical dependence is not expressed in the traditional logic form, e.g., as a propositional or first order logic expression, but as a copula, that is a specialized probability distribution. In fact, the copula corresponds to a set of logical expressions

$$\{(x_1 = a_1)\&(x_2 = a_2)\&\ldots\&(x_n = a_n) \text{ with probability } c(x_1, x_2, \ldots, x_n)\}$$

where c is a copula density.

The *interpretation* of copula as NTF-dependence is one of the key new points of this work showing how dependence in many-valued logic and probability theory can be fused in one concept by using copulas and marginal probabilities to clarify the relations between PT, FL and MVL. The next sections elaborate this approach.

3.3   Copula and Conditional Probability

Conditional probabilities capture some aspects of dependence between variables that include both frequency-based and logic based aspects. The relations between copula density and conditional probability are as follows as has been shown in Sect. 2:

$$p(x_2|x_1) = c(u_1, u_2)p(x_2) \quad p(x_1|x_2) = c(u_1, u_2)p(x_1)$$
$$p(x_2|x_1)/p(x_2) = c(u_1, u_2) \quad p(x_1|x_2)/p(x_1) = c(u_1, u_2)$$

This shows that copula density as a "frequency-free" dependence can be obtained from frequency-dependent conditional probabilities $p(x_1|x_2)$ by "normalizing" it. We also can see that

$$p(x_2|x_1) \leq c(u_1, u_2)$$
$$p(x_1|x_2) \leq c(u_1, u_2)$$

because $p(x_2) \leq 1$. In other words, frequency-free measure of dependence is no less than max of conditional probabilities:

$$\max(p(x_1|x_2), p(x_2|x_1)) \leq c(u_1, u_2)$$

In a logical form copula density represents a measure $c(u_1, u_2)$ of dependence for the conjunction,

$$(x_1 = a_1) \& (x_2 = a_2)$$

In contrast the conditional probabilities $p(x_1|x_2)$ and $p(x_2|x_1)$ represent measures of dependence for the implications:

$$(x_1 = a_1) \Rightarrow (x_2 = a_2), \qquad (x_2 = a_2) \Rightarrow (x_1 = a_1).$$

In a general case the relation between conditional probability and copula involves conditional density of copula as shown in Sect. 2.

$$p(x_n|x_1, \ldots, x_{n-1}) = c(x_n|x_1, \ldots, x_{n-1}) p(x_n)$$

This conditional density of copula also represents a frequency-free dependence measure.

3.4 Copulas, t-norms and Fuzzy Logic

Many copulas are t-norms [42], which are the basis of many-valued logics and fuzzy logic [31, 32]. Many-valued logics, classical logic, and fuzzy logic are *truth-functional*: "...the truth of a compound sentence is determined by the truth values of its component sentences (and so remains unaffected, when one of its component sentences is replaced by another sentence with the same truth value)" [31]. Only in very specific and simple for modeling cases, probabilities and copulas are made truth-functional. This is a major difference from t-norms that are truth-functional. The impact of the truth functionality on fuzzy logic is shown in the Bellman-Giertz theorem [43]. Probability theory, fuzzy logic and many-values logics all used t-norms, but not in the same way.

The concept of copula and t-norm are not identical. Not every copula is a t-norm [42] and only t-norms that satisfy the 1-Lipschitz condition are copulas [44]. Equating t-norms with copulas would be equating truth-functional fuzzy logic with probability theory that is not truth-functional. A joint probability density $p(x_1, x_2)$ in general cannot be made a function of only $p(x_1)$ and $p(x_2)$. A copula density or a conditional probability is required in addition due to properties $p(x_1, x_2) = c(x_1, x_2) \; p(x_1)p(x_1)$ and $p(x_1, x_2) = p(x_1|x_2)p(x_2)$. In this sense the *probability theory is more general than fuzzy logic*, because it covers both truth-functional and non-truth functional dependences. Thus, in some sense fuzzy logic with t-norms represents the *"logic of independence"*. On the other side, the copula gives a probabilistic interpretation of the fuzzy logic t-norms as well as interpretation of fuzzy logic theory as a many-valued logic.

Postulating a t-norm is another important aspect of the differences in the usage of copulas in FL and PT. In FL and possibility theory, a t-norm, e.g., $\min(x_1, x_2)$ is often postulated as a *"universal"* one for multiple data sets, without the empirical justification for specific data.

In PT according to Sklar's Theorem [24, 35, 36], the copula $C(u_1, \ldots, u_n)$ is *specific/unique* for each probability distribution F with uniform marginals. If this copula happened to be a t-norm, then it is a unique t-norm, and is also *specific* for that probability distribution.

This important difference was also pointed out in [45]: "In contrast to fuzzy techniques, where the same "and"-operation (t-norm) is applied for combining all pieces of information, the vine copulas allow the use of different "and"-operations (copulas) to combine information about different variables." Note that the copula approach not only allows different "and"-operations but it will actually get different "and"-operations when underlying joint probability distributions differ. Only for identical distributions the same "and"-operations can be expected.

The t-norm $\min(x_1, x_2)$ is a dominant t-norm in FL and possibility theory. This t-norm is known in the copula theory as a *comonotonicity copula* [46]. It satisfies a property

$$P(U_1 \leq u_1, \ldots, U_d \leq u_d) = P(U \leq \min\{u_1, \ldots, u_d\}) = \min\{u_1, \ldots, u_d\}, u_1, \ldots, u_d \in [0, 1]$$

where a *single* random uniformly distributed variable $U \sim U[0, 1]$ produced the random vector $(U_1, \ldots, U_d) \in [0, 1]^d$. The variables of the comonotonicity copula are called *comonotonic* or *perfectly positively dependent variables* [47].

3.5 Forms of Dependence Based on Common Cause and Direct Interaction

Forms of dependence. Dependence can take multiple forms. One of them is dependence with *common source/cause*, i.e., a third object S is a cause of the dependence between A and B. We will call it as *third-party dependence*. In dancing music, S serves as the third party that synchronizes dancers A and B. In physics such a type of dependence is often apparent in synchronization of events, e.g., in laser photon synchronization the source for the photon synchronization is interaction of the photons in a crystal.

Technically the dependencies between A and B in these cases can be discovered by using the correlation methods, which includes computing the copulas and the conditional probabilities $p(A|B)$, $p(B|A)$. However, correlation methods do not uncover the *type of dependency*, namely *third-party* dependence or dependence as a result of *direct local interaction* of objects A and B. The last one is a main type of laws in the classical physics.

Historically correlation based dependencies (e.g., Kepler's laws) were augmented later by more causal laws (e.g., Newton's laws). It was done via accumulation of more knowledge in the physics area and much less by more sophisticated mathematical methods for discovering the causal dependencies. This is an apparent now in constructing Bayesian networks (BN) where causal links are mostly built manually by experts in the respective field. Links build by discovering correlations in data are still may or may not be causal.

4 Physics and Expanding Uncertainty Modeling Approaches

This section provides physics arguments for expanding uncertainty modeling approaches beyond current probability theory and classical and non-classical logics to deal with dependent/related evens under uncertainty. After that we propose a way of expanding uncertainty modeling approaches. The first argument for expansion is from the area of duality of particles and waves and the second one is from the area of Bell's inequality [48, 49]. The dependencies studied in the classical physics can be described using the classical logic and the classical set theory.

Logic operates with true/false values assigned to states of particles. To establish the true state of a particle without involving fields it is sufficient to know the state of the individual particle without considering all the others particles. In this sense particles are independent one from the others. In this sense the classical logic and the associated set theory are the conceptual instruments to study classical physics. In quantum physics with the "entanglement" the states of all particles are "glued" in one global entity where any particle is sensitive or dependent to all the others. In this sense the information of any individual particle (local information) is not sufficient to know the state of the particle. We must know the state of all the other particles (global information) to know the state of one particle. The physical phenomena that are studied in the quantum physics have more complex (non-local) dependencies where one physical phenomenon is under the influence of *all* the other phenomena both local and non-local. These dependencies have both stochastic probabilistic component (that can be expressed by a joint probability) and a logic component that we express by copula.

4.1 Beyond Propositional Logic and Probability Theory

Particles and waves are classically considered as distinct and incompatible concepts because particles are localized and waves are not [50]. In other words, particles interact locally or have *local not global dependence*, but waves interact *non-locally* possibly with all particles of its media [51]. The classical propositional logic and set theory have no tools to deal with this issue and new mechanisms are needed to deal with it. The global dependence (interaction with non-local elements of the media) is a property of the structure of the media (whole system). Feynman suggested a concept of the negative probability that later was developed by Suppes and others [51–54] to address such issues.

Limitations of probability theory. Kolmogorov's axioms of PT deal only with *one type of dependence*: dependence of sets elementary events in the form of their *overlap* that is presence of the common elements in subsets. This is reflected in the formula for probability of the union $A \cup B$:

$$p(A \cup B) = p(A) + p(B) - p(A \cap B).$$

where dependence is captured by computing $p(A \cap B)$ and subtracting it to get $p(A \cup B)$.

While capturing overlap is a huge *advantage* of PT over truth-functional logics (including fuzzy logic and possibility theory) that do not take into account overlap $A \cap B$. However, axioms of the probability theory do not include other dependencies. In fact, PT assumes *independence* of each elementary event from other elementary events, requiring their overlap to be empty, $e_a \cap e_b = \emptyset$.

What are the options to deal with the restriction of the probability theory in the tasks where more dependencies must be taken into account?

We have two options:

- The first option is to *choose other elementary events* that will be really independent.
- The second one is *incorporating relations* into a new theory.

The first option has difficulty in quantum physics. Particles are dependent and cannot serve as independent elementary events. This creates difficulties for building a classical probabilistic theory of particles. Next, at a particular time, we simply may have no other candidates for elementary events that would be more elementary than particles.

Generalization of Probability The previous analysis shows that we need to explore the second option: adding relations between elementary events and their sets. Below we propose a generalization of the probability for dependent events that we call **relation-aware probability**.

Let $\{R_i^+(A, B)\}$ be a set of "positive relations" and $\{R_j^-(A, B)\}$ be a set of "negative relations" between sets A and B. Positive relations increase probability $p(A \cup B)$ and negative relations decreases probability $p(A \cup B)$.

Next we introduce probability for relations $\{p(R_i^+(A, B))\}$, $\{p(R_j^-(A, B))\}$ and define the **relation-aware probability** of the union:

$$p(A \cup B) = p(A) + p(B) + \sum_{i=1}^{n} p(R_i^+(A, B)) - \sum_{j}^{m} p(R_j^-(A, B))$$

The classical formula

$$p(A \cup B) = p(A) + p(B) - p(A \cap B)$$

is a special case of this probability with the empty set of positive relations and with a single negative relation, $R(A, B) \equiv A \cap B \neq \emptyset$ that is intersection of A and B is not empty. A negative relation can work as "annihilator". A positive relation between A and B can work as a catalyst to increase A or B. For instance, we can get

$$p(A \cup B) = p(A) + p(B) + 0.2p(A) + 0.1p(B) - p(A \cap B).$$

As a result, the relation-aware probability space can differ from Kolmogorov's probability space having the more complex additivity axioms.

Another potential approach is associated with changing the classical logic into logics that allow conflicts, e.g., fuzzy logic, agent model of uncertainty [8], and active sets [12]. This will make the paradox more *apparent* due to abilities to embed the conflict via gradual measures of uncertainty such as known in fuzzy logic, and many-valued logic.

4.2   Beyond the Set Theory: Bell Inequality

Consider sets S_1-S_8 formed by three circles A, B and C as shown in Fig. 4, e.g., $S_1 = A \cap B \cap C$ and $S_8 = A^C \cap B^C \cap C^C$ for complements of A, B and C. These sets have the following classical set theory properties that include the *Bell's inequality*

$$(A \cap B^C) \cup (B \cap C^C) = S_1 \cup S_7 \cup S_4 \cup S_6$$
$$A \cap C^C = S_7 \cup S_4$$
$$A \cap C^C \subseteq (A \cap B^C) \cup (B \cap C^C)$$
$$\left| A \cap C^C \right| \leq \left| (A \cap B^C) \cup (B \cap C^C) \right|$$

Next consider the *correlated* (*dependent*) events, for instance we can have an event with the property A, and another event with a negated property A^C. This is a type of dependence that takes place for particles. Consider an event with property $A \cap C^C$ that is with both properties A and C^C at the same time. We cannot measure the two properties by using one instrument at the same time, but we can use the correlation to measure the second property, if the two properties are correlated. We can also view an event with property $A \cap C^C$ as two events: event e_1 with property A and e_2 with the property C in the opposite state (negated). The number of pairs of events (e_1, e_2) is the same as the number of events with the superposition of A and C^C, $A \cap C^C$. In [49] d'Espagnat explains the connection between the set theory

Fig. 4 Set theory intersections or elements

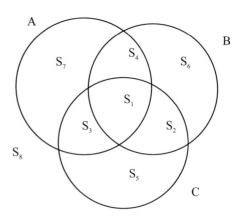

and Bell's inequality. It is known that the Bell's inequality that gives us the reality condition is violated [48].

How to reconcile the logic Bell's inequality and its violation in physics? The Bell inequality is based on the classical set theory that is consistent with the classical logic.

The set theory assumes *empty overlap* (as a form of *independence*) of elementary units. For dependent events the Bell's inequality does not hold. Thus the logic of dependence (logic of dependent events) must differ from the logic of independence (logic of independent events) to resolve this inconsistence. It means that we should use a theory that is *beyond* the classical set theory or at least use the classical set theory differently. The dependence logic [14, 15] is one of the attempts to deal with these issues.

4.3 Dependence and Independence in the Double Slit Experiment

The double slit experiment involves an electron gun that emits electrons through a double slit to the observation screen [55, 56]. Figure 5 [51] shows theoretical result of the double slit experiment when only the set theory is used to combine events: one event e_1 for one slid and another event e_2 for the second slid. In this set-theoretical approach it is assumed that events e_1 and e_2 are *elementary events* that do not overlap, $(e_1 \cap e_2) = \emptyset$, independent). In this case, the probability $p(e_1 \cap e_2) = 0$ and the probability that either one of these two events will occur is $p(e_1 \cup e_2) = p(e_1) + p(e_2)$. The actual distribution differs from distribution shown in Fig. 5 [51].

As an attempt to overcome this dependence difficulty and related non-monotonicity in quantum mechanics the use of *upper probabilities*, with the axiom subadditivity, has been proposed [51–54].

Below we explore whether the *copula* approach and *relation-aware probability* can be helpful in making this dependence *evident* and *explicit*. The relation-aware probability is helpful, because it explicitly uses the relations that are behind the dependences. This is its advantage relative to negative probabilities. The axioms of negative probability have no such components.

Fig. 5 [51]. Distribution of independent particles (events)

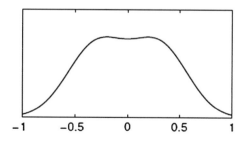

The value of the copula approach for this task needs more explorations. Copula allows representing the actual distribution F as a composition that includes the copula. The actual distribution is [51]

$$p(\alpha_1, \alpha_2) = k \cos^2(\alpha_1 - \alpha_2)$$

Respectively its copula is

$$C(F(\alpha_1), F(\alpha_2)) = k \sin^2[A \cdot B - C \cdot D] = C(u_1, u_2) \tag{4}$$

$$A = \frac{1}{2} \left(\frac{\frac{\pi}{2} \pm \sqrt{(\frac{\pi}{2})^2 + \frac{16}{\pi} \arccos(\frac{1}{k}\sqrt{F_1(\alpha_1)})}}{2} \right)^2$$

$$B = \frac{\frac{\pi}{2} \mp \sqrt{(\frac{\pi}{2})^2 + \frac{16}{\pi} \arccos(\frac{1}{k}\sqrt{F_2(\alpha_2)})}}{2}$$

$$C = \left(\frac{\frac{\pi}{2} \mp \sqrt{(\frac{\pi}{2})^2 + \frac{16}{\pi} \arccos(\frac{1}{k}\sqrt{F_2(\alpha_2)})}}{2} \right)^2$$

$$D = \frac{\frac{\pi}{2} \pm \sqrt{(\frac{\pi}{2})^2 + \frac{16}{\pi} \arccos(\frac{1}{k}\sqrt{F_1(\alpha_1)})}}{2}$$

This copula represents the dependence with the eliminated impact of marginal distributions via conversion of them into the uniform distribution. In this way the copula has the meaning of *graduate dependence between variables* resembling the many-valued logic, and t-norm compositions of variables. This is a valuable property. Thus while the copula is still a probability distribution it also represents a gradual degree of truth and falseness. The copula (5) is tabulated as follows:

$$M1 = \begin{pmatrix}
1 & 0.846 & 0.7 & 0.561 & 0.43 & 0.307 & 0.192 & 0.088 & 0 \\
0.846 & 1 & 0.969 & 0.899 & 0.805 & 0.692 & 0.56 & 0.401 & 0.125 \\
0.7 & 0.969 & 1 & 0.979 & 0.923 & 0.839 & 0.725 & 0.572 & 0.25 \\
0.561 & 0.899 & 0.979 & 1 & 0.982 & 0.93 & 0.843 & 0.708 & 0.375 \\
0.43 & 0.805 & 0.923 & 0.982 & 1 & 0.982 & 0.927 & 0.82 & 0.5 \\
0.307 & 0.692 & 0.839 & 0.93 & 0.982 & 1 & 0.98 & 0.909 & 0.625 \\
0.192 & 0.56 & 0.725 & 0.843 & 0.927 & 0.98 & 1 & 0.973 & 0.75 \\
0.088 & 0.401 & 0.572 & 0.708 & 0.82 & 0.909 & 0.973 & 1 & 0.875 \\
0 & 0.125 & 0.25 & 0.375 & 0.5 & 0.625 & 0.75 & 0.875 & 1
\end{pmatrix}.$$

It has the same form in its extreme values as classical logic equivalence relation ($\equiv$).

$$y = x_1 \equiv x_2 = \begin{bmatrix} 1\ 0 \\ 1\ 1\ 0 \\ 0\ 0\ 1 \end{bmatrix}$$

This link is also evident from [42, p. 182]:

$$C(x, 0) = C(0, x) = 0, C(x, 1) = C(1, x) = x$$

which is true for all x from [0, 1] including x = 1.

5 Conclusion and Future Work

This paper had shown the need for a theory *beyond* the classical set theory, classical logic, probability theory, and fuzzy logic to deal with dependent events including the physics paradox of particles and waves and the contradiction associated with the Bell inequality.

It is shown that the *copula* approach and *relation-aware probability* approach are promising approaches to go beyond the limitations of known approaches of PT, FL, MVL, and the classical set theory. Specifically *copula* and *relation-aware probability* approaches allow making dependences between variables in multidimensional probability distributions more *evident* with abilities to model the dependencies and relations between multiple variables more e*xplisit*.

It is pointed out that the concept of copula is *more general* than the concept of t-norm. Therefore the probability theory in some sense is *more general* than fuzzy logic, because it covers both truth-functional and non-truth functional dependences.

The *third-party dependence, direct local interaction* and causal dependence are analyzed concluding that discovering causal dependencies is far beyond current approaches in PT, FL, and MVL.

We introduced the concept of a Non-Truth-Functional *Logical Dependence* (**NTF-logical dependence**) derived as a copula from multidimensional probability distribution that does not depend on the frequencies of occurrence of variables. The interpretation of copula as NTF-dependence is one of the key of new points of this work. It shows how dependence in many-valued logic and probability theory can be fused in one concept by using the copulas and marginal probabilities to clarify the relations between PT, FL, and MVL to give a "frequency-free" dependence measure. It shows how, starting from probability and specifically from conditional probability, we introduce "logic" dependence.

We conclude that presence of t-norms in both probability theory (in copulas), and in fuzzy logic, creates a new opportunity to join them using t-norm copulas as a bridge.

Future directions are associated with linking the proposed approaches with dependence logic. Dependence logic proposed in 2007 [14] was later expanded to the contexts of modal, intuitionistic and probabilistic logic. Its goal is modeling the dependencies and interaction in dynamical scenarios. It is a type of first-order logic, and a fragment of second-order logic. While it claims applications in a wide range of domains from quantum mechanics to social choice theory, it still is in earlier stages of applications.

Another future opportunity to build the more realistic uncertainty modeling approaches is associated with the involvement of agents. Agents and teams of agents often are a source of input data, and the values of probabilities, copulas, and t-norms. The situation, when these agents are in a total agreement, can be modeled as a *total dependence* of their valuations, graduated or discrete (True/False). When a team of agents is in a conflict with each other in the valuation of the same attribute, they provide conflicting degrees of truth/falseness, which require adequate modeling approaches [8]. Active sets [12] that represent sets of agents can be a bridge between the many-valued logic, and fuzzy sets for providing the interpretable input data (valuations).

References

1. B. Kovalerchuk, and D.Shapiro, "On the relation of the probability theory and the fuzzy sets theory, Foundation Computers and Artificial Intelligence, Vol. 7, pp. 385–396, 1988.
2. D. Dubois, and H. Prade "Fuzzy sets, probability and measurement," Europ. J. of Operations Research, Vol. 40, pp. 135–154, 1989.
3. D.I. Blockley, "Fuzziness and probability: a discussion of Gaines' axioms," Civ. Engng Syst., Vol. 2, pp. 195–200, 1985.
4. P. Cheeseman, "Probabilistic versus fuzzy reasoning," in Uncertainty in Artificial Intelligence 1, L. Kanal, and J. Lemmer, Eds., North-Holland, Amsterdam, 1988, pp. 85–102
5. B.R. Gaines, "Fuzzy and probability uncertainty logics," Information and Control, Vol. 38, pp. 154–169, 1978.
6. G.J. Klir, and B. Parviz, "Probability-possibility transformations: a comparison," Int. J. of General Systems, Vol. 21, pp. 291–310, 1992.
7. B. Kosko, "Fuzziness vs. probability," Int. J. of General Systems, Vol. 17, pp. 211–240, 1990.
8. G. Resconi, B. Kovalerchuk: Agents' model of uncertainty. Knowl. Inf. Syst. 18(2): 213–229, 2009.
9. G. Resconi, B. Kovalerchuk: Agents in neural uncertainty. IJCNN 2009, pp. 2649–2656
10. G. Resconi, B. Kovalerchuk: Fusion in agent-based uncertainty theory and neural image of uncertainty. IJCNN 2008: pp. 3538–3544, 2007
11. G. Resconi, B. Kovalerchuk: Explanatory Model for the Break of Logic Equivalence by Irrational Agents in Elkan's Paradox. EUROCAST 2007: 26–33
12. G. Resconi, C. J. Hinde: Introduction to Active Sets and Unification. T. Computational Collective Intelligence 8: 1–36, 2012.
13. Graadel, E., Vaananen, J.: Dependence and independence. Studia Logica, 101(2), pp. 399–410, 2013.
14. J.Vaananen, Dependence Logic - A New Approach to Independence Friendly Logic, Cambridge University Press, 2007.
15. J. Ebbing, A. Hella, J-S. Meier, J. Muller, J. Virtema and H. Vollmer, Extended modal dependence logic, in: WoLLIC, 2013, pp. 126–137.
16. J. Ebbing, P. Lohmann, Complexity of model checking for modal dependence logic, in: M. Bieliková, G. Friedrich, G. Gottlob, S. Katzenbeisser and G. Turán, editors, SOFSEM, LNCS 7147, 2012, pp. 226–237.

17. J. Ebbing, P. Lohmann and F. Yang, Model checking for modal intuitionistic dependence logic, in: G. Bezhanishvili, S. Lobner, V. Marra and F. Richter, eds., Logic, Language, and Computation, Lecture Notes in Computer Science 7758, Springer, 2013 pp. 231–256.
18. P. Galliani, The dynamification of modal dependence logic, Journal of Logic, Language and Information 22 (2013), pp. 269–295.
19. J. Hintikka, and G. Sandu, Informational independence as a semantical phenomenon, in: Logic, methodology and philosophy of science, VIII (Moscow, 1987), Stud. Logic Found. Math. 126, North-Holland, Amsterdam, 1989 pp. 571–589.
20. J.-S. Müller, and H. Vollmer, Model checking for modal dependence logic: An approach through post's lattice, in: L. Libkin, U. Kohlenbach and R. Queiroz, editors, Logic, Language, Information, and Computation, Lecture Notes in Computer Science 8071, 2013 pp. 238–250.
21. F. Yang, "On Extensions and Variants of Dependence Logic," Ph.D. thesis, University of Helsinki, 2014.
22. I.R. Goodman, Fuzzy sets as equivalence classes of random sets, In: Yager, R., et al. (eds.), Fuzzy Sets and Possibility Theory, Oxford, UK: Pergamon Press, 327–432, 1982.
23. H. T. Nguyen, V. Kreinovich, "How to Fully Represent Expert Information about Imprecise Properties in a Computer System–Random Sets, Fuzzy Sets, and Beyond: An Overview", International Journal of General Systems, 2014, 43(5–6), pp. 586–609.
24. P. Embrechts, F.Lindskog, A. McNeil: Modelling dependence with copulas and applications to risk management. In *Handbook of heavy tailed distributions in finance*, Rachev ST, ed. Elsevier/North-Holland, Amsterdam, 2003.
25. P. Jaworski, F. Durante, W. K. Härdle, T. Rychlik (Eds), Copula Theory and Its Applications, Lecture Notes in Statistics, Springer, 2010.
26. J. C. Fodor and M. Roubens: Fuzzy Preference Modelling and Multicriteria Decision Support, Kluwer, Dordrecht 1994.
27. O. Hadžić, E. Pap: Fixed Point Theory in Probabilistic Metric Spaces. Kluwer, Dordrecht 2001.
28. B. Schweizer, A. Sklar: Probabilistic Metric Spaces. North-Holland, New York, 1983.
29. E. Klement, R. Mesiar and E. Pap: Triangular Norms. Kluwer, Dordrecht 2000.
30. E. Klement; R. Mesiar; E. Pap, Invariant copulas, Kybernetika, Vol. 38 2002, No. 3, 275–286.
31. Many-Valued Logic, Stanford Encyclopedia of Philosophy, 2015, http://plato.stanford.edu/entries/logic-manyvalued/#TNorBasSys
32. P. Hájek, Metamathematics of Fuzzy Logic, Dordrecht: Kluwer, 1998.
33. K. Aas, C. Czado, A. Frigessi, and H. Bakken. Pair-copula constructions of multiple dependencies. Insurance: Mathematics and Economics, 44:182–198, 2009.
34. R. Accioly and F. Chiyoshi. Modeling dependence with copulas: a useful tool for field development decision process. Journal of Petroleum Science and Engineering, 44:83–91, 2004.
35. A. Sklar. Fonctions de repartition a n dimensions et leurs marges. Publications de l'Institut de Statistique de L'Universite de Paris, 8:229–231, 1959.
36. A. Sklar: Random variables, joint distribution functions, and copulas. Kybernetika 9, 1973, 449–460.
37. J.-F. Mai, M. Scherer, "Simulating Copulas: Stochastic Models, Sampling Algorithms, and Applications", World Scientific, 2012, http://www.worldscientific.com/doi/suppl/10.1142/p842/suppl_file/p842_chap01.pdf
38. M.S. Tenney, "Introduction to Copulas", Enterprise Risk Management Symposium, 2003, Casualty Actuarial Society, http://www.casact.org/education/rcm/2003/ERMHandouts/tenney1.pdf
39. C. Klüppelberg, C. and S. Resnick, The Pareto copula, aggregation of risk and the Emperor's socks. Preprint, Technical University of Munich, 2009.
40. T. Mikosch, Copulas: tales and facts. Extremes, Vol. 9, Issue 1, pp. 3–20, Springer, 2006.
41. B. Kovalerchuk, E. Vityaev, J. Ruiz, Consistent and Complete Data and "Expert" Mining in Medicine, In: Medical Data Mining and Knowledge Discovery, Springer, 2001, pp. 238–280.
42. G. Beliakov, A. Pradera and T. Calvo, Aggregation Functions: A Guide for Practitioners, Springer, Heidelberg, Berlin, New York, 2007.

43. R. Bellman, M. Giertz. "On the Analytic Formalism of the Theory of Fuzzy Sets." Information Sciences 5 (1973): 149–156.
44. M. Navara, Triangular norms and conorms, Scholarpedia, 2(3):2398, 2007. http://www. scholarpedia.org/article/Triangular_norms_and_conorms
45. S. Sriboonchitta, J. Liu, V. Kreinovich, and H.T. Nguyen, "Vine Copulas as a Way to Describe and Analyze Multi-Variate Dependence in Econometrics: Computational Motivation and Comparison with Bayesian Networks and Fuzzy Approaches", In: V.-N. Huynh, V. Kreinovich, and S. Sriboonchitta eds.), Modeling Dependence in Econometrics, Springer, Berlin, 2014, pp. 169–187.
46. E. Jouini and C. Napp, Conditional comonotonicity, Decisions in Economics and Finance, 27 (2):153–166, 2004.
47. A. Patton, Applications of Copula Theory in Financial Econometrics, Dissertation, University of California, San Diego, 2002.
48. J. Bell, "On the Einstein Podolsky Rosen Paradox". Physics 1 (3): 195–200, 1964
49. D'Espagnat Conceptual Foundations of Quantum Mechanics, 2nd ed. Addison Wesley
50. R. Feynman, R., Leighton, and M. Sands, The Feynman lectures on physics: Mainly mechanics, radiation, and heat, volume 3. Basic Books, NY, 2011.
51. J.A. de Barros, G. Oas and P. Supper, Negative probabilities and counterfactual reasoning on the double – slit experiment, [quant-ph] 16 Dec 2014 arXiv:1412.4888v1.
52. P. Suppes, M. Zanotti, Existence of hidden variables having only upper probabilities Foundations of Physics 21 (12):1479–1499 (1991)
53. J. A. de Barros, P. Suppes, "Probabilistic Inequalities and Upper Probabilities in Quantum Mechanical Entanglement," Manuscrito, v. 33, pp. 55–71 (2010).
54. P. Suppes, S. Hartmann, Entanglement, upper probabilities and decoherence in quantum mechanics, In: M. Suaráz et al (ed.), EPSA Philosophical Issues in the Sciences: Launch of the European Philosophy of Science Association, Springer, pp. 93–103, 2010
55. P. Feynman, QED: The Strange Theory of Light and Matter. Princeton University Press, 1988.
56. Double-slit experiment, Wikipedia, http://en.wikipedia.org/wiki/Double-slit_experiment, 2015.

Note to the Polemics Surrounding the Second Gödel's Theorem

Klimentij Samokhvalov

Abstract An extensive philosophical debate continues for years about impact of arithmetization of metamathematics on the Hilbert program with many papers published. Many of these works claim a death-blow to Hilbert's Program by Gödel's Second Incompleteness Theorem. This note provides a short and understandable argument that in fact Gödel's Second Incompleteness Theorem does not deliver the notorious death-blow (the *coup de grace*) to Hilbert's Program.

Keywords Primitive recursive arithmetic · Gödel's numbering · Gödel's canonical Consis · Second Gödel's incompleteness theorem · Hilbert's program

1 Introduction

A philosophical debate about the impact of arithmetization of metamathematics on the Hilbert program intensified after publishing "Arithmetization of metamathematics in a general setting" by Feferman [1] in 1960. The general content and the course of this debate are presented in excellent reviews [2, 3] that include an extensive bibliography. However, in our opinion, this "drama of ideas" requires the additional analysis that is provided below.

Let S be a formal system in the first order language with signature

$$(+, \cdot, ', 0, \approx).$$

It is assumed that S includes Primitive Recursive Arithmetic (PRA). We say that S is *consistent* if and only if an arithmetic sentence X exists, for which it is not true that X is provable in S.

K. Samokhvalov (✉)
Institute of Mathematics, Russian Academy of Sciences, Novosibirsk, Russia
e-mail: kfsamochvalov@mail.ru

© Springer International Publishing AG 2017
V. Kreinovich (ed.), *Uncertainty Modeling*, Studies in Computational
Intelligence 683, DOI 10.1007/978-3-319-51052-1_16

Symbolically,

$$\text{CON}(S) \Leftrightarrow \exists X \text{ (it is not true that } \vdash_S X), \tag{1}$$

where $\text{CON}(S)$ is an acronym for "S is consistent", and $\vdash_S \ldots$ is an abbreviation for "… is provable in S". Due to (1), it is clear that for any particular arithmetical formula P, the implication

$$\text{(It is not true that } \vdash_S P) \Rightarrow \text{CON}(S) \tag{2}$$

takes place. It is also clear that the reverse implications, generally speaking, are not true. For example, the implication

$$\text{CON}(S) \Rightarrow \text{(it is not true that } \vdash_S 0 \approx 0)$$

is not true. In fact, the sentence $0 \approx 0$ is provable in PRA, and S includes PRA.

It is also clear that the second Gödel's incompleteness theorem can effectively find a certain special case, wherein the reverse implication is true. In fact, this theorem says:

$$\text{CON}(S) \Leftrightarrow \text{(it is not true that } \vdash_S \text{Con}_S), \tag{3}$$

where Con_S is a canonical Gödel's *consis* (definite arithmetical formula of some special type).

Note that only finite provable statements are credible for staunch Hilbertians. Therefore, if the second Gödel's theorem is generally intended to have at least some value for them, it must be acknowledged that the equivalence (3) is proved by finite means.

Moreover, in terms of Hilbertians, this finitely establishing equivalence (3) fully covers the content of the given theorem. So, for Hilbertians everything else that is usually attributed to Gödel's second theorem is idle talk.

This includes the following usual formulation of Gödel's second theorem:

if the system S is consistent, then the arithmetic sentence Con_S that *expresses consistency of S in S* is not provable in S.

In this formulation, the phrase in italics is redundant. For, whatever it meant, the actual finite proof of the theorem is independent of it.

In this finite proof, Con_S appears simply as a specific formal formula in language S, with respect to which it is important to know in advance only one thing: whether this formula is provable or not provable in S. It does not matter in this case at all, whether it expresses something or does not express anything, whether it is true or false, understandable or not, etc.

2 What Is "Expressed" by Gödel's Theorem?

When somebody says: "A formula Con_S 'expresses' consistency of S in S", we can ask: "What is this specific formula 'expresses' in a different choice of Gödel's numbering?". Then it becomes clear that in fact Con_S "expresses" not a property of the system S, which is the consistency of S, but the certain property of an ordered pair (g, S), where g is Gödel's numbering. Substituting one for the other is logically incorrect. In this circumstance one can see some inadequacy of *arithmetized* metamathematics to metamathematics envisioned by Hilbert initially.

On the other hand, finitely established equivalence (3) does not say anything about whether non-arithmetic meta-statement $CON(S)$ or non-arithmetic meta-statement (It is not true that $\vdash_S Con_S$) are finitely provable or not.

Therefore, the second Gödel's theorem is not "a fatal blow to Hilbert's program".

The natural context of both Gödel's incompleteness theorems is just the incompleteness of sufficiently strong consistent formal systems, rather than the question of the absence or presence of the finite proof of their consistency.

3 Afterword

Sadly, of course, a finite proof of the consistency of arithmetic is still not found. However, it should be remembered that Hilbert described the concept of a finite proof somewhat vaguely and never gave a precise definition of this concept, relying on the ability to immediately recognize, as soon as some reasoning is there, whether it is finite, or not [4].

It is not hard to guess here that Hilbert, like all rational people, was well aware that even a well-deserved trust in something is always an empirical psychological fact, which can be established in advance only as probable, and only within some empirical hypothesis. However, it was certainly distasteful for Hilbert, who was prone to some form of Kantian philosophy to base mathematics on the empirical hypotheses. Naturally, therefore, he described the concept of finite proof without giving exact definition.

References

1. S. Feferman, "Arithmetization of metamathematics in a general setting", *Fund. Math.*, 1960, Vol. 49, pp. 35–92.
2. R. Zach, "Hilbert's Program", In: E. N. Zalta (ed.) *The Stanford Encyclopedia of Philosophy* (Spring 2009 Edition), http://plato.stanford.edu/archives/spr2009/entries/hilbert-program/.
3. P. Raatikainen, "Gödel's Incompleteness Theorems", In: E. N. Zalta (ed.) *The Stanford Encyclopedia of Philosophy* (Winter 2013 Edition), http://plato.stanford.edu/archives/win2013/entries/goedel-incompleteness/.
4. A. Ignjatovic, "Hilbert's program and the omega rule", *Journal of Symbolic Logic*, 1994, Vol. 59, pp. 322–343.

Ontological Data Mining

Evgenii Vityaev and Boris Kovalerchuk

Abstract We propose the ontological approach to Data Mining that is based on: (1) the analysis of subject domain ontology, (2) information in data that are interpretable in terms of ontology, and (3) interpretability of Data Mining methods and their results in ontology. Respectively concepts of Data Ontology and Data Mining Method Ontology are introduced. These concepts lead us to a new Data Mining approach—Ontological Data Mining (ODM). ODM uses the information extracted from data which is interpretable in the subject domain ontology instead of raw data. Next we present the theoretical and practical advantages of this approach and the Discovery system that implements this approach. The value of ODM is demonstrated by solutions of the tasks from the areas of financial forecasting, bioinformatics and medicine.

1 Introduction

At the International Workshop on Philosophies and Methodologies for Knowledge Discovery (22–26 August 2005, Copenhagen, Denmark) it was pointed out that any KDD&DM (Knowledge Discovery in Data Bases and Data Mining) *method* has its own *ontology* [1–4]. Any KDD&DM method explicitly or implicitly assume:

E. Vityaev (✉)
Sobolev Institute of Mathematics SB RAS, Acad. Koptyug Prospect 4,
Novosibirsk, Russia
e-mail: vityaev@math.nsc.ru

E. Vityaev
Novosibirsk State University, Novosibirsk 630090, Russia

B. Kovalerchuk
Computer Science Department, Central Washington University,
Ellensburg, WA 98926-7520, USA
e-mail: BorisK@cwu.edu

© Springer International Publishing AG 2017
V. Kreinovich (ed.), *Uncertainty Modeling*, Studies in Computational
Intelligence 683, DOI 10.1007/978-3-319-51052-1_17

1. some types of input data;
2. some language for data interpretation and hypothesis class construction (knowledge space of the KDD&DM method);
3. confirmed hypothesis in this language.

By the ontology of KDD&DM & ML method we mean the language for constructing hypotheses. At the same time any **subject domain** also has its own **ontology** that includes the system of notions for description of objects. If we apply some KDD&DM & ML method for solving a task in the given subject domain, we want to get results that are interpretable in the ontology of that subject domain. For that purpose we need a following requirement.

Requirement. *For the interpretability of the KDD&DM & ML results, the ontology of KDD&DM & ML method must be interpretable in the ontology of the subject domain.*

For example, to apply a classification method that uses a language of spherical shapes for hypotheses, we need interpret spherical shapes in the ontology of the subject domain.

The knowledge extracted by a KDD&DM method on some data is a set of confirmed hypothesis that are interpretable in the ontology of the KDD&DM method and at the same time in the ontology of the subject domain.

The subject domain ontology induces **data ontology**. We need to emphasize that quantities in data are not numbers themselves, but numbers with their **interpretation**. For example, abstract numbers 200, 3400, 300, 500 have three different interpretations shown in Table 1.

For every quantity there are relations and operations that are meaningful for this quantity. This interpretation of quantities is a core approach of the Representational Measurement Theory (RMT) [5–7]. The RMT interprets quantities as *empirical*

Table 1 Values and their interpretation

Interpretation	Values	Meaningful operations
Abstract numbers	5, 3400, 360, 500	Meaning of $360 > 5$ is not clear. These numbers can be just labels
Abstract angles	5, 3400, 360, 500	360 meaningfully greater that 5. It is implicitly assumed that angles are rotational angles
Azimuth angles	5, 3400, 360, 500	Azimuth operations. 360 meaningfully less that 5, because Azimuth $360 =$ Azimuth 0
Rotational angles	5, 3400, 360, 500	Rotational angle operations. 360 meaningfully greater that 5 if the angle represents the rotation time

systems—algebraic structures defined on objects of subject domain with a set of relations and operations interpretable in the subject domain ontology.

More specifically, main statements of the measurement theory relative to data mining issues are as following [5–7]:

- numerical representations (scales) of quantities and data are determined by the corresponding empirical systems;
- scales are unique up to a certain sets of permissible transformations such as changing measurement units from meters to kilometers for ratio-scales;
- KDD&DM & ML methods need to be invariant relative to the sets of permissible transformations of quantities in data.

To obtain data ontology in accordance with the measurement theory, we need to transform data into many-sorted empirical systems. This transformation is described in [3, 8] for such data types as pair comparisons, binary matrices, matrices of orderings, matrices of proximity and an attribute-based matrix. Such transformation faces the following problem. Many physical quantities possess an interpretable operation • which has all formal properties of the usual addition operation. However, medicine and other areas may have no empirical interpretation for that operation. For example, empirical system of physical pressure, measured by a tonometer, have operation • as physical quantity, but have no this operation as medical quantity, because it is not interpretable in medicine. In that case, the • operation should be removed from the empirical system of pressure in medicine, and the corresponding scale should be reconsidered. The order relation for pressure is obviously interpretable in medicine, and it can be included in the empirical system of the pressure. Thus, data ontology and scales strongly depend on the subject domain ontology.

Consider another example from the area of finance. What is the ontology of financial time series? It also can be considered from points of view (ontologies) of: (1) a trader-expert, (2) one of the mathematical disciplines, (3) technical analysis, (4) trade indexes etc. We need to specify the ontology of a subject domain before determining the data ontology.

If a KDD&DM & ML method uses operations or relations that are not interpretable in the data ontology and hence in the subject domain ontology, then it may obtain non-interpretable results. For example, the average patients' temperature in the hospital has no medical interpretation. If all relations and operations, which are used in the method, are included in the data ontology, then the algorithm results will be interpretable in the subject domain ontology and hence invariant relative to the permissible transformations of scales of the used data.

Thus, to avoid the non-invariance of the method and non-interpretability of its results, we need to use only relations and operations from the data ontology. It means that hypotheses tested by the method must include only these relations and operations. However, as we pointed out above, scales of quantities depend on the data ontology. Hence, the invariance of the method cannot be established before we revise the scales of all quantities based on the data ontology.

2 Relational (Ontological) Approach to Data Mining

Since in any application of KDD&DM & ML methods we need to established data ontology and transform data into many-sorted empirical systems, we proposed a Relational Data Mining (RDM) approach [1–3, 9] to knowledge discovery that is working directly with the data presented as many-sorted empirical systems. We also developed a software system "Discovery" that implements this approach in a rather general form and extracts knowledge from the many-sorted empirical systems for various classes of hypothesis.

Thus, Relational Data Mining approach consists of the following stages:

1. transform data into many-sorted empirical systems according to data ontology;
2. use background knowledge expressed in the first-order logic for learning and forecasting;
3. determine, for corresponding Data Mining task, the hypothesis class in terms of many-sorted empirical system;
4. acquire knowledge, by testing hypotheses from the hypothesis class on many-sorted empirical systems;
5. use acquired knowledge for predictions, classifications, pattern recognition etc.

Relational Data Mining approach provides following possibilities that cannot be performed by other Data Mining methods:

- analyze data with unusual scales. Traditionally Data Mining methods use only few scale types, but in the measurement theory, there are known hundreds scale types;
- perform data exploration—when we can simultaneously vary data ontology, information extracted from data and hypothesis classes. The appropriate version of the system discovery was developed and applied for several tasks [10];
- acquire knowledge for data onthologies for which there are no Data Mining methods with appropriate set of relations and operations. Below we give examples of such data onthologies and corresponding hypotheses.

Relational Data Mining approach and Discovery system use logic-probabilistic techniques related to the Probabilistic Logic Programming for knowledge acquisition. However, there is a problem of probability and logic synthesis for this approach. This problem was discussed at the series of conferences with the title Progic (Probability+Logic, 2006–2013) [11]. In the framework of our approach, we propose a new solution to this problem based on special semantic probabilistic inference [9, 12–14]. The software system Discovery implements the semantic probabilistic inference in the process of hypotheses testing and prediction. We demonstrated in series of experiments that, in contrast with the Probabilistic Logic Programming the Discovery system can acquire knowledge from data with a significant noise level, for example, in financial forecasting [8]. Moreover, the Discovery system acquire maximal specific rules, for which we prove that they are predict without contradictions [9, 13].

3 Applications of Relational Data Mining

In the frame of Relational Data Mining approach, some problems were solved in the area of financial forecasting, bioinformatics, medicine, and other areas [15]. Here we present data onthologies and hypotheses classes for these tasks.

3.1 Financial Forecasting

As data ontology for the S&P500 (close) forecasting, we used information about local maxima and minima and their order interconnections. In some experiments we used information about weekdays with second and third differences between prices for these days [8, 16].

As objects, we used the periods for five consequent days for which the following relations and operations where determined:

1. function $wd(\mathbf{a})$ that display five consequent days, for example $wd(\mathbf{a}) = \langle 1, 2, 3, 4, 5 \rangle$ means 5 days beginning from Monday;
2. first order difference:

$$\Delta_{ij}(a_t) = (SP500(a_t^j) - SP500(a_t^i))/SP500(a_t^i), \ i < j, \ i,j = 1, \ldots, 5,$$

   where a_t—5 days, a_t^j, a_t^i—corresponding i-th and j-th day of the 5 days. This function represents the difference of SP500C between the i-th and j-th days, normalized relative the i-th day;
3. second order difference, i.e., the difference between two first order differences

$$\Delta_{ijk}(a_t) = \Delta_{jk}(a_t) - \Delta_{ij}(a_t).$$

In this ontology, we stated following classes of hypotheses:

1. $(wd(a) = wd(b) = \langle d_1, \ldots, d_5 \rangle) \& (\Delta(a) \le \Delta(b))^{\varepsilon 1} \Rightarrow (SP500(a^5) \le SP500(b^5))^{\varepsilon 0}$, where $\Delta(a)$, $\Delta(b)$—arbitrary first or second order differences for 5 days periods $\mathbf{a}$ and $\mathbf{b}$ with the same sequences of days; $\varepsilon 1$, $\varepsilon 0 = 1$ (0) if relation has no (has) negation, for example A^1 means simply A, while A^0 means "not A".
   The example of the discovered regularity on real SP500C data for that hypotheses class is:
   IF for any 5-day objects $\mathbf{a}$ and $\mathbf{b}$ beginning from Tuesday,
   AND the SP500 difference $\Delta_{12}(a_t)$ is smaller the difference $\Delta_{12}(b_t)$,
   THEN the target stock for the last day of $\mathbf{a}$ will be greater than for the last day of $\mathbf{b}$.

2. At the same time, hypotheses with two differences in the premise where tested:

$$(wd(a) = wd(b) = \langle d_1, \ldots, d_5 \rangle)\&(\Delta(a) \leq \Delta(b))^{\varepsilon 1}\&(\Delta(a) \leq \Delta(b))^{\varepsilon 2}$$
$$\Rightarrow (SP500(a^5) \leq SP500(b^5))^{\varepsilon 0},$$

3. The hypotheses with three and more relations at the premise also where tested:

$$(wd(a) = wd(b) = \langle d_1, \ldots, d_5 \rangle)\&(\Delta(a) \leq \Delta(b))^{\varepsilon 1}\& \ldots \&(\Delta(a) \leq \Delta(b))^{\varepsilon k}$$
$$\Rightarrow (SP500(a^5) \leq SP500(b^5))^{\varepsilon 0}.$$

For example, one of the tested statements was:
IF for any 5-day objects **a** and **b** with weekdays $\langle d_1, \ldots, d_5 \rangle$,
the SP500C difference $\Delta_{12}(a_t)$ is smaller than $\Delta_{12}(b_t)$
AND the SP500C difference $\Delta_{23}(a_t)$ is greater than $\Delta_{23}(b_t)$
AND the SP500C difference $\Delta_{123}(a_t)$ is greater than $\Delta_{123}(b_t)$
AND ...
THEN the target stock for the last day of **a** will be greater than for the last day
of **b**.

The example of rules discovered by the system Discovery is as follows:
*IF Current 5 days end on Monday and there are some other ("old") 5 days (from
the history of years 1984–1996) that end on Monday too*
*AND the relative SP500C difference between Tuesday and Thursday for the old
5 days is not greater than between Tuesday and Thursday for the current 5 days*
*AND the relative SP500C difference between Tuesday and Monday for the old 5
days is greater than between Tuesday and Monday for the current 5 days*
*AND the relative difference between SP500C differences for Tuesday, Wednesday
and Wednesday, Thursday for the old 5 days is not greater than for the pairs of days
for the current 5 days*
AND we omit linguistic description of $(\Delta_{245}(a_t) > \Delta_{245}(b_t))$, which is similar to
previous one
*THEN the target value for Monday from the current 5 days should be not greater
than the target value for the Monday from the old 5 days, i.e., we forecast that a
target stock 5 days ahead from the current Monday will grow not greater than it was
5 days ahead from the old Monday.*

3.2   DNA Regulatory Regions Analysis

We asked experts of the Institute of Cytology and Genetics SD RAS about the information that is needed for the solution of the analysis of regulatory regions. From their point of view this information is the following: distance between valuable DNA signals that may vary in some range, repetition of signals, some DNA intervals which

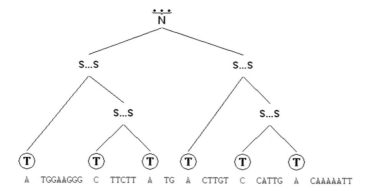

Fig. 1 Structure of complex signal

have a specific place relative to transcription start, and hierarchy of signals. Corresponding relations and operations where defined and hypothesis in these terms where formulated as a notion of a Complex Signals (CSs), which is defined recursively based on elementary signals and operations applied to them, see Fig. 1 [17, 18]. The recursive definition of the Complex Signal is following:

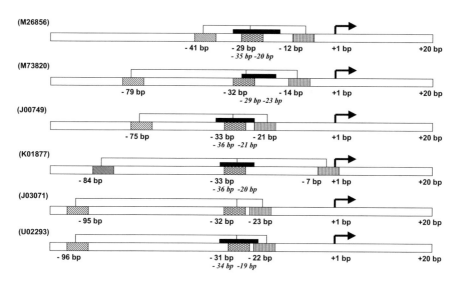

Fig. 2 Schematic localization of the complex signal $CWGNRGCN < NGSYMTAM < MAGKSHCN$ in promoters of endocrine system genes. The promoter sequences are aligned relative to the transcription start (position +1 bp), indicated by *arrows*. The EMBL identifiers of the promoters studied are given in parentheses to the *left*. The eight-bp oligonucleotide motifs composing the complex signal are shown as *dashed rectangles*; positions of the first nucleotides are indicated relative to the transcription start. The *black rectangles* mark experimentally defined positions of the TATA-box indicated in the TRRD database. Positions of its first and last nucleotides are italicized

1. the elementary signal is CS;
2. the result of the "repetition" or "interval" operation applied to CS is CS;
3. the result of the "distance" operation applied to a pair of CSs is CS.

Elementary signals are indivisible signals which are characterized by a name and location in the sequence. It may be nucleotide, motive, transcription factor binding site and any other signal. Figure 1 illustrates such signals. The examples of the complex signals are presented on the Fig. 2 [17, 18]. As elementary signals there where motifs of length 8 bp that were pre-selected as specific for the set of promoters by the program ARGO [19].

3.3 Breast Cancer Diagnostic System

Expert J.Ruiz (Baton Rouge Women hospital) defined 11 features, for developing the breast cancer diagnostic system [2, 20]. These features constituted the data ontology.

Table 2 Discovered breast cancer diagnostic rules

Diagnostic rule	F-criteria	Value of F-criteria			Diagnosis Round-Robin test (%)
		0.01	0.05	0.1	
IF NUMber of calcifications per cm^2 between 10 and 20	0.0029	+	+	+	93.3
AND VOLume $> 5\,cm^3$	0.0040	+	+	+	
THEN Malignant					
IF TOTal # of calcifications >30	0.0229	–	+	+	100.0
AND VOLume $> 5\,cm^3$	0.0124	–	+	+	
AND DENSITY of calcifications is moderate	0.0325	–	+	+	
THEN Malignant					
IF VARiation in shape of calcifications is marked	0.0044	+	+	+	100.0
AND NUMber of calcifications is between 10 and 20	0.0039	+	+	+	
AND IRRegularity in shape of calcifications is moderate	0.0254	–	+	+	
THEN Malignant					
IF variation in SIZE of calcifications is moderate	0.0150	–	+	+	92.86
AND variation in SHAPE of calcifications is mild	0.0114	–	+	+	
AND IRRegularity in shape of calcifications is mild	0.0878	–	–	+	
THEN Benign					

Using this ontology and data in terms of this ontology the set of regularities was discovered. Table 2 some of these regularities are presented.

4 Representative Measurement Theory

In accordance with the measurement theory, numerical representations of quantities and data are determined by empirical systems. In this section we present main definitions of the measurement theory [5–7].

An *empirical system* is a relational structure that contains a set of objects A, k (i)—ary relations $P_1, \ldots, P_n$ and k (j)—ary operations $\rho_1, \ldots, \rho_m$ defined on A,

$$\mathbf{A} = \langle A, P_1, \ldots, P_n, \rho_1, \ldots, \rho_m \rangle.$$

Every relation P_i is a Boolean function (a predicate) with k (i) arguments from A, and ρ_j is the k (j) argument operation on A. System $\mathbf{R}$

$$\mathbf{R} = \langle R, T_1, \ldots, T_n, S_1, \ldots, S_m \rangle$$

is called a *numerical system of the same type as a system* $\mathbf{A}$, if R is a subset of Re^m, $m \geq 1$, Re^m is a set of m-tuples of real numbers, every relation T_i has the same arity $k(i)$ as the corresponding relation P_i, and every real-value function S_j has the same arity $k(j)$ as the corresponding operation ρ_j.

A numerical system $\mathbf{R}$ is called a *numerical representation* of the empirical system $\mathbf{A}$, if a (strong) homomorphism $\phi : A \rightarrow R$ exists such that:

$$P_i(a_1, \ldots, a_{k(i)}) \Rightarrow T_i(\phi(a_1), \ldots, \phi(a_{k(i)})), \ i = 1, \ldots, n;$$

$$\phi(\rho_j(a_1, \ldots, a_{k(j)})) = S_j(\phi(a_1), \ldots, \phi(a_{k(j)})), j = 1, \ldots, m.$$

The strong homomorphism means that, if predicate $T_i(\phi(a_1), \ldots, \phi(a_{k(i)}))$ is true on $\langle \phi(a_1), \ldots, \phi(a_{k(i)}) \rangle$, then there exists tuple $\langle b_1, \ldots, b_{k(i)} \rangle$ in A, such that $P_i(b_1, \ldots, b_{k(i)})$ is true and $\phi(b_1) = \phi(a_1), \ldots, \phi(b_{k(i)}) = \phi(a_{k(i)})$. We will denote such homomorphism between the empirical system $\mathbf{A}$ and numerical system $\mathbf{R}$ as $\Phi : \mathbf{A} \rightarrow \mathbf{R}$. Thus, the numerical system $\mathbf{R}$ represents a relational structure in computationally tractable form with a complete retention of all the properties of the relational structure.

In the measurement theory, the following problems are considered:

(1) find a numerical representation $\mathbf{R}$ for an empirical system $\mathbf{A}$;
(2) prove a theorem that homomorphism $\Phi : \mathbf{A} \rightarrow \mathbf{R}$ exists;
(3) define the set of all possible automorphisms $f : R \rightarrow R$ (the uniqueness theorems), such that $f\Phi$ is also homomorphism $f\Phi : \mathbf{A} \rightarrow \mathbf{R}$.

Example A relational structure $A = \langle A, P \rangle$ is called a semi-ordering, if for all a, b, c, d $\in$ A the following axioms are satisfied:

$$\neg P(a, a);$$
$$P(a, b)\&P(c, d) \Rightarrow (P(a, d) \vee P(c, b));$$
$$P(a, b)\&P(b, c) \Rightarrow \forall d \in A(P(a, d) \vee P(d, c)).$$

Theorem 1 *If* $A = \langle A, P \rangle$ *is a semi-ordering and* $A/\approx$ *is finite, then there exists a function* $U : A \to Re$, *such that:*

$$P(a, b) \Leftrightarrow U(a) + 1 < U(b).$$

There are hundreds of numerical representations known in the measurement theory with few most commonly used. The strongest one is called the absolute data type (*absolute scale*). The weakest numerical data type is the nominal data type (*nominal scale*). There is a spectrum of data types between them. They allow us comparing, ordering, adding, multiplying, dividing values and so on. The classification of these data types is presented in Table 3. The basis of this classification is a transformation group. The strongest absolute data type does not permit transformations of data at all, and the weakest nominal data type permits any one-to-one transformation. Intermediate data types permit different transformations such as positive affine, linear and others (see Table 3).

The transformation groups are used to determine the invariance of a regularity. The regularity expression must be invariant to the transformation group; otherwise it will depend not only on the nature, but on the subjective choice of the measurement units.

Table 3 Classification of data types

Transformation	Transformation group	Data type (scale)
$X \to f(x)$,	F:Re $\to$ (onto) Re, 1 $\to$ 1 transformation group	Nominal
$X \to f(x)$,	F:Re $\to$ (onto) Re monotone transformation group	Order
$X \to rx + s, r > 0$	Positive affine group	Interval
$X \to tx^r, t, r > 0$	Power group	Log-interval
$X \to x + s$	Translation group	Difference
$X \to tx, t > 0$	Similarity group	Ratio
$X \to x$	Identity group	Absolute

5 Data Ontology in Different Subject Domains

From the measurement theory point of view, data is a many-sorted empirical system
A with the sets of relations and operations interpretable in the domain theory. For
instance, a "stock price" data type can be represented as a relational structure $\mathbf{A} =$
$\langle A; \{\leq, =, \geq\}\rangle$ with nodes A as individual stock prices and arcs as their relations
$\{\leq, =, \geq\}$. As we pointed out scales are strongly depend on the subject domain
ontology.

Let us consider the specificity of domain anthologies for different subject domains.
We consider six different cases:

Physical data in physical domains.
Physical data in non-physical domains.
Non-physical data in non-physical domains.
Nominal discrete data.
Non-quantitative and non-discrete data.
Mix of data.

1. *Physical data in physical domains*. Data contain only physical quantities, and
the subject domain is physics. This is a realm of physics with well-developed data
ontology and measurement procedures. In this case, the measurement theory [6]
provides formalized relational structures for all physical quantities and KDD&DM
methods can be correctly applied.

2. *Physical data for non-physical domains*. The data contain physical quantities,
but the subject domain is not physics. The ontology of the subject domain may refer to
finance, geology, medicine, and other areas. In these cases data ontology is not known,
as we pointed out above for the pressure in medicine. If the quantity is physical,
then we can define the relational structure using the measurement theory. However,
the physically interpretable relations of the relational structure are not necessarily
interpretable in ontology of subject domain. If a relation is not interpretable, it should
be removed from the relational structure.

3. *Non-physical data in non-physical domains*. For non-physical quantities, data
ontology is virtually unknown. There are two sub-cases:

Non-numerical data types. In [3, 8] it was developed a procedure for transformation the following representations of data into data ontology (many-sorted empirical
systems): pair-wise and multiple comparison data types, attribute-based data types,
and order, and coherence matrixes.

Numerical data types. Here, we have a measurer x(a), which produces a number as
a result of a measurement procedure applied to an object a. Examples of measurers are
psychological tests, stock market indicators, questionnaires, and physical measuring
instruments used in non-physical areas.

For this case let us define a data ontology as the set of *relations and operations*
for the measurer x(a). For any numerical relation $R(y_1, \ldots, y_k) \subset Re^k$ and operation
$S(x_1, \ldots, x_m) : Re^m \to Re$, where Re is the set of real numbers, an *relation* P^R on
A^k and an *operation* $\rho^S : A^m \to A$ can be defined as follows

$$P^R(a_1, \ldots, a_k) \Leftrightarrow R(x(a_1), \ldots, x(a_k)), \ \rho^S(a_1, \ldots, a_m) = s(x(a_1), \ldots, x(a_m)).$$

We should find such relations R and operations S that have interpretation in the subject domain ontology. The set of obtained interpretable relations is not empty, because at least one relation $(P^=)$ has an empirical interpretation: $P^=(a_1, a_2) \Leftrightarrow x(a_1) = x(a_2)$. In the measurement theory, many sets of axioms were developed for data, having only ordering and equivalence relations. For instance, given weak order relation $<_y$ (for the attribute y) and n equivalence relations $\approx_{x_1}, \ldots, \approx_{x_n}$ for the attributes $x_1, \ldots, x_n$ one can construct a complex relation $G(y, x_1, \ldots, x_n) \Leftrightarrow y = f(x_1, \ldots, x_n)$ (defined by the axiomatic system) between y and $x_1, \ldots, x_n$, such that $f(x_1, \ldots, x_n)$ is a polynomial [6]. A polynomial function uses multiplication, power and sum operations. Hence, these operations can be defined for $y, x_1, \ldots, x_n$ using only relations $<_y, \approx_{x_1}, \ldots, \approx_{x_n}$. Ordering and equivalence relations are usually empirically interpretable in the ontology of various subject domains.

4. *Nominal discrete data types*. Here, all numbers can be considered as names, and can be easily represented as predicates with a single variable. So, data are interpretable in the corresponding relational structures, because there is no difference between the numerical and empirical systems.

5. *Non-quantitative and non-discrete data types*. Data contain no quantities and discrete variables, but do contain ranks, orders and other non-numerical data types. This case is similar to the above item 3a.

6. *Mix of data types*. All the mentioned difficulties arise in this case.

6 Invariance of the KDD&DM Methods

The results of the KDD&DM & ML methods must not depend on the subjective choice of the measurement units, but usually it is not the case. Let us define the notion of invariance of a KDD&DM & ML method. To that end, we will use the common (attribute-based) representation of a supervised learning [21, 22] (see Fig. 3), where:

$W = \{w\}$ is a training sample;

$X(w) = (x_1, \ldots, x_n)$ is the tuple of values of n variables (attributes) for training sample w;

$Y(w)$ is the target function assigning the target value for each training example w;

The result of the learning of some KDD&DM & ML method M on the training sample $\{\langle X(w), Y(w) \rangle\}$, $w \in W$ is a rule J

$$J = M(\{\langle X(w), Y(w) \rangle\}),$$

that predicts values of the target function Y(w). For example, consider w with unknown value Y(w), but with known values of all attributes X(w), then

$$J(X(w)) = Y(w),$$

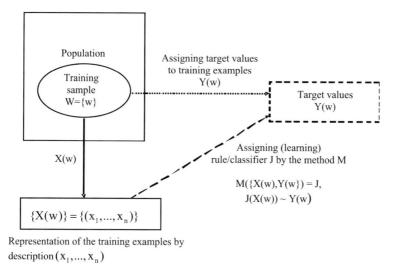

Fig. 3 Representation of a supervised learning

where $J(X(w))$ is a value generated by the rule J. The resulting rule J can be an algebraic expression, a logic expression, a decision tree, a neural network, a complex algorithm, or a combination of these models.

If attributes $(x_1, \ldots, x_n, y)$ are determined by the empirical systems $A_1, \ldots, A_n$, B having the transformation groups $g_1, \ldots, g_n, g$ respectively, then the transformation group G for all attributes is a product $G = g_1 \times \cdots \times g_n \times g$.

The KDD&DM & ML method M is invariant relative to the transformation group G iff for any $g \in G$ rules

$$J = M(\langle X(w), Y(w) \rangle), \ J_g = M(\langle gX(w), gY(w) \rangle),$$

produced by the method M, generate the same results for any $w \in W$

$$gJ(X(w)) = J_g(g(X(w))).$$

If the method is not invariant (that is the case for majority of the methods), then predictions generated by the method depend on the subjective choice of the measurement units.

The invariance of the method is closely connected to the interpretability of its results. The numerical KDD&DM & ML methods assume that operations such as $+, -, *, /$ can be used in an algorithm despite their possible non-interpretability. In this case, the method can be non-invariant and can deliver non-interpretable results. In contrast a KDD&DM & ML method M is invariant if it uses only information from empirical systems $A_1, \ldots, A_n, B$ as data and produces rules J that are logical expressions in terms of these empirical systems. This approach is a core of the Relational Data Mining.

7 Relational Methodology for the Analysis of KDD&DM Methods

A non invariant KDD&DM & ML method $M : \{X(w)\} \to J$ can be analyzed and another invariant method can be created based on M. Let us define a many-sorted empirical system A(W) that is a product of empirical systems $A_1, \ldots, A_n, B$ bound on the set W. Next we define transformation $W \to A(W)$ of data W into a many-sorted empirical system $A(W)$, and replace the numeric representation

$$W \to \{\langle X(w), Y(w)\rangle\}$$

by the transformation

$$W \to A(W) \to \{\langle X(w), Y(w)\rangle\} .$$

Based on method $M : \{\langle X(w), Y(w)\rangle\} \to J$, we define a new method $ML : A(W) \to J$ such that

$$ML(A(W)) = M(\{\langle X(w), Y(w)\rangle\}) = J,$$

using transformation $W \to A(W) \to \{\langle X(w), Y(w)\rangle\}$. Thus, method ML uses only interpretable information from data $A(W)$ and produces the rule J using method M.

Let us analyze the transformation of the interpretable information $A(W)$ into the rule J through the method M. If we apply only interpretable relations and operations from the method M (and change non interpretable ones to appropriate interpretable) to the interpretable information $A(W)$, we may extract some logical rule JL from rule J. This rule JL will contain only interpretable information from the rule J, expressed in terms of empirical system $A(W)$.

Let us define the next method

$$MLogic : A(W) \to JL,$$

where rule JL is a set of logical rules in terms of empirical system $A(W)$, produced by method M, and interpretable information $A(W)$.

The method MLogic is obviously invariant. If we consider all possible data for the method M, and all rules JL, that may be produced by the MLogic method, then we will obtain a class of rules (hypotheses) {JL} (knowledge space) of the method M.

As a result we obtain the ontology of the particular KDD&DM & ML method M as:

1. empirical system A(W) as a types of input data;
2. the set of relations and operations of the empirical system $A(W)$ as a language for data interpretation and hypothesis class {JL} construction (knowledge space);
3. discovered regularities as confirmed hypotheses.

Relational (ontological) approach to data mining includes (1) using an empirical system $A(W)$ as a form of the types of input data, (2) testing any hypotheses class {JL} in terms of empirical system $A(W)$ and (3) produce regularities as confirmed hypotheses. In this sense Relational (ontological) approach to data mining approximates any other Data Mining method.

Acknowledgements The work has been supported by the Russian Foundation for Basic Research (grant #15-07-03410-a) and Russian Federation grants (Scientific Schools grant of the President of the Russian Federation) #860.2014.1.

References

1. Kovalerchuk, B., Vityaev, E. Symbolic Methodology for Numeric Data Mining. Intelligent Data Analysis. Special issue on "Philosophies and Methodologies for Knowledge Discovery and Intelligent Data Analysis" eds. Keith Rennolls, Evgenii Vityaev. v.12(2), IOS Press, 2008, pp. 165–188.
2. Kovalerchuk, B., Vityaev, E., Ruiz, J.F. (2001). Consistent and Complete Data and "Expert" Mining in Medicine'. In: Medical Data Mining and Knowledge Discovery, Springer, pp. 238–280.
3. Vityaev, E.E. Knowledge discovery. Computational cognition. Cognitive process models. Novosibirsk State University Press, Novosibirsk, 2006, p. 293. (in Russian).
4. Vityaev, E., Kovalerchuk, B.Y. Relational Methodology for Data Mining and Knowledge Discovery. Intelligent Data Analysis. Special issue on "Philosophies and Methodologies for Knowledge Discovery and Intelligent Data Analysis" eds. Keith Rennolls, Evgenii Vityaev. v.12(2), IOS Press, 2008, pp. 189–210.
5. Fishbern PC (1970): Utility Theory for Decision Making. NY-London, J.Wiley&Sons.
6. Krantz, D.H., Luce, R.D., Suppes, P., Tversky, A. (1971, 1989, 1990). Foundations of measurement, Vol. 1,2,3—NY, London: Acad. press, (1971) 577 p., (1989) 493 p., (1990) 356 p.
7. Pfanzagl J. (1971). Theory of measurement (in cooperation with V. Baumann, H. Huber) 2nd ed. Physica-Verlag.
8. Kovalerchuk, B., Vityaev, E. (2000), Data Mining in finance: Advances in Relational and Hybrid Methods, Kluwer Academic Publishers, 308 p.
9. Vityaev E. The logic of prediction. In: Mathematical Logic in Asia. Proceedings of the 9th Asian Logic Conference (August 16–19, 2005, Novosibirsk, Russia), World Scientific, Singapore, 2006, pp. 263–276.
10. Demin A.V., Vityaev E.E. Universal system Discovery for knowledge acquisition development and its applications // Novosibirsk State University messenger, Information technologies, v.7, issue 1, Novosibirsk, 2009, p. 73–83. (in Russian).
11. The Progic series of conferences: http://www.kent.ac.uk/secl/philosophy/jw/progic.htm.

12. Smerdov, S.O., Vityaev, E.E. Logic, probability and learning synthesis: prediction formalization // Siberian electronic mathematical news. v.6, Institute of mathematics SD RAS, 2009, p. 340–365. (in Russian).
13. Vityaev, E., Kovalerchuk, B., Empirical Theories Discovery based on the Measurement Theory. Mind and Machine, v.14, #4, 551–573, 2004.
14. Vityaev, E., Smerdov, S. On the Problem of Prediction // K.E. Wolff et al. (Eds.): KONT/KPP 2007, LNAI 6581, Springer, Heidelberg, 2011, pp. 280–296.
15. Scientific Discovery website http://www.math.nsc.ru/AP/ScientificDiscovery.
16. Kovalerchuk, B., Vityaev, E. (1998): Discovering Lawlike Regularities in Financial Time Series. *Journal of Computational Intelligence in Finance 6 (3): 12–26.*
17. Kolchanov, N.A., Pozdnyakov, M.A., Orlov, Y.L., Vishnevsky, O.V., Podkolodny, N.L. Vityaev, E.E., Kovalerchuk, B.Y. Computer System "Gene Discovery" for Promoter Structure Analysis In: Artificial Intelligence and Heuristic Methods in Bioinformatics, Eds: P. Frasconi and R. Shamir, IOS Press, 2003, pp. 173–192.
18. Vityaev, E.E., Orlov, Y.L., Vishnevsky, O.V., Pozdnyakov, M.A., Kolchanov, N.A. Computer system "Gene Discovery" for promoter structure analysis, In Silico Biol. 2 (2002) 257–262.
19. Vishnevsky, O.V., Vityaev, E. E. Analysis and Recognition of Erythroid-Specific Gene Promoters Basing on Degenerate Oligonucleotide Motifs // Molecular Biology. November 2001, Volume 35, Issue 6, pp 833–840.
20. Kovalerchuk, B., Vityaev, E., Ruiz, J. (2000). Consistent Knowledge Discovery in Medical Diagnosis, IEEE Engineering in Medicine and Biology Magazine. Special issue: "Medical Data Mining", July/August 2000, pp. 26–37.
21. Pazzani, M., (1997), Comprehensible Knowledge Discovery: Gaining Insight from Data. First Federal Data Mining Conference and Exposition, pp. 73–82. Washington, DC.
22. Zagoruiko, N. G., Gulyaevskii, S. E., Kovalerchuk, B. Ya. Ontology of the Data Mining Subject Domain, Pattern Recognition and Image Analysis Journal, 2007, Vol. 17, No. 3, pp. 349–356. Pleiades Publishing, Ltd.

Ejectors for Efficient Refrigeration

Giuseppe Grazzini • Adriano Milazzo
Federico Mazzelli

Ejectors for Efficient Refrigeration

Design, Applications and Computational Fluid
Dynamics

 Springer

Giuseppe Grazzini
Department of Industrial Engineering
University of Florence
Florence, Italy

Adriano Milazzo
Department of Industrial Engineering
University of Florence
Florence, Italy

Federico Mazzelli
Department of Industrial Engineering
University of Florence
Florence, Italy

ISBN 978-3-030-09180-4   ISBN 978-3-319-75244-0   (eBook)
https://doi.org/10.1007/978-3-319-75244-0

This Springer imprint is published by the registered company Springer International Publishing AG part of Springer Nature.
The registered company address is: Gewerbestrasse 11, 6330 Cham, Switzerland

"To our families"

Preface: Past, Present, and Future of Ejector Refrigeration

The history of ejector refrigeration is tightly nestled in the broader history of refrigeration. When the steam jet chiller first appeared, many different refrigeration cycles had already been used for both refrigeration and air conditioning purposes. However, the steam jet refrigerator had the advantage that it could run using exhaust steam from any source (steam engines, industrial or chemical processes, etc.). Therefore, starting in 1910, steam jet refrigeration systems found use in breweries, chemical factories, theatres, ships, and trains.

Despite the promising start, the use of supersonic ejectors for refrigeration applications almost ended when Thomas Midgley Jr. and his associates introduced the first synthetic refrigerants during the 1930s. These gases could completely overcome the problems that hampered the large-scale commercialization of vapor compression systems. Meanwhile, steam plants became less common and electric energy was made broadly available in developed countries. Therefore, the steam jet cycles were gradually replaced by more efficient vapor compression systems (although some east European countries such as Czechoslovakia and Russia manufactured these systems as late as 1960[1]).

Despite the scarce success in refrigeration applications, the research on supersonic ejectors did not stop, given their widespread use in many other fields (as will be illustrated in Chap. 1). During the first half of the twentieth century, huge theoretical progresses were made in the understanding of the principles of aerodynamics applied to supersonic flows. The developments were pioneered by scientists like Ernst Mach, Ludwig Prandtl and his dynasty of brilliant students: Theodore von Karman, Theodore Meyer, Adolf Busemann, Hermann Schlichting and many others. By the end of the 1950s, Joseph H. Keenan and his colleagues at MIT had perfected the theory of mixing inside supersonic ejectors. Many design concepts have been developed since then, and systematic experimental activities have been performed to optimize system design.

---

[1] Arora, C.P., 2003. Refrigeration and Air Conditioning, Tata-McGraw-Hill.

   The oil crisis that followed the Yom Kippur War in 1973 and the rise in awareness of ozone depletion in 1974 laid out the groundwork for a revival of ejector systems. These two events, in conjunction with the increase in refrigeration demand and the appearance of stringent regulations on ozone depletion and global warming (Montreal and Kyoto Protocol in 1987 and 1992, respectively), prompted research toward new, economical, and environmentally safe technologies. As a result, ejector refrigeration experienced a renewed interest, and a great number of research centers worldwide started studies in this field.

   In recent years, thermodynamic optimization has been extensively pursued on ejectors and ejector chillers. Meanwhile, CFD has taken a leading role in the analysis of the internal flows within supersonic ejectors. Starting from a global analysis of the ejector behavior in terms of entrainment ratio and pressure lift, CFD simulations have been specialized on the study of complex internal phenomena, such as mixing and shock trains. These numerical results are now generating increasingly accurate results thanks to the appearance of new, sophisticated flow visualization studies. Meanwhile, a large number of configuration alternatives to the standard ejector cycle have been proposed, which include "passive" systems, cycles with multiple parallel ejectors, multiple nozzles, annular nozzles, and so forth.

   Yet, vapor compression and, to a lesser extent, absorption systems still completely dominate the refrigeration market. Quite surprisingly, the ejector design from a macroscopic point of view seems relatively unchanged in the last decades. Innovative proposals, like the above-cited unconventional configurations or improved design procedures, didn't impact the fundamental structure of the ejector chiller as would have been expected. Industries rely on their consolidated experience and proceed with understandable caution on the path of innovation, while the huge amount of scientific literature produced seems somewhat distant from practical application.

   To date, the use of ejectors to enhance the efficiency of conventional refrigeration cycles seems to be the most promising development. Ejector expansion cycles are emerging as enabling technology for CO_2 vapor compression cycles, as proven by the increased interest manifested by leading global players in the refrigeration arena (e.g., Danfoss or Carel). Other refrigerants (ammonia, hydrocarbons, or HFO) allow moderate losses in the expansion valve, and hence the insertion of additional components to increase the system efficiency could seem questionable. However, as the size increases, even a growth in efficiency below 10% may well be worth of a modest complication. In principle, the cost of a mass-produced ejector could be acceptable in comparison with other basic components of the refrigeration system. As will be shown in Sect. 1.3, the insertion of an ejector may be beneficial to the creation of an intermediate evaporation level or may allow liquid circulation in flooded evaporators. As soon as these new possibilities are fully pursued and the ejector effectively integrated with all other functions within the refrigeration system (e.g., regulation and control), we believe that the diffusion of ejectors for expansion recovery will become widespread. Also, the integration of ejectors in absorption systems seem to produce significant results without requiring too heavy modifications to the cycle even if, to the author's knowledge, there are no attempts of commercialization yet.

The future of supersonic ejectors in heat-powered refrigeration cycles, on the other hand, seems to be somewhat more uncertain. Research conducted over the years at our department suggests that ejector chillers can be easily manufactured with low-cost, off-the-shelf components (apart from the ejector itself, which must be tailored to the specific application under consideration). Unfortunately, the weak point of this type of system is still represented by the low thermodynamic efficiency. As an order of magnitude, one could set a target COP = 0.7, which is currently reached by commercial lithium bromide absorption chillers, but significantly higher than the typical values reported for ejector chillers in the same working conditions.

The use of solar energy to power ejector cycles seems a logical outcome but must be carefully evaluated. In a recent review, Kim and Infante Ferreira (2014)[2] make a comparison between solar thermal and solar electric cooling technologies, both in terms of thermodynamic performances and economic feasibility. The results show that, at present, the cheapest solution is represented by the PV panels coupled with commercial vapor compression chiller. This result is largely due to the recent dramatic decrease in PV cost and to the large production volumes that make vapor compression chillers very inexpensive.

However, solar thermal collectors have also seen a significant decrease of their cost, mainly due to the large amount of collectors manufactured and installed in China. In particular, evacuated tube collectors, thanks to their reduced heat loss, perform better at relatively high temperatures and have reached a high market share. These could be profitably adopted to power heat-driven refrigerators.

Waste heat recovery may represent another field for successful application. Heat recovery in industries, combined heating, power and cooling, or district heating and cooling may all potentially profit from the application of ejector chillers. Whenever waste heat is directly available in form of water vapor at moderate temperature, the steam ejector is undoubtedly the cheapest and simplest option.

An overturn of the current supremacy of absorption chillers on the heat-powered refrigeration market is unlikely and undesirable, but ejector refrigeration could offer an effective alternative in all cases where simplicity, reliability, and low investment costs are required. From a practical point of view, the key parameter in any refrigeration system is the total cost per unit cooling load. If the input thermal energy has low or zero cost (e.g., waste heat or solar power), the cost of cooling is mainly related to the investment, which therefore is the main objective function to minimize.

Compared with lithium bromide/water absorption refrigerators, ejector cycles may avoid problems of internal corrosion and crystallization of the solution. Ammonia/water absorption refrigerators, on the other hand, use a toxic fluid, while ejector chillers may use nontoxic, nonflammable, and environmentally safe refrigerants

---

[2]Kim, D.S., Infante Ferreira, C.A., 2014. Solar refrigeration options—a state-of-the-art review. International Journal of Refrigeration, 31, 3–15.

(e.g., water). These advantages potentially offer significant savings in terms of capital and life cycle maintenance costs.

Ejector refrigeration may also have a chance of playing a role in specific markets like developing countries, where access to electric power is limited and technical expertise for the maintenance and reparation of standard compression and absorption cycles is lacking. Other opportunities may hopefully arise in the future.

This book is an attempt to review ideas, techniques, results, and open issues, combining information from the open literature and from the industry. It is intended as a bridge between the extensive collection of scientific papers appearing on the relevant journals and the practical handbooks that have been published in the past, extending also toward the information published by the manufacturers. As such, it should promote and stimulate the discussion between the different players in the refrigeration arena and the experts of ejectors, including those who come from completely different fields.

23rd December 2017, Florence, Italy   Giuseppe Grazzini
                                                                      Adriano Milazzo
                                                                      Federico Mazzelli

Acknowledgments

The authors wish to thank Prof. Ian Eames for all the passionate and fruitful discussions about ejectors that inspired us while writing this book. Another very good friend, Prof. Yann Bartosiewicz, has shared with us many interesting ideas. Prof. Srinivas Garimella and Adrienne B. Little have also inspired us with some original proposals. Last but not least, Prof. Konstantin E. Aronson and his coworkers Dmitry V. Brezgin and Ilia Murmanskii have been involved in our work and contributed with original ideas.

Frigel S.p.A. has supported us technically and economically throughout all these years, and we are very grateful to Michele Livi, who has been responsible for the ejector project.

Since 2016, the authors have participated in a "PRIN" research project (funded by the Italian Ministry of University and Research for strategic national projects) named "Clean Heating and Cooling Technologies for an Energy Efficient Smart Grid" that groups eight Italian Universities working on refrigeration.

Special thanks also go to those people who belong or have been part of our research group and involved in our research on ejectors: Andrea Rocchetti, Francesco Giacomelli, Furio Barbetti, Jafar Mahmoudian, Simone Salvadori, Dario Paganini, and Samuele Piazzini.

Contents

List of Symbols: Global

Latin letters		Greek letters	
a	Speed of sound [m s^{-1}] or Van der Waals constant [m^5 kg^{-1} s^{-2}]	α	Volume fraction [−]
A	Cross section [m^2]	β	Mass fraction [−]
b	Van der Waals constant [m^3 kg^{-1}]	γ	Specific heat ratio [−]
B	Second virial coefficient [m^3 kg^{-1}] (4)	Γ	Liquid mass generation rate [kg m^{-3} s^{-1}]
$c_\mathrm{p}, c_\mathrm{v}$	Specific heat capacity [J kg^{-1} K^{-1}]	δ_ω	Vorticity thickness [m]
C	Third virial coefficient [m^6 kg^{-2}]	δ'	Mixing layer spreading rate
$\dot{C}$	Molecules condensation rate [n. molecules s^{-1}]	ε	Turbulence dissipation rate [m^2 s^{-3}]
d	Primary nozzle diameter [m]	η	Efficiency [−] or droplets per unit volume mixture [m^{-3}]
D	Mixer/diffuser diameter [m]	θ	Expansion ratio
e	Specific internal energy [J kg^{-1}]	Θ	Velocity ratio
E	Internal energy [J]	Λ	Square root of density ratio
$\dot{E}$	Molecules evaporation rate [n. molecules s^{-1}]	μ	Dynamic viscosity [kg m^{-1} s^{-1}]
$\dot{Ex}$	Exergy flux [W]	ν	Cinematic viscosity [m^2 s^{-1}]
f	Darcy friction factor [−]	ξ	Geometric ratio or Kantrowitz non-isothermal correction
F	Helmholtz free energy [J]	ρ	Density [kg m^{-3}]
g	Gravity acceleration [m s^{-2}] or specific Gibbs free energy [J kg^{-1}]	ς	Compression ratio
G	Gibbs free energy [J]	σ	Surface tension [J m^{-2}]
h	Specific enthalpy [J kg^{-1}]	τ	Shear stress [Pa]
h_lv	Latent heat [J kg^{-1}]	υ_l	Liquid kinematic viscosity [m^2 s^{-1}]
$\dot{I}$	Nucleation current [n. molecules s^{-1}]	φ_ss	Supersaturation degree
J	Nucleation rate [s^{-1} m^{-3}]	φ_sc	Supercooling degree

(continued)

k	Turbulence kinetic energy [m^2 s^{-2}] or thermal conductivity [W m^{-1} K^{-1}]	ϕ	Loss coefficient
k_b	Boltzmann constant [J K^{-1}]	Φ	Generic flux
K_a	Average Roughness Height [m]	ω	Entrainment ratio [−], specific dissipation rate [s^{-1}]
K_{rms}	Root mean squared roughness [m]	**Subscripts**	
K_{sg}	Equivalent sand-grain roughness [m]	∞	Undisturbed, isentropic flow region
l	Molecular mean free path [m]	A	Condenser exit
L	Duct length [m]	C	Condenser, discharge
m	Mass [kg]	c	Critical droplet
m_m	Mass of one molecule [kg]	crit	Fluid critical point, critical ejector pressure
$\dot{m}$	Mass flow rate [kg s^{-1}]	CS	Condenser source (external circuit)
$\dot{m}_v$	Mass velocity [kg m^{-2} s^{-1}]	d	Droplet
M	Mach number	D	Discharge
M_c	Convective Mach number	e	Nozzle exit (ESDU procedure)
M_{mol}	Molar mass	E	Entrained fluid, evaporator
n	Number of droplets per unit mass of mixture [kg^{-1}]	ES	Evaporator source (external circuit)
n_x	Droplet size distribution	eff	Effective = laminar + turbulent
P	Pressure [Pa]	f	Refrigeration
q	Specific heat power [W kg^{-1}], heat transfer rate per unit area [W m^{-2}]	ff	Flat film
q_C, q_E	Accommodation factors	g	Generator
Q	Heating or cooling energy [J]	G	Motive fluid, generator
$\dot{Q}$	Heating or cooling power [W].	GS	Generator source (external circuit)
r	Radial coordinate [m], radius [m]	in	Input
R	Specific gas constant [J kg^{-1} K^{-1}]	irr	Irreversible
s	Specific entropy [m^2 s^{-2} K^{-1}]	l	Liquid
S	Entropy [J kg^{-1}]	m	Mixing section, mixed flow, mixture, or molecule
S_x	Surface of a liquid cluster with x molecules [m^2]	p	Primary flow, primary nozzle
t	Time [s]	rev	Reversible
T	Temperature [K]	s	Secondary flow
u	Velocity [m s^{-1}]	sat	Saturation
U	Time-averaged velocity [m s^{-1}]	sh	Superheating
v	Specific volume [m^3 kg^{-1}]	sub	Subcooling
V	Volume [m^3]	t	Turbulent, total/stagnation condition
w	Specific shaft work [J kg^{-1} s^{-1}]	th	Throat
W	Shaft work [J]	v	Vapor
$\dot{W}$	Shaft power [W]	y	Virtual secondary choke section

(continued)

x	Axial coordinate [m] or vapor quality	'	Primary flow (ESDU procedure)
y	Transversal or radial coordinate [m]	"	Secondary flow (ESDU procedure)
z	Vertical coordinate [m]		
Acronym			
COP	Coefficient of performance		
CRMC	Constant rate of momentum change		
EEV	Electronic expansion valve		
ER	Entrainment ratio		
FP	Feed pump		
Gb	Gibbs parameter		
GWP	Global Warming Potential		
HTC	Heat transfer coefficient [$\mathrm{W\ m^{-2}\ K^{-1}}$]		
N.B.P.	Normal boiling point		
NXP	Nozzle exit plane		

Chapter 1
Introduction

1.1 Working Principle of Ejectors

The basic scheme of an ejector is shown in Fig. 1.1. The shape and proportioning of the parts are purely indicative. The motive (or "primary") fluid is fed through a nozzle which, in most cases, is shaped as a converging/diverging duct in order to accommodate a supersonic flow at the exit. The entrained (or "secondary") fluid is fed through the annular space that surrounds the primary nozzle. In this way, at the nozzle exit the two streams come in touch. Their velocities are highly different, and hence a transfer of momentum accelerates the secondary and decelerates the primary flow. We may imagine that a central core of primary flow and a lateral shell of secondary flow remain substantially unaffected, while the mixing takes place in an intermediate zone shaped as a cylindrical wedge, where turbulent shear stress produces a velocity distribution that grows steeply toward the ejector axis. Actually, if the primary flow is supersonic, a sequence of oblique shocks will form along the mixing zone, undergoing multiple reflections.

In modern applications, the secondary flow normally accelerates up to sonic speed, and hence the whole mixed stream is supersonic. This mixed stream must be decelerated in order to convert its kinetic energy and finally reach the exit pressure, intermediate between the high value featured by the motive fluid at inlet and the low value ("suction pressure") of the entrained flow. This happens in a supersonic diffuser which follows the mixing zone and features a convergent-divergent or cylindrical-divergent shape.

It may be worth to point out that, from a functional point of view, the ejector substitutes the much more complex assembly shown in Fig. 1.2. The ejector eliminates the transmission of mechanical work from the expansion to the compression via the connecting shaft. Flow energy is transmitted directly from the two flows, avoiding rotating blades, bearings, lubrication, etc.

Obviously, the direct interaction between streams at different velocities introduces some limitations and specific losses. For example, the transition between

© Springer International Publishing AG, part of Springer Nature 2018
G. Grazzini et al., *Ejectors for Efficient Refrigeration*,
https://doi.org/10.1007/978-3-319-75244-0_1

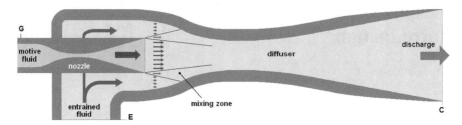

Fig. 1.1 Schematic section of a supersonic ejector

Fig. 1.2 Functionally
equivalent assembly

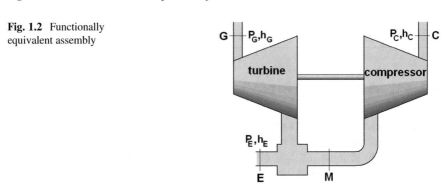

super- and subsonic flow should ideally occur at the diffuser throat, and the velocity
should decrease continuously. This ideal condition happens for a single combination
of inlet/exit conditions and is hence practically unfeasible. In practice, the supersonic
flow decelerates to subsonic velocity through a second shock train.

For stable operation the shock should take place downstream of the diffuser
throat. In this condition, the ejector flow rates are insensitive to any increase in the
discharge pressure. When discharge pressure increases, the shock moves toward the
inlet side of the ejector, and, as it reaches the throat, the ejector experiences its most
efficient working condition.

However, any further small increase in the discharge pressure causes the flow to
become subsonic in the throat, and hence the flow rate becomes dependent on the
discharge pressure. In this condition the ejector becomes unstable, i.e., an increase in
the discharge pressure produces a steep decrease in the ejector performance which,
in most applications, is unacceptable. Many ejectors feature a cylindrical zone
upstream of the conical diffuser. They work in a stable condition as far as the
shock occurs within this cylindrical zone.

A set of nondimensional parameters may be introduced:

- Entrainment ratio $\omega = \dot{m}_s/\dot{m}_p$, i.e., ratio between the secondary ($\dot{m}_s$) and the
  primary ($\dot{m}_p$) mass flow rates
- Compression ratio $\zeta = P_C/P_E$, i.e., ratio between the discharge (P_C) and entrained
  fluid (P_E) pressures
- Expansion ratio $\theta = P_G/P_E$, i.e., ratio between the motive (P_G) and entrained fluid
  (P_E) pressures

Fig. 1.3 Map of the operation for a supersonic ejector

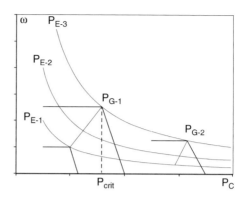

The aforementioned behavior may be described in terms of entrainment ratio as shown in Fig. 1.3. For a given combination of primary and secondary conditions, the line representing ω as a function of the discharge pressure P_C has a first horizontal part on the left and a second sharply decreasing part on the right. The dividing point corresponds to the *critical discharge pressure* P_{crit} (or maximum discharge pressure) that is commonly taken as the limit operating condition. This critical point conjugates the maximum entrainment ratio with the maximum compression ratio ζ_{crit}.

When the entrained fluid pressure is lowered, e.g., from P_{E-3} to P_{E-1} at constant primary fluid pressure P_{G-1}, the operating curve moves left and downward, i.e., both ζ_{crit} and ω decrease.

When the motive fluid pressure is raised, e.g., from P_{G-1} to P_{G-2} at constant entrained fluid pressure (θ increases), ω is lowered but ζ_{crit} increases.

Another fundamental parameter is the area ratio between primary nozzle and mixer/diffuser throat flow sections. In supersonic conditions, these sections limit the two flow rates and hence the entrainment ratio. We may hence introduce a ratio $\xi = d/D$ between the nozzle throat diameter d and the diffuser throat diameter D. As this ratio increases, the motive flow increases and, as a rule, the compression ratio grows. Correspondingly the entrainment ratio decreases, because the entrained flow remains constant or even decreases, as an increased portion of the diffuser throat section is occupied by the primary flow.

1.1.1 Ejectors as Components of Refrigeration Systems

An ejector may be used as a fluid-driven compressor in a refrigeration system. The motive fluid can be heated up in a "generator," where it boils at constant pressure. The entrained fluid is vaporized at low pressure in an evaporator, in order to extract the cooling load from a cold source. The mixed fluid at discharge is condensed and, once in the liquid state, is divided into two flows: the first goes to the evaporator through an expansion valve, and the second is pumped back at suitable pressure toward the generator (Fig. 1.4).

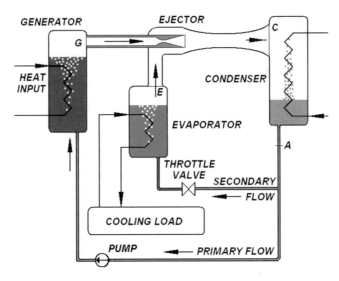

Fig. 1.4 Basic scheme of ejector-based refrigeration system

The ideal thermodynamic cycle, shown in Fig. 1.5 for steam on a pressure/enthalpy diagram, is actually comprised of two cycles sharing the condensation C-A: the motive cycle has a practically vertical left side (the pump absorbs very little work in the case of water) and then has a constant pressure heating, vaporization, and superheating (if present) up to point G. From this point, the vapor expands in the primary nozzle and mixes with the vapor exiting from the evaporator at state E. In this simplified representation, the mixing process is assumed at constant pressure, slightly below the evaporator exit pressure. Actually, in the real process, the expansion of the entrained fluid before mixing is scarcely significant, and the mixing pressure is hardly distinguishable on the diagram from the evaporator pressure. The mixed fluid is then compressed in the diffuser up to the condenser pressure and is discharged at state C. The entrained fluid exiting from the condenser in state A is expanded through a valve, and the process is assumed isoenthalpic.

The cycle efficiency may be calculated as a ratio between useful effect and input power:

$$\text{COP} = \frac{\dot{Q}_f}{\dot{Q}_m + \dot{W}_{\text{pump}}} = \omega \frac{h_E - h_A}{h_G - h_A} \tag{1.1}$$

where $\dot{Q}_f$ is the cooling power, $\dot{Q}_m$ the motive heat power, and $\dot{W}_{\text{pump}}$ the generator feed pump power.

The last part of Eq. 1.1 contains the entrainment ratio and a further ratio between the enthalpy differences $(h_E - h_A)$ and $(h_G - h_A)$ that depends on the fluid and operating conditions. Therefore, the global performance for fixed fluid and working

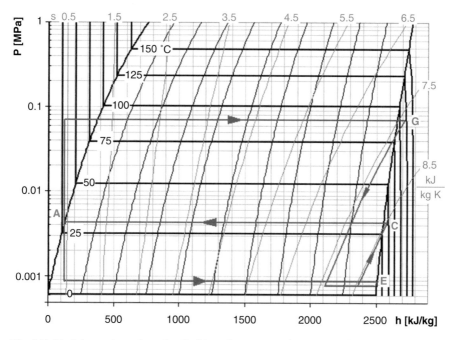

Fig. 1.5 Ideal thermodynamic cycle of refrigeration system using a steam ejector

conditions depend on the ejector entrainment ratio. This latter may be evaluated from an energy balance on the ejector and turns out to be

$$\omega = \frac{h_G - h_C}{h_C - h_E} \tag{1.2}$$

i.e., the ratio between motive $(h_G - h_C)$ and compression $(h_C - h_E)$ enthalpy differences. Ejector losses increase the enthalpy at condenser entrance h_C, decreasing numerator and increasing denominator in Eq. 1.2.

As a heat-powered refrigeration system, the machine sketched in Fig. 1.4 may be used in addition to a heat engine for combined heating, cooling, and power generation. For example, it may complement a district heating system in order to guarantee air conditioning in summer or produce refrigeration in civil or industrial environments from any form of waste heat. Alternatively, it may be used for "solar cooling" in conjunction with solar thermal panels having a suitably high temperature or use other forms of renewable energy.

1.2 Historical Background

An early application of a rudimentary jet device may be seen in the "blast pipe" found in the smokebox (the volume at the end of the boiler where smoke and ash are collected before going toward the chimney) since the very first steam locomotives (Fig. 1.6). This device directs exhaust steam from the cylinders through a nozzle at the bottom of the chimney, in order to reduce the flue gas pressure and increase the draught on the fire. The blast pipe significantly increased the locomotive power and produced the familiar intermittent flow of smoke from the chimney, synchronous with the alternate motion of the pistons.

The French inventor Henri Giffard, as a locomotive engineer, was familiar with the blast pipe. When he attempted to apply a steam engine to a dirigible, Giffard deserved a simple, light, and reliable means to pump water in the boiler. Mechanical feed pumps were heavy and complex. Since 1850 he understood the basic principle of the steam injector, writing down a momentum equation which is basically still valid (Kranakis 1982). The attempt to build a steam-powered flying vehicle was unsuccessful, but the injector survived and was patented in 1858, gaining immediate success.

In contrast to pumps, the injector was small and efficient, had no moving parts (which eliminated a major source of friction), needed no oiling, and even served to preheat the feedwater. Injectors had an additional advantage with regard to locomotive boilers because they would operate when the locomotive was at rest.

Because of its simplicity and efficiency, the injector made traditional pumps virtually obsolete. By 1860 several French railroad lines were regularly outfitting their locomotives with injectors, and they were also in use by the French Navy. At the London Exhibition of 1862, nearly one-third of the locomotives were equipped solely with injectors. Injectors were manufactured in America since 1860, and a year later nearly 1200 had been sold in the USA alone. At the turn of the century, the number had risen to over a half million. By that time, many locomotive engineers had never seen one of the older feed pumps.

Figure 1.7 shows a very refined construction, with a non-return valve at the water discharge port and an overflow exit that are necessary when the ejector is started. A further refinement is the needle that varies the nozzle exit section. Unfortunately, the properties of expanding steam were not completely known in that period, and hence

Fig. 1.6 Blast pipe in a steam locomotive

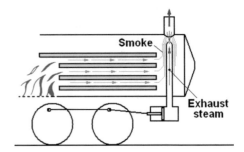

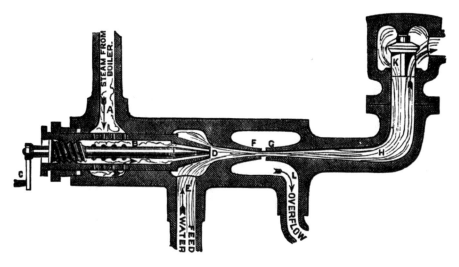

Fig. 1.7 Giffard's injector (Routledge 1876)

the need for a convergent-divergent nozzle was not realized. By the way, the modeling of an expanding flow of metastable, condensing steam is still quite troublesome nowadays.

It's worth to note that the injector, having a compression ratio equal to the expansion ratio, gives a first proof of the importance of a correct energy balance in the evaluation of these devices. The high enthalpy of the steam coming from the boiler is transformed in kinetic energy within the nozzle and, once transferred to the feed water flow, may overcome a relatively high discharge pressure. This was very surprising for the engineers of the middle nineteenth century and stimulated a lively discussion (Kranakis 1982), eventually promoting a widespread comprehension and acceptance of the newly born first law of thermodynamics.

The same principle successfully demonstrated by the injector was easily transferred to the steam ejector (Fig. 1.8) used on early locomotives as a vacuum pump for the brake circuit (Encyclopædia Britannica 1911). In this case, the large availability of steam and the simplicity and ruggedness of ejectors offered a convenient technical solution to the urgent need of a reliable braking system (the vacuum brake was the first intrinsically safe method because it automatically stops the train whenever the circuit is accidentally opened to atmosphere). The vacuum brake had major success in the UK, where it survived until the 1970s.

Another fundamental field of usage for ejectors was soon found in steam power plants, where incondensable gases had to be evacuated from the condenser and steam at various pressure levels was available along the expansion. An ejector was used for this purpose by Sir Charles Parson in 1901 (Chunnanond and Aphornratana 2004). Again, a steam ejector was cheaper and more reliable than any other kind of vacuum pump.

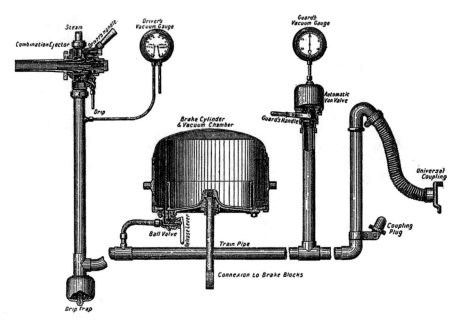

Fig. 1.8 Vacuum brake for trains – the ejector is on the upper left corner (Encyclopædia Britannica 1911)

Coming to the main topic of this book, we must finally mention the "Machines frigorifiques à vapeur d'eau et à éjecteur" invented by Maurice Leblanc (1911). The idea of using an ejector to produce a low-pressure reservoir wherein water could be evaporated and subtract heat at low temperature is even older, but the commonly cited reference year for this invention is 1910. The system is also known as "Westinghouse-Leblanc" due to the rapid commercialization in the USA and was very successful in this period, especially for use on ships. This is not surprising, as the ejector refrigerator must get rid of a substantial amount of heat (sum of the cooling load and motive heat) and is very sensitive to the heat sink temperature. Therefore, the availability of cool seawater was undoubtedly a key for success in this application. Simplicity and ruggedness were also very convenient for ship operators. Many different refrigeration cycles had already been used for both refrigeration and air conditioning purposes. However, the steam jet refrigerator had the advantage that it could run using exhaust steam from any source (steam engines, industrial or chemical processes, etc.). Hence, from 1910 to the early 1930s, steam jet refrigeration systems were successful in factories, for air conditioning of large buildings and on trains (Stoecker 1958; Arora 2003).

Despite this promising start, the use of supersonic ejectors for refrigeration applications almost disappeared when the first synthetic refrigerants were introduced during the 1930s. These gases could overcome the problems that hampered the large-scale commercialization of vapor compression systems. Furthermore, steam was losing its importance as heating fluid, while all buildings hosting refrigerators or air conditioners were served by the electric energy network. Therefore, however scarcely efficient and reliable initially, electrically operated compressors became increasingly common and soon virtually unrivalled. Steam ejector chillers survived (and are still in use) in those industrial plants where steam is available at low cost.

Much later, new perspectives opened for ejector chillers. In the 1970s, two events, the oil crisis and the stratospheric ozone depletion, suddenly interrupted the apparently unlimited growth of the conventional refrigeration market. On one side the consolidated working fluids were questioned and eventually phased out in a relatively short period. On the other hand, refrigeration and air conditioning were recognized as an important item in the inventory of electricity end users. This stimulated a renewed interest toward heat-powered refrigeration systems and environmentally safe working fluids. A significant amount of literature was published along the 1980s and the 1990s, reporting a number of thermodynamic analysis and various experimental results.

Unfortunately for ejector chillers, by that time absorption systems were ready to dominate the market of heat-powered refrigeration. Since their invention in 1858 by Ferdinand Carré, absorption machines enjoyed a strong research effort, and now, specially thanks to the good performance at low heat source temperature offered by lithium bromide systems, they are mass produced and leave few chances to competitors.

A possible breakthrough for ejector refrigeration systems was sought in the use of alternative working fluid. If steam is not provided and the heat source has a relatively low temperature, the circuit may be filled with any suitable refrigerant. The quest for alternatives started since the 1990s (Dorantes and Lallemand 1995) and is still open. Many authors propose synthetic refrigerants with the aim of overcoming the indisputable drawbacks of steam, i.e.:

- The high triple-point temperature that impedes to work below 0 °C
- The very low pressure at evaporator and condenser that poses sealing problems
- The high specific volume at low pressure that increases the system volume and cost
- The problem of condensation along the expansion that complicates the analysis and the design of the ejector

On the other hand, steam has various advantages; it is available everywhere at very low cost, is absolutely safe for the environment and operators, and requires a low pressure at generator.

In this last decade, another promising use of ejectors within refrigeration systems has emerged. Instead of trying to replace the mechanical compressor, the ejector can be used to complement it, working on the expansion side of the cycle. In this case, the motive fluid is the high-pressure liquid coming from the condenser, and the entrained fluid is the low-pressure vapor coming from the evaporator. The mixed flow, after pressure recovery, is sent to an intermediate pressure reservoir, e.g., a liquid separator that feeds the expansion valve (and hence the evaporator) and the compressor. With respect to other proposed means for expansion work recovery, the ejector has the advantage to have no moving parts. This idea dates back to the patent

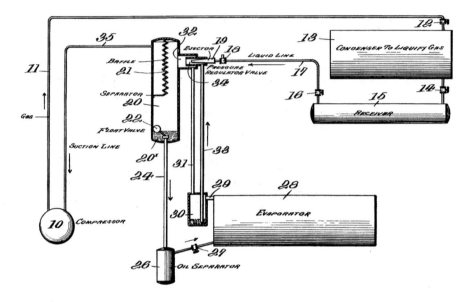

Fig. 1.9 Ejector for expansion work recovery (Gay 1931)

(Fig. 1.9) of Norman H. Gay (1931) but has known a rapidly increasing interest since
the rebirth of carbon dioxide refrigeration systems in the 1990s, due to the high
incidence of the expansion loss in CO_2 inverse cycles.

1.3 Applications

As pointed out, ejectors may have a number of applications and are referred to by
different names. A classification may be found, e.g., in Sokolov and Zinger (1989):

1. Ejectors using motive and entrained fluid in the same phase
2. Ejectors using motive and entrained fluid in different phases but having no phase
   change within their bodies
3. Ejectors using motive and entrained fluid in different phases and featuring a phase
   change

Clearly, the difficulty of analysis increases from case 1 to 3. For example, an
ejector used in a heat-powered chiller featuring a working fluid that does not
condense along the expansion belongs to the first category, while an ejector used
for expansion work recovery in a vapor compression cycle belongs to the third one.

Other sources (DIN 24290:1981-08) classify the jet devices according to the
nature of entrained and motive flows, as shown in Table 1.1.

For gaseous entrained flows, the code makes a further distinction according to the
entrained flow pressure level and compression ratio.

Table 1.1 Classification of jet devices (DIN 24290:1981-08)

	Motive flow		
Entrained flow	Gas	Steam	Liquid
Jet ventilator	Gas jet ventilator	Steam jet ventilator	Liquid jet ventilator
Jet compressor	Gas jet compressor	Steam jet compressor	Liquid jet compressor
Jet vacuum pump	Gas jet vacuum pump	Steam jet vacuum pump	Liquid jet vacuum pump
Jet liquid pump	Gas jet liquid pump	Steam jet liquid pump	Liquid jet liquid pump
Jet solid pump	Gas jet solid pump	Steam jet solid pump	Liquid jet solid pump

A list (not necessarily exhaustive) of industrially proven applications may include (Transvac 2017):

- *Steam ejectors for vacuum processing*, competitive whenever steam is available, and the need of high reliability suggests avoiding mechanical vacuum pumps
- *Water and wastewater treatment*, e.g., desalination, tank mixing and aeration, dosing of additives, ozone injection, and slurry pumping and heating
- *Fluid and solid handling* using liquid as the motive force to entrain, mix, and dilute a secondary solid (powder or granules), liquid, or gas, eliminating the need for mechanical dosing pumps and mixers or compressors
- *Scrubbing and pollution control*
- *Oil and gas*, e.g., for enhanced oil recovery, production boosting, flare gas recovery, and wastewater treatment
- *Nuclear power plants*, e.g., for slurry transport, tank mixing, and fluid pumping

For example, an ejector may enhance oil recovery from exhausting wells by using a small fraction of the high-pressure gas coming out from the compressor as a motive fluid within an ejector. The liquid separator that collects the outcome of the oil wells is connected to the entrained flow inlet of the ejector. In this way, the compressor is fed by the ejector discharge pressure and may continue to work at design conditions even if the oil well pressure reduces. In other cases, the motive fluid is high-pressure water, and the entrained fluid is a multiphase mixture (oil, gas, and water). The mixture components are then separated by gravity, downstream of the ejector discharge.

Ejectors are useful as tank mixers whenever the fluid at hand is corrosive or flammable. Mixing is hence performed without moving parts or electric motors, while the pump may be placed outside of the tank.

Vacuum is useful in innumerable processes (drying, distillation, extraction, etc.). Ejectors may reach high vacuum through multistage configurations and/or in combination with other kinds of vacuum pumps. A manufacturer (Shutte Koerting 2017) offers ejectors featuring up to six stages, with intermediate condensers, that may reach an absolute pressure of 3 μm Hg (0.4 Pa). Interestingly, manufacturers that offer hybrid systems (e.g., (Croll Reynolds 2017)) use the multistage ejector on the low-pressure end of the cascade, while the liquid-ring vacuum pump is used for discharging to atmosphere, in order to reduce the steam consumption for the process.

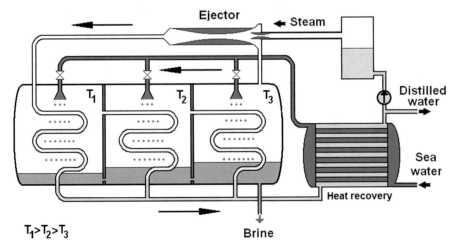

Fig. 1.10 Ejector on a MED desalination plant

High vacuum is required in distillation processes whenever high temperature may damage the product.

According to another manufacturer (Graham 2017), in vacuum oil refining the volume of gas to be evacuated is so large that ejectors are the only affordable device. The combination of ejectors and condensers produces and maintains subatmospheric pressure within the distillation column to permit fractionation of crude oil into its various components, such as light or heavy vacuum gas oils, and reduce the amount of lower valued residuum. The ejector system continually extracts from the distillation column cracked and inert gases along with associated saturated steam and hydrocarbon vapors.

Another relevant topic is multi-effect distillation (MED) (Fig. 1.10) that is one of the most important and widely used large-scale desalination methods. The energy consumption of MED seawater desalination systems may be reduced by the use of ejectors between the stages. Again, the ejector is perfectly fit for the sizes required by such plants.

The possibility to handle solids or corrosive fluids is enhanced by the relatively simple shape and by the absence of mechanical couplings and moving parts. These circumstances allow using a wide range of materials (stainless steel, bronze, ceramic, plastics, etc.) that may be tailored to withstand virtually any kind of flowing media. Moreover, even if erosion takes place on some parts of the ejector, these parts are easier and cheaper to substitute than a mechanical compressor or pump.

Another manufacturer (GEA Wiegand GmbH 2017) includes in their range of products and services evaporation plants, membrane filtration, distillation/rectification plants, alcohol production lines, condensation plants, steam jet cooling plants, steam vacuum plants, heat recovery plants, and gas scrubbers. Moreover, they offer a specific line of standardized steam jet cooling plants covering a range from 20 to 15,000 kW of refrigeration. These plants will be dealt with in the next subsection.

1.3.1 Refrigeration Applications

Coming specifically to refrigeration applications, the ejector, as seen in Sect. 1.1.1, substitutes the compressor and integrates the cooling cycle with a heat-powered engine cycle. In theory, one could think of a direct combustion system where a fuel is fed to the generator, but obviously the use of renewable or waste energy is much more appealing.

In the case of waste energy, normally the energy input is in the form of hot fluid. Thinking, e.g., to flue gases coming from an internal combustion engine or from a gas turbine, the generator simply takes the form of a recovery heat exchanger and may be designed with standard procedures. Obviously, the higher is the temperature of the waste heat, the higher is the exergy of the heat input, and potentially the useful effect grows. A careful selection of the working fluid is imperative, in order to maximize the heat recovery and the global efficiency.

A special case is that of steam as heat input fluid. In this case, the ejector may be dramatically simplified by feeding directly the hot steam to the primary nozzle, whence it acts as motive fluid (Power 1993).

The plants on the high side of the range typically feature a vertical construction, with condenser on top. These plants may be several meters high, and hence the low pressure required for condensation at ambient temperature is guaranteed by gravity (Fig. 1.11). This is made possible by the high density of liquid water. Note also in Fig. 1.11 the double-stage evaporator that directly feeds the chilled water circuit, using again gravity and a pump. Steam is produced elsewhere.

Typically, the performance of these plants is expressed in terms of steam consumption per unit cooling power, which makes unpractical and somewhat questionable the calculation of a COP in the usual nondimensional form. However, the efficiency generally has an increasing trend as the plant size grows.

These big plants have significant air infiltrations toward all the parts at low pressure. Therefore, a continuous air extraction is performed on top. Smaller plants have horizontal configuration and can have a closed circuit, in order to avoid air infiltration. A closed circuit is necessary when the working fluid is synthetic and must be kept within the circuit for long periods without refilling.

Another quite obvious application of ejector refrigeration is solar cooling. A well-known advantage of this practice is the temporal coincidence between heat input and cooling load, even if in reality some kind of energy storage may be needed.

The overall COP of the solar jet refrigeration cycle can be approximately expressed as (Chunnanond and Aphornratana 2004)

$$\text{COP}_{\text{overall}} = \eta_{\text{solar\_panel}} \cdot \text{COP}_{\text{ejector\_chiller}} \tag{1.3}$$

This shows that the performance and cooling capacity of the solar-powered system depend strongly on the type of solar collector. When the generator of the refrigeration system is operated at a temperature between 80 and 100 °C, a single glazed flat plate type collector with a selective surface is recommended, while the

Fig. 1.11 Steam jet
industrial cooling plant

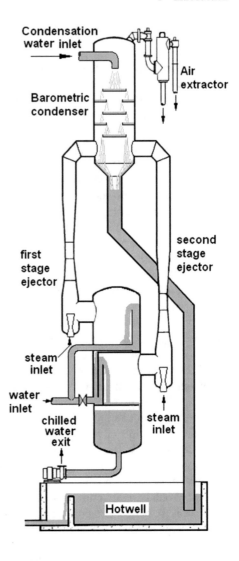

vacuum tube or parabolic solar concentrating collectors can provide higher operating
temperature when required (Chunnanond and Aphornratana 2004). Although the
installation of these types of high-efficiency collectors may result in increase in the
overall efficiency, a careful economic analysis is needed to understand whether the
relatively higher costs are worth the efficiency gains (Zhang and Shen 2002; Huang
et al. 2001; Nguyen et al. 2001).

In the simplest configuration, the solar collector acts as the generator of the
refrigeration system. However, this configuration poses some problems in terms of
pump sizing and regulation because the solar collector and refrigerator circuits may
require different flow rates and operating pressures (the control of the optimal flow
rate and boiler pressure becomes difficult with a single pump (Al-Khalidy 1997)). In

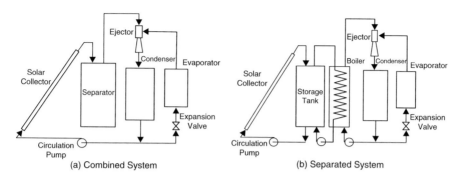

Fig. 1.12   Solar-powered ejector cycles. Left side: combined configuration. Right side: separated configuration (From Chunnanond and Aphornratana (2004))

order to eliminate this restriction, the solar and refrigeration systems are usually separated. Figure 1.12 shows both the integrated and separated configurations. In the separated configuration, the fluid circulating in the solar collector should have significant heat transfer properties, as well as a boiling point higher than the possible temperature occurring in the system (Chunnanond and Aphornratana 2004).

Many research groups worldwide have performed theoretical calculation, computer simulation, and experimental work on solar-powered ejector chillers (Abdulateef et al. 2009).

Up to now, we focused on "ejector-based" chiller configurations. However, supersonic ejectors can also be used to improve the performance of different types of conventional refrigeration systems by recovering throttling losses and boosting compressor efficiency.

A great variety of hybrid cycles can be found in the literature, from combined vapor compression/ejector systems to absorption or adsorption-ejector refrigerators (see (Besagni et al. 2016) for an extensive review). Among the various alternatives, the most promising solutions in terms of practical attainability and potential performance increase seem to be the enhancement of standard vapor compression and absorption cycles. In these types of applications, the ejector is used to recover the throttling loss inside any of the expansion valves present in the cycle. This allows appreciable efficiency increases without requiring significant modifications of the cycle configuration.

A large number of different systems for the throttling loss recovery have been proposed over the years: micro turbines, screw expanders, scroll expander, reciprocating engines, etc. However, due to the difficulties in the realization of ad hoc, reliable, two-phase flow devices, in most cases the efficiency gain is not worth of the costs of the new solution. In this context, the use of an ejector is particularly advantageous, as its operation and design are inherently robust with respect to two-phase flow (e.g., limited or absent problem of erosions) and the device is simple and inexpensive.

A first example of ejector for throttling loss recovery, as seen in Sect. 1.2, is the one introduced by Gay (1931) and is shown in Fig. 1.13a, where a liquid separator

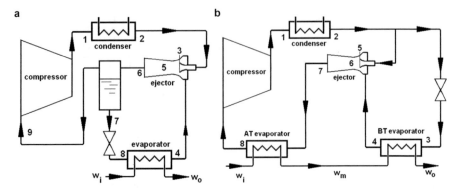

Fig. 1.13 Various configurations of expansion work recovery with ejectors

acts as intermediate pressure level and the recovered work is directly subtracted from the compressor that operates on a reduced compression ratio. The ejector receives high-pressure liquid coming from the condenser; the liquid expands inside a convergent-divergent nozzle following an (ideally) isentropic transformation. Inside the nozzle, the liquid refrigerant undergoes phase change and enters the mixing chamber as a vapor-liquid mixture with relatively low quality. Inside the mixing chamber, the primary flow entrains and accelerates the saturated vapor coming from the evaporator. The two streams mix and then are recompressed along the ejector diffuser. The flow coming out from the ejector, a high-quality mixture of vapor and liquid, enters the separator whence the vapor is sent to the compressor, while the liquid passes through a valve that reduces its pressure down to the evaporator level. Due to the small pressure gap between the separator and evaporator, the throttling loss of the combined chiller is much lower than in conventional cycles. The energy recovered from the liquid is spent to compress the vapor coming from the evaporator, thus reducing the pressure ratio and work of the compressor.

A drawback of the scheme shown in Fig. 1.13a is the presence of a liquid separator that hosts a relatively large mass of fluid and hence adds weight, cost, and risk, in case of refrigerants posing safety or environmental problems. The alternative scheme shown in Fig. 1.13b substitutes the liquid separator with a second evaporator, at intermediate pressure, that receives the partially evaporated fluid coming from the ejector. In this way, the availability of two temperature levels allows to face a large range of temperature on the cold fluid side. For example, this system has been applied to air conditioning on cars, where hot ambient air has to be brought down to below the dew point in order to perform dehumidification, and hence a single evaporation temperature introduces high heat transfer irreversibility. A slightly modified scheme has been employed in air conditioning systems for the automotive industry (Ishizaka et al. 2009), where the ejector may be just a few centimeters long and fits in the manifold of the evaporator.

The throttling loss increases as the cycle approaches or overcomes the critical temperature of the working fluid. Therefore, some authors (Hafner et al. 2014) concentrate their work specifically on transcritical CO_2 systems and propose to use several ejectors for expansion work recovery at the "gas-cooler" exit. One of these ejectors is dedicated to the liquid recirculation within the flooded evaporators, in lieu of a mechanical pump. In this way, the heat transfer efficiency of the evaporator is raised, and the evaporation temperature can be increased as well.

A further way through which ejectors can be profitably used is the enhancement of absorption chillers. In a conventional absorption refrigerator, two valves are commonly employed to separate the high-pressure side, consisting of the generator and condenser, from the low-pressure side, made up by the evaporator and absorber. The ejector can be used to recover the throttling losses from either the weak solution coming from the generator or the saturated liquid coming from the condenser. In many cases, the improved configuration can lead to COP values close to that of a typical double-effect absorption cycle machine (Chunnanond and Aphornratana 2004). Moreover, the hybrid ejector-absorption refrigeration machine features a relatively simple scheme that requires less investment costs with respect to other conventional, high-performance absorption systems.

Many different schemes have been proposed in order to combine the ejector to the absorption cycle. One possible configuration uses the ejector to raise the absorber pressure, thus reducing the solution concentration. Chung et al. (1984) and Chen (1998) investigated this configuration by using the high-pressure liquid solution returned from the generator as the ejector's motive fluid. The scheme is illustrated in Fig. 1.14. Experimental investigation showed that the use of the ejector allowed a pressure ratio of around 1.2 across the absorber and evaporator. The higher absorber pressure resulted in a lower solution mass flow rate, thus reducing the pump work.

Sözen and Özalp (2005) proposed a similar scheme, operated with aqua ammonia and powered by solar collectors. The COP improved by about 20% using the ejector (Abdulateef et al. 2009).

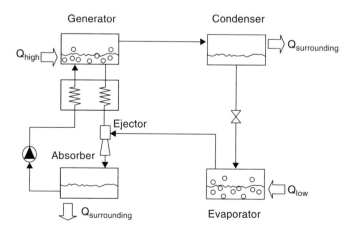

Fig. 1.14 Scheme of the absorption/ejector refrigerator with throttling loss recovery from the weak solution (Chunnanond and Aphornratana 2004)

1.4   Closure

The brief history and the list of applications reviewed in this chapter are surely not exhaustive. However, they may be sufficient to demonstrate that ejectors in general are fascinating devices and have a huge potential for further evolution.

Clearly, on the heat-powered refrigeration market, ejector chillers may have little chance to replace efficient and well-established systems like lithium bromide absorption machines. However, some specific pros may make it competitive in some niche markets.

From a practical point of view, the key parameter is the total cost per unit cooling load. If the input thermal energy has low or zero cost (e.g., waste heat or solar power), the cost of cooling is mainly related to the investment. Therefore, the most of our effort should be devoted to the reduction of the investment cost.

Ejectors are intrinsically simple, robust, and reliable. Probably more attention should be placed in their design and optimization, in order to increase their efficiency, but this shouldn't increase the system complexity and cost. If this attempt is successful, ejector-based refrigeration systems may contribute to global energy efficiency by opening new markets for heat-powered refrigeration.

References

Abdulateef, J., Sopian, K., Alghoul, M., & Sulaiman, M. (2009). Review on solar-driven ejector refrigeration technologies. *Renewable and Sustainable Energy Reviews, 13*, 1338–1349.

Al-Khalidy, N. (1997). Experimental investigation of solar concentrators in a refrigerant ejector refrigeration machine. *International Journal of Energy Research, 21*, 1123–1131.

Arora, C. (2003). *Refrigeration and air conditioning*. s.l.:Tata-McGraw-Hill.

Besagni, G., Mereu, R., & Inzoli, F. (2016). Ejector refrigeration: A comprehensive review. *Renewable and Sustainable Energy Reviews, 53*, 373–407.

Chen, L. (1998). A new ejector-absorber cycle to improve the COP of an absorption system. *Applied Energy, 30*, 37–41.

Chung, H., Hum, M. H., Prevost, M., & Bugarel, R. (1984). Domestic heating application of an absorption heat pump, directly fired heat pump. In *Proceedings International Conference University of Bristol*. s.l., s.n.

Chunnanond, K., & Aphornratana, S. (2004). Ejectors: Applications in refrigeration technology. *Renewable and Sustainable Energy Reviews, 8*, 129–155.

Croll Reynolds. (2017). [Online] Available at: www.croll.com. [Accessed 28 02 2017].

DIN 24290:1981-08. (1981). *Jet pumps (ejectors); terms, classification*. s.l.: s.n.

Dorantes, R., & Lallemand, A. (1995). Prediction of performance of a jet cooling system operating with pure refrigerants or non-azeotropic mixtures. *International Journal of Refrigeration, 18*, 21–30.

Encyclopædia Britannica. (1911). *Brake* (Vol. 4). s.l.: s.n.

Gay, N. (1931). *Refrigerating system*. US Patent, Patent No. 1,836,318.

GEA Wiegand GmbH. (2017). [Online]. Available at: http://produkte.gea-wiegand.de/GEA/index_en.html. [Accessed 28 02 2017].

Graham. (2017). [Online]. Available at: http://www.graham-mfg.com/. [Accessed 28 02 2017].

Hafner, A., Försterling, S., & Banasiak, K. (2014). Multi-ejector concept for R-744 supermarket refrigeration. *International Journal of Refrigeration, 43*, 1–13.

Huang, B., Petrenko, V., Samofatov, I., & Shchetinina, N. (2001). Collector selection for solar ejector cooling system. *Solar Energy, 7*, 269–274.

Ishizaka, N., et al. (2009). Next generation ejector cycle for car air conditioning systems. *ATZ Autotechnology, 111*, 34–38.

Kranakis, E. (1982). The French connection: Giffard's injector and the nature of heat. *Technology and Culture, 23*, 3–38.

Leblanc, M. (1911). *Notice sur lesMachines frigorifiques à vapeur d'eau et à éjecteur du système Westinghouse-Leblanc et sur leur application à la marine.* Gauthier-Villars, Paris: s.n.

Nguyen, V., Riffat, S., & Doherty, P. (2001). Development of a solar-powered passive ejector cooling system. *Applied Thermal Engineering, 21*, 157–168.

Power, R. (1993). *Steam jet ejectors for the process industries* (1st ed.). s.l: McGraw-Hill.

Routledge, R. (1876). *Discoveries and inventions of the nineteenth century.* London: George Routledge & Sons.

Shutte Koerting. (2017). [Online]. Available at: http://www.s-k.com/index.tpl. [Accessed 28 02 2017].

Sokolov, E., & Zinger, N. (1989). *Jet devices (in Russian).* Moscow: Energoatomizdat.

Stoecker, W. (1958). *Steam-jet refrigeration.* Boston: McGraw-Hill.

Sözen, A., & Özalp, M. (2005). Solar-driven ejector-absorption cooling system. *Applied Energy, 80*, 97–113.

Transvac. (2017). [Online]. Available at: http://www.transvac.co.uk/. [Accessed 28 02 2017].

Zhang, B., & Shen, S. (2002). *Development of solar ejector refrigeration system.* 1st International Conference on Sustainable Energy Technologies, s.l.

Chapter 2
Physics of the Ejectors

2.1 Influence of Fluid Properties on Ejector Behavior

Among the various physical factors influencing the ejector behavior and effectiveness within a refrigeration cycle, the nature of the working fluid undoubtedly deserves a careful discussion. All heat-powered cycles expose the working fluid to a wide range of temperatures, as it comes in touch with the hot thermal source as well as the refrigerated space. Some working fluids have a high saturation pressure at the hot thermal source temperature, increasing the system cost. Ejector cycles, as shown in Chap. 1, have a pump that receives the liquid from the condenser and feeds the vapor generator. If this latter works at high pressure, the pump may absorb a significant electric power.

Fluid charge is higher in comparison with vapor compression cycles, because heat-powered cycles combine a motive cycle and a cooling cycle. This exacerbates potential risks if flammable and/or toxic fluids are used.

From an environmental point of view, the large fluid charge per unit cooling power yields potentially high damage in case of accidental release. Therefore, a credible candidate fluid should at least have zero ODP and satisfy the current regulations in terms of GWP. In Europe, F-gas regulations ban the use of refrigerants with GWP > 2500 by 2017. This value may hence be set as a threshold.

Starting from the fundamental paper by Dorantes and Lallemand (1995), which used ten chloro- and hydrofluorocarbons, many authors have considered pure fluids as well as azeotropic or non-azeotropic mixtures as candidate fluids for ejector refrigerators. Sun (1999) compared 11 refrigerants including water, halocarbon compounds, a cyclic organic compound, and an azeotrope. Optimum ejector area ratios and COPs were computed for each fluid, concluding that ejectors using R134a and R152a perform well, regardless of operating conditions.

Cizungu et al. (2001) simulated a refrigeration system using a one-dimensional ideal gas model. Theoretical model validation was carried out on R11, and COPs were obtained for four fluids (R123, R134a, R152a, and R717) with different

© Springer International Publishing AG, part of Springer Nature 2018
G. Grazzini et al., *Ejectors for Efficient Refrigeration*,
https://doi.org/10.1007/978-3-319-75244-0_2

operating conditions and area ratios. For low-grade heat source, R134a and R152a achieved higher COP.

Five refrigerants (R134a, R152a, R290, R600a, and R717) were considered by Selvaraju and Mani (2004) within a one-dimensional computer simulation, thermodynamic fluid properties being obtained through REFPROP library. The best performance was found for R134a.

Using empirical correlation from the literature, a comparison of COPs for a solar-powered ejector refrigeration system operating with eight different working fluids was made by Nehdi et al. (2008), obtaining the best performance for R717.

Petrenko (2009) asserted that R245fa, R245ca, R600, and R600a offer low environmental impact, good performance, and moderate generator pressure. This last feature makes feed pump selection easier. Kasperski and Gil (2014) concentrated on hydrocarbons, showing that R600a yields good performance. Varga et al. (2013) confirmed the validity of this fluid.

Wang et al. (2015) introduced in the simulations the real properties of refrigerants calculated via REFPROP thermodynamic libraries. They compared R141b, R123, R600a, R142b, R134a, R152a, R290, and R717, concluding that the latter yields the highest COP.

Chen et al. (2014) compared R134a, R152a, R245fa, R290, R600, R600a, R1234ze, R430A, and R436B, accounting for the effect of superheating of the primary flow. According to their simulation, R245fa and R600 have the highest COP.

Fang et al. (2017) have shown a study on possible drop-in replacement of HFC by HFO. They used R1234yf and R1233ze(E) in the ejector experimented by Garcia del Valle et al. (2014) and showed by numerical simulation that the ejector performance was only slightly modified with respect to the previously used R134a. On the whole, the performance of the refrigeration system was however decreased (-4.2 and -26.6% reduction in COP and cooling capacity for R1233ze(E); -9.6 and -19.8% for R1234yf).

Summing up all these findings, desirable features of a working fluid for an ejector refrigerator are:

- Zero ODP and low GWP
- Low flammability and toxicity
- High latent heat and high density at generator, condenser, and evaporator temperatures, in order to reduce system size and cost

A set of candidate fluids is listed in Table 2.1.

Once CO_2 is excluded for its very low critical point, the first and rather obvious choice is water. Costless and absolutely safe, it has a very high latent heat throughout the typical range of temperatures encountered in ejector cycles. Major drawbacks are the very low pressure and density of steam at cold temperatures and the rather high triple point that impedes low-temperature applications.

A second possibility is given by hydrocarbons that play a central role in the domestic refrigeration market. Isobutane (R600a) is taken here as an example of this class of fluids, which share low GWP and rather high COP in vapor compression

Table 2.1 Relevant data for a set of zero-ODP fluids

Fluid	M_{mol} [kg/kmol]	T_{crit} [K]	P_{crit} [MPa]	N.B.P.[a] [K]	Expansion	GWP	Safety
Water	18.015	647.1	22.064	373.12	W	0	**A1**
R600a	58.122	407.81	3.629	261.4	D	20	**A3**
R134a	102.03	374.21	4.059	247.08	W	1300	**A1**
R143a	84.041	**345.86**	3.761	225.91	W	**4300**	A2L
R152a	66.051	386.41	4.517	249.13	W	120	**A2**
R218	188.02	**345.02**	2.64	236.36	D	**8600**	A1
R227ea	170.03	374.9	2.925	256.81	D	**3500**	A1
R236ea	152.04	412.44	3.502	279.34	D	1200	A1
R236fa	152.04	398.07	3.2	271.71	D	**9400**	A1
R245fa	134.05	427.16	3.651	288.29	D	950	**B1**
R32	52.024	**351.26**	5.782	221.5	W	550	A2L
R365mfc	148.07	460.0	3.266	313.3	D	890	–
R41	34.033	**317.28**	5.897	194.84	W	97	–
RC318	200.03	388.38	2.778	267.18	D	**10,000**	A1
R1234yf	114.04	367.85	3.382	243.66	D	4	A2L
R1234ze	114.04	382.52	3.636	254.19	D	6	A2L
R1233zd	130.5	438.75	3.772	195.15	D	<5	A2L

In bold: selected fluid or T_{crit} < 80 °C or GWP > 2500
M_{mol}, T_{crit}, P_{crit}, and N.B.P. from NIST REFPROP; GWP from (Calm and Hourahan 2001); safety from (ASHRAE 2008)
W wet, D dry expansion
[a]Normal boiling point ($P = 101.3$ kPa)

cycles. Isobutane has a "dry expansion," i.e., its entropy decreases along the upper limit curve, which is useful to avoid condensation within the ejector. The obvious burden of hydrocarbons is flammability, which may represent a serious problem as the fluid charge increases.

A third group includes the fluorocarbons. These fluids have zero or low flammability and favorable thermodynamic properties, but generally high GWP. Some of them have dry expansion. Among fluorocarbons, R134a has a reasonable cost and is well known in the refrigeration industry but has a high saturation pressure at generator temperature and a relatively high GWP.

The three fluids at the bottom are fluoro-olefins, promising alternatives to fluorocarbons with low GWP and generally low flammability, though likely to have a high cost in the near future.

After elimination of all fluids with GWP > 2500 and/or T_{crit} < 90 °C, we end up with the ten fluids marked in bold character in Table 2.1.

The selected fluids have remarkably different thermodynamic properties, as shown in the temperature-enthalpy diagram of Fig. 2.1. Enthalpy is set to 200 kJ kg^{-1} for all fluids at 0 °C (IIR reference state), which is unusual for water, but useful here for comparison purposes. Saturation curves are calculated through NIST REFPROP subroutines (Lemmon et al. 2013).

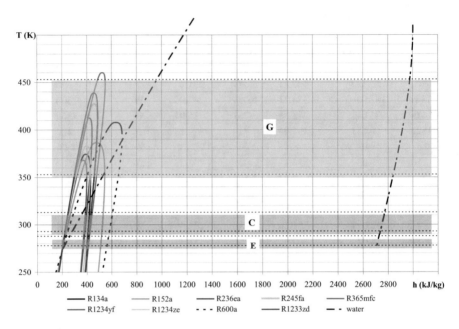

Fig. 2.1 *T-h* diagram for the selected fluids (Milazzo and Rocchetti 2015)

This diagram highlights the amount of latent heat that each fluid may exchange in the relevant temperature ranges. Water has 4–6 times latent heat values, leaving the isobutane far on the left. Fluorocarbons and fluoro-olefins form a bundle of hardly distinguishable curves that cover a very short interval on the h-axis (200–300 kJ kg^{-1}). Some fluids also show low critical temperatures.

The upper temperature range (gray area "G") represents the hot thermal source and encompasses relatively low values (from 80 °C as, e.g., solar collectors or district heating) as well as higher values (exhaust from internal combustion engines, up to 180 °C). The middle range, area "C," represents the ambient heat sink and, for stationary applications, may range from 25 to 40 °C. The lowest range, area "E," comprises the likely cold source temperature and, for air conditioning or industrial applications, goes from 5 to 15 °C.

When thermal sources with finite heat capacity (e.g., fluid streams) are accounted for, they act as boundaries for the cycle, as shown in Fig. 2.2. The diagram shown as an example refers to R245fa. Hot fluid enters at temperature T_{GS} which is higher than point G (exit from generator) by a suitable minimum difference ΔT_G. The same limit on the temperature difference must be obeyed in all sections of the heat exchanger. Likewise, the cold fluid is supposed to be delivered to the user at temperature T_{ES} which must be higher than the evaporation temperature by an amount ΔT_E. The same limit condition must be met at evaporator exit (point E). Finally, the ambient air (or water, if available) is fed to the condenser at T_{CS} and must keep a minimum temperature difference ΔT_C from the condensing fluid. All these data, as well as superheating and subcooling values, are specified in Table 2.2.

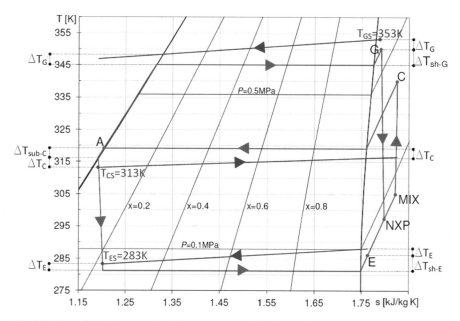

Fig. 2.2 Example of ejector cycle with temperatures of the thermal sources (Milazzo and Rocchetti 2015)

Table 2.2 Input data

T_{GS}	Hot fluid at generator inlet [°C]	80–180	ΔT_{sh-E}	Superheating at evaporator exit [°C]	5
T_{CS}	Ambient fluid at condenser inlet [°C]	25–40	ΔT_{sh-G}	Minimum superheating at generator exit [°C]	5
T_{ES}	Cold fluid at delivery [°C]	5–15	ΔT_{subC-}	Subcooling at condenser exit [°C]	3
ΔT_G	Minimum ΔT at generator [°C]	3	η_p	Primary expansion efficiency	0.95
ΔT_C	Minimum ΔT at condenser [°C]	3	η_{mix}	Mixing efficiency	0.91
ΔT_E	Minimum ΔT at evaporator [°C]	2	η_{pump}	Feed pump efficiency	0.75

These data have been used in a thermodynamic simulation code, as described by Milazzo and Rocchetti (2015). This code allows the use of various fluids, relying on NIST REFPROP for their thermodynamic properties, and calculates the system performance. The primary nozzle, the mixing process, and the feed pump are simulated via fixed values of efficiency. The supersonic diffuser is simulated by a one-dimensional model that accounts for friction losses via a friction factor and hence reflects the effect of the different velocities at the end of the mixing process for the various fluids. The model is rather simplified and has some assumptions that should be validated experimentally. Therefore, the results are not to be taken as absolute, but only as a comparison between the various fluids. It should be clearly

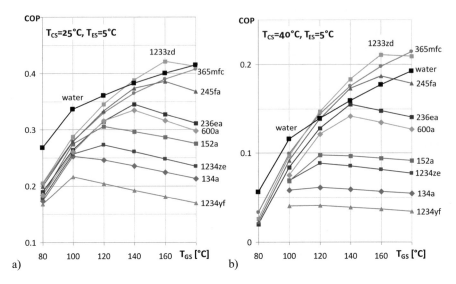

Fig. 2.3 Calculated COP values for the selected fluids - ambient temperature: (**a**) 25 °C; (**b**) 40 °C (Milazzo and Rocchetti 2015)

understood that the model produces a different ejector design for each fluid and working condition, corresponding to the maximum entrainment ratio achievable with the given exit pressure. It should also be mentioned that the saturation pressure at the generator has been chosen to be subcritical for all fluids. Therefore, at high values of T_{GS}, fluids featuring a relatively low critical pressure have been given a significant superheating at the generator. This may be a penalty for some fluids but serves also as a limit on generator pressure. Superheating at generator has also been applied to wet-expanding fluids, in order to avoid excessive liquid fraction at the primary nozzle exit (a lower limit of 0.85 has been posed on the vapor quality).

The results are shown in Fig. 2.3 for the two extreme values of T_{CS}. The lowest COP values pertain to R134a and its low GWP substitutes R1234ze and R1234yf. These fluids show peak COPs in the 100–120 °C generator temperature range. A second group of fluids collects R152a, R600a, and R236ea, showing peaks between 120 and 140 °C. R245fa and R1233zd have peaks at $T_{GS} = 160$ °C. For all these fluids, the COP starts declining as the constraint on saturation temperature at generator becomes effective. The two fluids with higher critical temperature (water and R365mfc) have increasing COPs through the whole graph.

The COP values shown in Fig. 2.3b are obviously lower than those of Fig. 2.3a, due to the increased condensation temperature. Furthermore, the four lowermost curves are interrupted at $T_{GS} = 100$ °C, because the fluids at hand cannot operate at lower values.

The curve pertaining to water has a markedly different shape, with a moderate but stable increase above $T_{GS} = 100$ °C, where the limit on nozzle-end quality comes into effect. Water turns out to be the best option at low generator temperatures, while at higher temperatures other fluids offer better performance. However, the COP obtained with water is still increasing $T_{GS} = 180$ °C, showing that at higher generator temperatures, water would exceed the result of any other fluid, given its very high critical temperature.

Other results reported by Milazzo and Rocchetti (2015) show that the COP increases, as expected, as the evaporator temperature is raised.

These results are quite different from others reported in the relevant literature. For example, Varga et al. (2013) place water invariably at the bottom in all COP diagrams. A possible reason for this evident discrepancy may be in their model, which recalls the classic one-dimensional, ideal gas scheme presented by Huang et al. (1999). In the case of steam, given the very high latent heat of condensation, even small amounts of condensed water may heavily affect the state of the expanding fluid. On the other hand, Milazzo and Rocchetti (2015) postulate an equilibrium state. This may be regarded as an opposite approach to the real fluid behavior: the ideal gas, on the one hand, postulates the fluid properties to be frozen throughout the expansion, while the equilibrium two-phase fluid is supposed to follow the expansion with no delay.

When these promising results on efficiency are combined with safety, cost, and availability of water, the importance of a careful study and optimization of steam ejectors turns out to be indisputable, even if the physics of the condensing ejector is complex, as will be shown in paragraph 2.4. Water also features a very low saturation pressure at generator, allowing a lighter and cheaper construction of the latter. Given the low generator pressure and the high density, water may even be fed by gravity, eliminating the pump (Grazzini and D'Albero 1998). Alternatively, the generator may be fed by an injector, whose merit has been discussed in Chap. 1.

High specific volume at evaporator and condenser may well increase the size of these heat exchangers, but this can be a minor problem for fixed installations. The high triple point of water may even out to be an opportunity, as shown by Eames et al. (2013) who integrated an ice storage system within an ejector chiller.

2.2 Supersonic Expansion and Compression

2.2.1 Nozzles

Any ejector contains variable area ducts specifically designed in order to expand or compress the fluid. The first law of thermodynamics, for a one-dimensional stream flowing with velocity u in the x direction, rising by dz. in vertical direction and exchanging a heat δq and a work δw, yields

$$udu + gdz + dh = \delta q - \delta w \qquad (2.1)$$

The second law of thermodynamics gives

$$\delta q = Tds - Tds_{irr} \tag{2.2}$$

Enthalpy, being a state variable, may be evaluated along any reversible transformation having the same endpoints of the one considered for Eqs. 2.1 and 2.2:

$$dh = Tds + dP/\rho \tag{2.3}$$

Once dh and δq are substituted in Eq. 2.1, the two terms Tds cancel out. Assuming null work transfer to and from the fluid (no moving devices) and neglecting gravity effects, the resulting equation reduces to

$$udu + dP/\rho + Tds_{irr} = 0 \tag{2.4}$$

Note that this equation is fully independent from Eq. 2.1, since it contains also Eq. 2.2. Therefore, we may combine it with Eq. 2.1 that may be further simplified assuming that heat exchange with the surrounding environment is negligible.

The continuity equation links the other variables to the area variation dA:

$$\frac{d\rho}{\rho} + \frac{dA}{A} + \frac{du}{u} = 0 \tag{2.5}$$

These equations may be complemented with a suitable equation of state that links the thermodynamic properties, e.g.,

$$T = T(P,h) \tag{2.6}$$
$$\rho = \rho(P,h) \tag{2.7}$$

If the irreversibilities consist of the sole effect of fluid viscous friction, the term Tds_{irr} may be expressed for a straight, constant section duct of length L and diameter D, in terms of a friction factor f, i.e.,

$$\int_L Tds_{irr} = f\frac{L}{D}\frac{u^2}{2} \tag{2.8}$$

The friction factor is commonly calculated by one of the well-known relations interpolating the experimental data (e.g., (Churchill 1977)). Obviously this approach is not strictly valid for a variable area duct, where the flow undergoes acceleration or deceleration. However, at high values of the Reynolds number, the friction factor tends to become constant and depends only on the roughness of the duct wall. Therefore, if the diameter variations along the duct length are sufficiently gradual, Eq. 2.8 may be used in differential form and substituted in Eq. 2.4. In any case, it may be useful as a definition of a convenient loss factor that should be determined experimentally for the case at hand and as a reminder of the inevitable increase of the friction with the kinetic energy of the stream.

In the design phase, it is easy to use the previous equations to define the geometry of the ducts. For example, one may assign a value for acceleration or deceleration

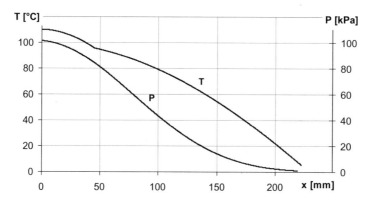

Fig. 2.4 Pressure and temperature along a steam nozzle

du/dx and substitute in Eq. 2.1 in order to have the enthalpy variation along the duct. Equation 2.4 may hence be used to calculate the pressure. All other parameters are calculated from pressure and enthalpy by NIST REFPROP functions, and the area is calculated by Eq. 2.5.

This approach forms the basis for the CRMC (Constant Rate of Momentum Change) method for designing the supersonic diffuser, introduced by Eames (2002). The CRMC method assumed ideal gas and isentropic flow, but these hypotheses are not strictly necessary and may be released.

As an example, we may design a nozzle for steam expanding from a boiler featuring a saturation temperature of 100 °C plus 10 °C superheating down to a receiver containing saturated steam at 5 °C, i.e., at pressure $P = 873$ Pa. Mass flow rate is set at 0.1 kg/s. Friction factor is constant, $f = 0.03$. Acceleration rate is set to a value $du/dx = 5000$ s^{-1}. The calculated expansion is shown in Fig. 2.4, where the temperature curve clearly shows the slope variation corresponding to the start of condensation. It must be stressed that this calculation was carried out assuming thermodynamic equilibrium throughout the expansion, which is quite unrealistic due to the significant steam supercooling as will be discussed later on. However, this result may be useful to highlight the significant error incurred when steam is modeled as an ideal gas. The nozzle shape resulting from this calculation is not practical, due to the high divergence angle at the exit (Fig. 2.5). Therefore, the nozzle should be stretched or a more refined criterion should be envisaged, e.g., reducing the acceleration rate in the diverging part of the nozzle.

Clearly, these calculations are valid only if the receiver pressure is strictly equal to the design value. If the receiver operates at higher pressure, the supersonic flow downstream of the nozzle throat will experience a shock that adjusts the pressure up to the desired value.

The shock is normally treated as a zero-thickness (Zucker and Biblarz 2002), adiabatic, and dissipative process that causes a sudden deceleration combined with a sharp increase in pressure and temperature. The flow area is thought to be constant, as the shock thickness has been neglected. Given the conditions immediately before

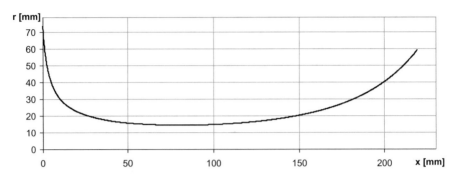

Fig. 2.5 Steam nozzle profile for $du/dx = 5000$ s^{-1}

the shock (point 1), the fluid state behind it (point 2) may be calculated imposing the conservation of mass, energy, and momentum:

$$\rho_2 u_2 = \rho_1 u_1 \tag{2.9}$$

$$h_2 + u_2^2/2 = h_1 + u_1^2/2 \tag{2.10}$$

$$P_2 + \rho_2 u_2^2 = P_1 + \rho_1 u_1^2 \tag{2.11}$$

Another relation is yielded by the equation of state Eqs. 2.7 and 2.8. These equations form a nonlinear system in four unknown variables and may be solved numerically. For example, the nozzle described in Fig. 2.4 would undergo a normal shock in the exit section when the receiver pressure is set to 9.45 kPa. This shock causes a significant increase in temperature (from 5 to 97.6 °C) and entropy (0.96 kJ kg^{-1} K^{-1}), the latter witnessing a strong irreversibility. If the receiver pressure is still higher, the shock will move to the left, toward the nozzle throat, and the divergent length downstream of the shock will see a pressure increase (subsonic flow in a divergent duct).

If the receiver pressure is somewhere between 873 Pa and 9.45 kPa, the shock will become conical and develop downstream of the nozzle exit. The flow through an oblique or conical shock does decelerate but may remain supersonic (even if the velocity component normal to the shock must be subsonic behind it). The oblique shock deviates the flow away from the normal to the shock wave, as shown, e.g., in Fig. 2.6, and brings the fluid in region 2 at the same pressure of the surroundings. However, as the ejector axis behaves like an impermeable wall, the flow that converges toward the axis, if supersonic, will originate another oblique shock that produces a further pressure increase (Zucker and Biblarz 2002). The pressure difference between region 3 and the surrounding fluid originates a Prandtl-Meyer expansion that propagates from the end of the oblique shock and lowers the pressure but turns the flow slightly away from the axis. The expansion fan will in turn reflect on the axis, and the reflected expansion will create another zone where the fluid stays

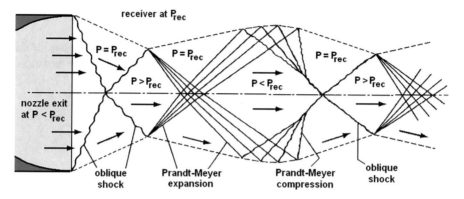

Fig. 2.6 Flow configuration at the exit of an over-expanded nozzle (Zucker and Biblarz 2002)

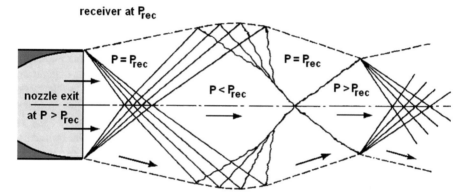

Fig. 2.7 Flow configuration at the exit of an under-expanded nozzle (Zucker and Biblarz 2002)

below the receiver pressure. A further Prandtl-Meyer may bring back the pressure in equilibrium with the surroundings but again will turn the flow toward the axis and cause another oblique shock and a further zone with high pressure. Eventually this sequence of expansions and compressions may repeat until the fluid viscosity dampens out the oscillations.

Therefore, the often cited mechanism that sees the primary nozzle to expand the fluid down to very low pressure in order to produce a "suction" for the secondary flow, in the case of supersonic ejectors, is oversimplified and misleading. When the nozzle has an exceedingly high "area ratio" (i.e., the ratio between the throat and the exit cross sections), it is said to be "over-expanded," and its effectiveness is impaired by the irreversibility suffered across the series of oblique shocks described above. This circumstance will be proven by numerical and experimental results in the following chapters.

If the nozzle, on the other hand, is "under-expanded," i.e., it is designed with an area ratio which is too low for the given receiver pressure, the situation is the one depicted in Fig. 2.7. The oblique shocks at the nozzle exit are replaced by Prandtl-

Meyer expansion fans, and the high-pressure region originates further downstream, where the shocks are weaker.

We may conclude that the nozzle should be operated at its design point or, if the working conditions are variable, it should preferably move toward under-expanded operation.

2.2.2 Diffusers

When it comes to the mixing chamber and diffuser, the problem is more complicated. Differently from supersonic nozzles, the dynamics of diffusers cannot be described by isentropic flow. This is due to the severe losses resulting from the interaction of compression waves or shocks with the boundary layer (Pope and Goin 1978). The study of this interaction is quite complex because of its unsteady and three-dimensional nature. Herein, we only illustrate some general features specific to supersonic ejector flows; more details on the shock-boundary layer interactions can be found in specialized readings (e.g., (Dolling 2001; Smits and Dussauge 2006)).

The configuration and intensity of the shocks inside generic channels depend on many factors: the Mach number upstream of the shock, the geometry of the duct, the intensity and direction of the pressure gradient, and, most importantly, the presence and interaction with the viscous boundary layer (Matsuo et al. 1999). This is normally stable when the pressure is decreasing in the direction of the boundary layer growth, i.e., for supersonic nozzles. However, it becomes unstable and tends to separate from the wall when the pressure is increasing in the direction of growth, i.e., for diffusers (Pope and Goin 1978).

Experimentally, it has been observed that with an increasing Mach number, the shock configuration changes from that of a single normal shock to a sequence of lambda shocks, called "shock train" (Shapiro 1953; Matsuo et al. 1999). This process is illustrated in Fig. 2.8. In the case of low freestream Mach number ($M_1 < 1.2$), the structure of the shock resembles that of an inviscid normal shock. In this regime, the interaction between the shock and the boundary layer is generally weak and no separation occurs. With increasing Mach number, the shock becomes more inclined with possible appearance of bifurcations near the wall. The interaction with the boundary layer becomes progressively stronger, and separation may occur at the foot of the shock. Finally, for freestream Mach numbers greater than approximately 1.5, one of more bifurcated shocks appears in conjunction with an extensive separation region.

The causes of the formation of the shock train are to be found in the complex interaction between the shocks and the viscous boundary layer (Shapiro 1953; Matsuo et al. 1999). A proof for this was produced by Carroll and Dutton (1990) who performed visual investigations of the shock train at various boundary layer thicknesses. The test was executed by forcing the shock to occur at different positions along the duct, where the boundary layer is at different stage of development. As shown in Fig. 2.9, a decreasing height of the boundary layer thickness

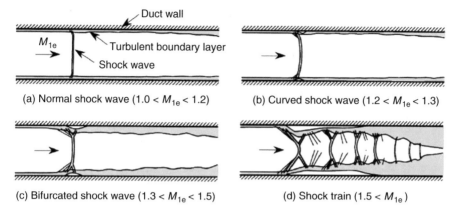

Fig. 2.8 Sketch of shock-boundary layer interaction leading to the shock train; (a) Normal shock wave; (b) Curved shock wave; (c) Bifurcated shock wave; (d) Shock train, (Matsuo et al. 1999)

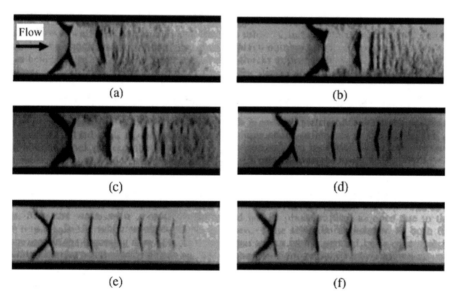

Fig. 2.9 Shock train formation at different levels of the boundary layer thickness, the upstream Mach = 1.6 and is constant in all tests. The boundary layer thickness progressively increases from (a) to (f), (Matsuo et al. 1999) (Copyright from Carroll and Dutton (1990))

reduces the formation of the shock train. In the ideal case of inviscid flow, the sonic shock would be perfectly normal. This information is important in ejector design because the shock train produces greater losses than an equivalent normal shock occurring at the same upstream Mach number (Matsuo et al. 1999). Hence, the presence of any obstacle or geometrical feature that produces an increase in the boundary layer thickness should be avoided.

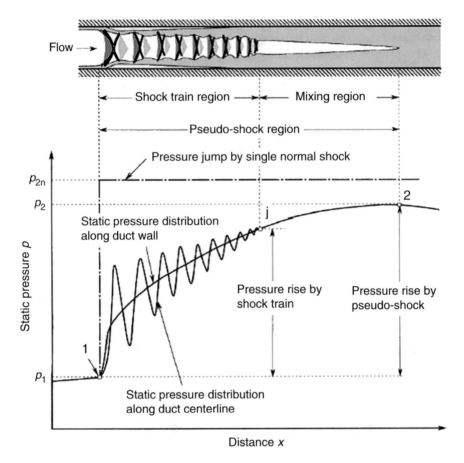

Fig. 2.10 Schematic static pressure distribution along the duct centerline and wall surface in constant-area duct (Matsuo et al. 1999)

In the case of a straight channel, downstream the region of the shock train, the central part of the duct is still occupied by supersonic flow. The mixing of the high-speed core with the surrounding subsonic flow leads to an increase of the static pressure until a maximum value is reached. This is followed by a zone of decreasing pressure due to friction losses, as illustrated in Fig. 2.10. The region consisting of both the shock train and the static pressure rise was named "pseudo-shock" by Crocco (1958). Matsuo et al. (1999) list several simplified theoretical models of the various fluids for the pseudo-shock region. Although these could be employed within an ejector model, the assumption of a normal shock generally predicts the pressure rise across the shock with an uncertainty that is approximately within 6% (Johnson III and Wu 1974). Consequently, the adoption of more complicated schemes may be avoided. Furthermore, none of the proposed models take into account the event of a nonuniform velocity profile. Nonuniform velocity fields, with an external region slower than the center of the duct, frequently occur in

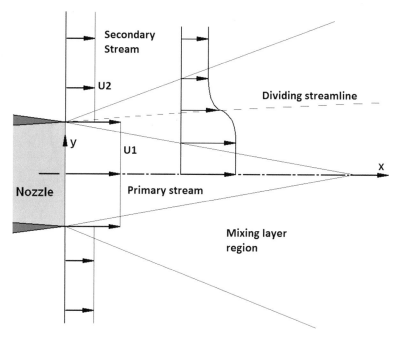

Fig. 2.11 Mixing layer inside an ejector (The dividing streamline is the virtual line that separates two regions having mass flow rates equal to that of the primary and secondary mass flow rates, respectively)

supersonic ejector due to the possibility of an incomplete mixing process. The presence of this type of flow field could be a further cause of formation of the shock train.[1] The length and effectiveness of the mixing process then becomes an important parameter to reduce shock losses in the diffuser: whereas longer diffusers may produce greater frictional losses, the increased uniformity of the flow could lead to lower shock losses.

2.3 Entrainment and Mixing

When the primary and secondary flows meet inside the mixing chamber, they give rise to a narrow region of strong mixing called "mixing layer."

Figure 2.11 shows a simplified scheme of the flow in the zone downstream of the nozzle exit, highlighting a "mixing layer region." Outside this region the primary (motive) and secondary (suction) streams flow isentropically. Inside the mixing

---

[1] As stated before, the presence of the boundary layer may be at the origin of the shock train. However, the boundary layer itself is nothing more than a particular nonuniformity in the velocity flow field.

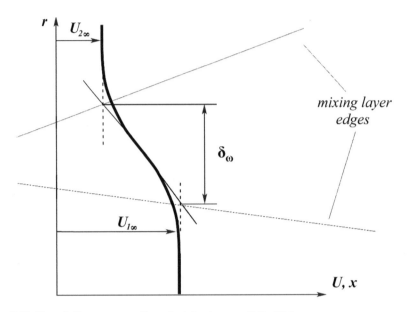

Fig. 2.12 Hyperbolic tangent profile and mixing layer vorticity thickness

layer, the time-averaged velocity smoothly varies from the value of the undisturbed primary stream to that of secondary stream. The radial extension of the shear region is usually measured by the definition of a "shear layer thickness." This is a fundamental quantity for the analysis and prediction of the mixing process. Unfortunately, there are many ways to define this thickness that are not completely equivalent and lead to difficulties in the comparison of experimental data.

In what follows, only one of these will be considered, i.e., the "vorticity thickness"; more details on the various alternative definitions are provided by Gatsky and Bonnet (2013).

The vorticity thickness is defined as the distance given by the velocity difference across the layer divided by the maximum slope of the velocity profile:

$$\delta_\omega = \Delta U_\infty / \left(\frac{\partial U}{\partial y}\right)_{\max} \tag{2.12}$$

where ΔU_∞ is the difference between the undisturbed primary and secondary stream velocities.

This definition is mostly useful whenever the mixing layer velocity profile can be described by an analytical function. For instance, if the velocity profile is approximately reproduced by a hyperbolic tangent (this is a recurrent choice in the literature), the knowledge of the vorticity thickness allows exact definition of the mixing layer edges. This concept is illustrated in Fig. 2.12.

The shear layer thickness is not the sole important parameter in the analysis of mixing layer. A further fundamental quantity is represented by the "spreading rate" of the shear layer along the longitudinal direction. This is simply defined as the axial variation of the shear layer thickness:

$$\delta'_\omega = \frac{d\delta_\omega}{dx} \qquad (2.13)$$

An interesting feature of most shear flows (jets, mixing layers, and wakes) is that the spreading rate is constant, i.e., the shear region grows linearly with distance (see Fig. 2.11).

For incompressible mixing layers, a widely used correlation for the spreading rate was devised by Brown (1974) and Brown and Roshko (1974) and is a function of only the velocity and density ratio across the layer:

$$\delta'_{\omega\_incomp} = 0.085 \cdot \left[\frac{(1 + \Lambda)(1 - \Theta)}{(1 + \Theta \cdot \Lambda)} \right] \qquad (2.14)$$

where $\Lambda = \sqrt{\rho_{\infty2}/\rho_{\infty1}}$ and $\Theta = U_{\infty2}/U_{\infty1}$.

The terms inside square brackets in Eq. 2.16 describe the effect of density and velocity difference across the layer, which is to increase the spreading rate for large velocity differences as well as when the density is greater on the low-speed side.

Although Eq. 2.14 and other alternative formulations (see, for instance, (Dimotakis 1986)) are strictly valid for incompressible mixing layers, they form the basis for the study of the mixing at high speeds. A key aspect that is of particular concern for supersonic ejector studies is that compressible mixing layers are affected by a significant reduction of the spreading rate with respect to equivalent incompressible configurations. This feature has been identified in several experimental investigations performed in the 1970s (e.g., (Brown and Roshko 1974; Ikawa 1973)) and severely reduces the effectiveness of mixing inside supersonic layers. The causes of this phenomenon have been the subject of studies for more than 50 years, and yet no clear explanation has been found. Early studies tried to explain the effect by the density variations resulting from the high expansion of the motive stream. However, the experimental work of Brown and Roshko (1974) demonstrated that this was not the main cause and that the reason could be the impact of compressibility on the turbulence structure of the flow. Papamoschou and Roshko (1988) later found that the decrease of mixing layer spreading rate may be described by means of a parameter called convective Mach number:

$$M_c = \frac{\Delta U_\infty}{a_{\infty1} + a_{\infty2}} \qquad (2.15)$$

where $U_{\infty1}$, $U_{\infty2}$, $a_{\infty1}$, and $a_{\infty2}$ are the velocities and sound speeds of the primary and secondary stream outside the mixing layer.

The convective Mach number roughly represents the Mach number of the relative motion of the large eddies within the mixing layer (Smits and Dussauge 2006). This parameter has been found to approximately correlate the experimental data for compressible mixing layer spreading rates. However, despite the many different correlations proposed (see (Smits and Dussauge 2006) or (Gatsky and Bonnet 2013)), none could really reproduce the experimental data with enough confidence (the uncertainty are usually well above 20%).

Among these, one of the most popular was provided by Papamoschou and Roshko (1988) and later readapted by Papamoschou (1993, 1996):

$$\delta'_{\omega} = \delta'_{\omega\_incomp} \cdot f(M_c) = 0.085 \cdot \left[\frac{(1 + \Lambda)(1 - \Theta)}{(1 + \Theta \cdot \Lambda)} \right] \cdot f(M_c) \qquad (2.16)$$

The term $f(M_c)$ is called "compressibility function" and is defined as the ratio of the compressible to incompressible spreading rate. In practice, the definition of this parameter is aimed at concentrating all the effect of compressibility in a single term. In the case of nearly incompressible mixing layers, the value of this function is unitary, and Eq. 2.16 can be used to approximately describe the spreading rate of low-speed mixing layers. When the mixing layer is highly compressible, the compressibility function is well below unity and brings about a significant reduction in the mixing layer spreading rate. From the correlation of several experimental data, $f(M_c)$ can be expressed by an exponential function, as follows (Papamoschou 1993):

$$f(M_c) = \frac{\delta'_{\omega}}{\delta'_{\omega\_incomp}} \approx 0.25 + 0.75e^{-3M_c^2} \qquad (2.17)$$

Unfortunately, Eq. 2.17 shows discrepancies of the order of 20% or more with respect to experimental data (Papamoschou, 1993). Many other correlations have been proposed that in some cases can provide better levels of agreement (see for instance (Gatsky and Bonnet 2013)). Nevertheless, when plotting $f(M_c)$ as a function of the convective Mach number, the spreading of experimental results remains substantial, regardless of the formulation employed.

This may be explained by considering that, in general, mixing layers may be influenced by blockage effects, thickness and surface conditions of the splitter plate, inlet turbulence level, and acoustic disturbances (Smits and Dussauge 2006). These effects are hardly captured by the use of a single parameter like the compressibility function. Therefore, a greater accuracy may be achieved by the introduction in Eq. 2.16 of some additional variables accounting for these factors. Unfortunately, at present it is very difficult to isolate and measure these "second-order" effects with experimental means, and all previous investigations lack information and data in this regard.

However, despite the low accuracy, Eq. 2.16 provides a mean to easily calculate the spreading rate by the knowledge of flow conditions in the isentropic region outside the layer. Moreover, the knowledge of the spreading rate allows deriving a fundamental equation for the maximum shear stress inside the mixing layer. By means of dimensional arguments, it can be demonstrated that the maximum shear stress is approximately given by (Papamoschou 1993):

$$\tau_{\max} \sim \rho_{\mathrm{avg}} U_{\mathrm{avg}} \Delta U_\infty \cdot \delta'_\omega = K \cdot \frac{1}{2}(\rho_{\infty 1} + \rho_{\infty 2})\Delta U_\infty^2 \left[\frac{(1+\Lambda)(1+\Theta)}{2(1+\Theta \cdot \Lambda)}\right] \cdot f(M_c)$$

$$(2.18)$$

where K is an empirical constant, obtained from subsonic constant-density experiments. Papamoschou suggests use of Wygnanski and Fiedler's value, $K = 0.013$ (Wygnanski and Fiedler 1970). The knowledge of the shear stress in some point of the mixing layer is important because it allows the calculation of the momentum balance without recurring to complex numerical procedure and turbulence models.

In conclusion, it should be mentioned that a large part of the literature on compressible mixing layers is related to the 2D planar case. By contrast, supersonic ejectors have generally circular cross sections, which implies the presence of mixing layers with annular shape. Nevertheless, many studies have confirmed that results for the spreading rate of planar mixing layers agree well with those for annular flows (Freund et al. 2000; Gatsky and Bonnet 2013). Consequently, the same correlations may be used for predicting both the planar and annular mixing layer growth rates.

2.4   Supersonic Condensation

Phase-change phenomena inside supersonic ejectors can occur in different ways depending on the application. In standard supersonic ejector cycles, condensation or freezing may occur due to the expansion of a "wet refrigerant" inside the two-phase dome. Conversely, throttling loss recovery in ejector expansion cycles requires the expansion of a saturated liquid or supercritical fluid that leads to evaporation of a substantial fraction of the primary flow.

Although the physical mechanisms involved are similar, the theory behind these phase-change phenomena has historically evolved in distinct directions and applications. In particular, for condensing high-speed flows, much of the work has been carried out in the context of steam turbine research. Low-pressure steam turbines are affected by problems of droplet formation that lead to thermodynamic losses and blade erosion (Gyarmathy 1962). Advances in this field have been mostly pioneered by the prominent work of Aurel Stodola at the beginning of the twentieth century (Stodola 1927).

In general, condensation phenomena inside supersonic ejectors are significantly more complex than in standard devices such as condensers. The high levels of speed, compressibility, and turbulence notably complicate the study that, in most cases, must rely on empiricism and experimental data.

In the ideal case of a reversible transformation, the condensation process follows a path of equilibrium states, and no losses occur. Inside supersonic ejectors, however, the very limited residence time and high cooling rates lead to a substantial departure from the equilibrium process. As the primary flow rapidly expands inside the motive nozzle, thermodynamic equilibrium is not maintained and, at a certain degree of expansion, the vapor state collapses and condensation takes place abruptly as a shock-like disturbance. This is generally called the "condensation shock."

This sudden change of state of aggregation leads to an instantaneous and localized heat release (heat of vaporization). The heat release alters the thermodynamic conditions along the motive nozzle by increasing pressure and temperature as well as reducing the Mach number. Downstream the condensation shock, the flow contains a considerable number of tiny liquid droplets (of the order of $10^{19}/dm^3$) that may affect the turbulence levels and the subsequent ejector dynamics. More than this, the condensation shock implies large gradients between the phases that result in thermodynamic irreversibilities.

In studying the condensation inside a supersonic nozzle, it is important to distinguish between two different stages of the process: the droplet formation stage or *nucleation* and the *droplet growth*.

Although in high-speed condensation these two processes occur almost simultaneously, the division is important because of the different tools that can be employed for the analysis: while the study of droplet growth can be handled by the familiar means of classical thermodynamics, the *prediction* of the nucleation process must consider the microscopic behavior of the fluid. At this scale, the usual continuum hypothesis does not hold, and the study must rely upon statistical mechanics or kinetic theory concepts. The statistical mechanics approach is not covered in this context and a review of this method is given by Ford (2004).

By contrast, the kinetic approach is at the base of the "classical nucleation theory," which still today represents the most common approach to predict droplet formation, at least for engineering purposes.

2.4.1 Phase Stability

Classical thermodynamics assumes that phase transition occurs immediately at the saturation line. In real systems, however, phase change usually takes place under non-equilibrium conditions (Carey 1992). Anytime the fluid crosses the saturation line without incurring in a phase transition, the system is said to be in a "metastable state." Under these conditions, the system is not thermodynamically stable, meaning that a perturbation can drive the system far from the initial conditions, i.e., to a different state of aggregation.

From a macroscopic point of view, these non-equilibrium conditions can be reached following an infinite number of different paths. Among these, two of the most common are the isothermal increase of pressure above the saturation line, i.e., the vapor *supersaturation*, and the isobaric cooling below the saturation temperature, viz., the vapor *supercooling*.

These two different paths allow the definitions of parameters that quantify the "degree of meta-stability" of the system, respectively, called the "degree of supersaturation" and the "degree of supercooling":

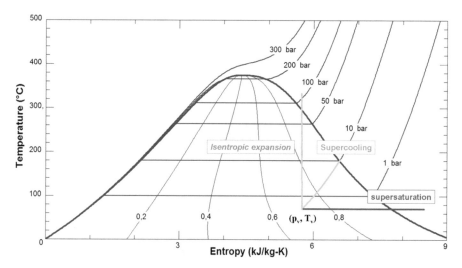

Fig. 2.13 T-s diagram for steam, showing the various processes that can lead to a metastable state

$$\varphi_{ss} = \frac{P_v}{P_{sat}(T_v)} \qquad \varphi_{sc} = T_{sat}(P_v) - T_v \qquad (2.19)$$

where both parameters are >1 in metastable conditions and the subscript v, which stands for *vapor*, represents the local static state of the fluid.

It is important to note that these two reference transformations do not describe the isentropic expansion of a vapor inside a supersonic nozzle, as shown in Fig. 2.13. In practice, however, it doesn't matter how the system reached the metastable condition as this represents a well-defined thermodynamic state. Therefore, both parameters can be equally used to locate the metastable state in the phase space.

From a microscopic point of view, the metastable condition is characterized by the continuous formation of liquid nuclei due to the random movement of the vapor molecules. The molecular fluctuations that create the liquid nuclei constitute a probabilistic phenomenon that must be studied by statistical concepts. The knowledge of the rate and magnitude of these density fluctuations is crucial for the prediction of the conditions under which condensation starts.

Once a liquid nucleus is formed, this can either collapse, grow, or stay in equilibrium with the surrounding vapor. In order to understand which one of these routes the nucleus will take, it is necessary to study its stability. In other words, we must ask if equilibrium is possible and whether this is stable or not.

Generally speaking, the stability of any liquid or vapor in metastable conditions can be analyzed by at least two different thermodynamic approaches. The first of these methods ignores the microscopic behavior of the fluid (i.e., the continuous formation of nuclei) and studies the thermal and mechanical stability of a pure, single-phase fluid. The analysis shows that there is a well-defined limit beyond which no metastability is possible and the system must undergo phase transition.

Table 2.3 Conditions for the occurrence of a spontaneous process and equilibrium state for different kinds of systems and surroundings

System constraints	Surrounding conditions	Spontaneous process	Equilibrium state
Isolated system No mass, work or heat exchange	/	$dS \geq 0$	Maximum S
Closed system No mass or work exchange	Constant T	$dF \leq 0$	Minimum F
Closed system No mass exchange	Constant P and T	$dG \leq 0$	Minimum G

This is called the "spinodal limit." Details of this approach can be found in Grazzini et al. (2011) or Carey (1992).

The second approach focuses on the stability analysis of a two-phase system composed of a droplet surrounded by an infinite mass of pure vapor. This second approach is crucial for the prediction of the nucleation process and will be illustrated next.

In general, the spontaneity of a process and its equilibrium state are described by different thermodynamic functions[2] depending on the system constraints and surrounding conditions. Some of these conditions are summarized in Table 2.3, where S is the entropy, F is the Helmholtz free energy, and G is the Gibbs free energy (a highly recommended discussion about these concepts can be found in (Ford 2013)).

In the stability analysis of the droplet-vapor problem, the surrounding is assumed to be at a well-defined state with constant pressure, P_v, and temperature, T_v (this may represent a "frozen" state condition of the isentropic expansion inside the nozzle). Under these constraints, the stability of the droplet can be analyzed by computing the variation of the system Gibbs energy from the initial condition of pure vapor to the state where the droplet has formed.

Following Bakhtar et al. (2005), the droplet formation can be conveniently subdivided into three main stages: a first isothermal expansion of the vapor down to the pressure of saturation, the formation of the liquid interface, and the recompression of the liquid to the local value of pressure. It is important to note that this is by no means the sequence that is really followed by the system; however, being G a state function, it is irrelevant what path is selected, as long as the initial and final states coincide.

The total Gibbs free energy variation is thus given by

---

[2]In analogy with mechanics, these are called thermodynamic potentials in view of their use in describing the direction of spontaneous process and the conditions for equilibrium of the system.

$$\Delta G_{\text{vapour}\rightarrow\text{droplet}} = \Delta G_{\text{vapour\_expansion}} + \Delta G_{\text{droplet\_formation}}$$
$$+ \Delta G_{\text{liquid\_compression}} \tag{2.20}$$

where G is the free energy of the total mass involved in the condensation, m_l.

By assuming ideal gas behavior of the vapor phase, the first term is calculated as follows:

$$\Delta G_{\text{vapour\_expansion}} = m_l R T_v \int\limits_{P_v}^{P_{\text{sat}}(T_v)} \frac{1}{P} dP = -\frac{4}{3}\pi r^3 \rho_l R T_v \ln \varphi_{\text{ss}} \tag{2.21}$$

The second term depends on the sole surface tension given by[3]

$$\Delta G_{\text{droplet\_formation}} = 4\pi r^2 \sigma(T_v) \tag{2.22}$$

Finally, the last term can be computed as follows:

$$\Delta G_{\text{liquid\_compression}} = \frac{m_l}{\rho_l} \cdot \int\limits_{P_{\text{sat}}(T_v)}^{P_v} dP = \frac{4}{3}\pi r^3 (P_v - P_{\text{sat}}(T_v)) \tag{2.23}$$

where it was assumed a constant density for the liquid phase. This last term is usually small and can be neglected without incurring in significant approximations (Bakhtar et al. 2005).

Summing up the various contributions, the total free energy variation is given by

$$\Delta G_{\text{vapour}\rightarrow\text{droplet}} = 4\pi r^2 \sigma(T_v) - \frac{4}{3}\pi r^3 \rho_l R T_v \ln \varphi_{\text{ss}} \tag{2.24}$$

The above expression is one of the fundamental equations that constitute the classical nucleation theory.

In general, the establishment of thermodynamic stable equilibrium under the considered constraints requires the system to reach a minimum of the total Gibbs free energy, as is briefly demonstrated in Appendix A.

Therefore, in order to understand whether equilibrium between the vapor and the droplet can exist, we must compute the derivative of Eq. 2.24 as a function of the radius[4] and impose it equal to zero:

$$\frac{dG}{dr} = 4\pi r \sigma - 4\pi r^2 \rho_l R T_v \ln \varphi_{\text{ss}} = 0 \tag{2.25}$$

Solving for the radius gives

---

[3]This is by definition the energy required to form an interface of unit area (Carey 1992).

[4]We are analyzing a fixed state of the droplet-vapor system having pressure and temperature equal to T_v and P_v. Hence, the free energy variation depends solely on the droplet radius.

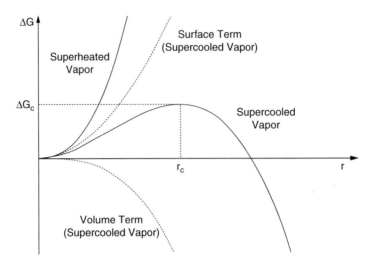

Fig. 2.14 Gibbs free energy variation for the process of droplet formation in the case of superheated and supercooled vapor state (Bakhtar et al. 2005)

$$r_{\mathrm{c}} = \frac{2\sigma}{\rho_{\mathrm{l}} R T_{\mathrm{v}} \ln \varphi_{\mathrm{ss}}} \qquad (2.26)$$

The above expression describes the radius for which the droplet is in equilibrium with the surrounding vapor. This is usually called the *critical radius*. The most important aspect that must be noticed in Eq. 2.26 is that the critical radius becomes smaller as the degree of supersaturation increases. We will come back to this aspect later.

Solving Eq. 2.26 for the denominator and inserting the resulting expression in Eq. 2.24 return a very compact equation for the Gibbs free energy required to form a droplet having precisely the critical radius:

$$\Delta G_{\mathrm{c}} = \frac{4}{3} \pi r_{\mathrm{c}}^2 \sigma \qquad (2.27)$$

The final step of the analysis consists in determining whether this equilibrium is stable or not. In order to discover this, the Gibbs free energy expression, Eq. 2.24, can be plotted to study its behavior as the radius varies. This is shown in Fig. 2.14.

As a first remark, Eq. 2.24 is composed of a surface and a volume term: while the first term is always positive, the sign of the second term depends on the supersaturation ratio. In the case of superheated vapor (i.e., $0 < \varphi_{\mathrm{ss}} < 1$), the volume term is positive and ΔG becomes a monotonically increasing function of the radius. This means that whenever a nucleus is formed inside a superheated vapor, this will spontaneously tend to collapse because the minimum of the Gibbs free energy is for a radius equal to zero.

By contrast, when the vapor is in metastable conditions, the two terms have opposite signs. Due to the different exponents, the surface term dominates for small radius, while the opposite is true for larger radius. Consequently, the Gibbs free energy increases up to a maximum occurring at the critical radius and then decreases down to negative values. The presence of this maximum indicates that the equilibrium is *unstable*: nuclei smaller than the critical size must collapse, while those with radii greater than the critical have the tendency to grow.

From a *micro*scopic point of view, ΔG can be considered like an energy barrier. This can be understood by considering that the formation of a droplet requires the enclosure of a large number of molecules into a very small and confined region. Due to the repulsive forces between molecules, a potential barrier form and the establishment of the nucleus require some work to be done on the system. This work must be provided by the kinetic energy of the same molecules constituting the surrounding vapor.

From a *macro*scopic point of view, the potential barrier can be correctly interpreted by investigating further the significance of the Gibbs free energy. By assuming constant pressure and temperature of the surrounding vapor, the droplet formation can be approximately regarded as an isothermal and isobaric process. Under these constraints, the Gibbs free energy variation between the initial (pure vapor) and final state (liquid droplet) of the system becomes equivalent to its exergy variation[5]:

$$\Delta G = \Delta E + P_v \Delta V - T_v \Delta S = \Delta Ex \qquad (2.28)$$

where it is assumed that the reference state for exergy is at T_v and P_v.

Therefore, in the ideal case of a reversible process, the Gibbs free energy variation corresponds to the *maximum reversible work* that can be extracted from the system or, else, the *minimum reversible work* required to create the droplet (the aforementioned energy barrier):

$$\Delta G_{\mathrm{rev}} = \Delta E + P_v \Delta V = W_{\min} \qquad (2.29)$$

Inserting Eq. 2.26 into Eq. 2.27 returns the minimum work required to form a critical cluster as a function of the flow parameters:

$$\Delta G_c = \frac{16}{3} \frac{\pi \sigma^3}{(\rho_l R T_v \ln \varphi_{ss})^2} = W_c \qquad (2.30)$$

In Eq. 2.30 it is interesting to note the strong influence of the surface tension. This, in turn, depends on the system temperature and is equal to zero at the fluid critical point. Therefore, lower values of supersaturation should be expected for flow conditions that are close to the critical point. Moreover, as the supersaturation increases, the W_{eq_min} decreases, but the barrier never disappears completely (Bakhtar et al. 2005). In other words, a metastable state becomes increasingly less stable as the degree of supersaturation (or supercooling) grows.

---

[5]Rigorously, this is the availability of the system; see (Carey 1992).

2.4.2   Nucleation

In the preceding section, it was shown that the condensation of a supersaturated vapor requires the formation of droplets with radius greater than r_c. Although this information is essential to model the phase transition, the correct prediction of the nucleation stage requires also the knowledge of how many "critical-sized" nuclei form in the vapor stream.

There are two main mechanisms through which critical clusters can form. The first is due to the presence of foreign particles within the vapor or surface vacancies at the solid walls containing the flow. Qualitatively speaking, these impurities and surface imperfections constitute the primordial sites where the molecules aggregate to form an embryo. This process is called *heterogeneous nucleation* and is typical of phase transitions inside conventional condenser (Carey 1992). The second mechanism is called *homogeneous nucleation* and originates from random density fluctuations due to thermal agitation of the vapor molecules.[6] Although it can be observed in any system, this type of nucleation is the primary mechanism through which droplets form inside high-speed nozzles (Wegener and Mack 1958).

The homogeneous nucleation is thus a stochastic phenomenon that must be addressed by means of statistical and probabilistic evaluations. As it might be guessed, the probability that a cluster forms depends on the ratio between the potential barrier and the average kinetic energy of the vapor molecules. This ratio defines a dimensionless parameter called the Gibbs number:

$$Gb = \frac{\Delta G}{k_b T_v} \tag{2.31}$$

where k_b is the Boltzmann constant and ΔG is given by Eq. 2.24.

In the classic approach of the homogeneous nucleation theory, the formation of clusters is described through the so-called *nucleation rate*, J, defined as the number of nucleation events occurring in a unit volume per unit time (Brennen 1995). Many formulations exist for J, but almost all of them assume the general form:

$$J = J_0 e^{-Gb} \tag{2.32}$$

where J_0 is some factor of proportionality that will be described next.

The presence of the exponential in Eq. 2.32 is indicative of the shock-like nature of the homogeneous condensation phenomenon. We will come back to this aspect later on.

Over the years, a great number of different expressions and corrections have been devised for both J_0 and the argument of the exponential. Many of these are carefully reviewed by Bakhtar et al. (2005) and will not be detailed here. Herein, the focus is

---

[6]The reason for these two names comes from the fact that in the first case nucleation occurs at well-defined spots whose location is not uniformly distributed, while the second occurs homogeneously all over the volume of fluid.

only on the classical formulation and its demonstration, which is useful in understanding the main assumptions that lie behind the theory. In deriving the explicit expression for J, we will follow a procedure similar to that outlined by Bakhtar et al. (2005), but an analogous derivation can be found in Carey (1992).

Before to begin, it is important to stress the fact that the analysis will not deal with the probability of formation of individual clusters, but rather on the time evolution of their *size distribution*:

$$\frac{dn_2}{dt}, \frac{dn_3}{dt}, \frac{dn_4}{dt}, \cdots \cdots \frac{dn_x}{dt} \tag{2.33}$$

where n_x represents the number of clusters composed of "x-molecules" that are found every instant in a unit volume of gas.

The analysis considers a metastable vapor in supersaturation conditions ($\varphi_{ss} > 1$). Due to the thermal agitation of the molecules, clusters are continuously formed and disrupted within the volume of fluid. Therefore, the identities and position of the individual clusters change instantaneously within the vapor volume. However, due to the large numbers involved, the average population of each cluster group is more stable in time.

In order to find a simple mathematical expression for the evolution of the size distribution, it is assumed that the passage of a cluster from one size to another occurs only by the acquisition or loss of single molecules. From a simple molecules balance, it follows that the population of the "x-sized" cluster group, n_x, depends solely on the rate of condensation and evaporation from the groups of smaller and larger sizes:

$$\frac{dn_x}{dt} = \left(\dot{C}_{x-1} n_{x-1} - \dot{E}_x n_x \right) - \left(\dot{C}_x n_x - \dot{E}_{x+1} n_{x+1} \right) = \dot{I}_{(x-1) \leftrightarrow x} - \dot{I}_{x \leftrightarrow (x+1)} \tag{2.34}$$

where $\dot{E}$ and $\dot{C}$ represent here the average evaporation and condensation rates of an entire cluster group, measured in [n. molecules/s].

The variable $\dot{I}$ represents the so-called *nucleation current*, i.e., the number of clusters that move from one size group to another in the unit of time and volume. Figure 2.15 represents schematically the concept of the nucleation current assumption.

In general, liquid embryos can form in several different ways. For instance, an embryo composed of x molecules may form by the aggregation of two smaller clusters, by the disruption of a larger cluster, or even by the collision of x single molecules. Hence, the correct evaluation of the nucleation rate, J, would require the calculation of the probability connected to each of this possible way of cluster formation. However, the assumption of the existence of a simple nucleation current greatly simplifies the calculation of the *nucleation rate* by implying that $J = I$.

Moreover, the goal of the analysis is now simplified to the evaluation of the nucleation current for the only "critical-sized" clusters:

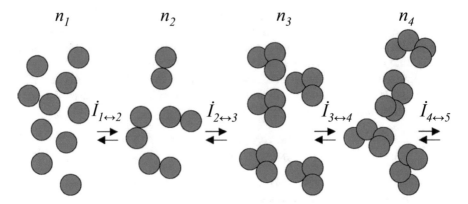

$$n_1 \qquad n_2 \qquad n_3 \qquad n_4$$

$$\dot{I}_{1\leftrightarrow2} \qquad \dot{I}_{2\leftrightarrow3} \qquad \dot{I}_{3\leftrightarrow4} \qquad \dot{I}_{4\leftrightarrow5}$$

Fig. 2.15 Scheme of the nucleation current (Adapted from Ford (2004))

$$J_c = \dot{I}_{c\leftrightarrow(c+1)} \tag{2.35}$$

In order to further simplify the calculations, it is assumed that the size distribution inside the vapor volume is steady:

$$\frac{dn_x}{dt} = 0 \tag{2.36}$$

This assumption is reasonable because it is known that n_x reaches a steady distribution in few µs, whereas the active nucleation period in a nozzle flow lasts typically 10–50 µs (Bakhtar et al. 2005). Based on this, the transient stage is usually ignored, and it is assumed that the steady distribution is attained instantaneously. Inserting Eq. 2.36 into Eq. 2.34 leads to

$$\dot{I}_{(x-1)\leftrightarrow x} = \dot{I}_{x\leftrightarrow(x+1)} = \dot{I} \tag{2.37}$$

This means that the nucleation current is equal for each size group and, in particular, we need to focus on the nucleation current of the critical-sized group of clusters. Now, this is simply given by

$$J_c = \dot{I} = \dot{C}_x n_x - \dot{E}_{x+1} n_{x+1} \tag{2.38}$$

In order to find J_c, we must find expressions for the average evaporation and condensation rates in Eq. 2.38. The kinetic theory of gases serves this purpose. In practice, $\dot{E}$ and $\dot{C}$ can be obtained by calculating the rate at which vapor and liquid molecules impact the cluster surface from both sides. By assuming ideal gas behavior and spherical droplets, the evaporation and condensation rates are given by

$$\dot{C}_x = q_C \frac{S_x}{m_m} \frac{P_v}{\sqrt{2\pi R T_v}}$$
$$\dot{E}_x = q_E \frac{S_x}{m_m} \frac{P_l}{\sqrt{2\pi R T_l}} \tag{2.39}$$

where S_x is the cluster surface, m_m is the mass of one molecule,[7] and q_C and q_E are the so-called *accommodation coefficients*.

These are defined as the ratio between the number of molecules that actually cross the interface and the total number of those impacting the surface. In practice, these coefficients quantify the fraction of impacts that, from both sides, concretely results in a condensation or evaporation of molecules.

Unfortunately, the direct substitution of Eq. 2.39 into Eq. 2.38 does not lead to a simple solution. In the original derivation, a different approach is used that requires considering the hypothetical situation of an equilibrium cluster distribution given by

$$n'_x = n_v e^{-Gb_x} = n_v \exp\left(-\frac{\Delta G_x}{kT_v}\right) \tag{2.40}$$

where n'_x is the equilibrium population of the "critical-sized" cluster group.

This expression is the well-known Boltzmann distribution and represents the equilibrium cluster distribution in a superheated gas ($\varphi_{ss} < 1$). Due to the monotonically increasing trend of Gibbs free energy (see again Fig. 2.14), droplet growth in a superheated gas is prohibited, but cluster formation is nevertheless active. In this condition, the cluster distribution reaches a steady distribution where, for each size group, the number of created clusters is equal to those destroyed. Hence, under these hypotheses, the nucleation current is zero:

$$\dot{I} = 0 \Rightarrow \dot{C}_x n'_x = \dot{E}_{x+1} n'_{x+1} \tag{2.41}$$

Inserting Eq. 2.41 into Eq. 2.38 to eliminate $\dot{E}_{c+1}$ results in

$$J_c = \dot{C}_x n'_x \left(\frac{n_x}{n'_x} - \frac{n_{x+1}}{n'_{x+1}}\right) \tag{2.42}$$

This expression can be rearranged in differential form to give

$$J_c = \dot{C}_x n_x \frac{\partial(n/n')}{\partial x} \tag{2.43}$$

Inserting Eq. 2.40 and Eq. 2.39 into Eq. 2.43 and integrating the resulting expression over the whole range of x-groups finally return the classical formulation

---

[7]The molecular mass or weight, m [kg/molecule], is not to be confused with the molar mass, M [kg/kmol]. The mass of one molecule is found by dividing the molar mass by the Avogadro number: $m = M/N_A$. In (Carey 1992) M is mistakenly reported as the molecular weight.

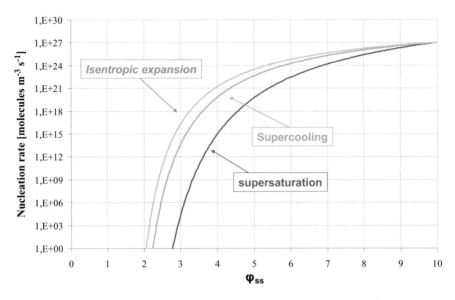

Fig. 2.16 Comparison of nucleation rates for different processes, as a function of the supersaturation ratio

for the critical nucleation rate (details of the integration passages can be found in (Bakhtar et al. 2005)):

$$J_c = q_C \frac{\rho_v^2}{\rho_l} \left(\frac{2\sigma}{\pi m_m^3} \right)^{1/2} \exp(-Gb_c) \qquad (2.44)$$

where J_c is measured in [n. molecules/m^{3}s].

To develop further Eq. 2.44, Gb_c can be equally expressed as a function of the critical radius or the thermodynamic state of the system by, respectively, making use of Eq. 2.27 and Eq. 2.30:

$$Gb_c = \frac{\Delta G_c}{k_b T_v} = \frac{4\pi r_c^2 \sigma(T_v)}{k_b T_v} = \frac{16}{3} \frac{\pi \sigma(T_v)^3}{k_b T_v^3 (\rho_l R \ln \varphi_{ss}(T_v))^2} \qquad (2.45)$$

In particular, the last expression is of practical use for thermodynamic calculations.

Although Eq. 2.45 was developed by considering an isothermal compression of the vapor, its use is not limited to this process because all the quantities in the expression are state variables. For instance, Fig. 2.16 shows a comparison of nucleation trends for different types of processes. These were selected in order to have the final state in common ($T_v = 300\ K$, $P_v = 35,000\ Pa$) and qualitatively represent the transformations depicted in Fig. 2.13.

Most interestingly, Fig. 2.16 illustrates that the isentropic expansion has the steepest trend among the three types of process. Indeed, we will see in Chap. 4

that during the rapid expansion inside de Laval nozzles, condensation takes the semblance of a dynamic shock.

The classical nucleation theory is remarkable in the way it provides for a very simple analytic expression for the nucleation rate. However, many assumptions were needed in order to achieve a simple closed solution, which may compromise the accuracy as well as the physical correspondence with the real phenomenon.

For instance, the classical nucleation rate assumes that thermal equilibrium exists between the vapor and the growing cluster (i.e., the liquid temperature is always equal to T_v). However, the attainment of thermal equilibrium requires either slow transformations or a very effective transport of heat from the cluster to the surrounding vapor. In the absence of any of these conditions, the temperature in the cluster increases to value greater than T_v due to the release of the latent heat of condensation. This localized heat release makes the molecules of the liquid cluster more energized, thus enhancing the rate at which they evaporate from the cluster surface. Therefore, the net rate of condensation on each cluster is reduced, and, consequently, the critical nucleation rate J_c is partially suppressed.

In order to account for this effect, Kantrovitz (1951) calculated a correction factor for the critical nucleation rate:

$$J_{c\_NISO} = \frac{J_c}{1 + \xi} \tag{2.46}$$

where J_c is the classical nucleation rate from Eq. 2.44 and ξ is given by

$$\xi = q_C \frac{2(\gamma - 1)}{(\gamma + 1)} \frac{h_{lv}}{RT_v} \left(\frac{h_{lv}}{RT_v} - \frac{1}{2} \right) \tag{2.47}$$

where γ is the specific heat ratio, h_{lv} is the latent heat of condensation, and it was assumed that nucleation occurs under conditions that are not too close to the critical point. Typically, Kantrowitz' correction reduces the critical nucleation rate for water by a factor of 50–100 (Bakhtar et al. 2005).

In addition to a reduction of the nucleation rate, the uncertainty connected to the liquid phase temperature directly impacts the value of the surface tension which, in turn, can significantly affect the critical nucleation rate (σ is present in Eq. 2.44 both in the pre-exponential factor and in the exponential argument, where it is elevated at the third power!).

The surface tension is generally a linearly decreasing function of temperature. For instance, a semiempirical expression from Eötvös and Ramsay-Shields gives the "flat-film surface tension" as a function of the fluid critical temperature (Wegener and Mack 1958):

$$\sigma_{ff} = c v_l (T_c - T_v) \tag{2.48}$$

where v_l is the liquid cinematic viscosity and c is a constant which for many liquids is equal to 2,12.

Unfortunately, the uncertainty on the liquid temperature is not the only source of potential error for σ. Apart from the effect of surface impurities,[8] the surface tension is generally believed to depend on the curvature of the cluster surface. At the very high supersaturation ratios achieved in condensing nozzle flows, a critical droplet is composed of 10–50 molecules (Bakhtar et al. 2005). At these very small curvature radii, the value of surface tension may depart consistently from the conventional flat-film value.

For instance, Tolman (1949) arrived by thermodynamic methods at a first approximation for the surface tension of a drop of radius r:

$$\sigma(r) = \sigma_{ff}\frac{r}{r + 2C} \tag{2.49}$$

where C is a constant of the order of the free molecular path.

Although the surface tension is generally believed to diminish with the radius of curvature (like in Eq. 2.49), many other theories have been developed which are in marked contrast between each other, even on the sign of the variation (Bakhtar et al. 2005). Consequently, despite the large uncertainty that this may introduce, the surface tension is in many cases calculated by considering the conventional flat-film value.

In conclusion of this section, it may be worth to spend some words on the expression for pre-exponential factor in the critical nucleation rate equation:

$$J_0 = q_C\frac{\rho_v^2}{\rho_l}\left(\frac{2\sigma}{\pi m^3}\right)^{1/2} \tag{2.50}$$

Although the effect of an error in J_0 may be small compared with the effect on the exponent, nevertheless, a seemingly never-ending debate on the accommodation factor has accompanied the research on nucleation theory since perhaps its beginning.

In general, it is believed that q_C is of the order of the unity (by definition, the accommodation factor can't be greater than one). However, it should be noted that for very small clusters, no reliable way to measure the accommodation factors has been devised which, ultimately, are empirically tuned to make the theoretical trends coincide with experimental data (see, for instance, (Young 1982)). However, whenever the Kantrowitz non-isothermal correction is adopted, this has the fortuitous effect of making J_c almost insensitive to the accommodation factor, at least for values in the range 0,1–1,0 (Bakhtar et al. 2005).

---

[8] Although these dramatically change the value of surface tension, they should not be present inside a freshly formed droplet by vapor condensation. By contrast, this is a big issue in bubble nucleation or cavitation phenomena (see (Carey 1992) or (Brennen 1995)).

2.4.3 Droplet Growth

In high-speed condensations, the mass of the critical-sized nucleus is very much smaller than the mass of liquid that condenses upon it (Hill 1966). Indeed, it is the growth of the droplets that produces the macroscopic changes on the nozzle and ejector dynamics. Consequently, it is very important to accurately calculate this final stage of the condensation process in order to determine the trends of the mixture flow variables (Mach, temperature, pressure, and entropy).

In general, the growth of a droplet can be evaluated by computing the fluxes of mass, momentum, and energy that cross its surface:

$$\frac{dm_d}{dt} = \Phi_{m\_in} - \Phi_{m\_out}$$
$$\frac{dm_d u}{dt} = \Phi_{mu\_in} - \Phi_{mu\_out} \qquad (2.51)$$
$$\frac{dm_d E}{dt} = \Phi_{mE\_in} - \Phi_{mE\_out}$$

where Φ_m, Φ_{mu}, and Φ_{mE} are, respectively, the mass, momentum, and energy fluxes entering or leaving the drop surface. Although some authors have proposed general formulation, in common practice, these fluxes are calculated differently depending on the size of the droplet.

In particular, the analysis is generally subdivided into three main regimes that depend on the ratio between the droplet radius and the molecular mean free path,[9] i.e., the Knudsen number:

$$Kn = \frac{\lambda}{2r_d} \qquad (2.52)$$

During the initial phase of the droplet growth, the liquid nucleus is generally much smaller than the mean free path, i.e., $Kn \gg 1$. Under these conditions, named as *free molecular regime*, the continuum hypothesis does not hold, and the calculation of the droplet growth must be accomplished by means of kinetic theory or statistical mechanics concepts. At the other extreme is the situation where $Kn \ll 1$. In this case the droplet is large enough to apply the macroscopic balances for heat, mass, and momentum. In between these two conditions is what is called the transition regime ($Kn \sim 1$). This is the most difficult to analyze and is usually handled by means of interpolation formulae that connect the continuum and free molecular regimes.

In the case of the free molecular regime ($Kn \gg 1$), the mass transfer is calculated by evaluating the rate of molecule collision with the droplet surface. Thus, the mass

---

[9]This is the average distance that a molecule or cluster can travel without incurring in collision with other gas particles.

conservation equation follows from the balance between the evaporation and condensation rates:

$$\frac{dm_d}{dt} = 4\pi r^2 \rho_1 \frac{dr}{dt} = m_m \left(\dot{C} - \dot{E} \right)$$ (2.53)

where m is the mass of one molecule and r is the droplet radius.

Substitution of Eq. 2.39 into Eq. 2.53 leads to an expression for the time derivative of the droplet radius, which is the quantity of interest:

$$\rho_1 \frac{dr}{dt} = \left(q_C \frac{P_v}{\sqrt{2\pi R T_v}} - q_E \frac{P_s(T_1)}{\sqrt{2\pi R T_1}} \right)$$ (2.54)

where the assumption was made that the droplet pressure is equal to the saturation pressure at the droplet temperature (Hill 1966).

In the case of nozzle flows, it was recognized by many authors (e.g., (Wegener and Mack 1958)) that, to a very good approximation, the droplet velocity can be considered equal to that of the surrounding vapor. This is particularly true in the free molecular regime, where the dimensions of the droplets are so small that their inertia can be considered negligible. Consequently, the momentum exchange between the phases is zero, and the momentum balance needs not to be calculated for this regime.

The energy balance can be derived by considering the energy transport of each of the molecules condensing or evaporating from the droplet surface. Young (1982) provides for a simple expression of the energy balance, which was derived by the work of Hill (1966):

$$\rho_1 h_{lv} \frac{dr}{dt} = \frac{P_v}{\sqrt{2\pi R T_v}} \frac{c_p + c_v}{2} \cdot (T_1 - T_v)$$ (2.55)

If the vapor temperature and pressure variations are known, Eq. 2.54 and Eq. 2.55 constitute a set of two equations that can be integrated numerically to give the two unknowns of the system, namely, T_1 and r. However, in order to reduce further the computational requirements, the prescriptions of a formulation for the droplet temperature, T_1, dispense from the resolution of the mass conservation equation. For instance, in many studies the liquid temperature is assumed to be at the saturation conditions corresponding to the vapor pressure, i.e., $T_1 = T_s(P_v)$.

Although this imposition avoids calculating a further equation, it also implicates that the only driving potential for the droplet growth is the temperature difference between the vapor and liquid phase, i.e., the supercooling degree (it should be noted that the pressure of the two phases is equal in this case, $P_1 = P_v$). In general, this is a crude simplification because the droplet growth depends both on thermal gradient and on the pressure difference between the two phases (Young 1991). Moreover, any effect of pressure increase due to the droplet surface tension is neglected.

For the continuum regime, $Kn \ll 1$, a similar set of equations can be found by considering the conservation of mass and momentum energy of the vapor surrounding the droplet (Young 1991). By considering a local reference system of spherical coordinates at the center of the droplet, the only nonzero velocity component is the radial velocity, and the conservation equations, Eq. 2.51, reduce to[10]

$$\frac{d}{dr}(r^2\rho u) = 0$$
$$(r^2\rho u)\frac{du}{dr} = r^2\frac{dP}{dr}$$
$$(r^2\rho u)\frac{d(h+u^2)}{dr} = \frac{d}{dr}\left(r^2 k\frac{dT}{dr}\right)$$

(2.56)

where k is the thermal conductivity of the liquid and it was assumed a steady condensation process. Integration of the above equation from the droplet surface to the far-field leads to equivalent expressions of Eq. 2.54 and Eq. 2.55 in the continuum regime (Young 1982, 1991):

$$\rho_1\frac{dr}{dt} = \sqrt{2\rho_1(P_v - P_d)}$$
$$\rho_1 h_{lv}\frac{dr}{dt} = k\frac{dT}{dr} \approx \frac{k}{r}(T_1 - T_v)$$

(2.57)

where it was assumed a linear temperature variation and P_d is the pressure at the droplet surface.

Things get really involved when Kn approaches unity because none of the two limiting situations described before can be applied without incurring in significant errors. Many interpolation formulae have been proposed to cover this range of conditions (see, e.g., (Young 1982)), but their accuracy has never been assessed rigorously (Young 1991).

Nevertheless, it should be noticed that in most nozzle experiments, the average drop size is usually smaller than one mean free path so that it is possible to use the results of the kinetic theory for predicting mass and energy fluxes to and from the drop surface (Hill 1966). This however may not be true for simulations of complete ejectors, as will be discussed in Chap. 4.

2.4.4 Condensing Nozzle

In general, there are two ways of testing nucleation and droplet growth theories: the first, more appropriate for fundamental physics investigations, involves condensation in cloud and expansion chambers; the second, which is more suited for engineering studies, deals with supersonic expansions in de Laval nozzles. As it was shown by many authors (e.g., (Wegener and Mack 1958; Hill 1966)), the converging/diverging nozzle is a remarkable test bench for wet-steam model theories. The

---

[10]The complete set of conservation equation in spherical coordinates can be found in Appendix B of (Bird et al. 2002).

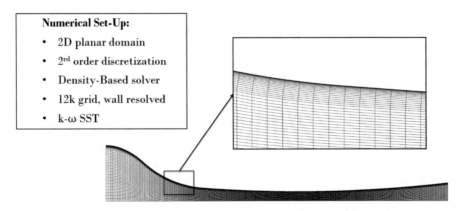

Numerical Set-Up:

- 2D planar domain
- 2rd order discretization
- Density-Based solver
- 12k grid, wall resolved
- k-ω SST

Fig. 2.17 Numerical scheme and computational domain for the Moses and Stein nozzle

advantages of this type of experiment are numerous: first of all, the "simplicity" of the steady isentropic flow, which can be easily reproduced by Q1D calculations. In turn, this allows marking the effects of condensation by simple pressure measurements, as will be illustrated shortly. Moreover, it has been shown by Stodola that for this type of expansion the effects of dust particles are entirely insignificant, meaning that the condensation is of the homogeneous type (Hill 1966). The only drawback of nozzle experiments is that nucleation and droplet growth are tightly coupled and is hard to validate the theories separately (Bakhtar et al. 2005).

Among the many nozzle experiments that can be found in the literature, that of Moore et al. (1973) and Moses and Stein (1978) appear to be the most popular test cases for validating wet-steam models (Starzmann et al. 2016).

In what follows, the ANSYS Fluent wet-steam model (ANSYS Inc. 2016) is used to simulate the nozzle experiment from Moses and Stein (1978) in order to illustrate the main impact of non-equilibrium condensation on supersonic flow. The numerical scheme adopted for these simulations is described in Fig. 2.17.

Figures 2.18 and 2.19 show the results for two sets of experiments performed by keeping fixed the inlet pressure while varying the inlet temperature (Fig. 2.18) and vice versa (Fig. 2.19). The figures present the normalized pressure trend along the nozzle axis and focus on the region downstream the nozzle throat where the condensation shock takes place.

As can be seen, for all the tested cases, the agreement is satisfactory both in terms of condensation starting position and asymptotic pressure trend. The two sets of curves show the main features and effects caused by the non-equilibrium condensation. Due to the concentrated heat release, the pressure curves deviate significantly from the hypothetical dry isentropic trend (the dotted curve on each figure). In turn, this brings about some undesirable effects. Firstly, the non-equilibrium heat transfer between the two phases causes the entropy to increase, thus producing losses and reducing the nozzle efficiency. Secondly, the trends of pressure and Mach number are altered. This has consequences on the subsequent development of the mixing layer because increasing the nozzle exit pressure impacts the correct expansion of the

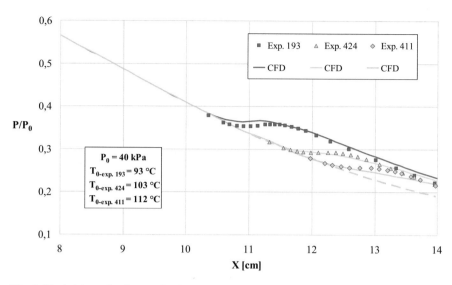

Fig. 2.18 Axial trends of normalized static pressure for three experiments with increasing inlet temperature

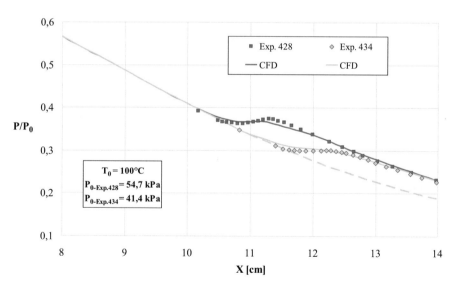

Fig. 2.19 Axial trends of normalized static pressure for two experiments with decreasing inlet pressure

primary flow. In practice, if the primary nozzle of a supersonic ejector is designed to be perfectly expanded under a specific set of conditions, the pressure increase would cause the primary jet to become under-expanded. Consequently, in addition to direct losses caused by the heat transfer, condensation can cause indirect losses due to the

Table 2.4 Onset conditions for various experiments, comparison between numerical and experimental data (in blue is the vapor onset temperature below 0 °C)

	T_0 [°C]	P_0 [Pa]	T_{onset} [°C]	P_{onset}[Pa]	φ_{ss_onset}	x_{Exp} [cm]	x_{CFD} [cm]	Error
Exp. 191	96.1	17,812	−15.8	3906	21.7	13.26	13.37	−0.8%
Exp. 193	92.9	43,023	12.8	15,252	10.3	10.74	10.56	1.7%
Exp. 234	97.9	34,957	6.4	11,012	–	11.49	11.64	−1.3%
Exp. 244	110.4	26,944	−6.0	6199	–	13.18	13.37	−1.4%
Exp. 248	108.8	19,492	−17.5	3840	–	14.1	14.2	−0.7%
Exp. 252	101.2	40,050	9.6	12,292	10.3	11.5	11.47	0.3%

increase of the shock diamonds intensity and to a different flow behavior within the mixing chamber.

The agreement that is seen at condensation onset mostly depends on the accuracy of the nucleation rate equation. In particular, the point at which the pressure differs from the isentropic value by 1 percent is commonly referred to as the "onset of condensation." In their paper, Moses and Stein report data for this quantity in a wide range of conditions.

Table 2.4 shows a comparison of numerical and experimental data for the onset of condensation. From inspection of the different cases, it can be seen that the difference between CFD and experiments is always around 1%. Moreover, it is interesting to note that the accurate matching holds even for experiments where the minimum vapor temperature was below 0 °C.[11]

In order to check the accuracy of the droplet growth formulation, the numerical trend for the liquid mass fraction can be compared with those obtained by light scattering measurement. Unfortunately, Moses and Stein report only one of such profiles. Nevertheless, even for this one case, the agreement between theory and experiments is accurate, as shown in Fig. 2.20.

Nevertheless, it should be noted that the comparisons are made only on macroscopic variables, whereas a better assessment of the numerical simulations must be done by comparing the results on the population of droplets and their size distributions. In this case, the agreement is much more difficult to achieve, and different combinations of models and calibrating parameters can produce comparable levels of accord with experiments (Bakhtar et al. 2005).

---

[11]It should be noted that vapor temperatures below zero do not necessarily imply ice formation. This is because the droplet temperature is always higher than that of the vapor and close to the liquid saturation temperature.

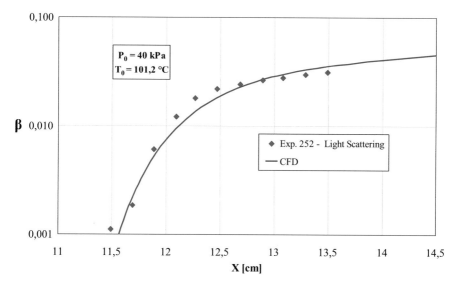

Fig. 2.20 Axial trends of liquid mass fraction, comparison of experimental and CFD trends

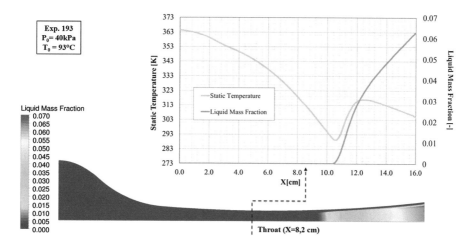

Fig. 2.21 Liquid mass fraction and temperature axis trends (top); liquid mass fraction contour (bottom) (the two images are not on the same scale)

Finally, Fig. 2.21 illustrates the axis trend and contour of the liquid mass fraction within the nozzle. As can be seen, the condensation occurs well after the nozzle throat and reaches values of around 6/7% of the total mass. Despite this small percentage, the impact on the temperature trend is considerable, with an increase of more than 30 °C in less than 2 cm. After this abrupt change, the temperature starts to decrease again, although at a lower rate.

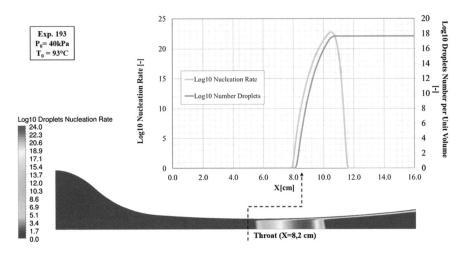

Fig. 2.22 Log_{10} nucleation rate and droplet number axis trends (top); Log_{10} nucleation rate contour (bottom) (the two images are not on the same scale)

Figure 2.22 further illustrates that the nucleation process occurs just in a limited region of the nozzle and arrives at values of the order of 10^{25} nuclei per unit time and unit volume. Downstream the condensation shock, the liquid continues to form via condensation on the droplet surface, but the total number of liquid nuclei remains constant. In this region, the large availability of condensing sites ($\sim 10^{18}$ nuclei/m^3) practically restores the thermodynamic equilibrium conditions, and the condensation process follows a path with a reduced influence of the metastability effects.

2.5   Flashing Ejectors

The idea behind the ejector expansion cycle is very promising and apparently simple (see Chap. 1). Unfortunately, the underlining physics is actually very complex. The liquid coming from the condenser (or, in the case of CO_2, the transcritical fluid coming from the gas cooler) undergoes a phase change while expanding in the supersonic nozzle. Due to the high velocity, the phase change takes place in non-equilibrium conditions, i.e., the fluid crosses a metastable region.

A first indication of the metastable region boundary is given by the "spinodal" curves (Bejan 1988; Carey 1992). Spinodal curves can be seen as a consequence of the van der Waals theory of liquid-gas phase transition that aims to account for the finite size of molecules and for the attractive force between them. The van der Waals equation of state

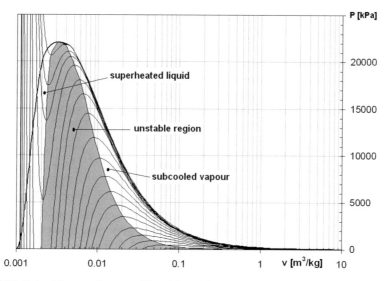

Fig. 2.23 Spinodal curves for water (Grazzini et al. 2011)

$$\left(P + \frac{a}{v^2}\right)(v - b) = RT \tag{2.58}$$

can be written as

$$v^3 - \left(b + \frac{RT}{P}\right)v^2 + \frac{a}{P}v - \frac{ab}{P} = 0 \tag{2.59}$$

showing that, for a given value of P and T, there can be up to three values of v. This happens below the critical point, where the shape of the isothermal curve is far from the equilibrium behavior of real fluids. All states between the saturated liquid and the saturated vapor curves appear unrealistic. However, some of these states can actually be reached in metastable conditions (Bejan 1988; Carey 1992). This is true for the isothermal segments comprised between the saturated liquid curve and the minimum and between the saturated vapor curve and the maximum, while the curve segment between the minimum and maximum represents an unstable condition (Bejan 1988) because in this area it would be $(\partial P/\partial v)_T > 0$. Therefore, the loci of the minima and maxima on the isothermals mark the boundary between the two metastable areas (on the sides) and the unstable area (in the middle). These two loci are named "spinodal curves" due to their resemblance with the spine of a vertebrate animal when they are plotted on the isothermals (Fig. 2.23).

In order to build the spinodal curves, instead of calculating as usual the van der Waals coefficients a, b, and R as a function of the critical point, we may calculate them for each temperature imposing that the van der Waals isothermal crosses the saturation curves in the saturation points, i.e.,

$$P_{sat} = \frac{RT_{sat}}{(v_1 - b)} - \frac{a}{v_1{}^2} = \frac{RT_{sat}}{(v_v - b)} - \frac{a}{v_v{}^2} \qquad (2.60)$$

These two equations may be complemented by the "Maxwell condition" (Bejan 1988), that is, the null value of work for the thermodynamic cycle comprised between the van der Walls isothermal and the constant pressure vaporization line. This means that the two areas between the $P = P_s$ horizontal line and the S-shaped van der Waals isotherm must be equal. Hence,

$$\int_{v_1}^{v_v} \left[\frac{RT_{sat}}{(v - b)} - \frac{a}{v^2} - P_{sat} \right] dv = 0 \qquad (2.61)$$

This procedure has been used in the paper by Grazzini et al. (2011) to build the spinodal curves for water (Fig. 2.23). This limit however has proven rather unrealistic. The classic nucleation theory, recalled in Sect. 2.4, gives more realistic results. Nucleation of bubbles in a rapidly expanding liquid is dealt with, e.g., by Elias and Chambré (1993). If the rate of pressure recovery due to the vapor formation balances (or exceeds) the rate of expansion imposed by the nozzle shape, the fluid undergoes a "flashing." The limit of homogeneous nucleation, however, may still be unattainable in practice, because any disturbance (effect of walls, dirt particles carried with the expanding fluid, gasses dissolved in the liquid, etc.) will cause the phase transition to start well in advance with respect to the spinodal curve.

Once the phase change has started, the fluid motion is influenced by the strong density difference between the two phases. Even for very low values of the vapor mass fraction (or vapor quality) x, the vapor volumetric fraction (or "void" fraction) α_v reaches close-to-unity values. This may be easily seen considering the velocities of the two phases:

$$u_v = \frac{\dot{Q}_v}{A_v} = \frac{\dot{m}}{\rho_v} \frac{x}{\alpha_v} \qquad\qquad u_l = \frac{\dot{Q}_l}{A_l} = \frac{\dot{m}}{\rho_l} \frac{1 - x}{1 - \alpha_l} \qquad (2.62)$$

In the simple case of homogeneous flow, i.e., $u_v = u_l$, α_v turns out to be

$$\alpha_{vH} = \frac{1}{1 + \left(\frac{1-x}{x}\right) \frac{\rho_v}{\rho_l}} \qquad (2.63)$$

As it may be seen in Fig. 2.24, which is referred to carbon dioxide at 275 K, for $x = 0.2$ we have $\alpha_{vH} \approx 0.7$, i.e., the vast majority of the flow section is occupied by vapor. Obviously this is just a limit, as usually the two phases proceed at significantly different velocity. This is accounted for by a "slip factor" $S = u_v/u_l$ that appears behind the second term at denominator in Eq. 2.63.

The flow patterns within the primary nozzle are very difficult to predict. Bubbles are formed by a nucleation process (boiling) that is substantially similar to the droplet nucleation described in Sect. 2.4. A bubble of diameter D is stable if its

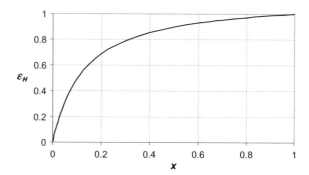

Fig. 2.24 Vapor volumetric fraction as a function of quality – CO_2 at 275 K

internal pressure is sufficient to balance the pressure of the surrounding liquid plus the surface tension σ (Young-Laplace equation):

$$P_v = P_l + \frac{4\sigma}{D} \tag{2.64}$$

Nucleation is followed by bubble growth, as evaporation takes place at the liquid interface. Coalescence between the bubbles could form a zone in the central part of the flow section with vapor moving at high speed, while the liquid could stay on the walls (annular flow) as what happens in any tubular evaporator at high mass velocity. However, the present case is completely different, because the flow is substantially adiabatic and the velocity may be supersonic. If the vapor flows at high speed on a liquid film, the interface becomes unstable and may form droplets which will be carried on by the vapor (mist flow).

At the nozzle exit, the liquid may either travel in form of droplets or create ligaments, as the liquid film is thorn away by the interaction with the surrounding vapor phase. The formation of drops from a continuous liquid phase, either jet or film, is called "breakup." Afterward, the droplet traveling along the flow path may become unstable, incurring in a "secondary" breakup. The dynamics of the droplet and the stability of the equilibrium between internal and external forces are governed by the Rayleigh-Plesset equation (Plesset 1949). Note that this process is completely different from the droplet nucleation and growth which was described in Sect. 2.4.

As a general rule, the mean drop size is roughly proportional to the square root of the liquid jet diameter or sheet thickness (Lefebvre 1989). In any case, the formation of droplets (atomization) dramatically increases the liquid interface area. The minimum energy required for atomization is equal to the surface tension multiplied by the increase in liquid surface area. Whenever atomization occurs under conditions where surface tension forces are important, the Weber number, which is the ratio of the inertial force to the surface tension, is a useful dimensionless parameter for correlating drop size data.

$$We_1 = \frac{\dot{m}_v^2 d_{eq}}{\rho_1 \sigma} \tag{2.65}$$

In Eq. 2.65 $\dot{m}_v$ is the mass velocity, i.e., the mass flow rate divided by the flow area, and d_{eq} is the hydraulic diameter of the core liquid flow. The surface tension decreases as the fluid approaches the critical point.

Liquid viscosity is important as well, as it affects not only the drop size distributions but also the nozzle flow rate and spray pattern. An increase in viscosity lowers the Reynolds number and also hinders the development of any natural instabilities in the jet or sheet. The combined effect is to delay disintegration and increase the size of the drops in the spray.

One of the main problems is the lack of reliable experimental results that impedes the development of accurate theoretical and numerical tools for the analysis of the ejector. Several works about flashing jets are available in literature, but most of them concern industrial safety and are focused on the understanding of jet behavior after the expansion through an orifice.

Laser diagnostic techniques for measuring particle size and velocity in sprays are commonly available, but, to the best of our knowledge, they have not been used on ejectors. High-speed pulsed imaging is being used to study drop size distributions and spray structure. A convenient method for assessing and comparing sprays is diffraction particle sizing, which measures the diffusion of a laser beam when it crosses the spray. The analysis of the light distribution as a function of the distance from the line-of-sight is elaborated via software and gives an estimate of the droplet size distribution in real time. In principle, it should be possible to open two optical accesses near the nozzle exit plane in order to have a laser beam across the flow and perform a diffraction particle sizing. The presence of different peaks in the size distribution would give an indication of the various droplet forming mechanisms.

For more details about topics like primary and secondary breakup, flow entrainment, jet angles, droplet distribution, and interactions with turbulence, the reader is referred to (Lefebvre 1989; Linn and Reitz 1998, or Polanco et al. 2010).

The flash boiling phase-change mechanism is often analyzed distinguishing between "mechanical" and "thermodynamic" effects, even if actually these two aspects are strictly connected. Aamir and Watkins (2000) analyzed the behavior of a liquid propane jet through an orifice and suggested that the thermodynamics part governs the process before the nozzle exit, while the mechanical effects are of major importance in the evolution of the spray.

Cavitation is a further topic that has some resemblance with flashing ejectors. Cavitation can be considered as a "mechanical" process, as the simplest cavitation models consider the mass transfer as driven only by the difference between the actual pressure and the vapor pressure (Singhal et al. 2002). The use of cavitation models is one of the most common approach to analyze the mass transfer between phases for flashing devices; in these models the mass exchange rate is obtained from the Rayleigh-Plesset model (Singhal et al. 2002; Giese and Laurien 2000).

Several models have been proposed for the calculation of the critical mass flux of a flashing jet. The simplest one is the Homogeneous Frozen Model (HFM), which assumes that vapor-liquid mixture is in mechanical and thermodynamic equilibrium (the phases have the same temperature and velocity) and the quality (vapor mass fraction) remains constant through the whole length of the nozzle (Fire Science Center 1994). The second model is the Homogeneous Equilibrium Model (HEM), where the phases have the same temperature and velocity and the mass transfer between phases happens in equilibrium conditions. This model has been implemented in 1D model and 2D/3D CFD calculations, for flashing nozzles and ejectors working with various fluids (Giacomelli et al. 2016; Biferi et al. 2016; Palacz et al. 2015).

Angielczyk et al. (2010) have proposed the formulation of a Homogeneous Relaxation Model (HRM) in order to enhance the performance of the HEM accounting for the non-equilibrium phase change. Another possibility is yield by the Moody's model, in which the hypothesis of an equal velocity between the phases is abandoned and a slip facto is introduced at the exit of the orifice (Fire Science Center 1994).

We finish this resume (without pretending to be exhaustive) citing the Henry and Fauske model (Henry and Fauske 1971), which is based on the assumption of a non-equilibrium phase-change process but sees the phases flowing at the same velocity (Fire Science Center 1994). Lawrence and Elbel (2016) used the same correlation to determine the throat diameter of the motive nozzle of a two-phase ejector working with R410a.

Appendix A

Spontaneity of an Isothermal and Isobaric Process

By definition, the specific Gibbs free energy is given by

$$g = e + pv - Ts \qquad (2.\text{A}1)$$

For a generic transformation, its variation is computed as follows:

$$dg = de + vdp + pdv - sdT - Tds \qquad (2.\text{A}2)$$

In the case of isothermal and isobaric transformation, the equation simplifies to

$$dg = de + pdv - Tds \qquad (2.\text{A}3)$$

By using the first and second law, it is easy to show that the quantity on the RHS must be either minor or equal to zero:

$$dg = de + pdv - Tds \leq 0 \qquad (2.\text{A4})$$

where the equal sign is strictly valid for reversible transformations.

Therefore, under the current settings, a reversible transformation leaves the Gibbs free energy of the system unaltered. If the transformation is not reversible, g must diminish, i.e., a spontaneous (irreversible) process that occurs *inside* an isothermal and isobaric system is always accompanied by a decrease in Gibbs free energy.

Hence, the minimum of the Gibbs free energy corresponds to a situation where no other internal transformations are allowed, i.e., to an equilibrium state.

References

Aamir, A., & Watkins, A. (2000). Numerical analysis of depressurization of highly pressurized liquid propane. *International Journal of Heat and Fluid Flow, 21*, 420–431.

Angielczyk, W., Bartosiewicz, Y., Butrimowicz, D., & Seynhaeve, J. (2010). *1-D Modeling of supersonic carbon dioxide two-phase flow through ejector motive nozzle.* s.l., International Refrigeration and Air Conditioning Conference.

ANSYS Inc. (2016). *ANSYS fluent theory guide.* Canonsburg, PA: release 18.0.

ASHRAE (2008). *Addenda to designation and safety classification of refrigerants,* s.l.: ANSI/ASHRAE Standard 34-2007.

Bakhtar, F., Young, J. B., White, A. J., & Simpson, D. A. (2005). *Classical nucleation theory and its application to condensing steam flow calculations.* s.l., s.n.

Bejan, A. (1988). *Advanced engineering thermodynamics.* s.l.:John Wiley and Sons.

Biferi, G., et al. (2016). *CFD Modeling of high-speed condensationin supersonic nozzle, part II: R134a.* Atlanta: Fourth International Conference on Computational Methods for Thermal Problems.

Bird, R., Stewart, W., & Lightfoot, E. (2002). *Transport phenomena* (2nd ed.) s.l.:John Wiley and Sons.

Brennen, C. (1995). *Cavitation and bubble dynamics.* s.l.:Oxford University Press.

Brown, G. (1974). *The entrainment and large structure in turbulent mixing layers, Proceedings of the 5th Australasian Conf. on Hydraulics and Fluid Mechanics.* New Zealand, s.n.

Brown, G., & Roshko, A. (1974). On density effects and large structure in turbulent mixing layers. *Journal of Fluid Mechanics, 64*, 775–816.

Calm, J., & Hourahan, G. (2001). Refrigerant data summary. *Engineered Systems, 18*(11), 4–88.

Carey, V. (1992). *Liquid-vapor phase-change phenomena: An introduction to the thermophysics of vaporization and condensation processes in heat transfer equipment.* New York: s.l.:Taylor & Francis Series Group, LLC.

Carroll, B., & Dutton, J. (1990). Characteristics of multiple shock wave/turbulent boundary-layer interactions in rectangular ducts. *Journal of Propulsion, 6*, 186–193.

Chen, J., Havtun, H., & Palm, B. (2014). Screening of working fluids for the ejector refrigeration system. *International Journal of Refrigeration, 47*, 1–14.

Churchill, S. (1977). Friction factor equations spans all fluid-flow regimes. *Chemical Engineering Journal, 84*, 91–92.

Cizungu, K., Mani, A., & Groll, M. (2001). Performance comparison of vapour jet refrigeration system with environment friendly working fluids. *Applied Thermal Engineering, 21*, 585–598.

Crocco, L. (1958). One-dimensional treatment of steady gas dynamics. In: H. Emmons (Ed.), *Fundamentals of gas dynamics* (pp. 110–130). s.l.:Princeton University Press.

Dimotakis, P. E. (1986). Two-dimensional shear layers. *AIAA Journal, 24*(11), 1791–1796.

Dolling, D. S. (2001). Fifty years of shock-wave/boundary-layer interaction research: What's next? *AIAA Journal, 39*(8), 1517–1531.

Dorantes, R., & Lallemand, A. (1995). Prediction of performance of a jet cooling system operating with pure refrigerants or non-azeotropic mixtures. *International Journal of Refrigeration, 18*, 21–30.

Eames, I. (2002). A new prescription for the design of supersonic jet-pumps: The constant rate of momentum change method. *Applied Thermal Engineering, 22*, 121–131.

Eames, I., Worall, M., & Wu, M. (2013). An experimental investigation into the integration of a jet-pump refrigeration cycle and a novel jet-spay thermal ice storage system. *Applied Thermal Engineering, 53*, 285–290.

Elias, E., & Chambre, P. L. (1993). Flashing Inception in water during rapid decompression. *Journal of Heat Transfer, 115*, 231–238.

Fang, Y., et al. (2017). Drop-in replacement in a R134 ejector refrigeration cycle by HFO refrigerants. *International Journal of Refrigeration, 121*, 87–98.

Fire Science Center (1994). *The blowdown of pressurized containers,* s.l.: Fire Science Center, University of New Bruswick.

Ford, I. (2004). Statistical mechanics of water droplet nucleation. *Journal of Mechanical Engineering Science, 218*(C8), 883–899.

Ford, I. (2013). *Statistical physics an entropic approach.* s.l.:John Wiley & Sons, Ltd.

Freund, J., Lele, S., & Moin, P. (2000). Compressibility effects in a turbulent annular mixing layer. Part 1: Turbulence and growth rate. *Journal of Fluid Mechanics, 421*, 229–267.

Garcia del Valle, J., Saiz Jabardo, J., Castro Ruiz, F., & San Jose Alonso, J. (2014). An experimental investigation of a R-134a ejector refrigeration system. *International Journal of Refrigeration, 46*, 105–113.

Gatsky, T., & Bonnet, J.-P. (2013). *Compressibility, turbulence and high speed flow* (2nd ed.). Oxford: Academic Press.

Giacomelli, F., Mazzelli, F., & Milazzo, A. (2016). *Evaporation in supersonic CO2 ejectors: Analysis of theoretical and numerical models.* Firenze: International Conference on Multiphase Flow.

Giese, T., & Laurien, E. (2000). *A three dimensional numerical model for the analysis of pipe flows with cavitation.* Bonn: Proceedings of the Annual meeting on nuclear technology.

Grazzini, G., & D'Albero, M. (1998). *A Jet-Pump inverse cycle with water pumping column.* June 2–5, Oslo, Norway, Proceedings of Natural Working Fluids '98.

Grazzini, G., Milazzo, A., & Piazzini, S. (2011). Prediction of condensation in steam ejector for a refrigeration system. *International Journal of Refrigeration, 34*, 1641–1648.

Gyarmathy, G. (1962). *Bases for a theory for wet steam turbines (translated from German: "Grundlagen einer Theorie der Nassdampfturbine").* Doctoral Thesis No. 3221, ed. s.l.:ETH Zurich.

Henry, R., & Fauske, H. (1971). The two-phase critical flow of one-component mixtures in nozzles, orifices, and short tubes. *Journal of Heat Transfer, 93*(2), 179–187.

Hill, P. G. (1966). Condensation of water vapour during supersonic expansion in nozzles. *Journal of Fluid Mechanics, 25*(3), 593–620.

Huang, B., Chang, J., Wang, C., & Petrenko, V. (1999). A 1-D analysis of ejector performance. *International Journal of Refrigeration, 22*, 354–364.

Ikawa, H. (1973). *Turbulent mixing layer in supersonic flow.* s.l.:Ph.D. thesis, California Institute of Technology.

Johnson III, J., & Wu, B. (1974). *Pressure recovery and related properties in supersonic diffusers: A review,* s.l.: Report of the National Technical Information Service.

Kantrovitz, A. (1951). Nucleation in very rapid vapour expansions. *The Journal of Chemical Physics, 19*, 1097–1100.

Kasperski, J., & Gil, B. (2014). Performance estimation of ejector cycles using heavier hydrocarbon refrigerants. *Applied Thermal Engineering, 71*, 197–203.

Lawrence, N., & Elbel, S. (2016). Experimental investigation on the effect of evaporator design and application of work recovery on the performance of two-phase ejector liquid recirculation cycles with R410A. *Applied Thermal Engineering, 100*, 398–411.

Lefebvre, A. (1989). *Atomization and sprays*. USA: s.l.:Emisphere Publishing Corporation.

Lemmon, E., Huber, M., & McLinden, M. (2013). *NIST Standard reference database 23: Reference fluid thermodynamic and transport properties-REFPROP, Version 9.1*, s.l.: National Institute of Standards and Technology.

Linn, P., & Reitz, R. D. (1998). Drop and spray formation from liquid jet. *Annual Review of Fluid Mechanics, 30*, 85–105.

Matsuo, K., Miyazato, Y., & Kim, H. (1999). Shock train and pseudo-shock phenomena in internal gas flows. *Progress in Aerospace Sciences, 35*, 33–100.

Milazzo, M., & Rocchetti, A. (2015). Modelling of ejector chillers with steam and other working fluids. *International Journal of Refrigeration, 57*, 277–287.

Moore, M. J., Walters, P. T., Crane, P. I., & Davidson, B. J. (1973). *Predicting the fog-drop size in wet-steam turbines*. s.l., s.n.

Moses, C. A., & Stein, G. D. (1978). On the growth of steam droplets formed in a Laval nozzle using both static pressure and light scattering measurements. *Journal of Fluids Engineering, 100*, 311–322.

Nehdi, E., Kairouani, L., & Elakhdar, M. (2008). A solar ejector air-conditioning system using environment-friendly working fluids. *International Journal of Energy Research, 32*, 1194–1201.

Palacz, M., et al. (2015). Application range of the HEM approach for CO2 expansion inside two-phase ejectors for supermarket refrigeration systems. *International Journal of Refrigeration, 59*, 251–258.

Papamoschou, D. (1993). Model for entropy production and pressure variation in confined turbulent mixing. *AIAA Journal, 31*(9), 1643–1650.

Papamoschou, D. (1996). Analysis of partially mixed supersonic ejector. *Journal of Propulsion and Power, 12*(4), 736–741.

Papamoschou, D., & Roshko, A. (1988). The compressible turbulent shear layer: An experimental study. *Journal of Fluid Mechanics, 197*, 453–477.

Petrenko, V. (2009). *Application of innovative ejector chillers and air conditioners operating with low boiling refrigerants in trigeneration systems*. Louvain-la-Neuve: International Seminar on ejector/jet-pump technology and application.

Plesset, M. (1949). The dynamics of cavitation bubbles. *Journal of Applied Mechanics, 9*, 277–282.

Polanco, G., Holdøb, A., & Munday, G. (2010). General review of flashing jet studies. *Journal of Hazardous Materials, 173*, 2–18.

Pope, A., & Goin, K. (1978). *High-speed wind tunnel testing*. s.l.:Wiley.

Selvaraju, A., & Mani, A. (2004). Analysis of an ejector with environment friendly refrigerants. *Applied Thermal Engineering, 24*, 827–838.

Shapiro, A. H. (1953). *The dynamics and thermodynamics of compressible fluid flow* (Vol. II). New York: Ronald Press.

Singhal, A., Athavale, M., Li, H., & Jiang, Y. (2002). Mathematical basis and validation of the full cavitation model. *Journal of Fluids Engineering, 124*, 617–624.

Smits, A., & Dussauge, J.-P. (2006). *Turbulent shear layers in supersonic flow* (2nd ed.). New York: Springer.

Starzmann, J., et al. (2016). *Results of the International Wet Steam Modelling Project, Wet Steam Conference*. Prague, s.n.

Stodola, A. (1927). *Steam and gas turbines*. s.l.:McGraw-Hill.

Sun, D.-W. (1999). Comparative study of the performance of an ejector refrigeration cycle operating with various refrigerants. *Energy Conversion and Management, 40*, 873–884.

Tolman, R. (1949). Effects of droplet size on surface tension. *The Journal of Chemical Physics, 17*, 333.

Varga, S., Lebre, P., & Oliveira, A. (2013). Readdressing working fluid selection with a view to designing a variable geometry ejector. *International Journal of Low Carbon Technologies, 10*, 1–11.

Wang, F., Shen, S., & Li, D. (2015). Evaluation on environment friendly refrigerants with similar normal boiling points in ejector refrigeration system. *Heat and Mass Transfer, 51*(7), 965–972.

Wegener, P., & Mack, L. (1958). *Condensation in supersonic and hypersonic wind tunnels*. New York: Academic Press Inc..

Wygnanski, I., & Fiedler, H. (1970). The two-dimensional mixing region. *Journal of Fluid Mechanics, 41*, 327–362.

Young, J. (1991). The condensation and evaporation of liquid droplets in a pure vapour at arbitrary Knudsen number. *International Journal of Heat and Mass Transfer, 34*, 1649–1661.

Young, J. B. (1982). The spontaneous condensation in supersonic nozzles. *Physico Chemical Hydrodynamics, 3*(1), 57–82.

Zucker, R., & Biblarz, O. (2002). *Fundamentals of gas dynamics* (2nd ed.). Hoboken: John Wiley & Sons, Inc..

Chapter 3
Ejector Design

3.1 Zero-Dimensional Design

Traditionally, the ejector design criteria have been classified into two categories, according to the position of the nozzle (ESDU 1986): if the nozzle exit is located within the cylindrical part of the ejector and the mixing of primary and secondary flow occurs herein, we have a "constant-area mixing ejector"; if the nozzle exit is located in a "suction chamber" (usually conical) in front of the cylindrical part, we have a "constant-pressure mixing ejector." This latter denomination is justified by experimental evidence: if the mixing process takes place in a sufficiently large volume, the secondary flow touching the suction chamber wall remains at substantially constant pressure until completion of the entrainment, i.e., until it reaches the sonic speed. This point of view was introduced by Keenan et al. (1950) in one of the first attempts to analyze the ejector behavior. A refinement of this concept was presented by Munday and Bagster (1977), who assumed that the expanding primary flow and the suction chamber wall form a converging duct where the secondary flow (subsonic) accelerates until it reaches the sonic velocity. The actual mixing should begin from this point onward.

Most authors (e.g., Huang et al. (1999), Chunnanond and Aphornratana (2004), Besagni et al. (2016)) agree on the superior performance of the constant-pressure design over the constant-area one, which is now basically abandoned. Actually, this distinction must be taken as a general reference and the interaction between the primary and secondary flow at the nozzle exit is much more complex, as was shown in Chap. 2.

Even if the nozzle exit is placed well before the entrance of the cylindrical part of the ejector, the mixing process will continue throughout the cylinder, and a markedly nonuniform velocity will be recognizable even in the conical diffuser. What is really important when dealing with supersonic ejectors is the existence of a clearly defined section where the flow is fully supersonic. If not so, the ejector is said to work at "off-design" condition, as was mentioned also in the previous chapters.

© Springer International Publishing AG, part of Springer Nature 2018   71
G. Grazzini et al., *Ejectors for Efficient Refrigeration*,
https://doi.org/10.1007/978-3-319-75244-0_3

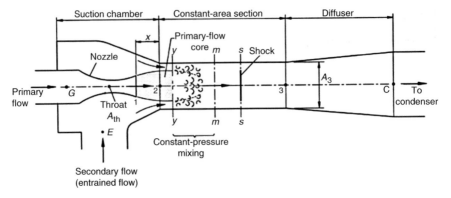

Fig. 3.1 Scheme of constant-pressure mixing ejector (Huang et al. 1999)

3.1.1 Huang et al. Model

Even if the model introduced by Munday and Bagster (1977) and improved by Huang et al. (1999) has some rather artificial assumptions, it is still used by many authors (Huang et al. have been cited by about 400 authors till May 2017) and, by careful calibration of some parameters, may be brought in good agreement with the experimental results. The symbols for the model and the reference locations within the ejectors are defined in Fig. 3.1.

Note that in the paper by Huang et al. (1999), the model is defined as "1D," i.e., one dimension. Actually, given that it calculates the flow characteristics in a finite number of sections, without specifying the distance between them, we prefer to classify this model as "0D."

The assumptions are:

- The working fluid is an ideal gas with constant properties c_p and $\gamma = c_p/c_v$.
- The ejector operates at steady state.
- The kinetic energy at ejector inlet/outlet ports is negligible.
- The flow is modeled as isentropic, and irreversibilities are taken into account by using experimentally determined coefficients.
- Mixing starts at section y–y when the secondary flow has reached sonic speed and proceeds at constant pressure within the constant-area section until the shock that occurs at section s–s.
- The ejector wall is adiabatic.

The first equation relates the primary mass flow $\dot{m}_p$ to the fluid thermodynamic properties (γ, R) and conditions at the generator (P_G, T_G). The mass flow is proportional to the nozzle throat area A_{th}. An isentropic efficiency η_p introduces a first parameter to be found experimentally.

$$\dot{m}_p = \frac{P_G A_{th}}{\sqrt{T_G}} \sqrt{\frac{\gamma}{R} \left(\frac{2}{\gamma + 1}\right)^{\frac{\gamma+1}{\gamma-1}}} \sqrt{\eta_p} \qquad (3.1)$$

Assuming that the nozzle efficiency is close to unity, the conditions at nozzle exit (section p1) are evaluated by classic gas-dynamic isentropic relations:

$$\frac{A_{p1}}{A_{th}} \approx \frac{1}{M_{p1}} \left[\frac{2}{\gamma+1}\left(1 + \frac{\gamma-1}{2}M_{p1}^2\right)\right]^{\frac{\gamma+1}{2(\gamma-1)}} \tag{3.2}$$

$$\frac{P_G}{P_{p1}} \approx \left(1 + \frac{\gamma-1}{2}M_{p1}^2\right)^{\frac{\gamma}{\gamma-1}} \tag{3.3}$$

From section p1, the primary flow is thought to continue its expansion without mixing until it reaches section y. At this section, primary and secondary flows have the same pressure $P_{py} = P_{sy}$ and start to mix. P_{sy} may be found considering that at section y the entrained flow reaches sonic velocity, that is, for an approximately isentropic flow,

$$\frac{P_E}{P_{sy}} \approx \left(\frac{\gamma+1}{2}\right)^{\frac{\gamma}{\gamma-1}} \tag{3.4}$$

Once P_{py} is known, we may go back to the nozzle exit assuming again an approximately isentropic expansion:

$$\frac{P_{py}}{P_{p1}} \approx \left(\frac{1 + \frac{\gamma-1}{2}M_{p1}^2}{1 + \frac{\gamma-1}{2}M_{py}^2}\right)^{\frac{\gamma}{\gamma-1}} \tag{3.5}$$

The flow area, on the other hand, is evaluated introducing in the isentropic relation a coefficient ϕ_p that should account for viscous loss at the boundary between primary and secondary flow:

$$\frac{A_{py}}{A_{p1}} = \frac{\phi_p M_{p1}}{M_{py}}\left[\frac{2}{\gamma+1}\frac{1 + \frac{\gamma-1}{2}M_{py}^2}{1 + \frac{\gamma-1}{2}M_{p1}^2}\right]^{\frac{\gamma+1}{2(\gamma-1)}} \tag{3.6}$$

The secondary mass flow rate is

$$\dot{m}_s = \frac{P_E A_{sy}}{\sqrt{T_E}}\sqrt{\frac{\gamma}{R}\left(\frac{2}{\gamma+1}\right)^{\frac{\gamma+1}{\gamma-1}}}\sqrt{\eta_s} \tag{3.7}$$

The isentropic efficiency of the secondary flow expansion is a third parameter to be specified.

Given that the section y has been assumed to stay in the cylindrical portion of the ejector, the area of this portion is now specified as

$$A_3 = A_{py} + A_{sy} \tag{3.8}$$

The temperatures of the two streams are related to the Mach numbers and stagnation conditions:

$$\frac{T_G}{T_{py}} = 1 + \frac{\gamma - 1}{2} M_{py}^2 \tag{3.9}$$

$$\frac{T_E}{T_{sy}} = 1 + \frac{\gamma - 1}{2} M_{sy}^2 \tag{3.10}$$

The mixing process ends at a section m. Introducing a mixing loss coefficient ϕ_m, a momentum balance between y and m yields

$$\phi_m \left(\dot{m}_P u_{Py} + \dot{m}_S u_{Sy} \right) = \left(\dot{m}_P + \dot{m}_S \right) u_m \tag{3.11}$$

An energy balance gives

$$\dot{m}_P \left(h_{py} + \frac{u_{py}^2}{2} \right) + \dot{m}_s \left(h_{sy} + \frac{u_{sy}^2}{2} \right) = \left(\dot{m}_p + \dot{m}_s \right) \left(h_m + \frac{u_m^2}{2} \right) \tag{3.12}$$

Primary, secondary, and mixed stream velocities are also obviously related to Mach numbers:

$$u_{py} = M_{py} \sqrt{\gamma R T_{py}}; \quad u_{sy} = M_{sy} \sqrt{\gamma R T_{sy}}; \quad u_m = M_m \sqrt{\gamma R T_m} \tag{3.13}$$

After mixing, the flow undergoes a normal shock in section s. Assuming an isentropic flow before and after the shock, the pressure rise is concentrated in this section and brings the stream from the mixed flow pressure $P_m = P_{py} = P_{sy}$ to the final value P_3 at the end of the cylindrical duct. The classic gas dynamic relations give

$$\frac{P_3}{P_m} = 1 + \frac{2\gamma}{\gamma + 1} \left(M_m^2 - 1 \right) \tag{3.14}$$

$$M_3^2 = \frac{1 + \frac{\gamma-1}{2} M_m^2}{\gamma M_m^2 - \frac{\gamma-1}{2}} \tag{3.15}$$

Finally, the conical diffuser produces a further pressure recovery up to

$$\frac{P_C}{P_3} = \left(1 + \frac{\gamma - 1}{2} M_3^2 \right)^{\frac{\gamma}{\gamma-1}} \tag{3.16}$$

The authors suggest to use the above described set of equations to calculate the entrainment ratio and the diameter D_3 of the ejector from the knowledge of the boundary conditions and of the nozzle geometry. The procedure works as follows:

1. The stagnation conditions at generator P_G, T_G and the area of the primary nozzle throat A_{th} are introduced in Eq. (3.1) to calculate the primary mass flow rate $\dot{m}_p$.
2. $\dot{m}_p$ and the nozzle exit area A_{p1} are used in Eqs. (3.2) and (3.3) to calculate the Mach number M_{p1} and the pressure P_{p1}.

3. The stagnation conditions at the evaporator are used in Eq. (3.4) to calculate the pressure P_{sy} at section y where the secondary flow reaches sonic speed.
4. The primary flow must have the same pressure in the section y, and hence, from Eqs. (3.5) and (3.6), we may calculate the primary flow Mach number M_{py} and area A_{py}.
5. A tentative value of area A_3 is used to calculate the secondary flow area A_{sy}.
6. The secondary flow area yields the mass flow rate $\dot{m}_s$ through Eq. (3.7).
7. Primary and secondary flow temperatures T_{py}, T_{sy} are evaluated through Eqs. (3.9) and (3.10).
8. Primary and secondary flow velocities u_{py}, u_{sy} are calculated through Eq. (3.13).
9. Mixed flow velocity u_m is calculated by Eq. (3.11).
10. Mixed flow enthalpy (and temperature) is calculated by Eq. (3.12).
11. Mixed flow Mach number M_m is calculated by Eq. (3.13).
12. Pressure P_3 and Mach number M_3 after the shock are evaluated by Eqs. (3.14) and (3.15).
13. Pressure P_C at end of the conical diffuser is evaluated by Eq. (3.16).
14. If the ejector exit pressure is known, a comparison is made with the calculated value. If this latter is higher than the real exit pressure, the tentative value of A_3 must be decreased and vice versa. A new value of A_3 is hence provided at step 5 of the procedure, until convergence.
15. When convergence is reached, the entrainment ratio $\omega = \dot{m}_s/\dot{m}_p$ comes out as the final result of the procedure.

The above described procedure has been calibrated using an experimental setup that uses R141b as the working fluid. Two nozzles and eight mixers/diffusers have been combined in various ways, producing 11 configurations. Four saturation temperatures at generator are combined with two saturation temperatures at evaporator, giving eight working conditions. A total of 50 experimental points has been produced. For each point, the authors have found the critical exit pressure, the entrainment ratio, and the area ratio A_3/A_t.

The discrepancy between calculated and measured values is within $\pm 10\%$ in 75% of the cases, with a maximum error of 23%. The best values of the empirical parameters are $\eta_P = 0.95$, $\eta_S = 0.85$, and $\phi_P = 0.88$. The last parameter, the mixing loss coefficient ϕ_M, has a value that increases from 0.8 to 0.84 as the area ratio A_3/A_t decreases from 8.3 to 6.9. An empirical linear relation is proposed. The authors explicitly recommend to use this model only for dry-expanding fluids, because condensation would severely affect the results.

The procedure may be inverted in order to switch input data and results. For example, the primary and secondary mass flow rates may be input and the diameters of the nozzle and diffuser may be output.

Unfortunately, the working fluid R141b is now phased out in the majority of industrialized countries. Obviously, the flow velocities are variable with the fluid, and hence the loss coefficients cannot be simply transferred from R141b. Even for the same fluid, the efficiencies vary with operating conditions.

3.1.2   ESDU Procedure

An older but more detailed tool for the design and verification of ejectors is provided by ESDU, a society originated from the Technical Department of the UK Royal Aeronautical Society in 1940, currently owned by IHS Markit Inc. ESDU offers various tools concerning ejectors, comprising computer programs, design hints, and worked examples. The most recent versions are:

- ESDU 94046: issued in December 1994, supersedes ESDU 86030, and covers steam-driven gas ejectors, introducing the program ESDUpac A9446
- ESDU 92042: revised in June 2011, introduces the program ESDUpac A9242 for design and performance of ejectors using compressible gases

Here we report a short description of the first version, ESDU 94046, as an example of the general structure of these procedures. Practical design hints are reported as well.

Figure 3.2 shows a general scheme of an ejector, with a mixing duct that may have a conical part at inlet and a subsequent cylindrical part. The diffuser follows the mixing duct and recovers a further amount of pressure. Note that in this section, for the sake of concision, the nozzle exit section is named "e," instead of NXP as elsewhere in the book.

The longitudinal profile of the primary nozzle may have a circular inlet portion with radius $r > 0.3\ d_{th}$ and a conical divergent portion. The throat joining the convergent and divergent portions must be short and provide a smooth transition, i.e., have no discontinuity of slope or curvature. Included angle of the divergent part is a compromise between friction losses and flow separation and usually is around $10°$. The nozzle exit must have a sharp edge as far as possible. The outer surface should be smooth and cylindrical or slightly conical. These two features guarantee a narrow wake from the nozzle edge and improve the mixing between primary and secondary flows. All surfaces should have high quality finish. The supply line should guarantee a velocity below 50 m/s.

Secondary inlet and mixing chamber entry must be smooth and short. Sharp constrictions or expansions must be avoided. A bell-mouth inlet is recommended.

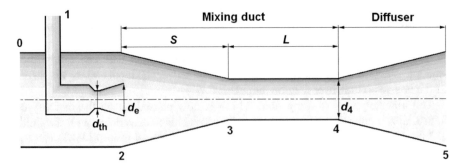

Fig. 3.2   Scheme of ejector for ESDU procedure

The flow passage between the secondary inlet and the nozzle outer surface must be constantly converging. A conical entry is simpler, but produces more losses. The maximum velocity of the secondary stream must be limited throughout the passage (<10–20 m/s for liquid and <100 m/s for gases). For liquids, cavitation must be accounted for.

The mixing chamber, as already explained, may be formed as a straight cylinder (constant-area design) or a converging cone (constant-pressure design). Actually the profile needed for a strictly constant-pressure flow would be much more complicated than a cone. ESDU sees no reason to suppose that either design is more efficient. Normally, steam-liquid ejectors have constant-area design, and steam-gas ejectors have constant-pressure design. The mixing process should be finished before the diffuser. If not so, the ejector efficiency decreases. If, however, the mixing chamber is excessively long, pressure losses increase. A constant-area mixing chamber should have a length in the range 5–10 d_4. For constant-pressure design, the effective mixing length depends on the nozzle position. In this case, the distance between nozzle exit and diffuser outlet $S + L$ should be from 5 to 10 d_4. The conical inlet of the mixing chamber should have an included angle of 2–10°. The length of the cylindrical portion L should be adjusted accordingly. Normally this results in 2–4 d_4. Surface finish must be as smooth as economically feasible.

The primary nozzle position should be movable for performance optimization. This is because the primary flow repeatedly converges and diverges due to expansion waves and compression shock waves. The secondary mass flow rate depends on the cross-sectional area occupied by the primary flow at the mixing duct entry, which varies with the nozzle position. A movable primary nozzle would allow to adjust the position as a function of operating conditions, but would also complicate the ejector and reduce its reliability. A simpler solution is a fixed primary nozzle whose position may be modified before commissioning. In principle, retracting the nozzle should increase the area available for the secondary flow, increasing the entrainment ratio at the expense of the maximum discharge pressure. In some cases, however, the secondary flow may separate from the walls and cause instability. On the other hand, moving the nozzle into the mixing chamber should reduce the secondary flow area, but increase the discharge pressure. Typically, movements as small as 1 mm cause noticeable changes in the performance. For constant-area mixing, the nozzle should be placed 0.5–1 d_4 from the mixing chamber entrance. For constant-pressure mixing, there is no exact rule because many geometrical factors influence the system behavior. An experimental optimization is recommended. The nozzle should be carefully aligned on the ejector axis.

The diffuser should have a moderate divergence angle (3–4°; never >7°) to avoid flow separation, especially if the mixing chamber is short and there is a risk of imperfect mixing at diffuser entrance. If the downstream pipe has a smaller diameter, a contraction may be introduced, and usually it doesn't produce heavy losses.

The objective function when designing a steam-driven ejector is to minimize the steam consumption for a given secondary flow rate and compression ratio.

The ESDU design procedure assumes an ideal gas behavior ($\gamma = 1.315$ e $R = 461.5$ J kg^{-1} K^{-1}) for steam as well as for the secondary flow. This is reasonable for superheated steam, but is commonly applied also to saturated steam, assuming that condensation does not occur due to metastable behavior. (In practice, condensation does occur and even freezing may be encountered, causing significant operation problems.)

Parameters related to the primary and secondary flow are marked with (') and ("), respectively. This holds also for fluid thermodynamics properties γ and R. The flow sections are named as in Fig. 3.2.

The throat area is calculated assuming unitary Mach number:

$$A_{\text{th}} = \dot{m}' \frac{\sqrt{RT_{\text{t1}}}}{\kappa P_{\text{t1}}} \tag{3.17}$$

where

$$\kappa = \sqrt{\frac{\gamma}{\left(\frac{\gamma+1}{2}\right)^{\frac{\gamma+1}{\gamma-1}}}} \tag{3.18}$$

Primary nozzle losses are accounted for in terms of static pressure by a coefficient

$$C_{\text{D}} = \frac{\left(P'_{\text{e}}\right)_{\text{actual}}}{\left(P'_{\text{e}}\right)_{\text{isentropic}}} \tag{3.19}$$

Well-designed nozzles can reach $C_{\text{D}} > 0.95$.

Primary and secondary flows are in static pressure equilibrium at nozzle exit plane:

$$P''_{\text{e}} = \left(P'_{\text{e}}\right)_{\text{actual}} \tag{3.20}$$

A nondimensional parameter x links the total pressures of the two streams:

$$x = C_{\text{D}} \frac{P'_{\text{te}}}{P''_{\text{te}}} \tag{3.21}$$

We can introduce a generic function for any parameter μ as

$$f_1(\mu, \gamma) = \left(1 + \frac{\gamma - 1}{2}\mu^2\right)^{\frac{\gamma}{\gamma-1}} \tag{3.22}$$

This form is analogous, for instance, to the isentropic relation:

$$\frac{P_{\text{t}}}{P} = \left(1 + \frac{\gamma - 1}{2}M^2\right)^{\frac{\gamma}{\gamma-1}} \tag{3.23}$$

Therefore, given the assumption (3.20), the parameter x can be written as

$$x = \frac{f_1\left(M'_e, \gamma'\right)}{f_1\left(M''_e, \gamma''\right)} \tag{3.24}$$

The Mach number of the secondary flow at nozzle exit plane must be comprised between 0 and 1, i.e.,

$$f_1\left(M'_e, \gamma'\right) \leq x \leq \frac{f_1\left(M'_e, \gamma'\right)}{\left(\frac{\gamma'+1}{2}\right)^{\frac{\gamma'}{\gamma'-1}}} \tag{3.25}$$

Introducing the speed of sound $a = \sqrt{\gamma P/\rho}$, the mass flow rate $\dot{m} = \rho A u$ can be written as $\dot{m} = A M \sqrt{\gamma P \rho}$. Introducing also the ideal gas relation $\frac{P}{\rho} = RT$, we have

$$\dot{m} = PAM \sqrt{\frac{\gamma}{RT}} \tag{3.26}$$

This relation may be written for the primary as well as the secondary flows:

$$\dot{m}' = \left(P'_e\right)_{\text{actual}} A'_e M'_e \sqrt{\frac{\gamma'}{R'T'_e}} \; ; \quad \dot{m}'' = P''_e A''_e M''_e \sqrt{\frac{\gamma''}{R''T''_e}} \tag{3.27}$$

The entrainment ratio $\omega = \dot{m}''/\dot{m}'$, remembering the pressure equilibrium, hence is

$$\omega = \frac{A''_e M''_e}{A'_e M'_e} \sqrt{\frac{\gamma'' R' T'_e}{\gamma' R'' T''_e}} \tag{3.28}$$

The static temperature may be related to the stagnation temperature as follows:

$$\frac{T_t}{T} = 1 + \frac{\gamma - 1}{2} M^2 \tag{3.29}$$

Another generic function of any variable μ is introduced:

$$f_2(\mu, \gamma) = \mu \sqrt{1 + \frac{\gamma - 1}{2} \mu^2} \tag{3.30}$$

and hence the entrainment ratio becomes

$$\omega = \frac{A''_e f_2\left(M''_e, \gamma''\right)}{A'_e f_2\left(M'_e, \gamma'\right)} \sqrt{\frac{\gamma'' R' T'_{te}}{\gamma' R'' T''_{te}}} \tag{3.31}$$

Heat exchange is assumed to be negligible; hence $T''_{te} = T_{t0}$; $T'_{te} = T_{t1}$.

The primary nozzle expansion ratio may be calculated as follows. The mass flow rate depends on the throat cross section where $M = 1$:

$$\dot{m}' = P_{\text{th}} A_{\text{th}} \sqrt{\frac{\gamma'}{R' T_{\text{th}}}} \tag{3.32}$$

The same equation may be applied to the nozzle exit:

$$\dot{m}' = C_{\text{D}} \left(P'_{\text{e}}\right)_{\text{isentropic}} A'_{\text{e}} M'_{\text{e}} \sqrt{\frac{\gamma'}{R' T'_{\text{e}}}} \tag{3.33}$$

Equating these two expressions yields

$$C_{\text{D}} \frac{A'_{\text{e}}}{A_{\text{th}}} = \frac{1}{M'_{\text{e}}} \frac{P_{\text{th}}}{\left(P'_{\text{e}}\right)_{\text{isentropic}}} \sqrt{\frac{T'_{\text{e}}}{T_{\text{th}}}} \tag{3.34}$$

Gas dynamic relations for isentropic flow yield

$$C_{\text{D}} \frac{A'_{\text{e}}}{A_{\text{th}}} = \frac{1}{M'_{\text{e}}} \left[\frac{\gamma' + 1}{2\left(1 + \frac{\gamma'-1}{2}\right) M'^{2}_{\text{e}}}\right]^{\frac{\gamma'+1}{2(1-\gamma')}} \tag{3.35}$$

This relation shows how the factor C_{D} that represents the expansion losses may be seen as a contraction of the effective flow area at the nozzle exit.

The mixing duct may be analyzed by writing its energy equation. Defining

$$\tau' = \frac{T'_{\text{te}}}{T_{\text{t4}}} \quad \text{and} \quad \tau'' = \frac{T''_{\text{te}}}{T_{\text{t4}}} \tag{3.36}$$

and setting $\dot{m}_4 = \dot{m}' + \dot{m}''$, an enthalpy balance yields

$$c'_{\text{p}} \tau' + \omega c''_{\text{p}} \tau'' = (1 + \omega) c_{\text{p4}} \tag{3.37}$$

The conservation of the total temperature gives

$$\frac{\tau'}{\tau''} = \frac{T'_{\text{te}}}{T''_{\text{te}}} = \frac{T_{\text{t1}}}{T_{\text{t0}}} \tag{3.38}$$

Substitution in Eq. (3.37) yields

$$\tau' = \frac{(1 + \omega) c_{\text{p4}}}{c'_{\text{p}} + \omega c''_{\text{p}} (T_{\text{t0}}/T_{\text{t1}})} \quad \text{and} \quad \tau'' = \frac{(1 + \omega) c_{\text{p4}}}{c'_{\text{p}} (T_{\text{t1}}/T_{\text{t0}}) + \omega c''_{\text{p}}} \tag{3.39}$$

These new parameters may be used in the expression for the entrainment ratio:

$$\omega = \frac{A''_{\text{e}} f_2 \left(M''_{\text{e}}, \gamma''\right)}{A'_{\text{e}} f_2 \left(M'_{\text{e}}, \gamma'\right)} \sqrt{\frac{\gamma'' R' \tau'}{\gamma' R'' \tau''}} \tag{3.40}$$

The continuity equation may be written for the mixing duct as follows:

$$\rho' A'_e u'_e + \rho'' A''_e u''_e = \rho_4 A_4 u_4 \tag{3.41}$$

Other nondimensional parameters may be introduced:

$$y = \frac{P_{t4}}{P''_{te}} \tag{3.42}$$

$$PP = \frac{C_D}{A_R} \frac{P'_{te}}{P''_{te}} \quad \text{(primary parameter)} \tag{3.43}$$

$$SP = \frac{A_R - 1}{A_R} \frac{P''_{te}}{P_4} \quad \text{(secondary parameter)} \tag{3.44}$$

plus a further general function:

$$f_3(\mu, \gamma) = \frac{f_2(\mu, \gamma)}{f_1(\mu, \gamma)} = \mu \left(1 + \frac{\gamma - 1}{2} \mu^2\right)^{-\frac{\gamma+1}{2(\gamma-1)}} \tag{3.45}$$

The continuity equation hence becomes

$$PP\sqrt{\frac{\gamma'}{R'\tau'}} f_3(M'_e, \gamma')$$

$$+ SP \frac{A''_e}{A_4 - A'_e} \sqrt{\frac{\gamma''}{R''\tau''}} f_3(M''_e, \gamma'') = M_4 \sqrt{\frac{\gamma_4}{R_4} \left(1 + \frac{\gamma_4 - 1}{2} M_4^2\right)} \tag{3.46}$$

but from the definitions of PP and SP

$$PP\left(1 + \frac{\gamma_4 - 1}{2} M_4^2\right)^{\frac{\gamma_4}{1-\gamma_4}} = C_D \frac{P'_{te}}{P_{t4}} \frac{A'_e}{A_4} = \frac{x A'_e}{y A_4} \tag{3.47}$$

$$SP\left(1 + \frac{\gamma_4 - 1}{2} M_4^2\right)^{\frac{\gamma_4}{1-\gamma_4}} = \frac{A_4 - A'_e}{y A_4} \tag{3.48}$$

and so Eq. (3.46) may be further simplified:

$$\frac{1}{y}\left[x\frac{A'_e}{A_4}\sqrt{\frac{\gamma'}{R'\tau'}} f_3(M'_e, \gamma') + \frac{A''_e}{A_4}\sqrt{\frac{\gamma''}{R''\tau''}} f_3(M''_e, \gamma'')\right] = \sqrt{\frac{\gamma_4}{R_4}} f_3(M_4, \gamma_4) \tag{3.49}$$

An area ratio $A_R = A_4/A'_e$ is defined. Furthermore, the ratio A''_e/A'_e can be found from Eq. (3.32). Finally, x is written as in Eq. (3.24) and so Eq. (3.49) becomes

$$y = \frac{1}{A_R} \frac{f_2(M'_e,\gamma')}{f_3(M_4,\gamma_4)f_1(M''_e,\gamma'')}(1+\omega)\sqrt{\frac{R_4\gamma'}{\gamma_4 R'\tau'}} \tag{3.50}$$

The mixing duct momentum equation

$$P'_e A'_e + \dot{m}'u'_e + P''_e A''_e + \dot{m}''u''_e = P_4 A_4 + \dot{m}_4 u_4 \tag{3.51}$$

can be made nondimensional as was done for the energy equation.

$$\frac{xf_4(M'_e,\gamma') + \frac{A''_e}{A'_e}f_4(M''_e,\gamma'')}{yA_R} = \frac{1+\frac{\gamma_4}{K}M_4^2}{f_1(M_4,\gamma_4)} \tag{3.52}$$

where $f_4(\mu,\gamma) = \frac{1+\gamma\mu^2}{f_1(\mu,\gamma)}$. Substituting x, y, and A''_e/A'_e,

$$\frac{\sqrt{\frac{R'\tau'}{\gamma'}}f_5(M'_e,\gamma') + \omega\sqrt{\frac{R''\tau''}{\gamma''}}f_5(M''_e,\gamma'')}{1+\omega} = \sqrt{\frac{R_4}{\gamma_4}}\frac{1+\frac{\gamma_4}{K}M_4^2}{f_2(M_4,\gamma_4)} \tag{3.53}$$

where $f_5(\mu,\gamma) = \frac{1+\gamma\mu^2}{f_2(\mu,\gamma)}$.

The last part of the ejector, the diffuser, obeys the continuity equation $\dot{m}_4 = \dot{m}_5$ and has an efficiency η_d; hence

$$f_3(M_5,\gamma_5) = f_3(M_5,\gamma_5)\frac{A_4}{\eta_d A_5} \tag{3.54}$$

The solution method may differ according to the available data. Introducing the pressure ratios

$$r_1 = \frac{P_{t1}}{P_{t5}}, \quad r_2 = \frac{P_{t5}}{P_{t0}} \quad \text{and} \quad r_3 = \frac{P_{t1}}{P_{t0}} \tag{3.55}$$

two of them may be found once one is known. For example, if r_1 is known,

$$\frac{x}{y} = C_D\frac{P'_{te}}{P_{t4}} = C_D\frac{P_{t1}}{P_{t4}} = C_D\eta_d\frac{P_{t1}}{P_{t5}} = C_D\eta_d r_1 \tag{3.56}$$

Combining this equation with (3.24), (3.31), and (3.50), we have

$$\frac{A_4}{A_e}d_h f_3(M_4,\gamma_4)\left(\sqrt{\frac{R'\gamma_4}{R_4\gamma'}} + \sqrt{\frac{R''\gamma_4}{R_4\gamma''}}\frac{c_h}{f_2(M''_e,\gamma'')}\right) = C_D\eta_d r_1 \tag{3.57}$$

where $c_h = \omega f_2(M'_e,\gamma')\sqrt{\tau''/\tau'}$ and $d_h = \dfrac{\sqrt{\tau'}}{(1+r_m)f_3(M'_e,\gamma')}$ are known quantities.

Equations (3.57) and (3.53) form a set in the two variables M''_e and M_4 and may be solved iteratively.

3.2 One- and Two-Dimensional Design

The two examples of design procedure presented in Sect. 3.1 show quite clearly the limits of this kind of approach:

- Ideal gas with constant thermodynamic properties must be assumed as the working fluid.
- Only a few sections are calculated, and the other geometric elements are at most suggested by empirical rules.
- Some coefficients, either in terms of loss factors or efficiencies, must be assumed, and their values must be calibrated for each case as a function of the working fluid, geometry, etc.
- Complicated algebra makes the solution quite time-consuming, and iterative solving processes are required.

Adding a longitudinal coordinate to the procedure is a quite simple operation and makes the description of the flow much more intuitive.

3.2.1 Nozzles

In common practice, nozzles and secondary adduction ducts are designed according to the results of 0D models. In many cases, the resulting geometry consists of straight ducts divided by sharp corners. These simple configurations facilitate the manufacturing of the component and allow the definition of a finite number of geometric parameters that can be easily optimized (see Sect. 3.4).

However, the presence of profile discontinuities inside the primary nozzle may increase the risk of shock formation due to the overlapping of multiple compression waves. With regard to these aspects, Pope and Goin (1978) remark the great importance of a smooth and gradual transition between the converging and diverging part: "experience has shown that a low tolerance in the actual coordinate of a nozzle is of considerably less importance than low tolerances on the smoothness and continuity of curvature downstream of the initial expansion at the throat."

A smooth profile of the primary nozzle supersonic region may be achieved by using the Method Of Characteristics (MOC; see, e.g., Zucrow 1976). This method allows the design of any continuous internal or external geometry, as long as the governing equations are hyperbolic and the flow is isentropic (as in supersonic nozzles). To date, the MOC technique is ordinarily employed to design the nozzles of supersonic and hypersonic wind tunnels with both planar and axial symmetries (Pope and Goin 1978). Several different MOC techniques were proposed over the years depending on the different design purposes. For instance, some of these aimed at generating the most uniform flow in the test section of a supersonic wind tunnel (Pope and Goin 1978; Shope 2006). Other types were intended to yield minimum length nozzles (Brown and Argrow 1999) or optimal thrust nozzles for space propulsion applications (Hoffman et al. 1972). In addition, a number of empirical techniques were developed to design the contraction of the nozzle (i.e., the subsonic part) and to control the evolution and stability of the boundary layer (Pope and Goin 1978).

In order to choose among all these options, the specific requirements of supersonic ejector applications must be understood. Clearly, the design of the nozzle should not only consider friction losses but also account for the indirect losses generated within the mixing chamber due to heat exchange and shocks. These losses originate from pressure and temperature gradients that can be minimized if the temperature and pressure levels of the two streams are matched at the mixing chamber inlet (i.e., the static state of the motive and suction stream must coincide). In turn, this can be achieved by a careful choice of the cycle operating temperatures and a proper design of the primary nozzle pressure ratio.

One further feature that must be considered when designing ejector nozzles and secondary inlet ducts is that the pressure in the mixing chamber should be always as high as possible. This concept is better understood by looking at the T-s diagram shown in Fig. 3.3. When the secondary inlet duct is designed in a way that accelerates the secondary flow, the pressure in the mixing chamber is decreased (the path represented by the points from 2 to 5 in Fig. 3.3). This implies that the primary flow must accelerate to a higher Mach number in order to reach the pressure equilibrium with the suction flow, leading to stronger shock trains and friction losses. Moreover, the pressure lift increases, causing an augmented "recompression work" that reduces the diffuser efficiency.

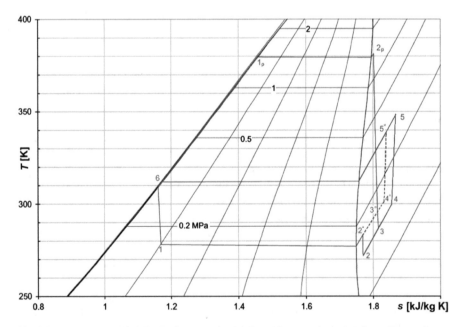

Fig. 3.3 T-s diagram of an R245fa ejector cycle, the dotted lines are the path followed by an ejector having no acceleration of the secondary flow

In order to avoid these irreversibilities, the secondary inlet duct should be large enough to limit the acceleration of the secondary stream, so that its pressure is almost equal to the stagnation pressure (the path represented by the points from 2* to 5* in Fig. 3.3).

3.2.2 CRMC Diffuser

The project of the mixing chamber/diffuser regions is perhaps the most delicate and challenging. This is because the preeminent sources of dissipation inside ejectors, i.e., mixing, friction, and shocks, are all consistently present in these regions.

Despite the long tradition in wind tunnel testing, the design of supersonic or transonic diffusers is still a very complex task. Differently from supersonic nozzles, the dynamics of diffusers is not described by irrotational flow. This is due to the severe losses originating from the interaction of shocks or compression waves with the boundary layer (for more detail on these aspects, the reader may refer to (Smits and Dussauge 2006)). Hence, the method of characteristics does not hold anymore, and design is usually performed following empirical rules (e.g., ESDU 1986).

An attempt to reduce the shock intensity within the mixing chamber/diffuser is provided by the CRMC method devised by Eames (2002). A modified version of this idea was already presented in Sect. 2.2 when dealing in general with supersonic expansion and compression. Here we present a more detailed description of the original procedure devised in (Eames 2002), and we discuss a few possible improvements.

The claim of this procedure was to eliminate the normal shock that takes place in the supersonic ejectors. To this aim, the diffuser is shaped in a way that allows a continuous rise of the static pressure by imposing a gradual decrease of the momentum. For simplicity, this gradual decrease is assumed to happen at constant rate α:

$$\dot{m}_p(1+\omega)\frac{du}{dx} = \alpha \tag{3.58}$$

The following further assumptions are made:

- The same fluid is assumed for both primary and secondary flows.
- The fluid is assumed to behave as an ideal gas with constant specific heat ratio γ.
- The primary mass flow $\dot{m}_p$ and the required entrainment ratio ω are specified.
- The total pressures and temperatures of the primary and secondary flows are known.
- The flow is assumed to be adiabatic throughout.
- The velocity of the secondary flow at entry to the entrainment region $u_{s,NXP}$ is specified.
- The velocity at diffuser exit u_2 is specified.
- The entrainment process is carried out at constant static pressure P_E.

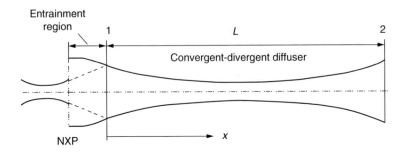

Fig. 3.4 CRMC diffuser

- At design condition the mixed stream is compressed within the diffuser so that at its throat the local Mach number is unitary.

Mixing takes place in the entrainment region that starts at nozzle exit (section NXP) and ends at section 1, where the zone hosting the secondary flow ends. The boundary conditions for Eq. (3.58), using the numbering of Fig. 3.4, are

$$u_x = u_1 \text{ at } x = 0 \quad \text{and} \quad u_x = u_2 \text{ at } x = L \tag{3.59}$$

Solving Eq. (3.58) with these boundary conditions yields

$$u_x = u_1 - \frac{u_1 - u_2}{L} x \tag{3.60}$$

The length L should be chosen as a compromise between the desire to decrease friction losses and the need to avoid recirculation. An appendix in Eames (2002) shows a simple geometric reasoning that guarantees a suitable average included angle (e.g., $8°$–$10°$) for the divergent part of the diffuser.

The velocity at diffuser inlet may be calculated from the conservation of momentum across the constant-pressure entrainment region:

$$u_1 = \frac{u_{p,\mathrm{NXP}} + \omega\, u_{s,\mathrm{NXP}}}{1 + \omega} \tag{3.61}$$

The exit velocity should be as low as possible, in order to reduce energy loss. An energy balance across the entrainment region gives

$$T_{t,1} = \frac{T_{tp} + \omega T_{ts}}{1 + \omega} \tag{3.62}$$

The static temperature at diffuser inlet is hence

$$T_1 = T_{t,1} - \frac{u_1^2}{2\, c_p} \tag{3.63}$$

The total pressure at diffuser inlet, given the assumption of constant static pressure across the entrainment region, is

$$P_{t,1} = P_{NXP} \left(\frac{T_{t,1}}{T_1} \right)^{\frac{\gamma}{\gamma-1}} \tag{3.64}$$

The secondary flow velocity at inlet of the entrainment region is rather low, so that the density variation can be neglected:

$$P_{NXP} = P_{t,s} - \frac{\rho_s u_{s,NXP}^2}{2} \tag{3.65}$$

Once the velocity is known from Eq. (3.60), the diffuser cross section may be determined from the fluid density, which can be calculated from the knowledge of static pressure and temperature:

$$T_x = T_{t,1} - \frac{u_x^2}{2 c_p} \tag{3.66}$$

$$P_x = P_{t,1} \left(\frac{T_x}{T_{t,1}} \right)^{\frac{\gamma}{\gamma-1}} \tag{3.67}$$

$$\rho_x = \frac{P_x}{R T_x} \tag{3.68}$$

and finally we have the diffuser local diameter:

$$D_x = 2 \sqrt{\frac{\dot{m}_p (1 + \omega) R T_x}{\pi P_x u_x}} \tag{3.69}$$

Summing up, this method allows the design of the convergent-divergent diffuser, i.e., its diameter at any point of its length, from the following input data:

- Individual gas constant
- Specific heat at constant pressure c_p
- Ratio of specific heats γ
- Stagnation pressure and temperature of fluid supplied to the primary nozzle $P_{t,p}$, $T_{t,p}$
- Stagnation pressure and temperature of fluid supplied at the secondary inlet $P_{t,s}$, $T_{t,s}$
- Entrainment ratio ω
- Primary nozzle mass flow rate $\dot{m}_p$
- Secondary flow velocity at inlet u_s
- Diffuser exit velocity u_2

Eames (2002) presents an experimental comparison between a conventional and a CRMC design. Reference data are listed in Table 3.1. The diffuser profile calculated by Eq. (3.69) is shown in Fig. 3.5 and closely resembles Fig. 3 in Eames (2002).

Table 3.1 Data for CRMC profile

Gas constant	R	462	J kg^{-1} K^{-1}
Specific heat ratio	γ	1.3	
Primary mass flow rate	m_p	0.001	kg s^{-1}
Entrainment ratio	ω	0.42	
Secondary flow velocity at inlet	u_s	50	m s^{-1}
Velocity ad diffuser outlet	u_2	50	m s^{-1}
Stagnation pressure, primary flow	P_{tp}	198	kPa
Stagnation temperature, primary flow	T_{tp}	393	K
Stagnation pressure, secondary flow	P_{ts}	872	Pa
Stagnation temperature, secondary flow	T_{ts}	278	Pa

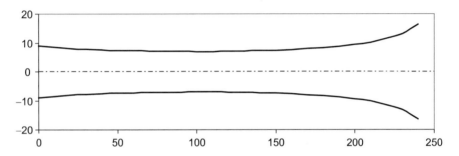

Fig. 3.5 Calculated CRMC profile

Actually, in order to complete the calculation, the primary flow velocity at nozzle exit is needed. This velocity is easily calculated from the stagnation conditions when a suitable efficiency of the primary nozzle is assumed. In our calculations we used $\eta_{isentropic} = \Delta h_{nozzle}/\Delta h_{isentropic} = 0.85$, obtaining 14 mm as the diameter of the diffuser throat.

Note that the divergence angle increases toward the exit and reaches quite high values. However, according to Eames (2002), the parameter that controls the formation of reverse flow is the global angle of the divergent part, i.e., the angle of the secant to the profile between the throat and the exit. In the case of Fig. 3.5, this angle is 7.65°.

The fundamental claim of this method is that, avoiding (or at least reducing) the intensity of the normal shock in the diffuser, the loss in total pressure is also reduced and the flow may overcome a higher pressure at the condenser. In other words, a diffuser designed according to the CRMC criterion should guarantee, ceteris paribus, a higher compression ratio. For example, using the data in Table 3.1, Eames (2002) calculated the compression ratio of the CRMC ejector to be 50% higher with respect to a conventional design. Experimental results by Worall (2001) showed a 48% increase.

Another example of CRMC design is the ejector for R245fa installed in our laboratory and described in detail in Chap. 5. In this case, the last part of the profile has been replaced by a straight cone in order to maintain a maximum divergence angle below 10°.

3.2.3   Improvement of the CRMC Procedure

A first limitation of the CRMC procedure concerns the lack of any indication about the design of the entrainment region. In a previous work (Eames et al. 2013), we presented an ejector for R245fa designed with a modified version of the CRMC profile. The entrainment zone was designed with a 1D model that confines the mixing within a truncated cone originating from the primary nozzle edge (Fig. 3.6). Around this cone, the secondary stream flows undisturbed. Any slice of the entrainment region absorbs a portion of the secondary flow, and the characteristics of the mixed flow are calculated accordingly. The outer profile of the entrainment region is shaped in order to host the unmixed secondary flow in the space between the solid wall and the cone. The entrainment region ends when all the secondary flow has been entrained.

Unfortunately, no indication is available about the entrainment cone angle, which does not coincide, even if may be related, with the primary nozzle exit angle. Furthermore, the ejector described in (Eames et al. 2013) proved rather unsuccessful, and this technique was abandoned. A more realistic approach to the analysis of the entrainment zone has been presented in Sect. 3.2.3.

As shown in Sect. 3.2.2, the assumptions of ideal gas and isentropic flow are easily removed from the CRMC design procedure. The procedure may also be extended to comply with an assigned discharge pressure, by a trial-and-error loop on the entrainment ratio. This modified procedure has been used since the first work by Grazzini et al. (2012) and lately by Milazzo and Rocchetti (2015). Though rather empirical, this procedure allows to substitute the fixed values of nozzle and diffuser

Fig. 3.6 1D model of the entrainment zone

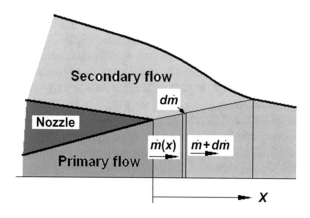

efficiencies assumed in most 0D models with a friction factor. Henceforth the friction losses are correlated to the velocity in the various zones of the ejector. Clearly, further experimental investigation is needed in order to estimate the friction factor.

3.3   Alternative Configurations

In this section we analyze a few alternative configurations to the classic ejector configuration referred to elsewhere in the book. These configurations include annular or multiple nozzles, multistage ejectors, and passive configurations.

3.3.1   Nozzle Modifications

As explicitly mentioned in ESDU 94046, the primary nozzle placed on the axis of the ejector may be substituted by an annular passage offering a converging/diverging section to the motive flow. This configuration may offer an improved mixing (i.e., require a shorter mixing length, typically around 2/3) due to the increased contact area between the primary and secondary stream. However, the nozzle surface wetted by the fast primary fluid increases, especially if the primary annulus is tangent to the mixing duct surface, and hence friction losses are expected to worsen. An experimental study of this type of configuration was investigated by Kim et al. (2006).

Another possibility is offered by the multi-nozzle configuration that has basically the same effect of the annular nozzle, i.e., increases the contact area between the two streams. Multiple nozzles were commonplace since the 1960s in steam ejectors for extraction of non-condensable gases from the condensers within steam power plants. A drawback of this configuration may be the increased blockage of the secondary inlet cross section due to the multi-nozzle assembly.

An exploration into the potential of multiple nozzles was presented in Grazzini et al. (2015). The method was already mentioned in Chap. 2 when dealing with the mixing process. The simulation showed that splitting the primary flow into two nozzles produces an entrainment ratio increased by 12%, while splitting into four nozzles produces an increase of 28%. A further investigation of multiple nozzle ejector was also carried out by Kracík and Dvořák (2015).

As a last option (although many more exist), we consider the "petal" nozzle, in which the diverging part of the nozzle is shaped into a number of lobes (generally between 3 and 6). The adoption of this peculiar shape modification can enhance the mixing rate by increasing the shearing surface between the primary and secondary streams and by providing higher turbulence levels and more effective vortices within the mixing layer. A further advantage stemming from a more rapid mixing is represented by the possibility of reducing the mixing chamber length and, consequently, the amount of friction losses.

Petal nozzle configurations have been investigated by Srikrishnan et al. (1996), Chang and Chen (2000), Opgenorth et al. (2012), and Rao and Jagadeesh (2014). Most of these studies highlighted an increase in the petal nozzle entrainment rate with respect to the conical shape. However, this increase was generally accompanied by a simultaneous decrease in the compression ratio (except Opgenorth et al. (2012) who found an increase in the critical pressure, with a nearly constant ER). Unfortunately, none of these studies clearly demonstrated whether the increase in ER offsets the augmented pressure losses. In this respect, the use of a clearly defined second-law efficiency may prove useful in shedding some light on this debated point (see Sect. 3.4).

3.3.2 Multistage Configurations

If the inlet pressure P_0 is very low, as is the case in many applications, multistage ejectors may be employed. Usually a single stage can reach a compression ratio from 5 to 10, i.e., a maximum vacuum around 10 kPa. With multistaging, pressures below 1 Pa can be reached. The stages may usefully be fitted with intermediate condensers (Fig. 3.7) that remove a significant amount of vapor in form of liquid water and hence reduce the flow to be compressed, decreasing the overall steam consumption. Intermediate condensers may be direct contact or surface. Their efficiency influences the overall efficiency of the system. An optimum compression ratio exists for each stage, and a careful optimization may greatly reduce the steam consumption.

A peculiar case of study was presented by Grazzini and Mariani (1998). The proposed multistage configuration was unconventional, due to the annular configuration of the primary nozzle in the second stage (Fig. 3.8). The objective being the transfer of momentum between the primary and secondary streams, there is no convenience in slowing down the mixed flow exiting from the first stage through a diffuser. The annular primary stream of the second stage entrains the flow directly

Fig. 3.7 Three-stage ejector with intermediate condensers

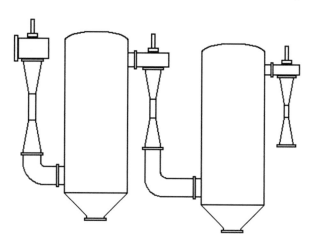

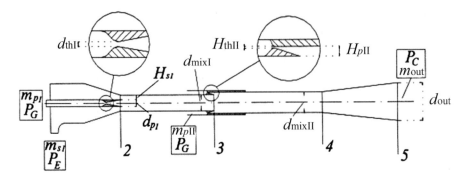

Fig. 3.8 Two-stage ejector with annular primary nozzle at second stage (Grazzini and Rocchetti 2002)

after the mixing phase, in order to reduce at minimum the irreversibility related to the interaction between streams flowing at different velocity. Furthermore, this configuration is very compact and easy to manufacture. The only drawback, in the case of condensable flow such as steam, is the impossibility to reduce the flow rate by interstage condensation. Hence this scheme, though originally proposed for steam, would probably be more adequate for dry-expanding fluids.

Grazzini and Mariani (1998) built a simple model for this kind of ejector, based on ideal gas and isentropic relations, and validated it against experimental data. A prototype was also built at our laboratory, following a design optimization that will be described in Sect. 3.5.

3.3.3   Passive Configurations

In passive systems configurations, the elimination of the pump allows the construction of an off-grid refrigerator. In addition, the absence of any rotary device increases the system reliability and reduces investment and maintenance costs. However, depending on the type of configuration, the passive system may incur in performance penalties.

One of the simplest ways to replace the liquid feed pump is represented by gravitational systems. These were first theoretically investigated by Grazzini and D'Albero (1998) and later by Nguyen et al. (2001), who built a solar-powered prototype to provide 7 kW of cooling to an office building. The scheme of the plant is depicted in Fig. 3.9a. The pump is eliminated by establishing a large gravity head between the condenser and the generator. Using water as the working fluid, the height difference required to produce a suitable pressure gradient is found to be around 7 m.

The system successfully ran at a boiler temperature of 80 °C, varying ambient temperatures, and 1.7 °C evaporator temperature, with a COP of 0.3. The authors

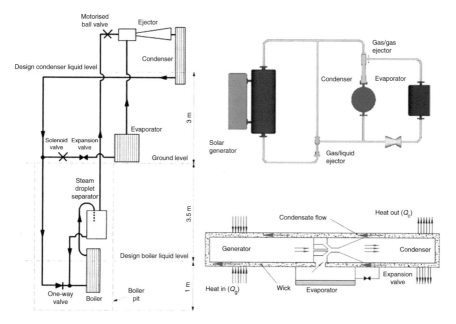

Fig. 3.9 Passive cycles – (**a**) scheme of a gravitational ejector chiller (Nguyen et al. 2001); (**b**) scheme of the bi-ejector refrigeration system from Shen et al. (2005); (**c**) the heat pipe/ejector refrigeration system from Ziapour and Abbasy (2010)

state that from an operating perspective, the system has several advantages, including very long lifetime, minimal maintenance requirement, low risk of breakdown, and no associated noise or vibration. On the other hand, the main drawbacks include the necessity to operate at subatmospheric pressures and large thermal inertia of the system.

Nguyen et al. studied also the economic feasibility of the system. From a cost-benefit analysis, they concluded that over a 30-year lifetime, the ejector cooling system is more expensive than an equivalent vapor compression system. The payback time for the ejector cooling system is *33 years*, and, consequently, a reduction in capital cost is necessary if commercial viability is to be achieved (Nguyen et al. 2001).

Gravitational systems were also theoretically studied by Kasperski (2009) by means of a simulation model. The analyses conducted by the authors revealed that the main limitation of this system lies in its requirement of great height differences and length of pipe work, which increases friction and heat losses.

In order to limit or eliminate the requirement for height difference, an injector may be used to substitute the feed pump by exploiting the motive energy of the vapor coming from the generator itself. Shen et al. (2005) refer to these cycles as the bi-ejector refrigeration systems. Their scheme is presented in Fig. 3.9b. Shen et al. studied this type of cycles numerically and found that the overall COP of the system is mainly affected by the gas-gas ejector entrainment ratio in the refrigeration loop. Different refrigerants also impact the system performance. Under the same operating conditions, the entrainment ratio of water is high, but the best overall system COP (~0.26) is achieved using ammonia as the refrigerant.

A problem in all these systems is that some active components are necessarily needed to run the auxiliary circuits (e.g., the cooling and hot water circuits, power controllers, and valves). Although these may require small amounts of electrical input, the use of pumps or fans to circulate the external streams and the eventual need of a solar source of electricity (i.e., a solar PV) may nullify the very "raison d'être" of the passive systems. A possible workaround would be to project heat exchanger cooled by natural convection. However, these may result in very large, expensive, and not-so-easily controllable components. An example of completely passive system is represented by the heat pipe/ejector refrigerator.

This type of system has first been devised by Riffat (1996) and later studied by Riffat and Holt (1998) and Ziapour and Abbasy (2010). Integration of the heat pipe with an ejector results in a compact system that can utilize solar energy or waste heat sources. The scheme of the cycle is shown in Fig. 3.9c and consists of a heat pipe coupled with an ejector, evaporator, and expansion valve. Heat is added to the generator where the working fluid evaporates and flows through the primary nozzle of the ejector. The flow exiting the nozzle entrains the secondary fluid and decreases the pressure in the evaporator. The mixed fluid is then condensate in the condenser. A part of the condensate is returned to the generator via the capillary tube, while the remainder is sent to the evaporator through the expansion valve.

Ziapour and Abbasy (2010) performed an energy and exergy analysis of the heat pipe/ejector system in order to find the optimum operating conditions. The results showed that COP could reach about 0.3 with a generator temperature of 100 °C, condenser temperature of 30 °C, and evaporator temperature of 10 °C. Unfortunately, to date there is no experimental prototype supporting these theoretical results, and some doubts arise about the practical attainability and control of these systems.

Other alternatives to the use of circulation pumps have been explored by Srisastra and Aphornratana (2005), Huang et al. (2006), Srisastra et al. (2008), and Wang et al. (2009). These systems try to replace the mechanical feed pump by means of particular configuration of storage tank and set of valves that work with a cyclic process. Although these systems require little external electrical or mechanical energy input, the efficiency is generally very low (Little and Garimella 2016).

3.4   Optimization

Before discussing any attempt for ejector optimization, a rigorous analysis of the optimization method in terms of objective functions and design variables must be done. Therefore, a first subsection will be devoted to the various possible definitions of ejector efficiency.

3.4.1 Ejector Efficiency

The definition of the efficiency for the supersonic ejector has been the cause of much debate and confusion in the literature. As will be shown, this is mainly due to the possibility of selecting several different efficiency definitions depending on the arbitrary choice of the useful system output. The following analysis builds upon a previous work carried out by McGovern et al. (2012).

The common way to define a second-law efficiency[1] is to envision a thermodynamically reversible reference process against which the real processes may be compared. In order to effectuate this comparison, the following four steps are necessary:

1. Identify the "physical" input and output quantities (flows) and equations describing the process at hand.
2. Define a thermodynamically reversible reference process which can potentially substitute the real system (this means that it must have the same type of input and output parameters).
3. Choose a "useful product" or "useful output" among the different parameters at hand. The choice usually depends on the specific task that the device must perform and need not necessarily be the physical output of the system.
4. Develop a performance metric based on a comparison between the real and the reversible "useful outputs." This could simply be chosen as the ratio between the two quantities.

Within this framework, efficiency is thought as a parameter that compares the *desired output* of a real system with the *ideal output* of a reversible system that can potentially substitute it.

Following the procedure outlined above, the first step requires to identify the "physical" input and output quantities (flows), as well as the equations describing the process at hand. This is done by considering a black-box system like that shown in Fig. 3.10.

For this type of black-box system, it is possible to apply the mass, energy, and entropy balances regardless of the spatial dimension and geometry (0D analysis). This type of analysis entirely overlooks the internal dynamics of the device (i.e., the momentum equation is not analyzed). At inlets and outlets, equilibrium states are assumed, and the fluid is considered to be in stagnation condition. For the case of an ejector, the equations are

$$\dot{m}_p h_G + \dot{m}_s h_E + Q_0 = \left(\dot{m}_p + \dot{m}_s\right) h_D$$
$$\dot{m}_p s_G + \dot{m}_s s_E + \dot{S}_{irr} = \left(\dot{m}_p + \dot{m}_s\right) s_D \tag{3.70}$$

---

[1]The first-law efficiency cannot be defined for an adiabatic ejector because energy is conserved.

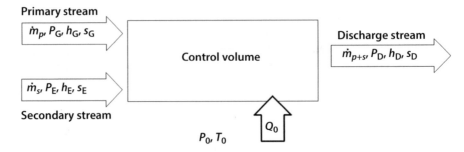

Fig. 3.10 Simplified ejector scheme

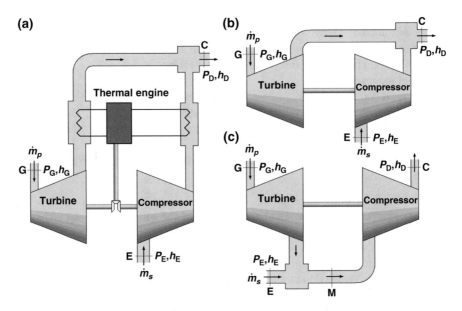

Fig. 3.11 Ideal reference systems for a supersonic ejector: reversible turbocharger coupled with heat engine (**a**) and simple reversible turbocharger with mixing after (**b**) or before the compression (**c**)

Normally the heat exchanged with the environment at P_0 and T_0 is neglected. Discharge pressure coincides with the condenser pressure P_C, while enthalpy and entropy are different.

In the second step, an ideal reference process must be identified. As explained by McGovern et al. (2012), for an ejector this can be envisioned as an ideal turbocharger coupled with a reversible engine. This ideal machine, shown in Fig. 3.11, is able to transform the pressure and temperature gradients between the primary and secondary streams into work. This work is reversibly delivered to the secondary flow in such a way that the thermodynamic state of the two mixing streams matches prior to getting in contact. Ideally, this should avoid production of mixing irreversibilities.

In the case of zero temperature difference between primary and secondary fluid, the ideal system is simplified by the absence of the thermal engine. As shown by Chunnanond and Aphornratana (2004), in this isothermal case, the ideal machine becomes a simple turbocharger, and two different configurations are possible (Fig. 3.11b or c). The first of the two options allow mixing of the two currents after the compression of the suction flow (Fig. 3.11b). In this case the primary stream expands just until the exhaust pressure.[2] In the alternative configuration (Fig. 3.11c), the mixing occurs before the recompression. This option is analogous to what happens inside ejectors, where the primary flow expands until the evaporator pressure and then both the primary and secondary flow are recompressed up to the condenser pressure.

The last two steps of the procedure consist in the choice of the useful output and of the performance metric.

Unfortunately, the arbitrariness in the choice of both the useful output and performance metric gives rise to a great number of different efficiency definitions. For instance, one could equally adopt the ratio between the ideal and real discharge pressure, the ratio between the ideal and real entrainment ratios, the ratio of secondary fluid mass flow rates, and so forth. Some of these options are discussed below but more can be found in McGovern et al. (2012).

In order to better understand these last two steps, it is useful to make a simple example that will introduce to the somewhat more complicated case of the ejector efficiency. The example considers the efficiency definition of a gas turbine, whose "black-box" scheme is shown in Fig. 3.12.

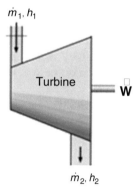

Fig. 3.12 Gas turbine "black-box" scheme

$\dot{m}_1, h_1$

Turbine

$\dot{W}$

$\dot{m}_2, h_2$

---

[2]This should be the best configuration for loss reduction in a real machine, because the compressor deals just with the secondary flow and works with a lower pressure ratio. For an ideal cycle, it doesn't matter which of the two options is selected, because there are no losses in any case.

The equations of the real and reversible systems are the following:

$$\dot{m} = \text{const} \qquad \dot{m}^{\text{rev}} = \text{const}$$
$$\dot{m}\left(h_2(P_2, s_2) - h_1(P_1, s_1)\right) = \dot{W} \quad \dot{m}^{\text{rev}}\left(h_2^{\text{rev}}\left(P_2^{\text{rev}}, s_2^{\text{rev}}\right) - h_1^{\text{rev}}\left(P_1^{\text{rev}}, s_1^{\text{rev}}\right)\right) = \dot{W}^{\text{rev}}$$
$$s_1 + s_{\text{irr}} = s_2 \qquad s_1^{\text{rev}} = s_2^{\text{rev}}$$

$$(3.71)$$

where the superscript "rev" over the variables indicates quantities of the ideal system.

In order to define the efficiency, all the ideal and real variables must be equated except for the identified useful output. In this way the two systems operate under the same boundary conditions, and the comparison is meaningful. A common choice in the case of a gas turbine is to consider the work output as the quantity of interest and to impose equal the inlet states and all other variables:

$$\dot{m} = \dot{m}^{\text{rev}}; P_1 = P_1^{\text{rev}}; s_1 = s_1^{\text{rev}}; P_2 = P_2^{\text{rev}} \qquad (3.72)$$

By substituting (3.72) into (3.71), the two systems of equations become

$$\dot{m} = \text{const} \qquad \dot{m} = \text{const}$$
$$\dot{m}(h_2 - h_1) = \dot{W} \qquad \dot{m}\left(h_2^{\text{rev}} - h_1\right) = \dot{W}^{\text{rev}} \qquad (3.73)$$
$$s_1 + s_{\text{irr}} = s_2 \qquad s_1 = s_2^{\text{rev}}$$

where the only different quantities between the real and ideal systems are the work outputs and exit state. At this point, the performance metric can be simply defined as the ratio between the ideal and real useful output:

$$\eta_{\text{turbine}} = \frac{\dot{W}}{\dot{W}_{\text{rev}}} = \frac{h_2 - h_1}{h_2^{\text{rev}} - h_1} \qquad (3.74)$$

which is the well-known "isentropic efficiency" for a turbine. This is usually described by means of an h-s diagram as shown in Fig. 3.13. The larger the difference between the outlet entropies, the lower the efficiency.

Fig. 3.13 h-s Diagram showing the isentropic efficiency for a gas turbine

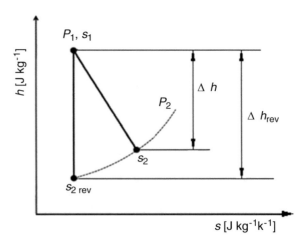

It is important to recognize that the "isentropic efficiency" is just one of the many possible alternatives that could be devised. In particular, there are six independent variables in common between the two systems ($\dot{m}$, P_1, P_2, s_1, s_2, $\dot{W}$). In order to define the efficiency and boundary conditions, only one "desired output" must be specified while leaving a further variable floating (s_2 in the preceding example). Therefore, there are 15 (6-choose-2, i.e., 6!/4!/2!) different ways in which the efficiency could be defined. In practice, the choices are lower due to additional constraints. One of these is that the inlet and outlet entropy cannot be both imposed equal to the real system, as this would imply that the entropy generation is zero. This reduces the option to 12 different alternatives (e.g., the ratio of mass flow rates, the ratio of outlet pressures, etc.).

In the case of an adiabatic ejector, the ideal reference system can be any one of the three shown in Fig. 3.11. In any of the three systems, the conservation equations for the ideal process can be written as follows:

$$\begin{aligned} \dot{m}_p h_G + \dot{m}_s h_E &= \left(\dot{m}_p + \dot{m}_s\right) h_D \\ \dot{m}_p s_G + \dot{m}_s s_E &= \left(\dot{m}_p + \dot{m}_s\right) s_D \end{aligned} \tag{3.75}$$

where the superscript rev is omitted for clarity.

The definition of the ejector efficiency follows the same procedure outlined for the gas turbine. Rearranging,

$$\begin{aligned} \dot{m}_p(h_G - h_D) &= \dot{m}_s(h_D - h_E) & \dot{m}_p^{\text{rev}}\left(h_G^{\text{rev}} - h_D^{\text{rev}}\right) &= \dot{m}_s^{\text{rev}}\left(h_D^{\text{rev}} - h_E^{\text{rev}}\right) \\ \dot{m}_p(s_G - s_D) &= \dot{m}_s(s_D - s_E) + \dot{S}_{\text{irr}} & \dot{m}_p^{\text{rev}}\left(s_G^{\text{rev}} - s_D^{\text{rev}}\right) &= \dot{m}_s^{\text{rev}}\left(s_D^{\text{rev}} - s_E^{\text{rev}}\right) \end{aligned} \tag{3.76}$$

There are eight independent variables in common with the real system. We must choose one variable as the useful output and let another float, so there are totally 28 different alternatives (8 choose 2). Among these, at least two seem best suited for supersonic ejector applications. The first identifies the entrainment ratio as the useful output and may be called the *ejector entrainment ratio efficiency*, η_{ER}.[3] The second considers the outlet pressure as the useful variable and may be called the *discharge pressure ratio efficiency*, η_{PR}.

In order to define η_{ER}, the link between the real and ideal system boundary conditions must be as follows:

$$\begin{aligned} \dot{m}_p &= \dot{m}_p^{\text{rev}}; \dot{m}_s \neq \dot{m}_s^{\text{rev}}; P_G = P_G^{\text{rev}}; s_G = s_G^{\text{rev}}; P_E = P_E^{\text{rev}}; s_E = s_E^{\text{rev}}; \\ P_D &= P_D^{\text{rev}}; s_D \neq s_D^{\text{rev}} \end{aligned} \tag{3.77}$$

The equations for the ideal system thus become

---

[3]McGovern et al. (2012) call this *reversible entrainment ratio efficiency*, η_{RER}.

$$\dot{m}_p\left(h_G - h_D\left(P_D, s_D^{\text{rev}}\right)\right) = \dot{m}_s^{\text{rev}}\left(h_D\left(P_D, s_D^{\text{rev}}\right) - h_E\right)$$
$$\dot{m}_p\left(s_G - s_D^{\text{rev}}\right) = \dot{m}_s^{\text{rev}}\left(s_D^{\text{rev}} - s_E\right) \tag{3.78}$$

By comparing (3.78) with the equations of the real system, the "*reversible entrainment ratio efficiency*" can be defined as

$$\eta_{\text{ER}} = \frac{\dot{m}_s}{\dot{m}_s^{\text{rev}}} = \frac{\dot{m}_s}{\dot{m}_s^{\text{rev}}}\frac{\dot{m}_p}{\dot{m}_p} = \frac{\omega}{\omega_{\text{rev}}} \tag{3.79}$$

The procedure for the definition of the discharge pressure ratio efficiency, η_{PR}, is analogous and returns the following definitions (more details can be found in Mazzelli (2015)):

$$\eta_{\text{PR}} = \frac{P_D}{P_D^{\text{rev}}} \tag{3.80}$$

McGovern et al. (2012) provide a very useful graphical representation on a Mollier diagram for both these two efficiencies.

It is important to note that the two definitions are equally valid even if they provide different information. While the first shows the distance from the maximum entrainment ratio attainable by a reversible machine operating at the given pressure levels, the second refers to the maximum discharge pressure at the given mass flow rates.

Although the choice of one efficiency over the other may be dictated by the specific application of the ejector, the joint adoption of both these efficiencies gives an indication on how the ejector is operating, by roughly answering to this question: is the ejector putting more energy into drawing secondary fluid or into compressing it?

Unfortunately, the ejector entrainment ratio efficiency, η_{ER}, and the discharge pressure ratio efficiency, η_{PR}, are only 2 among 28 different alternatives. Although additional constraints may lower this number, different ways to define the performance metric can add many more options. For instance, an analogous version of the discharge pressure ratio efficiency was proposed by Arbel et al. (2003) and is named the *reversible discharge pressure efficiency*:

$$\eta_{\text{RDP}} = \frac{P_D - P_E}{P_D^{\text{rev}} - P_E} \tag{3.81}$$

In this case the useful output is the same as for η_{PR}, but the performance metric is different (a ratio of differences, instead of a simple ratio). Indeed, the possible definitions of the ejector second-law efficiency seem to be countless.

3.4.2 Ejector Irreversibilities

We have seen in the previous section that the definition of an ejector second-law efficiency leads to a real plethora of different expressions. Nevertheless, it is possible to devise approaches that are independent from the arbitrary definition of the useful output. This is possible by focusing on the analysis of the irreversibilities inside the ejector.

A first example is the exergy analysis that, in the specific case of the supersonic ejector, leads to a simple efficiency definition as the ratio between the outlet and inlet exergy fluxes:

$$\eta_{ex} = \frac{\dot{E}x_{out}}{\dot{E}x_{in}} = \frac{\dot{E}x_{out}}{\dot{E}x_{primary\_in} + \dot{E}x_{secondary\_in}} \tag{3.82}$$

where, in the case of absence of chemical reactions, the exergy flux of a power or cooling system can be defined as follows:

$$\dot{E}x = \dot{m}\,(h - T_0 s) \tag{3.83}$$

where the T_0 is the reference state temperature.

The exergy definition allows an easy calculation of the efficiency by simply evaluating the conditions at the ejector's inlets and outlet. The exergy analysis measures how much of the available energy entering the device has been destroyed or lost in the unit of time. Hence, this type of analysis is independent from the definition of an arbitrary "useful output" (however, the choice of the reference temperature still remains arbitrary).

The exergy analysis can also be adapted to evaluate the efficiency of the various ejector's parts, provided that a sensible partition of the system is achieved. For instance, the ejector could be divided in four parts as shown in Fig. 3.14. According to such subdivision, four different exergy efficiencies could be defined as follows:

$$\eta_{primary\_nozzle} = \frac{\dot{E}x_{2'}}{\dot{E}x_{1'}} \tag{3.84}$$

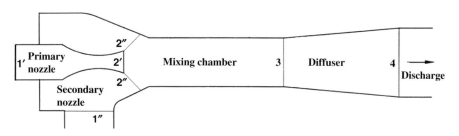

Fig. 3.14 Schematic subdivision of the ejector regions

$$\eta_{\text{secondary\_nozzle}} = \frac{\dot{E}x_{2''}}{\dot{E}x_{1''}} \tag{3.85}$$

$$\eta_{\text{mixing}} = \frac{\dot{E}x_3}{\dot{E}x_{2'} + \dot{E}x_{2''}} \tag{3.86}$$

$$\eta_{\text{diffuser}} = \frac{\dot{E}x_4}{\dot{E}x_3} \tag{3.87}$$

This type of approach highlights the regions where the losses are more intense and allows the designer to understand where the optimization is more needed.

Unfortunately, a clear distinction of the different regions may not always exist as each region may be associated with multiple flow processes. For instance, the primary expansion can proceed inside the mixing chamber or the mixing process can continue inside the diffuser.

Hence, a different approach is to estimate the impact of the various dissipation mechanisms on the ejector efficiency. This can be done by making use of the "indirect formulation" of the exergy efficiency:

$$\eta_{\text{ex}} = \frac{\dot{E}x_{\text{out}}}{\dot{E}x_{\text{in}}} = \frac{\dot{E}x_{\text{in}} - \dot{E}x_{\text{lost}}}{\dot{E}x_{\text{in}}} = 1 - \frac{T_0\dot{S}_{\text{irr}}}{\dot{E}x_{\text{in}}} \tag{3.88}$$

By making use of the above expression, the calculation of the efficiency reduces to the estimation of the entropy generation within the ejector.

An interesting study in this regard was proposed by Sierra-Pallares et al. (2016), who performed a CFD-based entropy generation analysis of a R134a ejector operating with three different mixing chambers. Although this can be done during the postprocessing phase, the analysis of the local entropy generation using CFD is not straightforward. This comes from the fact that irreversibilities arise from both mean and fluctuating gradients of the velocity and temperature field.

In particular, the time averaging of the entropy transport equation leads to four distinct production terms (Kock and Herwig 2004):

1. Entropy production by viscous dissipation due to mean velocity gradients
2. Entropy production by heat conduction due to mean temperature gradients
3. Entropy production by viscous dissipation due to fluctuating velocity gradients (turbulence dissipation)
4. Entropy production by heat conduction due to fluctuating temperature gradients (turbulent heat transport)

The first of the two sources is calculated from CFD results of the mean temperature and velocity fields, whereas the third term can be directly related to the turbulence dissipation rate, ε (Kock and Herwig 2004). The last term is the most critical because it requires the knowledge of the fluctuating temperature field, which is obtained resorting to a Boussinesq-like approximation coupled with a Strong Reynolds Analogy (i.e., it is assumed that a constant value of turbulent Prandtl number exists throughout the flow field). Moreover, a set of dedicated wall functions must be implemented in order to properly model the large entropy generation found at wall (more details are given by Kock and Herwig (2004)).

Sierra-Pallares et al. (2016) implemented the method devised by Kock and Herwig into ANSYS Fluent and used it for the closure of the Favre-averaged entropy transport equation. The method allowed the authors to reveal the areas where the irreversibilities were most intense and to identify the different mechanism responsible for the entropy generation.

In particular, peak regions of entropy production were found in the turbulent mixing layer downstream the primary nozzle and across the shock train region in the mixing chamber/diffuser. Most interestingly, the rate of entropy generation within the boundary layer was found to be relatively small compared with the bulk flow. However, it should be noted that the entropy production at wall was calculated by considering the boundary layer up to the point where the wall functions are applied (i.e., up to some point of the log-law region). Therefore, all the dissipation occurring in the outer (or wake) region of the boundary layer was allocated to the bulk flow (this also includes all the losses occurring in the recirculation regions, which are significant).

In terms of loss mechanisms, the CFD simulations showed that the fluctuating viscous dissipation accounted for more than 75% of the total entropy production, whereas the irreversibilities induced by heat transfer (due to both mean gradients and fluctuating ones) accounted for less than 2% of the total amount. This point is important because it demonstrates that most of the losses are connected to the large turbulence levels originating from the mixing of the two streams as well as from the shock-laden deceleration of the mixed stream in the diffuser. Whereas the first source is necessary for an effective entrainment of the suction stream, the reduction of the shock intensity in the diffuser would provide benefits worth of significant optimization efforts.

3.4.3 Ejector Optimization

The literature of ejector studies is abundant of attempts to optimize the system efficiency by changing some of the most relevant geometrical parameters (e.g., Yapici et al. 2008; Yadav and Patwardhan 2008; Zhu et al. 2009; Kong and Kim 2016). However, most of these studies are essentially carried out through manual trial-and-error strategies, and no attempt is made to implement systematic optimization procedures.

In principle, an optimization (or search) algorithm can be easily coupled to an ejector analytical model in order to explore the design space and find optimal geometrical parameters. However, this procedure is mainly suited for a first-attempt design and does not allow a true optimization of ejector performance. Indeed, we have seen in previous sections that irreversibilities inside ejectors are mostly located in the mixing chamber and diffuser and that a shape optimization of these components may lead to significant efficiency improvements. In this regard, the use of

Fig. 3.15 Aerodynamic
profile optimization process
(Milli 2006)

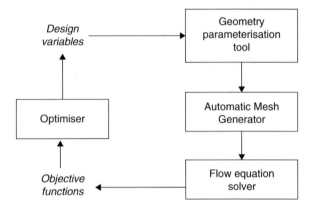

predictive tools, such as high-fidelity CFD simulations, is indispensable for the
reliability of the optimization outcomes.

Generally speaking, the main steps of a CFD-based optimization procedure
consist in the geometry parameterization, automatic grid generation, flow equation
solving, and search of the optimal objective function, as sketched in Fig. 3.15.

The first step of the process involves the definition of an adequate geometry
parameterization. According to Shahpar (2004), three different ways of parameterization can be identified:

- Engineering global parameters
- Local shape perturbation
- Spline and their generalization

The first category represents the most basic approach and consists in a sequential
modification of a finite number of key design parameters. In the case of supersonic
ejectors, these would be the nozzle throat and exit diameters, the diffuser throat
diameter, the NXP, the diffuser length and diverging angle, etc.

The second class of techniques is based on the application of bump functions to
perturb a prescribed aerodynamic shape. In this approach the independent variables
consist of a limited number of shape perturbation parameters. This method is
particularly useful when the design must evolve from an existing configuration
and only slight modifications are needed. The main drawback of these parameterization techniques lies in the difficulty of controlling profile modifications and
defining proper ranges for design variables (Milli 2006).

Finally, the third category is represented by piecewise polynomial curves that are
controlled so as to reproduce the desired aerodynamic profiles. Bezier curves
represent perhaps the most renowned example of the spline-based parameterization.
The advantage of this method is that it is possible to represent complex engineering
shapes by changing a small number of shape control points.

Depending on the method selected for the parameterization and the type of
problem, the generation of the computational grid can follow different approaches:

- Perturbation of a base computational grid
- Computation of a new grid for each explored configuration

The first method is usually very rapid because only limited modification to the mesh is needed. Moreover, this approach has the notable advantage of preserving grid topology. This point is important because it allows to restart the flow field simulation from a previously computed solution without using any interpolation algorithm. On the other hand, the second approach is usually more robust and tolerates larger geometry shape modifications. In addition, the mesh resolution is usually higher when the grid is generated from scratch instead of being adapted from an already computed distribution of nodes (Milli 2006).

Regarding the flow simulation step, it is obvious that the selection of the numerical setup must derive from a trade-off between the needs for accuracy and computational time savings. This would entail a careful choice of the modeling approximation (e.g., symmetric domains, simple turbulence models) and grid characteristics (type, size, wall resolution, etc.).

Finally, the most crucial aspect of the whole optimization process is represented by the choice of the most suitable search algorithm.

In the last decades, a plethora of different optimization algorithms and strategies have been developed, the review of which is beyond the scope of this book (the interested readers may refer to Nocedal and Wright (2006), for a global overview of the subject, or to more specific references for aerodynamic optimization, such as Shyy et al. (2001)).

Herein, the focus is only on the main features that must be considered in order to choose the best algorithm. Among these, the most relevant aspects that must be accounted are:

- The computational cost of the simulations
- The structure of the design space

In terms of computational costs, the use of CFD for the exploration of the design space generally requires several CPU hours for each analyzed point. When dealing with these very expensive objective functions, it is natural to try to avoid as much as possible unnecessary evaluations.

An interesting approach to achieve this goal consists in the use of the information collected from previous simulations in order to build a surrogate model (or metamodel) of the objective function (Locatelli and Schoen 2013). The surrogate model, also called response surface, is a function (generally a polynomial of low order) that fits the data points so far simulated and creates an approximated representation of the objective function in the design space (this is a hypersurface with dimensions equal to number of independent variables). After the surrogate model is built, it can be used to explore the behavior of the virtual objective function at points which are different from the observed ones. For instance, one may consider as the best guess for the next simulation the point for which the surrogate model presents a minimum. The exploration of the response surface is generally performed by using an optimization algorithm (e.g., a genetic algorithm).

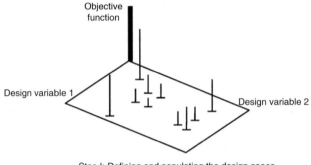

Step I: Defining and populating the design space

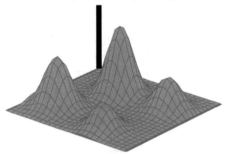

Step II: Interpolating with response function

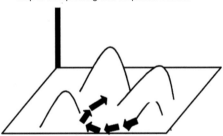

Fig. 3.16 Optimization process using the response surface methodology (Shyy et al. 2001)

A sketch of the response surface optimization methodology is given in Fig. 3.16 for a case with only two design variables.

The characteristics of the design space are also crucial in the choice of the most suitable optimization algorithm. In general, these characteristics are strictly related to the type of application under study. In particular, the optimization of supersonic ejectors represents a very difficult case.

Due to the largely compressible, 3D, viscous flow, the configuration of the design space is highly nonlinear. Indeed, even small changes in the wall profile can lead to significant modification of the field. For instance, a change in the curvature of the mixing chamber/diffuser may cause the appearance of a strong compression wave leading to a deterioration of the flow performance (i.e., a steep decrease of the objective function).

Moreover, the ejector behavior is characterized by abrupt changes in the flow pattern when the operating conditions are close to the critical point. This "on/off-like" behavior is reflected in a discontinuous design space that is very difficult to handle. Besides, the conditions near the critical point are generally characterized by the highest levels of efficiency, which implies that the optimization algorithm is most likely to concentrate the explorations in the vicinity of this unstable flow conditions.

An interesting approach to deal with this complex phase space is the use of a particular response surface methodology where the surrogate model is not constructed on standard polynomial functions but rather exploiting Neural Networks (NNs). Due to the higher flexibility of the NN functional form, this approach is particularly suited to problems where the physical system changes from one regime to another due to the presence of critical parameters (Shyy et al. 2001).

In any event, all these reasonings demonstrate that the shape optimization of supersonic ejectors is a process that must be studied with care.

Examples of CFD-based optimization studies are not frequent in the ejector literature.

One of the first works was performed by Dvorak (2007) who made a shape optimization of the mixing chamber of an axisymmetric ejector. The shape of the mixing chamber was parameterized using a cubic spline, and modification of the baseline mesh was obtained by using a mesh morphing approach. The objective function was the ejector ER which was maximized for different values of the backpressure. The results showed that the optimized configuration leads to better ER than the simple ejector with constant-area mixing chamber and conic diffuser. Most interestingly, the final shapes of the ejector present two throats, with the first throat placed right downstream the primary nozzle exit plane.

A few years later, Eves et al. (2012) building upon a previous work of Fan et al. (2011), performed a parametric optimization of a supersonic ejector design. The baseline geometry was obtained according to the CRMC criterion and then modified by describing the diffuser profile according to a hyperbolic tangent. Three parameters were considered for the optimization (nozzle throat diameter, diffuser throat, and exit diameter). New grids were generated by mesh morphing of the baseline grid. A response surface approach was used coupled to a genetic algorithm to explore optimal solutions in terms of maximal entrained mass flow rate.

The result of the optimization revealed the presence of an optimal solution front in terms of motive and entrained flow rates, as illustrated in Fig. 3.17.

Palacz et al. (2016) performed a CFD-based optimization of the geometry of a CO_2 flashing ejector for supermarket refrigeration. The CFD model was based on a homogeneous equilibrium model (HEM) calibrated according to experimental mass flow rate data ($\pm 10\%$ accuracy).

A genetic algorithm and an evolutionary algorithm were used to maximize the ejector ER by changing three main parameters of the mixing chamber (length, diameter, and entrance duct length). The results of the optimization showed a strong relation between ejector performance and the mixing chamber diameter, whereas the

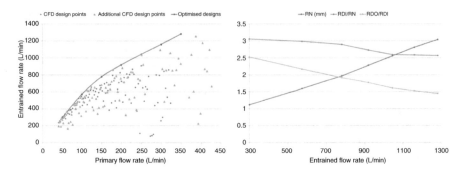

Fig. 3.17 Optimal design solutions (Rn is the nozzle radius, RDI is the diffuser inlet radius, RDO is the diffuser outlet radius) (Eves et al. 2012)

influence of the mixing section length was lower. Unfortunately, the efficiency of the optimized design was only 2% higher than that of the baseline design. In a subsequent study, Palacz et al. (2017) extended the previous results by considering six design variables and four operating conditions for each tested ejector. The average efficiency increase was 6% in this case.

Lee et al. (2016) performed another optimization study of a R600a flashing ejector for commercial applications. In their work, the CFD model was based on a homogeneous relaxation model, whose mass generation term was described according to an empirical correlation based on the primary jet Weber number (this represents the ratio of the inertial force to the surface tension, and it is a crucial parameter in determining the two-phase flow regime and the bubble or droplet sizes; see Polanco et al. 2010). Five geometric design parameters (mixing chamber diameter and length, diffuser length and diverging angle, nozzle length) were optimized in order to find optimal performance according to both the ejector ER and pressure lift (bimodal optimization). A Multi-Objective Genetic Algorithm (MOGA)-based Online Approximation-Assisted Optimization (OAAO)[4] technique was used to explore the ejector design space in order to find the Pareto front of optimal solutions.

As shown in Fig. 3.18, the optimization revealed the presence of trade-off between the pressure lift and entrainment ratio (and this should cause no surprise). By keeping fixed the ER of the baseline geometry, the pressure lift could be increased up to 10,379 Pa, while the ER could be enhanced up to 0.782 for the same pressure lift.

Finally, an additional ejector optimization study was carried out by Husain et al. (2016), who used a surrogate model methodology coupled with an evolutionary algorithm to find the optimal Pareto solutions for the pressure lift and motive pressure ratio (i.e., the ratio between the primary inlet and diffuser outlet pressures).

---

[4]This is simply a particular kind of surrogate modeling approach in which the response surface is explored by means of a genetic algorithm.

Fig. 3.18 Pareto front for the ejector optimization (Lee et al. 2016)

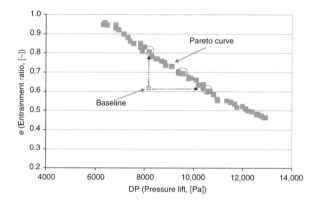

The results showed that both the pressure lift and efficiency of the jet pump could be significantly enhanced through the multi-objective optimization.

3.5   Ejector Chiller Efficiency and Optimization

Thermodynamic optimization of refrigeration systems in general is described in a vast amount of literature, from textbooks (Gordon and Ng 2000) to advanced research papers (Bejan et al. 1995). In the specific case of refrigeration systems based on an ejector, this latter obviously plays the main role, just like the compressor in a vapor compression refrigerator, and this justifies the prominence given in the previous sections to the ejector efficiency and optimization.

However, it's worth to underline that in general the optimal design of a complex system does not necessarily coincide with the optimum of each of its components. Moreover, the ejector is only a minor part in the global cost of the refrigeration system. The heat exchangers, as will be shown in Chap. 5 for our prototype chiller, form about one half of the system cost. This circumstance is important because in many cases the investment cost has a greater impact on the economic balance than the input energy cost.

For instance, in heat-powered refrigeration systems, the heat input may be renewable energy (solar) or waste heat (combined power-heat-cooling systems): in both cases the energy input is virtually free, and it affects the operation cost only for its impact on the capital cost (cost of the solar panels, cost of heat recovery exchanger, etc.). Therefore, a careful choice and sizing of the heat exchangers is clearly fundamental. On the other hand, the size of the heat exchangers does depend on the ejector efficiency, as inefficient systems need larger generators and condensers for a given cooling power.

The pump should receive some attention as well, as it represents the sole consumer of electric energy within the ejector chiller and a non-negligible investment cost. Small systems have relatively small mass flow rates and efficient, small size pumps may be difficult to find. The pump is also the sole moving mechanism and hence may determine the global system reliability.

An example of global system optimization that includes the effect of heat exchanger and pump performance is given by Grazzini and Rocchetti (2002) who analyze the efficiency of chiller with the same two-stage ejector presented by Grazzini and Mariani (1998) and three shell-and-tube heat exchangers. The optimization accounts for the pressure losses encountered by the external water flow in the heat exchangers. Therefore, the objective function is the system COP defined as

$$\text{COP} = \frac{\dot{Q}_E}{\dot{Q}_G + \dot{W}_{Pump} + \dot{W}_{PLG} + \dot{W}_{PLC} + \dot{W}_{PLE}} \tag{3.89}$$

where the subscript PL means pressure losses. The fixed input data are the inlet external water temperature at the generator, condenser, and evaporator plus the required cooling power. The independent variables are the external water flow rates, the number and inner diameter of the heat exchanger tubes, the steam flow rate at generator, the boiling and superheating temperature at evaporator and generator, and the condensing temperature. Tubes' inner diameter and thickness are chosen in the standard series. The code evaluates the ejector geometrical design and the flow rate needed at first and second stage primary nozzles, in order to obtain the requested pressure lift for the secondary stream. A cooling power of 5 kW and 12, 30, and 120 °C as inlet water temperatures at the evaporator, condenser, and generator are considered.

The results show that optimum heat exchanger sizes are quite large, particularly at the evaporator. Lowering the temperature difference at evaporator has a stronger effect on COP than at generator and condenser. The most important result, however, is the significant departure of the global optimum from the component optimum, i.e., optimizing all components separately does not necessarily imply the best result for the complete system. Therefore, when searching the optimum design for an ejector chiller, the simulation must involve all components.

The same optimization technique was applied by Grazzini et al. (2004) to a comparison between a two-stage and a CRMC steam ejector. The comparison shows that the CRMC design offers significantly higher performance.

A further work by Grazzini and Rocchetti (2008) has investigated the effect of the objective function used for optimization on the resulting design. Several possible objective functions have been tested: the first is the system COP defined by Eq. (3.89), while the second and the third divide this same COP by the sum of heat exchange surfaces and volumes, respectively. These two criteria are meant to weight the result on the basis of an investment cost parameter. A fourth objective function overcomes the problem suffered by the classic COP definition, i.e., the summation of quantities of different thermodynamic value (e.g., thermal and mechanical energy). To this aim, the thermal powers at numerator and denominator are multiplied by their Carnot factors $(1 - T/T_{amb})$ using the water at condenser inlet as ambient temperature. This introduces exergy as evaluation parameter in lieu of energy.

Another objective function introduces the Life Cycle Analysis (LCA), in order to account for the "embodied exergy" in the materials used for building the refrigeration system. In this way, the exergy delivered by the system is divided by the exergy used over the entire lifetime of the plant. Finally, a last criterion is the "entropy generation minimization" introduced by Bejan (1996).

The obtained results show how sensitive is the design on the objective function used and how the inclusion of the system cost in the evaluation may orient the choice toward different solutions.

The previous study highlights the inadequacy of the use of the COP alone to study the performance of refrigeration cycle or thermal systems in general. In particular, first-law arguments do not allow a proper comparison of refrigerators working under different operating temperatures, as the quality (or exergy) content of the energy sources is completely neglected.

In order to overcome this limitation, it is possible to define a second-law efficiency for the supersonic ejector cycle. As illustrated in Sect. 3.4, this is done by comparing the useful output of the real system with that of a corresponding ideal device.

The ideal thermodynamic cycle of a supersonic ejector chiller can be thought as composed by two reversible parts, the motive and the refrigeration cycle, as shown in Fig. 3.19. Under stationary conditions the work output of the motive cycle must equate that demanded by the chiller.[5]

Fig. 3.19 Ideal ejector refrigeration cycle

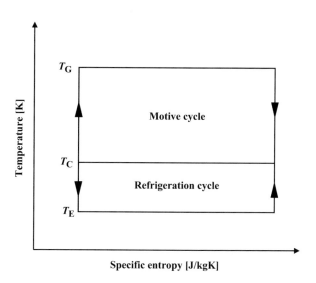

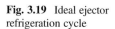

---

[5]Despite this, the two parts of the system can produce and require different "work per cycle" and have different areas in a *T-s* diagram depending on the corresponding mass of the operating fluid.

For the ideal motive cycle, the formulation of its efficiency is the well-known Carnot efficiency, which is derived by using the Clausius theorem:

$$\eta_{m\_ideal} = \frac{W_m}{Q_G} = \frac{Q_G - Q_C}{Q_G} = \frac{T_G - T_C}{T_G} \tag{3.90}$$

where the subscript C stands for condenser.

The efficiency or COP of the ideal refrigeration cycle is obtained in the same way, provided that the useful output and input energy source are changed according to refrigeration purposes:

$$COP_{r\_ideal} = \frac{Q_E}{W_r} = \frac{Q_E}{Q_C - Q_E} = \frac{T_E}{T_C - T_E} \tag{3.91}$$

The ideal efficiency of the complete ejector cycle is derived based on the previous relations. In this case, the useful output is the cooling load, while the only energy input is the heat transferred at the boiler. The ideal efficiency is then easily obtained by considering that the work output of the motive cycle must equate that required by the refrigeration cycle (although the specific work can be different as stated before). Moreover, the condenser temperature is the same:

$$COP_{ec\_ideal} = \frac{Q_E}{Q_G} = \frac{Q_E}{W_r}\frac{W_m}{Q_G} = COP_{r\_ideal} \cdot \eta_{m\_ideal} = \frac{T_E}{T_C - T_E}\frac{T_G - T_C}{T_G} \tag{3.92}$$

The ideal efficiency is useful in many ways. Firstly, it provides an easy tool to understand efficiency trends that are approximately followed by the real cycle. Secondly, it forms the basis for the definition of the second-law efficiency, which is given by the ratio of the real to the ideal first-law efficiencies:

$$\eta_{II\_ec} = \frac{COP_{ec\_real}}{COP_{ec\_ideal}} = \frac{Q_E}{Q_G + W_{pump\_real}}\frac{Q_G + W_{pump\_ideal}}{Q_{E\_ideal}} \approx \frac{Q_E}{Q_{E\_ideal}} \tag{3.93}$$

where it has been assumed that the difference between the ideal and real pump work is negligible.

It is important to underline that although it may seem natural to select the cooling load as the useful output of the system, the choice is actually arbitrary. As an example, the ejector cycle could be used in reverse mode as a heat pump (or in both ways simultaneously). The efficiency definition then would change, despite the system is the same. Fortunately, for power and refrigeration systems, this problem is really marginal, as the number of different possibilities is few and the useful output is usually well defined. By contrast, this is not the case when trying to define the efficiency of the supersonic ejector alone, as detailed in Sect. 3.4.1.

The use of the second-law efficiency provides reasons to understand why the ejector (and chiller) performance should be maximum at the critical point.

When a real system is working in the vicinity of the critical conditions, a decrease in condenser temperatures moves the operating point inside the plateau of the characteristic curve, and the COP remains constant. However, at the same time, the COP of the ideal cycle increases due to the lower condenser temperature, and the second-law efficiency decreases.

This efficiency reduction implies that a certain amount of available energy (or exergy) is being wasted by the system. This exergy destruction occurs through a progressive increase of shock intensity and turbulence dissipation in the ejector diffuser (upstream the shock the flow remains unaltered).

References

Arbel, A., et al. (2003). Ejector irreversibility characteristics. *Transactions of the ASME. Journal of Fluids Engineering, 125*, 121–129.

Bejan, A. (1996). Entropy generation minimization: the new thermodynamics of finite-size devices and finite-time processes. *Journal of Applied Physics, 79*(3), 1191–1218.

Bejan, A., Vargas, J., & Sokolov, M. (1995). Optimal allocation of a heat-exchanger inventory in heat driven refrigerators. *International Journal of Heat and Mass Transfer, 38*, 2997–3004.

Besagni, G., Mereu, R., & Inzoli, F. (2016). Ejector refrigeration: a comprehensive review. *Renewable and Sustainable Energy Reviews, 53*, 373–407.

Brown, B., & Argrow, B. (1999). Calculation of supersonic minimum length nozzle for equilibrium flow. *Inverse Problem in Engineering, 7*, 66–95.

Chang, Y.-J., & Chen, Y.-M. (2000). Enhancement of a steam-jet refrigerator using a novel application of the petal nozzle. *Experimental Thermal and Fluid Science, 22*, 203–211.

Chunnanond, K., & Aphornratana, S. (2004). An experimental investigation of a steam ejector. *Applied Thermal Engineering, 24*, 311–322.

Dvorak, V. (2007). Shape optimization and computational analysis of axisymmetric ejector. *Proceedings of the 8th International Symposium on Experimental and Computational Aerothermodynamics of Internal Flows*. Lyon.

Eames, I. (2002). A new prescription for the design of supersonic jet-pumps: the constant rate of momentum change method. *Applied Thermal Engineering, 22*, 121–131.

Eames, I., Milazzo, A., Paganini, D., & Livi, M. (2013). The design, manufacture and testing of a jet-pump chiller for air conditioning and industrial application. *Applied Thermal Engineering, 58*, 234–240.

ESDU. (1986). *Ejectors and jet pumps, data item 86030*. London, UK: ESDU International Ltd.

Eves, J., et al. (2012). Design optimization of supersonic jet pumps using high fidelity flow analysis. *Structural and Multidisciplinary Optimization, 45*, 739–745.

Fan, J., et al. (2011). Computational fluid dynamic analysis and design optimization of jet pumps. *Computers & Fluids, 46*, 212–217.

Gordon, J., & Ng, K. (2000). *Cool thermodynamics*. Cambridge, UK: Cambridge International Science Publishing.

Grazzini, G. & D'Albero, M. (1998, June 2–5). A Jet-Pump inverse cycle with water pumping column. *Proceedings of natural working fluids '98*. Oslo.

Grazzini, G., & Mariani, A. (1998). A simple program to design a multi-stage jet-pump for refrigeration cycles. *Energy Conversion and Management, 39*, 1827–1834.

Grazzini, G., & Rocchetti, A. (2002). Numerical optimization of a two-stage ejector refrigeration plant. *International Journal of Refrigeration, 25*, 621–633.

Grazzini, G., & Rocchetti, A. (2008). Influence of the objective function on the optimisation of a steam ejector cycle. *International Journal of Refrigeration, 31*, 510–515.

Grazzini, G., Rocchetti, A. & Eames, I. (2004). *A new ejector design method discloses potential improvements to the performance of jet-pump cycle refrigerators*. Heat Powered Cycle Conference, Larnaca.

Grazzini, G., Milazzo, A., & Paganini, D. (2012). Design of an ejector cycle refrigeration system. *Energy Conversion and Management, 54*, 38–46.

Grazzini, G., Mazzelli, F. & Milazzo, A. (2015, May 18–19). *Constructal design of the mixing zone inside a supersonic ejector*. Constructal Law & Second Law Conference, Parma.

Hoffman, J., Scofield, M., & Thompson, H. (1972). Thrust nozzle optimization including boundary layer effects. *Journal of Optimization Theory and Applications, 10*, 133–159.

Huang, B., Chang, J., Wang, C., & Petrenko, V. (1999). A 1-D analysis of ejector performance. *International Journal of Refrigeration, 22*, 354–364.

Huang, B., Hu, S., & Lee, S. (2006). Development of an ejector cooling system with thermal pumping effect. *International Journal of Refrigeration, 29*, 476–484.

Husain, A., Sonawat, A., Mohan, S., & Samad, A. (2016). Energy efficient design of a jet pump by ensemble of surrogates and evolutionary approach. *International Journal of Fluid Machinery and Systems, 9*, 265–276.

Kasperski, J. (2009). Two kinds of gravitational ejector refrigerator stimulation. *Applied Thermal Engineering, 29*, 3380–3385.

Keenan, J., Neumann, E., & Lustwerk, F. (1950). An investigation of ejector design by analysis and experiment. *Journal of Applied Mechanics, 17*, 299–309.

Kim, S., Jin, J., & Kwon, S. (2006). Experimental investigation of an annular injection supersonic ejector. *AIAA Journal, 44*(8), 1905–1908.

Kock, F., & Herwig, H. (2004). Local entropy production in turbulent shear flows: a high-Reynolds number model with wall functions. *International Journal of Heat and Mass Transfer, 47*, 2205–2215.

Kong, F., & Kim, H. (2016). Optimization study of a two-stage ejector–diffuser system. *International Journal of Heat and Mass Transfer, 101*, 1151–1162.

Kracík, J. & Dvorák, V. (2015). *Experimental and numerical investigation of an air to air supersonic ejector for propulsion of a small supersonic wind tunnel*. EPJ Web of Conferences. EFM14 – Experimental Fluid Mechanics, s.l.

Lee, M., et al. (2016). Optimization of two-phase R600a ejector geometries using a non-equilibrium CFD model. *Applied Thermal Engineering, 109*, 272–282.

Little, A., & Garimella, S. (2016). A critical review linking ejector flow phenomena with component- and system-level performance. *International Journal of Refrigeration, 70*, 243–268.

Locatelli, M., & Schoen, F. (2013). *Global optimization; theory, algorithms, and applications*. s.l.: MOS-SIAM.

Mazzelli, F. (2015). *Single & two-phase supersonic ejectors for refrigeration applications* (Ph.D. thesis). Florence.

McGovern, R., Narayan, G., & Lienhard, J. (2012). Analysis of reversible ejectors and definition of an ejector efficiency. *International Journal of Thermal Sciences, 54*, 153–166.

Milazzo, M., & Rocchetti, A. (2015). Modelling of ejector chillers with steam and other working fluids. *International Journal of Refrigeration, 57*, 277–287.

Milli, A. (2006). *Development and application of numerical methods for the aerodynamic design and optimisation of turbine components* (Ph.D. thesis). Università degli Studi di Firenze, s.l.

Munday, J. T., & Bagster, D. F. (1977). A new ejector theory applied to steam jet refrigeration. *Industrial & Engineering Chemistry Process Design and Development, 164*, 442–449.

Nguyen, V., Riffat, S., & Doherty, P. (2001). Development of a solar-powered passive ejector cooling system. *Applied Thermal Engineering, 21*, 157–168.

Nocedal, J., & Wright, S. (2006). *Numerical optimization*. s.l.: Springer.

Opgenorth, M., Sederstroma, D., McDermott, W., & Lengsfeld, C. (2012). Maximizing pressure recovery using lobed nozzles in a supersonic ejector. *Applied Thermal Engineering, 37*, 396–402.

Palacz, P., et al. (2016). CFD-based shape optimisation of a CO_2 two-phase ejector mixing section. *Applied Thermal Engineering, 95*, 62–69.

Palacz, P., et al. (2017). Shape optimisation of a two-phase ejector for CO_2 refrigeration systems. *International Journal of Refrigeration, 74*, 212–223.

Polanco, G., Holdøb, A.E., & Mundayc, G. (2010). General review of flashing jet studies. *Journal of Hazard Material, 173*, 2–18.

Pope, A., & Goin, K. (1978). *High-speed wind tunnel testing*. s.l.: Wiley.

Rao, S., & Jagadeesh, G. (2014). Novel supersonic nozzles for mixing enhancement in supersonic ejectors. *Applied Thermal Engineering, 71*, 62–71.

Riffat, S. B. (1996). International, Patent No. PCT-GB96-00855.

Riffat, S., & Holt, A. (1998). A novel heat pipe/ejector cooler. *Applied Thermal Engineering, 18*, 93–101.

Shahpar, S. (2004). *Automatic aerodynamic design optimisation of turbomachinery components – an industrial perspective, von Karman lecture series 2004–7*. s.l.: American Institute of Aeronautics and Astronautics.

Shen, S., et al. (2005). Study of a gas-liquid ejector and its application to a solar-powered bi-ejector refrigeration system. *Applied Thermal Engineering, 25*, 2891–2902.

Shope, F. (2006, June 5–8). *Contour design techniques for super/hypersonic wind tunnel nozzles*. 24th applied aerodynamics conference. AIAA, San Francisco.

Shyy, W., Papila, N., Vaidyanathan, R., & Tucker, K. (2001). Global design optimization for aerodynamics and rocket propulsion components. *Progress in Aerospace Sciences, 37*, 59–118.

Sierra-Pallares, J., García del Valle, J., García Carrascal, P., & Castro Ruiz, F. (2016). A computational study about the types of entropy generation in three different R134a ejector mixing chambers. *International Journal of Refrigeration, 63*, 199–213.

Smits, A., & Dussauge, J.-P. (2006). *Turbulent shear layers in supersonic flow* (2nd ed.). New York: Springer.

Srikrishnan, A., Kurian, J., & Sriramulu, V. (1996). Experimental study on mixing enhancement by petal nozzle in supersonic flow. *Journal of Propulsion and Power, 12*(1), 165–169.

Srisastra, P., & Aphornratana, S. (2005). A circulating system for a steam jet refrigeration system. *Applied Thermal Engineering, 25*, 2247–2257.

Srisastra, P., Aphornratana, S., & Sriveerakul, T. (2008). Development of a circulating system for a jet refrigeration cycle. *International Journal of Refrigeration, 31*, 921–929.

Wang, J., Wu, J., Hu, S., & Huang, B. (2009). Performance of ejector cooling system with thermal pumping effect using R141b and R365mfc. *Applied Thermal Engineering, 29*, 1904–1912.

Worall, M. (2001). *An investigation of a jet-pump thermal (ice) storage system powered by low-grade heat* (Ph.D. thesis). University of Nottingham, s.l.

Yadav, R., & Patwardhan, A. (2008). Design aspects of ejectors: effects of suction chamber geometry. *Chemical Engineering Science, 63*, 3886–3897.

Yapici, R., et al. (2008). Experimental determination of the optimum performance of ejector refrigeration system depending on ejector area ratio. *International Journal of Refrigeration, 31*, 1183–1189.

Zhu, Y., Cai, W., Wen, C., & Li, Y. (2009). Numerical investigation of geometry parameters for design of high performance ejectors. *Applied Thermal Engineering, 29*, 898–905.

Ziapour, B., & Abbasy, A. (2010). First and second laws analysis of the heat pipe/ejector refrigeration cycle. *Energy, 35*, 3307–3314.

Zucrow, M. (1976). *Gas dynamics*. s.l.: Wiley.

Chapter 4
Ejector CFD Modeling

4.1 Single-Phase Ejectors

Starting from the beginning of the last decade, the number of CFD studies regarding supersonic ejectors has continuously increased. A general feature that emerges from analyzing these works is that discrepancies between CFD and experiments are strongly related to operating conditions. In particular, many authors have shown that the prediction of the Entrainment Ratio (ER) at off-design conditions is significantly more challenging than that at on-design regime, where the discrepancies with experiments are generally of the order of few percent.

Previous studies blamed this low accuracy on turbulence modeling approximations. Although this may represent a source of error, it is by no means the only cause of possible inconsistencies with experimental data. Many aspects must be considered in order to set up an accurate numerical setting, and the selection of the turbulence model represents just one piece of the puzzle. Among these are compressibility effects, 3D effects, wall friction, heat transfer, and multiphase phenomena, which will be detailed in the following sections.

4.1.1 Numerical Aspects

A general discussion of theoretical and practical aspects for reaching high-quality CFD simulation results is certainly beyond the scope of this book. However, few comments are necessary in order to discuss some general guidelines that can help increase the confidence in the numerical results (and which are sometimes neglected in the specific literature concerning supersonic ejectors).

One of the first, critical issue of CFD simulations is the inevitable loss of accuracy due to nonuniform cell grid distributions (a uniform grid is one in which the distance

© Springer International Publishing AG, part of Springer Nature 2018
G. Grazzini et al., *Ejectors for Efficient Refrigeration*,
https://doi.org/10.1007/978-3-319-75244-0_4

between adjacent mesh points is constant, whereas in nonuniform grids the distance between mesh points changes along the domain).

It is a general property of finite difference approximation that discretization schemes can lose at least one order of accuracy, and sometimes two, on general nonuniform grids (Hirsch 2007). This problem is particularly detrimental with currently used discretization schemes, which are generally of second-order accuracy on uniform grids. Therefore, in order to achieve second-order accuracy on arbitrary grids, one has to consider discretization schemes that are of higher order on uniform grids (Hirsch 2007). Alternatively, the errors due to the grid nonuniformity can be minimized if the mesh is constructed in such a way that the size variation between consecutive cells is smooth and continuous. Therefore, it is important to always use prescribed analytical laws for defining grid size variation, for instance, by setting a constant clustering factor r:

$$\Delta x_{i+1} = r \cdot \Delta x_i$$

where Δx_i is the ith cell length along the x direction and typical values for r are between 1.1 and 2 (Hirsch 2007).

Nonuniform grid size distribution is not the sole parameter affecting the accuracy of CFD simulation. In particular, the distortion of grid elements from the ideal quadrilateral/cubic shape can be detrimental for the accuracy of the simulations. This distortion can be measured by many topological parameters (e.g., cell aspect ratio, skewness factor, etc.), although its impact is generally difficult to quantify precisely. Nevertheless, based on this consideration, it is possible to set some general guidelines that can help improve the quality of the simulations (Hirsch 2007):

- Avoid absolutely discontinuities in grid cell size. Any sudden jump in grid size could reduce the local accuracy to order zero.
- Ensure that the grid sizes vary in a continuous way in all directions.
- Minimize grid distortion, avoiding concave cells or cells with angles between adjacent edges that are too far away from orthogonality. If these angles are reduced to a few degrees, poor accuracy is guaranteed.
- Avoid cells with one or more very short edges, except in boundary layers where high aspect ratios are acceptable, provided the cells are sufficiently close to orthogonality to the solid surface.

In addition to these aspects, it is essential to notice that the same mesh distribution can lead to greater discretization errors in regions where the gradients of the flow properties are higher. In these regions, one should consider increasing the mesh refinement and avoiding any form of grid distortions. In ejector flow, for instance, two critical zones are those near the nozzle throat and at the primary nozzle trailing edge, as illustrated in Fig. 4.1.

In the throat zone, the problem is often caused by the abrupt slope variation in the converging part of the nozzle. In the specific case of structured grids with quadrilateral/hexahedral elements, the requirement of mesh continuity between the wall

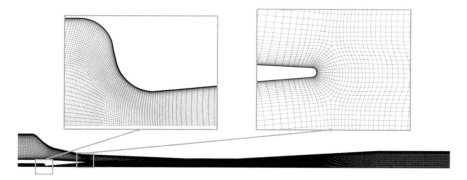

Fig. 4.1 Mesh details at the primary nozzle throat and trailing edge (grid with 70 k cells)

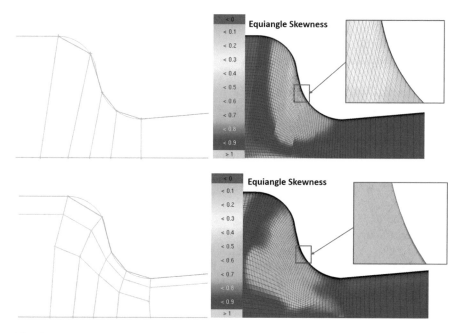

Fig. 4.2 Blocking structure and mesh skewness near the throat of a De Laval nozzle for two different block topologies

and axis boundaries results in a large distortion of the grid elements as well as a high non-orthogonal elements at the nozzle wall.

One way to alleviate the distortion of the cells is to construct a proper multi-block structure that allows a smoother transition from the curvature of the wall to that of the axis. This is illustrated in Fig. 4.2 which shows the mesh skewness near the throat of a nozzle for two different block structures (the skewness is a parameter that quantify the level of non-orthogonality of the grid cells). As can be seen, the inclusion of a

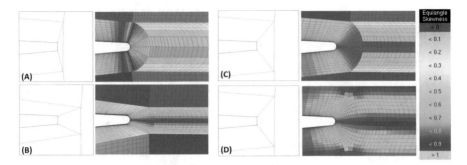

Fig. 4.3 Blocking structure and mesh skewness at the nozzle trailing edge: (**a**) case with high orthogonality at wall but high skewness in the mixing region; (**b**) case with higher orthogonality in the mixing region but high skewness at wall; (**c**) optimized case; (**d**) optimized case after mesh smoothing

number of additional blocks allows lower levels of skewness, particularly at the wall, where the mesh lines are more orthogonal to the wall (the resulting mesh is shown in Fig. 4.1).

In the region near the primary nozzle trailing edge, problems may arise when trying to simulate the real thickness of the nozzle fillet. Due to the very small curvature radii, the mesh in this region must be highly refined, and this refinement is unavoidably carried along the whole domain. However, this feature may turn to be an advantage. Indeed, a high mesh resolution in the region downstream of the nozzle trailing edge is important to accurately predict the wake formation and mixing layer development, which are key phenomena in determining the ejector entrainment ratio.

Therefore, great care should be paid in the optimization of the block structure as well as the node distribution in this region. Figure 4.3 illustrates an example of blocking structure and mesh skewness at the nozzle trailing edge. The quality of the mesh is highly sensitive to variations in this zone due to the opposed requirements of maintaining high orthogonality at the nozzle wall and downstream of the fillet. Thus, a compromise is generally needed, although a post-smoothing of the grid can improve the global quality of the final distribution (the resulting mesh is illustrated in Fig. 4.1).

One further aspect concerning grid quality is connected to the grid-flow alignment and the choice of the type of mesh. As a general rule, one should remember that meshes with quadrilateral/hexahedral elements will lead to a higher accuracy than grids with triangular/tetrahedral cells. A reason for this is to be found in what is called false diffusion. The false diffusion is a numerical error that appears in multidimensional problem when convective schemes are discretized by means of upwind schemes (for more details see (Patankar 1980)). The error arises when the flow is not perfectly aligned with the normal to the cell faces and has an excess in the diffusivity of the quantity being transported (e.g., momentum). The problem can effectively be reduced either by increasing the grid refinement, by choosing a higher discretization order, or by aligning the mesh elements to the flux direction. Due to

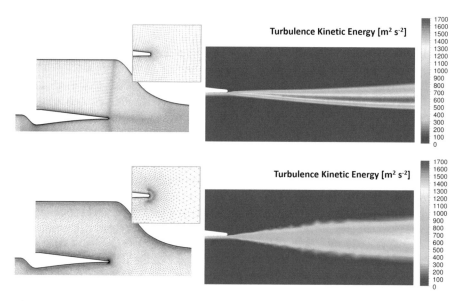

Fig. 4.4 Effect of the false diffusion on the transport of turbulence kinetic energy along the mixing layer for a structured grid with 80 k quadrilateral cells (top) and a hybrid grid (triangular plus quadrilateral at wall) containing 200 k cell (bottom)

their very nature, the last of these points is unfeasible in unstructured grids which are thus more affected by the numerical diffusivity.

As an example, Fig. 4.4 shows the effect of the false diffusion on the transport of turbulence kinetic energy across the mixing layer of a 2D axisymmetric ejector. The simulations are performed for a structured grid with 80 k quadrilateral cells and for a hybrid grid with 200 k cells (a hybrid grid is one with triangular cells in the core and quadrilateral elements at the wall). The difference in the mesh size is due to the fact that the hybrid mesh requires a much larger number of nodes to reach a mesh-independent solution for the mass flow rates. In both cases the convective term is discretized with a second-order upwind scheme.

Despite the larger number of elements, Fig. 4.4 clearly reveals an increased diffusivity for the unstructured mesh leading to a more smeared profile of the turbulence kinetic energy. This increase in turbulent momentum transport brings along an increased entrainment effect that, although small (the suction mass flow rate is 2% higher for the hybrid mesh), should be avoided.

Consequently, the use of structured grids should be preferred whenever possible as these require a much lower number of grid points to achieve the same level of accuracy of an unstructured grid. This is especially true for supersonic ejectors, where the strong "directionality" of the flow (axial velocity component always much greater than the transversal component) allows the structured grid to be properly aligned with the momentum fluxes.

Finally, a few words are in order about the choice of the flow solver. To date, the dichotomy "pressure-based solvers are suited for low-speed incompressible flow while density-based solvers are ideal for compressible flows" is quite surpassed. Many strategies have been devised to extend the range of applicability of both methods, which are generally implemented in commercial codes (for more details see (Miettinen and Siikonen 2015; Keshtiban et al. 2004)). As a consequence, the choice of the solver is not as restricted as in the past and should be made based on considerations of stability and computational efficiency.

As a general indication, the use of density-based solvers should be preferred in highly compressible flows or in any flow in which the conservation equations and the equation of state are highly coupled (e.g., non-isothermal gas flows). For these flows, the use of a pressure-based segregated approach may result in numerical instabilities and higher computational times. Using a pressure-based coupled algorithm may sometimes provide an alternative option to the density-based schemes. On the contrary, pressure-based solvers may be more efficient in cases where the compressible flow regions are present in a limited part of the computational domain.

In terms of accuracy, it is important to always use high-order discretization schemes in order to reduce the issue of numerical diffusivity (at least second-order schemes). Moreover, accurate prediction of shock-laden flows (like supersonic ejectors) generally requires the use of high-resolution schemes appropriately devised for shock-capturing problems. These schemes avoid the occurrence of spurious oscillations on the thermodynamic properties upstream and downstream of the shock discontinuity. One of these methods is the Monotone Upstream-Centered Scheme for Conservation Laws (MUSCL) (Van Leer 1979) which allows a sharp reconstruction of the shock by locally increasing the scheme diffusivity (this is done by using flux limiters; for more details, see (Hirsch 2007)). Other advanced approach may be used, like the Weighted Essentially Non-Oscillatory (WENO) schemes (Shu 2009) or different types of Riemann solvers (Toro 2009), but these are seldom implemented in commercial codes.

In any event, sensitivity analyses to the discretization order as well as to the mesh refinement should always be performed, possibly by employing error estimation procedures that quantitatively assess the numerical uncertainty of the computational setup (e.g., Richardson extrapolation). This is what is called the verification of a simulation (see (Roache 1997)). However, the accuracy of a simulation depends not only on numerical errors but also on physical modeling approximations. The assessment of the error connected to the physical model is what is called the validation of the model. In the next sections, we will analyze the impact of commonly used modeling approximations (more details on verification and validation procedures in CFD can be found in (Oberkampf and Trucano 2002; Roache 1997)).

4.1.2   Turbulence Modeling

The sensitivity of CFD results to the turbulence model used has been demonstrated by several authors in the field of ejector studies.

One of the first extensive studies was made by Bartosiewicz et al. (2005) who compared the results of six different turbulence models against axial pressure distributions and laser tomography pictures. The validation concentrated on the shock location, shock strength, average pressure recovery, and mixing length predictions. Unfortunately, no comparison was made in terms of mass flow rates or ER. From the investigation, it appeared that the k-ω SST gave a better agreement with experiments.

Hemidi et al. (2009a, b) performed a number of calculations on a supersonic air ejector using the standard k-ε and the k-ω SST turbulence models. The results illustrated that the k-ε model performed better in terms of global parameter predictions. Moreover, the analyses revealed that in many cases the two schemes produced similar results in terms of ER, although the predicted local flow features were highly dissimilar.

Sriveerakul et al. (2007) studied a supersonic steam ejector and compared the CFD results using a k-ε-realizable model with data on mass flow rates and wall pressure profiles. More recently, Besagni and Inzoli (2017) used the same data to study the performance of seven different turbulence models. The extensive analyses showed that the k-ω SST model had better agreement both in terms of global and local parameter prediction. Unfortunately, neither these two works considered the impact of two-phase phenomena in their simulations, and the condensing steam was simulated by means of the ideal gas equation of state.

Zhu and Jiang (2014) compared 3D CFD simulations using four different turbulence models with Schlieren optical measurements. They found that the RNG k-ε turbulence model better reproduced experimental data for entrainment ratio and shock wave structures.

In all these studies, the validation was made without a clear distinction between the main mechanisms that cause uncertainties in the modeling of the turbulent flow field. Perhaps, the only one exception in this regard is represented by the study performed by Garcia del Valle et al. (2015), who validated four different turbulence models with three different test cases including the simulation of the compressible turbulent mixing layer between two coaxial flows, the shock wave-turbulent boundary layer interaction, and the real gas flow behavior in a shock tube experiment.

The main purpose was to check the turbulence models independently and later check their suitability on complete ejector flow simulations. The investigation showed that the standard k-ε turbulence model yielded the most accurate solution when simulating the "single-feature" test cases, whereas the k-ω SST returned the poorest results. However, when comparing the results for the complete ejector, the k-ω SST turned out to perform the best.

More examples could be proposed which, however, would just add to the significant spreading of the results presented above. However, this scatter should cause no surprise on considerations that turbulence models are, in actual facts, very sophisticated correlations that achieve a good level of agreement only for the flow types for which they were previously calibrated. In this respect, the dynamic behavior of supersonic ejectors results from the simultaneous presence and complex interaction of many of these flow types, which altogether contribute to the global uncertainty of the CFD simulations.

Nevertheless, it may be useful to review some of the characteristics of commonly used turbulence models in order to devise desirable features for the simulations of ejector flows.

4.1.2.1 ε-Based Models

Among the two-equation RANS turbulence models, the standard k-ε model by Launder and Spalding (1972) is by far the simplest and oldest. This model has proven to provide accurate results for free shear layers and attached boundary layers with zero or favorable pressure gradient. However, due to its inability to correctly match the law of the wall in the near-wall region, the performance of this model is unsatisfactory for general boundary layer flows, especially those with strong adverse pressure gradients (Wilcox 2006).

In order to improve the accuracy, one may use more advanced wall function approaches. These however, are generally calibrated for specific types of wall flows (e.g., with adverse pressure gradient, heat exchange, or wall curvature) and may not be able to reproduce all the flow features occurring within ejector's boundary layers. A different approach is to change the turbulence-governing equations in order to allow the proper integration of momentum and energy equations through the viscous sublayer. One example is the so-called two-layer approach (e.g., (Rodi 1991)), which is based on a subdivision of the boundary layer into a viscous-dominated region near the wall and a fully turbulent region closer to the free stream. The standard k-ε formulation is used in the fully turbulent region, while a simpler model with only one transport equation for the turbulence kinetic energy is employed in the viscous region (Wolfshtein 1969). This particular near-wall treatment should improve predictions for general boundary layer applications, including those with adverse pressure gradient effects.

A further known issue of the standard k-ε model is related to the specific manner in which turbulent viscosity is defined with the Boussinesq approximation, that is:

$$\mu_t = C \cdot \rho \frac{k^2}{\varepsilon} \tag{4.1}$$

where ρ is the density, k is the turbulence kinetic energy, ε is the dissipation, and C is a constant.

It has been found in several studies (e.g., (Moore and Moore 1999)) that specifying a constant value in Eq. 4.1 can lead to unrealistically high levels of turbulent viscosity in flows with a high shear rate (such as impinging jets, flows approaching stagnation points, and shock-induced separations). These high levels of μ_t are "not realizable" in real turbulent flow. The k-ε-realizable model (Shih et al. 1995) improves upon the standard k-ε model by introducing a different formulation of the turbulent viscosity that ensures the "realizability" of the model. The formulation suggested by Reynolds (1987) basically substitutes the constant in Eq. 4.1 with an algebraic function that makes the values of μ_t match those obtained from experiments with various flows. In addition, the realizable model uses a different formulation for the ε transport equation, derived from the exact transport equation of fluctuating vorticity (Tennekes and Lumley 1972). Such a formulation should better represent the spectral energy transfer from large eddies to the small dissipative structures (Shih et al. 1995).

The k-ε-realizable model has been extensively validated and improves upon the standard k-ε model for a wide variety of flows, including round jets, mixing layers, rotating homogeneous shear flows, channel flows, and boundary layer flows. However, because the modification to the standard k-ε model is only valid for high Reynolds numbers, the model cannot be expected to provide any significant improvement in near-wall regions. Consequently, the near-wall approach that must be used for this model is the same as the one described for the standard version.

4.1.2.2 ω-Based Models

An alternative way to work around the near-wall issues of ε-based models is to adopt a different independent variable in place of the dissipation rate, namely, the specific dissipation rate, ω. Indeed, models based on ω have the ability to correctly reproduce the law of the wall without the need for any particular modification. This allows for a simple integration along the boundary layer, and results typically show much better agreement for boundary layer simulations, including those with adverse pressure gradients (Wilcox 2006). However, despite all recent improvements, ω-based models, and in particular the standard k–ω model, generally do not achieve the same level of agreement as the k-ε model for free mixing layers. Moreover, they suffer from an exaggerated sensitivity to the free stream value of ω, which can sometimes require a sensitivity analysis to be performed in order to find the correct solution. In order to reduce these negative characteristics, Menter (1994) devised a method that smoothly blends from a k–ω formulation, used in the near-wall region (viscous sublayer and log layer), and a k-ε model, which is active elsewhere (boundary layer outer wake region and free stream). The model is named k–ω Shear Stress Transport (SST) and employs a "stress limiter" to comply with the "realizability" requirements (Menter 1994). The stress limiter is simply a parameter that limits the eddy viscosity, μ_t, to a maximum value in cases of flows with large strain. Within the k–ω SST model, the stress limiter has been calibrated specifically to provide good results for transonic flow up to moderate supersonic speeds. As a result, the model achieves a satisfactory level of agreement with a wide range of flows, including those with adverse pressure gradients and transonic shock waves, as are seen in ejector flows.

4.1.2.3   Reynolds Stress Models

Despite these improvements, models based on the Boussinesq approximation are inherently limited in application. By definition, the turbulent viscosity is affected by the total turbulence kinetic energy, and not by any of its components. This implies that turbulence is modeled as an isotropic process. Consequently, common two-equation turbulence models fail to yield reliable results for flows where turbulence is largely anisotropic. Examples include flows with sudden geometrical changes, flow over curved surfaces, rotating and three-dimensional flows, and flows in noncircular ducts.

The Reynolds Stress Models (RSMs) tackle this problem by solving a transport equation for each independent component of the Reynolds stress tensor. In this way, it is possible to describe the evolution of the different components of the velocity fluctuations, which potentially leads to a greater accuracy in complex flows where turbulence anisotropy plays a role.

Unfortunately, the reliability of RSM simulations is somewhat limited by the numerous closure assumptions that must be employed (and calibrated) to model various terms in the transport equations for the Reynolds stresses. In addition, in much the same way as for two-equation models, RSMs describe the turbulence dissipation by means of a transport equation for the dissipation rate, ε, or specific dissipation rate, ω. Consequently, they inherit the same deficiencies resulting from the assumptions underlying these equations. For instance, ε-based RSMs fail to predict the law of the wall satisfactorily, and the adoption of complicated viscous damping functions or wall function approach may increase the stiffness of the equations system (Wilcox 2006).

Therefore, the use of the RSMs should always be evaluated with care because the potential benefits in terms of accuracy may not always justify the additional computational effort.

4.1.3   Compressibility and 3D Effects

Apart from turbulence modeling approximations, a key aspect for supersonic ejector studies is related to the impact of the flow compressibility on the mixing process. As stated in Chap. 2, compressible mixing layers are affected by a significant reduction of the spreading rate with respect to equivalent low-speed configurations. This issue has consequences on ejector flows because it severely reduces the entrainment and mixing rate inside the mixing chamber.

Due to the absence of any convincing theoretical explanation, turbulence models corrections have been proposed in order to predict this decrease empirically. For ω-based models, Wilcox (1992) proposes a correction to the turbulence kinetic energy equation based on the turbulent Mach number:

$$Ma_t = \frac{\sqrt{2k}}{a} \tag{4.2}$$

where k is the local turbulence kinetic energy and a is the local speed of sound.

This correction reduces the mixing layer entrainment by increasing the dissipation of turbulence kinetic energy through so-called dilatation-dissipation (Sarkar and Balakrishnan 1990). A value of $M_t \sim 0.25$ was found to be the threshold for compressibility to have any impact on the mixing layer (Wilcox 2006). Below this threshold the correction has no impact on the solution and its use may be avoided. This correction has demonstrated improved accuracy for compressible mixing layers. However, its use can negatively affect predictions for wall boundary layers at transonic and supersonic speeds and is especially detrimental for ε-based models (Wilcox 2006). Because of this, the application of the correction should be carefully evaluated on a case by case basis.

The impact of 3D effects is a further issue that must be considered to increase the fidelity of CFD predictions. This may be of particular relevance in case of non-axisymmetric ducts with possible presence of secondary flows.

Unfortunately, very few examples of fully 3D simulations of ejectors can be found in the literature. Pianthong et al. (2007) performed 3D simulations on an axisymmetric geometry in order to check the possible effects of a non-axially symmetric suction inlet (the suction port was on the bottom side of the ejector). Due to very low velocity at the suction inlet, the authors found that 3D simulation results were very similar to those from 2D simulations (however, the simulations were performed using a k-ε-realizable model which is based on the isotropic assumption for turbulence; therefore, the potential effect of circumferential turbulent transport, although probably minor, was not considered). Bouhanguel et al. (2009) also performed 3D axisymmetric simulations with different turbulence models. They compared the results with 2D simulations in the case of zero suction flow and found that 3D calculations were in better agreement with the experimental data.

More recently, Mazzelli et al. (2015) performed numerical and experimental analyses on a rectangular supersonic air ejector to evaluate the impact of turbulence modeling on the accuracy of CFD simulations. Three series of experimental curves at motive pressures of 2.0, 3.5, and 5.0 bar were compared with 2D and 3D simulations, using the four turbulence models described in the previous section: k-ε, k-ε-realizable, k-ω SST, and the stress-ω RSM from Wilcox (2006).

Figure 4.5 shows the comparison between the experimental ERs and those predicted with the different turbulence models. Clearly, the figure reveals that the differences between CFD and experiments are generally much lower at on-design conditions. In particular, when moving to off-design conditions, 2D simulations tend to miss the critical point and subsequent off-design points, whereas 3D calculations more closely match the experimental results. This is due to the friction losses caused by the side walls of the ejector test section that prevents the motive jet from transferring some momentum to the suction flow, thus reducing ejector ER and

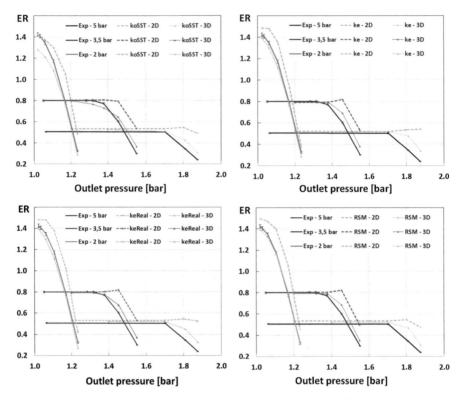

Fig. 4.5 Comparison of experimental and numerical characteristic curves for a rectangular supersonic air ejector (Mazzelli et al. 2015)

decreasing the critical pressure. This effect can only be captured by 3D calculations, whose results match the experimental curves more closely.

In terms of turbulence model accuracy, it was found that different models perform differently depending on the specific operating condition. Globally, the k-ω SST model performs the best, although ε-based models are more accurate at low motive pressures. The RSM model shows predictions comparable to those of the k-ω SST, but the model suffers from numerical stiffness and convergence issues that make its use inconvenient (a more detailed discussion of the results is provided in (Mazzelli et al. 2015)).

Apart from the results of simulations, it is important here to stress the method followed for the assessment of the CFD results and turbulence models' accuracy. First of all, the maximum values of turbulent Mach number were checked to understand the impact of compressibility on mixing layers. Inspections of CFD results revealed that peaks of the Ma_t were always less than 0.25 so that no correction was needed in the simulations. Additionally, many numerical and experimental trials were made in order to evaluate the impact of sources of error different from turbulence modeling approximations.

In particular, close examinations of the numerical errors revealed that the primary mass flow rate was systematically overpredicted. This was likely due to small discrepancies between the real and computational dimensions of the primary nozzle throat. Although the differences were not large, the overprediction of the motive flow rate may have affected the calculation of the critical pressure, as well as the estimation of suction flow entrainment. Additional trials were made to understand the influence of uncertainties in the nozzle exit position and of the presence of water condensation inside the air stream. However, these were found to have negligible impact on determining the mass flow rates. In contrast, a numerical exploration of the role of surface roughness showed that this parameter may have an influence on the transition to off-design conditions.

In conclusion, the study illustrated that while the inaccuracies of different turbulence models were comparable, the correct evaluation of the shear losses at the wall would impact significantly the accuracy of the simulations. This is further investigated in the next section.

4.1.4 Wall Friction and Heat Transfer

As stated at the beginning of this chapter, the inability of CFD models to predict the transition to the off-design regime stems mostly from inaccuracies in the evaluation of the total pressure losses inside the ejector. Due to the high levels of speed, these losses depend strongly on the kinetic energy dissipation within the boundary layers. Hence, the achievement of an adequate level of accuracy requires an accurate description of the friction losses, recirculations, and heat transfer at the ejector walls.

For historical reasons, the commonly used roughness definition in fluid dynamics is the "uniform sand-grain roughness height," K_{sg} (for details see (Taylor et al. 2006)). This is defined as the mean diameter of virtual sand grains that cover the surface, as illustrated in Fig. 4.6. In general, this particular roughness definition is what must be input in most CFD codes (e.g., (ANSYS Inc. 2016)). However, the "uniform sand-grain roughness height" is a quantity that is not measured by common profilometers. These latter usually return some average of the surface vertical displacement, e.g., the arithmetic average height, K_a, or the root mean square height,

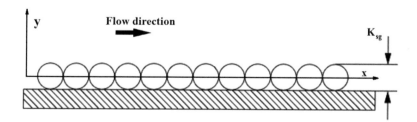

Fig. 4.6 Sand-grain roughness height

K_{rms}. Consequently, some conversion factors are necessary to compare measured roughness heights with values employed for numerical simulations.

Unfortunately, there is no exact conversion factor to transform a measured average roughness (arithmetic or root mean square) into an equivalent value of uniform sand-grain roughness. A recent work from Adams et al. (2012) estimated theoretically the conversion factors and found that these may be calculated by considering $K_{sg} \sim 3\ K_{rms}$ (also found by (Zagarola and Smits 1998)) and $K_{sg} \sim 5.9\ K_a$. By comparison with experimental data, they showed that conversion factors are subject to large uncertainty and should always be regarded as indicative values. Nevertheless, they concluded that using the conversion factor is always a better approximation than to use none.

The influence of wall roughness on numerical accuracy was investigated by Mazzelli and Milazzo (2015). The analysis that follows extends these results by further considering the impact of heat transfer across the ejector walls. It will be clear that the usual assumptions of hydrodynamically smooth and adiabatic surfaces lead in many cases to error at least comparable, if not greater, than those ascribed to turbulence models.

The numerical scheme and computational domain used for the simulations is illustrated in Fig. 4.7. Details of the numerical setup and validation process can be found in (Mazzelli and Milazzo 2015).

The main results of this analysis are condensed in Fig. 4.8. All at once, this chart shows the importance of the correct evaluation of the momentum and heat transfer at the ejector walls. In observing Fig. 4.8, the attention should be focused on the curve representing the numerical scheme with smooth and adiabatic walls (the fuchsia curve). This is the setup that is commonly adopted by most of the studies in ejector research. Notably, while this scheme correctly reproduces the ER results for the on-design regime, the same model is far from being accurate at off-design conditions.

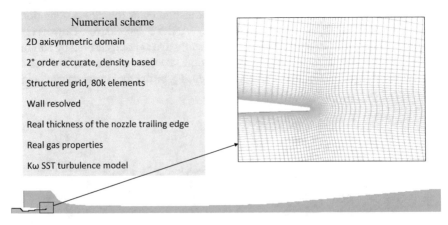

Numerical scheme

2D axisymmetric domain

2° order accurate, density based

Structured grid, 80k elements

Wall resolved

Real thickness of the nozzle trailing edge

Real gas properties

Kω SST turbulence model

Fig. 4.7 Numerical scheme and mesh characteristic for the CFD simulations (Milazzo and Mazzelli 2017)

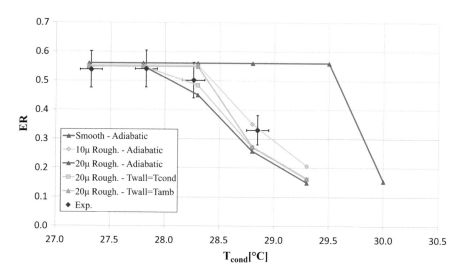

Fig. 4.8 ER for different values of wall roughness and temperature; $T_{gen} = 89$ °C, $T_{eva} = 5$ °C
(Milazzo and Mazzelli 2017)

By including even a small amount of wall roughness, the results for the ER curve
change dramatically. In particular, the gold and light-blue curves correspond to sand-
grain roughness heights of 10 and 20 μm. Clearly, as the condenser pressure
increases, higher values of friction cause the critical point to appear in advance.
This result is indeed expected, as greater friction translates into larger amounts of
total pressure losses, thus reducing the capability of the flow to withstand high values
of back pressure.

By inspection of Fig. 4.8, it appears that the curve with 10 μm roughness over-
estimates the critical pressure, which corresponds to about 28 °C of saturation
temperature. By contrast, the curve relating to 20 μm roughness height seems to
capture well the transition point, though it underestimates the ER at higher temperatures. The green and purple curves in Fig. 4.8 represent two numerical schemes
with 20 μm roughness height and two values of constant wall temperatures. These
are set equal to the condenser and ambient temperature correspondingly (for this it
was considered $T_{amb} = T_{cond} - 5$ °C). Although imposing a constant temperature
along the external wall is clearly a simplification, nonetheless, some interesting
aspects can be understood by this approximate analysis.[1] In particular, Fig. 4.8
shows that the lower the wall temperature, the higher becomes the critical pressure.
Therefore, it appears that a net heat loss toward the ambient produces a positive
effect in terms of flow stability.

---

[1]It should be noticed that the ejector's walls are made of aluminum which has high values of
longitudinal conductivity. Therefore, the hypothesis of constant wall temperature may be
reasonable.

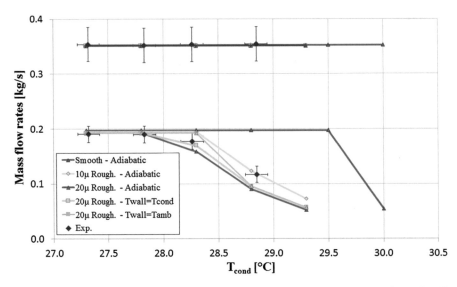

Fig. 4.9 Primary (top) and secondary (bottom) mass flow rates for different values of wall roughness and temperature; $T_{gen} = 89\ °C$, $T_{eva} = 5\ °C$

In Fig. 4.9, the same results of Fig. 4.8 are reported by separating the curves related to the primary and secondary mass flows. As one could expect, the numerical discrepancy with experiments is mainly due to the prediction of the secondary flow rate while the primary nozzle flow is correctly predicted. Nonetheless, it is important to always report data of mass flow rates alone because, in many cases, the error in ER may be lowered or augmented by compensation effect that hides the real accuracy of numerical simulations (e.g., when both the primary and secondary mass flow rates are overestimated, see (Mazzelli et al. 2015)).

Figure 4.10 shows the comparison between the static pressures measured along the diffuser wall and the corresponding profiles obtained by numerical simulations. Clearly, the calculated pressure profiles are highly dissimilar for different roughness heights. In particular, the curves corresponding to smooth surfaces are very distant from the experimental data.

By focusing on the case with $T_{cond} = 28.3\ °C$, it can be noted that, in much the same way as for the ER, the curve of 20 μm roughness height is the one thatg more closely reproduces the experimental data. Conversely, the curves with lower roughness still predict choked flow for this case, which is revealed by the presence of the shock in the diffuser (i.e., the steep rise in wall pressure seen in the figure).

Hence, it can be inferred that the equivalent "sand-grain roughness height" of the ejector, as predicted by numerical analysis, should be close to 20 μm. Converting this value through the aforementioned conversion factors gives an estimated arithmetic roughness height, K_a, of around 3.5 μm. The roughness of the ejector surface was subsequently measured in different locations by means of a Mahr contact

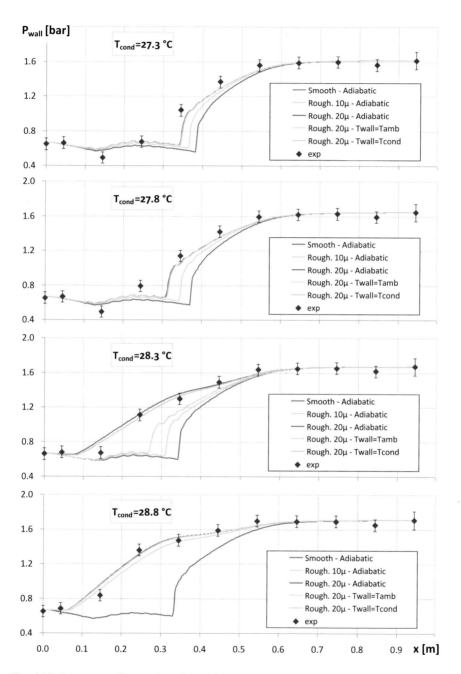

Fig. 4.10 Pressure profiles at the mixing chamber/diffuser wall, for different values of wall roughness and temperature; $T_{gen} = 89\ °C$, $T_{eva} = 5\ °C$, varying T_{cond}

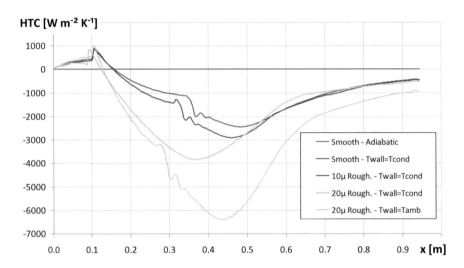

Fig. 4.11 Heat transfer coefficient profiles on the mixing chamber/diffuser wall for smooth and rough walls; $T_{gen} = 89\,°C$, $T_{eva} = 5\,°C$, $T_{cond} = 28.3\,°C$ (Milazzo and Mazzelli 2017)

surface profilometer. Resulting values of K_a ranged from 4 to 6 microns depending on the different measurement sites.

By comparing this value with that predicted by CFD, it seems that numerical analysis underestimates somewhat the experimental datum. Although this error may be partly due to numerical and experimental approximations, one must not forget the large uncertainty connected to the definition of conversion factor between different roughness heights (Adams et al. 2012). In any event, the check of the CFD predictions against experimentally measured roughness should always be carried out in order to avoid this analysis becoming a mere adjustment to the experimental results.

Figure 4.11 presents the profiles of Heat Transfer Coefficient (HTC) at the ejector external wall. As expected, the heat transfer increases with increasing surface roughness. Moreover, the location of the maximum peak in HTC is always after the mixing section throat ($x \sim 272$ mm) and downstream of the diffuser shock (which is visible by the presence of wiggles in the HTC curves). This may be due to the strong mixing occurring after the supersonic shock. In any event, it is clear that even in the most conservative case (i.e., smooth wall and wall temperature equal to the condenser temperature), the ejector surface cannot be considered adiabatic. Indeed, values like those reported in Fig. 4.11 are typical of the heat transfer of liquids in forced convection. This effect may impact the accuracy of the numerical simulations by changing the starting position and shape of the transition process, as can be seen in Figs. 4.8 or 4.9.

In addition to these numerical considerations, the heat loss toward the environment is a feature that must be taken into account for a correct sizing of the condenser. Overall, the heat loss through the ejector external wall is always between 2 and 5 kW in our simulations. This can hardly be considered a negligible quantity as it represents the 2–4% of the total heat rejected at condenser.

As a concluding remark, this analysis highlights the importance of the manufacturing process on the performance of a supersonic ejector refrigerator. Friction losses are obviously neither the only nor the greatest source of losses inside the ejector. Nevertheless, with a relatively small economic outlay (i.e., by polishing the internal surfaces), significant advantage could be gained in terms of efficiency. In this respect, it is important to note the large size of the ejector under study, which is designed to produce around 40 kW_f of nominal cooling power. These dimensions are larger than those commonly found in the literature. Therefore, the impact of wall roughness on ejector efficiency should be expected to be even greater for systems with smaller size.

4.2 Condensing Ejectors

As seen in Chap. 3, design techniques for ejectors are still a matter of discussion. Ideal gas models are usually employed to easily obtain a first set of basic dimensions. However, in the case of ejectors operating with steam (or other condensing fluids), the ideal gas behavior is far from being physically consistent. In order to produce a more refined design, some studies have attempted to include real fluid behavior inside a thermodynamic, 1D model of the ejector (e.g., (Cardemil and Colle 2012)). However, this was made by postulating thermodynamic equilibrium conditions, which implies overlooking of any non-equilibrium effects, especially the condensation shock. As a consequence, the design is ultimately conducted by making use of empirical prescriptions or correlations of experimental data (e.g., (ESDU 1986)).

Although this method can provide for a suitable sizing of the main ejector dimensions, its empirical nature does not permit optimization of the ejector performances. Consequently, the use of multiphase CFD simulations becomes necessary in order to numerically evaluate the condensing ejector physics and to refine first-attempt designs obtained by approximated techniques.

4.2.1 Numerical Approaches

In the past decades, several methods have been devised to simulate wet-steam flows, with different levels of complexity and accuracy. The simplest and perhaps most used is the so-called single-fluid approach. This is basically a fully Eulerian scheme that assumes the liquid phase to be uniformly dispersed within the vapor volume. The mass, momentum, and energy conservation equations are written for a homogeneous mixture, and two additional equations describe the conservation of the droplets number and liquid mass within the computational domain. This method is commonly exploited by commercial codes (e.g., ANSYS Fluent or CFX) and has been used by many research teams (Ariafar et al 2015; Wang et al. 2012; Giacomelli et al. 2016; Mazzelli et al. 2016).

A second method, called the "two-fluid" approach, is similar to the previous one with the exception that the conservation equations are solved for the two phases separately. This method can result in greater accuracy thanks to the possibility of describing the energy and force exchange between the phases. On the other hand, modeling the interphase interaction is a complex task that can lead to larger uncertainties than with the use of simpler models. This may be especially true in ejector applications, where the wet-steam flow must traverse shear layer and shock regions. The two-fluid method has been investigated by many authors, who have either adapted commercial codes (Gerber and Kermani 2004) or developed in-house solvers (Dykas and Wróblewski 2011, 2013).

The two approaches described above have in common the assumption of a monodispersed population of droplets, i.e., it is assumed that the droplet distribution can be adequately represented by a single equivalent size (e.g., the average droplet diameter). An intrinsic limit of this method is that it can't reproduce multiple nucleation phenomena. If nucleation occurs at other distinct locations in the flow path (e.g., in a subsequent flow expansion), additional sets of equations are needed to represent each droplet population formed in a specific region. Gerber et al. (2007) first developed this method called the "multi-fluid" approach, which proved to be useful for simulating multiple stages of steam turbines (Gerber 2008; Starzmann and Casey 2010; Starzmann et al. 2016).

In addition to the multi-fluid approach, there are at least two further methods capable to account for the polydispersed nature of wet-steam flows: the Population Balance Methods (PBM) and the Eulerian/Lagrangian approach. The population balance methods are fully Eulerian schemes that describe the evolution and transport of the spectrum of droplet size along the flow domain. They do so by solving the liquid phase conservation equations in the form of the size distribution and a number of its low order moments. White (2003) and Hughes et al. (2016) provide a detailed review of these methods.

In the Eulerian/Lagrangian schemes, the complete spectrum is represented by homogeneous groups of liquid particles that are generated during a prescribed time-step and are tracked along the flow trajectories. This last approach is perhaps the most accurate in modeling the droplet spectrum (Gerber 2002) and is sometimes considered as a reference value to test more approximate schemes (Hughes et al. 2016). Different examples of this technique are provided by (Young 1992; Gerber 2002; Sasao et al. 2013).

Despite this plethora of different modeling approaches, to date, commercial CFD codes generally feature models based solely on the single-fluid approach. Moreover, although these codes dispense from developing complex in-house solvers, the use of built-in models does not allow freedom in the change of model parameters and settings. In order to overcome this limitation, it is possible to implement user-defined wet-steam model in suitable commercial CFD codes. In what follows, we present the governing equations of a scheme based on the single-fluid approach. Details of more complex schemes can be found in the literature cited above.

The governing equations for a single-fluid model are written for a homogeneous mixture of water vapor and droplets, and they assume the form of the conventional Navier-Stokes equations for compressible flows:

$$\frac{\partial \rho_m}{\partial t} + \frac{\partial \rho_m u_{mj}}{\partial x_j} = 0$$

$$\frac{\partial \rho_m u_{mi}}{\partial t} + \frac{\partial \rho_m u_{mi} u_{mj}}{\partial x_j} = -\frac{\partial p}{\partial x_j} + \frac{\partial \tau_{ij\_eff}}{\partial x_j} \qquad (4.3)$$

$$\frac{\partial \rho_m E_m}{\partial t} + \frac{\partial \rho_m u_{mj} H_m}{\partial x_j} = \frac{\partial q_{j\_eff}}{\partial x_j} + \frac{\partial u_{mi} \tau_{ij\_eff}}{\partial x_j}$$

In Eq. 4.3, the properties of the mixture are described by means of mass or volume-weighted averages:

$$\varsigma_m = \beta \varsigma_l + (1 - \beta)\varsigma_v$$
$$\chi_m = \alpha_l \chi_l + (1 - \alpha_l)\chi_v \qquad (4.4)$$

where ς_m represents mixture thermodynamic properties like enthalpy, entropy, total energy, etc.; χ_m is the mixture density, molecular viscosity, or thermal conductivity; β is the liquid mass fraction; and α_l is the liquid volume fraction. The connection between these last two quantities is straightforward:

$$\beta = \frac{m_l}{m_l + m_v} = \frac{\alpha_l \rho_l}{\alpha_l \rho_l + (1 - \alpha_l)\rho_v} \qquad (4.5)$$

The evaluation of the mixture speed of sound requires special considerations (Brennen 1995) and is calculated here by means of an harmonic average:

$$a = \sqrt{\frac{1}{(\alpha_l \rho_l + \alpha_v \rho_v)\left(\frac{\alpha_l}{\rho_l a_l^2} + \frac{\alpha_v}{\rho_v a_v^2}\right)}} \qquad (4.6)$$

The transport equation for the mixture is coupled with the two equations for the conservation of the liquid mass and the droplets number:

$$\frac{\partial \rho_m n}{\partial t} + \frac{\partial \rho_m u_{mj} n}{\partial x_j} = \alpha_v J_c \qquad (4.7)$$

$$\frac{\partial \rho_l \alpha_L}{\partial t} + \frac{\partial \rho_l u_{mj} \alpha_L}{\partial x_j} = \Gamma \qquad (4.8)$$

where n is the number of droplets per unit mass of the mixture, and it is assumed that the two phases move at the same speed (no-slip condition).

The term J in Eq. 4.7 is the nucleation rate per unit volume of vapor and is expressed here through the classical nucleation theory modified with the Kantrowitz non-isothermal correction (Kantrovitz 1951) (see Chap. 2):

$$
J_c = \frac{q_C}{(1 + \xi)} \frac{\rho_v^2}{\rho_1} \left(\frac{2\sigma}{\pi m^3} \right)^{1/2} exp \left(-\frac{\Delta G_c}{k_b T_v} \right)
$$
$$
\xi = q_C \frac{2(\gamma - 1)}{(\gamma + 1)} \frac{h_{lv}}{RT_v} \left(\frac{h_{lv}}{RT_v} - \frac{1}{2} \right)
$$

(4.9)

where the Gibbs free energy needed to form a stable liquid cluster ΔGb^* is given by:

$$
\Delta G_c = \frac{4}{3} \pi r_c^2 \sigma = \frac{16}{3} \frac{\pi \sigma^3}{(\rho_1 RT_v \cdot \ln \varphi_{ss})^2}
$$

(4.10)

Equation 4.9 gives the rate at which liquid nuclei spontaneously form within the vapor stream.

In order to close the set of governing equations, it is necessary to provide the law for the liquid mass generation rate per unit volume of mixture, Γ, in Eq. 4.8. This quantity stems from two different sources:

$$
\Gamma = \Gamma_{nuc} + \Gamma_{grow} = \alpha_v m_d^* J + \rho_m n \frac{dm_d}{dt}
$$

(4.11)

where m_d is the mass of a generic liquid droplet and m_d^* is its value when the liquid nucleus first forms. By assuming a spherical shape for all liquid droplets, these are given by:

$$
m_d^* = \frac{4}{3} \pi \rho_1 r_c^3
$$
$$
m_d = \frac{4}{3} \pi \rho_1 r_d^3 = \frac{\rho_1 \alpha_1}{\rho_m n}
$$

(4.12)

The first of the two terms in the RHS of Eq. 4.11 describes the mass generated from freshly nucleated droplets. This term is significant only in the first stages of the condensation process, and it gets rapidly overtaken by the second term, Γ_{grow}, which represents the droplet growth.

In this model we use the formulation derived by Hill (1966) and described in Chap. 2:

$$
\frac{dr_d}{dt} = \frac{p_v}{\rho_1 h_{lv} \sqrt{2\pi RT_v}} \frac{c_p + c_v}{2} \cdot (T_s(p_v) - T_v)
$$

(4.13)

Equations from 4.3, 4.4, 4.5, 4.6, 4.7, 4.8, 4.9, 4.10, 4.11, 4.12, and 4.13 form a closed system of equations that can be solved as long as the vapor and liquid equations of state and thermodynamic properties are provided.

In this regard, calculations of the non-equilibrium phase change of steam necessarily requires the description of the fluid properties in metastable conditions, meaning that common tabulated properties cannot be used for this purpose. Unfortunately, there is a serious lack of experimental data for the properties of steam in supercooled conditions, which is regularly testified by reports of the International Association for the Properties of Water and Steam (IAPWS 2011). Consequently, it is necessary to extrapolate a generic equation of state outside its normal range of validity in order to describe metastable states within the saturation curve.

In the present model, the steam properties are calculated following the work of Young (1988) who derived a virial equation of state truncated at the third term of the expansion:

$$p = \rho_v R T_v \cdot \left(1 + B\rho_v + C\rho_v^2\right) \tag{4.14}$$

where B and C are the second and third virial coefficients. These are function of the temperature, and their expressions were calibrated to match steam data in the range between 273.16 and 1073 K. Moreover, formulations for the enthalpy, entropy, and specific heats are derived from the virial equations based on a procedure described by Young (1988). The steam thermal conductivity and dynamic viscosity are given by low order polynomial functions of the vapor temperature obtained from interpolation of NIST dataset (Lemmon et al. 2013). The liquid phase properties (viz., liquid density, specific heat capacity, thermal conductivity, and viscosity) are calculated assuming saturation conditions and are again expressed through empirical correlations obtained from NIST (Lemmon et al. 2013). Finally, the water surface tension is a function of the sole temperature and is expressed following Young (1982).The present model has been tested on an ejector test case that will be described next.

4.2.2 Ejector Flow

As was shown in Chap. 2, a good level of accuracy is generally achieved for nozzle flows. However, this may not hold true in ejector applications where the interaction between droplets, shocks, and shear layers may introduce many unpredictable effects. In this respect, most of the numerical studies on condensing steam ejectors have been accomplished through single-phase, ideal-gas simulations (e.g., (Sriveerakul et al. 2007)), and very few examples of CFD using wet-steam models exist (see, for instance, (Ariafar et al. 2015; Wang et al. 2012)).

In order to accurately validate numerical simulation on ejector applications, the comparison with experimental data should be made by considering both global and local parameters. Unfortunately, articles reporting these types of data appear to be very few. In particular, the study of Chunnanond and Aphornratana (2004) and the subsequent work of Sriveerakul et al. (2007) provide results for the entrainment ratios and pressure profiles along the ejector walls. However, no information on the separate primary and secondary mass flow rates is given, which makes difficult the assessment of numerical results.

Table 4.1 Summary of ejector boundary conditions

Stream	Total temperature [K]	Total pressure [kPa]
Motive	403	270
Suction	287	1.6
Discharge		From 4.2 to 7.5

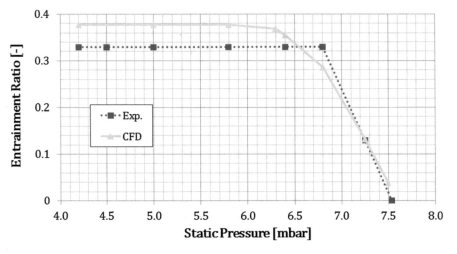

Fig. 4.12 Comparison of experimental (Al-Doori 2013) and numerical ER

In this section, the presented model is compared against data from the supersonic steam ejector studied by Al-Doori (2013) and Ariafar et al. (2015). The set of boundary conditions analyzed are summarized in Table 4.1.

The described model has been implemented within the commercial CFD package ANSYS Fluent v18.0 (ANSYS Inc. 2016). A k-ω SST turbulence model is selected for all the simulations because of the specific calibration for transonic applications (Menter 1994). In addition, due to the high Mach reached within the ejector mixing chamber, two additional UDFs were built to endow the turbulence model equations with the correction for compressible mixing layer. The computational domain used in this work is that illustrated in Fig. 4.1.

Figure 4.12 shows the comparison between the experimental and numerical ER curve. The numerical simulations produce a higher value of ER at on-design with a percent difference of about 14%. Moreover, the CFD results somewhat anticipate and smoothen the transition toward the off-design regime.

Despite the large discrepancy, Fig. 4.13 illustrates that when results for the motive and suction flows are analyzed separately, differences are always smaller than the corresponding value of ER. This is due to a summation of the errors when dividing the two quantities and should caution authors from reporting results in terms of ER alone. In particular, the greatest error is achieved for the data of the suction flow rates, with a percent error of about 7% at on-design, whereas the

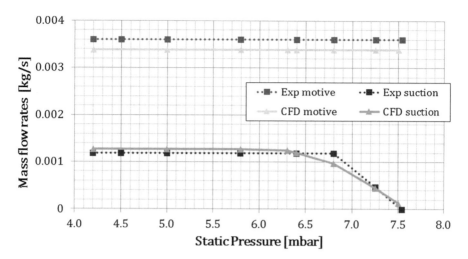

Fig. 4.13 Comparison of experimental (Al-Doori 2013) and numerical mass flow rates

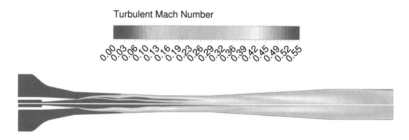

Fig. 4.14 Ma_t contour for the cases with compressibility correction (bottom) and without compressibility correction (top) ($P_{out} = 4.2$ kPa)

discrepancy for the motive flow rate is slightly less than 6%. For these two quantities, Al-Doori (2013) reports uncertainties of 0.6% and 1–2% for the primary and secondary mass flow rate, respectively.

In order to predict the entrainment rate of the suction flow with sufficient accuracy, it was necessary to account for the impact of compressibility effects on the mixing layer.

Figure 4.14 shows a comparison of the turbulent Mach number field between a simulation with the compressibility correction active and one without correction. As can be seen, Ma_t reaches very high levels within the mixing layer and downstream of the shock in the diffuser. In particular, a substantial part of the mixing layer presents $Ma_t > 0.25$, which is the threshold for compressibility to have any impact on the mixing layer (Wilcox 2006). The use of the correction limits these peak values and reduces the mixing layer spreading rate. In turn, the suppression of the spreading rate would result in a drop of the suction flow rate of nearly 17%, leading to a difference with the experimental mass flow rate of about 22% (as opposed to the 7% obtained with the correction).

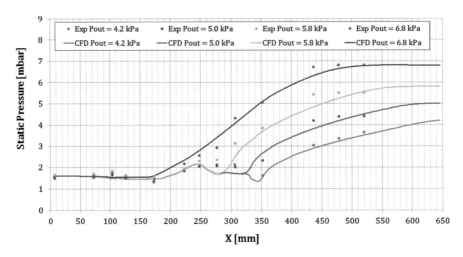

Fig. 4.15 Comparison of experimental (Al-Doori 2013) and numerical wall pressure profiles

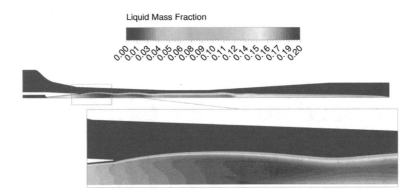

Fig. 4.16 Liquid mass fraction contour ($P_{out} = 4.2$ kPa)

Figure 4.15 presents the results for the pressure profiles along the ejector wall. Overall, the comparison with experimental data shows a good agreement, especially along the mixing chamber and entrance regions. The accord with experiments decreases as the flow approaches the diffuser, where large recirculations are found. These are notoriously hard to capture by common two-equation turbulence models and could partly account for the differences with experiments.

Furthermore, the discrepancies in the prediction of mass flow rates may also contribute to the discrepancies of the pressure trends. This is because the energy budget of the mixed stream is altered due to the different proportions of motive and suction flows. In turn, this may change the positions of the shock within the diffuser as well as the pressure trends at the wall.

Figure 4.16 shows the contour of the liquid mass fraction for the case with $P_{out} = 4.2$ kPa. As can be seen, the condensed phase reaches value up to 20% of

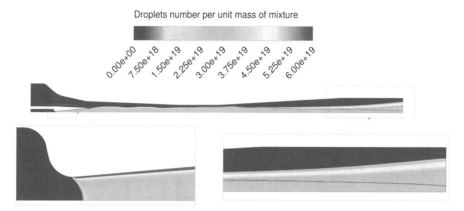

Fig. 4.17 Contour of the droplet number per unit mass of mixture (in purple is the line where the liquid mass fraction is zero; case with $P_{out} = 4.2$ kPa)

the total mass, with peak levels in the region downstream of the nozzle exit plane. This is due to the further acceleration caused by the primary jet under-expansion. The absence of any superheating at the motive inlet exacerbates this problem that can lead, in some extreme cases, to the formation of ice inside the ejector (as discussed later).

Figure 4.16 further illustrates that the liquid mass fraction goes to zero at the nozzle wall, due to the heat recovery induced by the fluid deceleration and viscous dissipation within the boundary layer.

Figure 4.17 shows the contour of the droplet number per unit mass of mixture, n. The contour reveals the presence of a radial distribution of the droplet number. This stems from the significant curvature of the nozzle profile at the throat, which induces a region of low pressure near the wall and causes a stronger nucleation. The nucleated droplets are then convected along streamlines, and the radial distribution persists almost unaltered till the outlet.

The figure displays also the line representing the boundary where the liquid mass fraction is zero. In general, the droplet number contour follows closely that of the liquid mass fraction all along the ejector. This is a direct consequence of the assumption of equal velocity between the phases. Yet, the comparison reveals also that near the ejector outlet, where the condensed mass evaporates completely, the liquid droplets survive in the form of nuclei with zero mass and volume. The reason for this error is to be found in the absence of a "nucleation sink" in the droplet transport equation, Eq. 4.7.

This issue is common to most of the single-fluid approaches and prevents their use in applications where secondary nucleation occurs (e.g., multistage steam turbine cascades). Although in principle it could be possible to add a sink term to the droplet number equation, in practice, the assumptions implicit in the single-fluid approach would anyhow lead to significant approximations (e.g., the sizes of droplets originating from different nucleation sites are averaged out, (Hughes et al. 2016)). Therefore, these complex cases require the adoption of more advanced approaches, such as the multi-fluid model or Lagrangian schemes that naturally account for the removal of droplets from the computational domain and can provide more accurate results than the single-fluid method used in this work.

4.2.3   Modeling Assumptions

In this section, we review certain specific limits connected with the developed model and discuss possible implications deriving from them. In doing so, we focus on those assumptions that seem particularly restrictive with respect to ejector applications.

A first limit of the model relates with the droplet growth regime. As detailed in Chap. 2, the growth rate formulation for a liquid droplet is calculated differently depending on the value of the Knudsen number:

$$Kn = \frac{l}{2r_{\mathrm{d}}} \tag{4.15}$$

Specifically, Hill's droplet growth law adopted in the present model is only valid for the free molecular regime, $Kn \gg 1$. Therefore, some doubts may arise about its applicability to ejector flows.

Figure 4.18 shows the trend of the Knudsen number along the ejector axis, for one of the simulated cases (but all cases present similar trends). Clearly, Kn is always greater than one except for a very small region near the primary nozzle throat. In this zone, the vapor temperature is still high, and the mean free path is of the same order of magnitude of the droplet diameters (around 10^{-9} m). Nevertheless, the assumption of free molecular regime appears to be appropriate for this study.

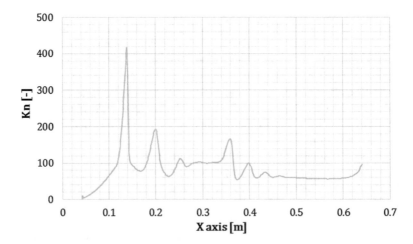

Fig. 4.18 Kn along the ejector axis (case with $P_{\mathrm{out}} = 4.2$ kPa)

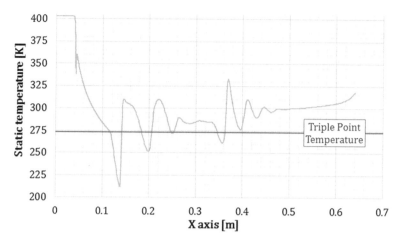

Fig. 4.19 Temperature trend along the ejector axis, in red the line corresponding to the triple-point temperature (case with $P_{out} = 4.2$ kPa)

A further key aspect of concern for steam ejector studies is related to the very low temperature levels attained by the expanding stream. This problem is particularly critical for ejector refrigeration applications, where efficiency considerations impose the use of low or no level of superheating at the motive inlet (as in this study). Consequently, the motive jet can reach temperature levels that go well below the triple point, causing the possible appearance of ice.

Figure 4.19 shows the mixture temperature trend along the ejector axis. Clearly, the temperature goes well below the limit of the triple point, and the presence of ice cannot be excluded (especially downstream of the nozzle exit plane where the mixture temperature reaches values close to 210 K). Nevertheless, ice crystal formation, in much the same way as for droplet nucleation, is fundamentally a time-dependent phenomenon, and some degrees of supercooling usually exists before the water vapor or liquid starts to solidify.

In particular, experiments in cloud chambers with pure water vapor indicates that the homogeneous nucleation of ice usually occurs with around 30–40 K of supercooling (Pruppacher 1995). By contrast, recent investigations in supersonic nozzles have shown that for the high cooling rates and small cluster sizes that are achieved in these devices, the supercooling can be as high as 90 K (i.e., supercooled water temperatures of nearly 190 K) (Wölk et al. 2013). However, these tests were conducted with ultrapure water and may not be directly applicable to the present study (this is because impurities in the water droplets or vapor stream can greatly anticipate crystal formation). As a result, the presence of ice cannot be excluded in the steam ejector under investigation.

The appearance of water ice crystal may induce substantial modifications to the mixture flow behavior. For instance, the change in the latent heat release (from the value of condensation to that of solidification) may modify the nozzle Mach and pressure profiles. Moreover, phenomena such as crystals agglomeration and

deposition may be important and could lead to modifications in the ejector geometrical profiles. On the other hand, the presence of shocks immediately after every
steam expansion may lead to the sudden melting of the ice, limiting its impact on the
global flow dynamics. In this case, numerical simulations may still incur in significant discrepancies due to the uncertain extrapolation of the supercooled water
properties below the triple-point temperature (most of these, including viscosity,
specific heat, surface tension, and others, present exponential variations with
decreasing temperatures (Pruppacher 1995)). In view of these many aspects, it is
important that future experimental investigations properly address the analysis of ice
formation inside steam ejectors.

4.3   Concluding Remarks: The Need for Validation

In this final section, we would like to discuss briefly the needs for a proper validation
approach in the field of ejector studies. Indeed, it is the authors' opinion that some
deficiencies in the assessment of the numerical error and physical model approximations contribute to the significant scatter of CFD results in this field.

   In general, it is useful to distinguish between two different sources of error: those
connected to the numerical approach (spatial discretization errors, equations trunca-
tion errors, convergence errors) and uncertainties stemming from physical modeling
(a precise distinction between error and uncertainties is provided by Oberkampf and
Trucano (2002)).

   Although the assessment of the numerical error is a well-established practice,
with most studies presenting details of grid sensitivity, convergence criteria,
discretization order, etc., the analysis of physical modeling approximations is less
common, and it prevents, in some cases, to reach a clear interpretation of the results.

   In particular, it seems that in most of the literature on ejectors, the analyses are
limited to evaluation of the impact of turbulence modeling. As stated at the beginning of this chapter, this should not be considered the only cause of possible
inconsistencies with experiments. For instance, the large discrepancies in ER that
are seen at off-design mainly arise from the failure of the CFD models to reproduce
the transition toward the unchoked regime. The correct prediction of this process
requires accurate evaluation of the kinetic energy dissipation across shocks, mixing
layers, and boundary layers. In turn, this necessitates the evaluation of the discussed
effects of compressibility, wall friction, multiphase phenomena, etc. Generally, all
these phenomena are highly coupled in ejector flows, and each of these contribute to
the global approximation of the physical model.

   As a consequence, some efforts should be done in order to properly validate the
numerical approach. At best, this entails a hierarchical approach to decouple the
effects of each source of error (for details see (Oberkampf and Trucano 2002)). At
least, a sensitivity analysis should be performed on each modeled phenomenon to
understand the relative impact on the solution.

This issue of physical model validation is particularly relevant to the field of wet-steam simulations. Very recently, the research group of the University of Cambridge has organized an interesting numerical challenge with the purpose of reviewing the ability of computational methods to predict condensing steam flows (Starzmann et al. 2016). Despite some physiological spreading of the results, the project has shown a general good agreement between the different numerical schemes and the experimental data. If, from the one hand, this is a pleasing confirmation of the significant modeling progress made, on the other hand, it appears quite clearly a sort of general convergence toward the use of a restricted number of well-known experimental test case and modeling settings (the model presented in previous sections does not represent an exception in this regard).

In particular, among the several calibration constants and empirical correlations, the computation of the surface tension has a chief impact on the phase change, since its expression, raised to the third power, appears within the exponential term of the nucleation rate equation (see Eq. 4.9). Therefore, small changes in its value can strongly influence the final number of nucleated droplets as well as the whole flow conditions downstream of the nucleation region.

Unfortunately, a general expression for the surface tension of nanodroplets is still to be found, and most of the previous "calibration work" has been accomplished with the assumption that the surface tension is solely dependent on temperature, whereas any possible effects of surface curvature (which are most likely present for very small droplets) are neglected. In view of this, any change in the formulation for the surface tension, although physically reasonable, could produce the paradoxical effect of a reduction in the level of agreement between simulations and experiments.

This fact may provide a possible explanation for the aforementioned convergence to similar numerical settings and implies that any substantial change on a "first-order-impacting" parameter should be followed by a recalibration of all other model constants and coefficients. This is, needless to say, a difficult and tedious task. Moreover, the uncertainty of experimental data, especially that connected to the measurement of droplets spectra and metastable steam properties, adds to these problems by extending the domains of valid solutions and making the validation of the condensation model even more uncertain (Wróblewski et al. 2009).

Consequently, in consideration of these many issues, the attempt to achieve a perfect matching with experiments seems to be a fruitless effort, at least until new sets of test cases with reliable data on surface tension, nucleated droplets spectra, and metastable fluid properties will be available.

References

Adams, T., Grant, C., & Watson, H. (2012). A simple algorithm to relate measured surface roughness to equivalent sand-grain roughness. *International Journal of Mechanical and Mechatronics Engineering, 1*, 66–71.

Al-Doori, G. (2013). *Investigation of refrigeration system steam ejector performance through experiments and computational simulations.* s.l.:Ph.D. thesis, University of Southern Queensland.

ANSYS Inc. (2016). *ANSYS fluent theory guide.* Canonsburg, release 18.0.

Ariafar, K., Buttsworth, D., & Al-Doori, G. (2015). Effect of mixing on the performance of wet steam ejectors. *Energy, 93,* 2030–2041.

Bartosiewicz, Y., Aidoun, Z., Desevaux, P., & Mercadier, Y. (2005). Numerical and experimental investigations on supersonic ejectors. *International Journal of Heat and Fluid Flow, 26,* 56–70.

Besagni, G., & Inzoli, F. (2017). Computational fluid-dynamics modeling of supersonic ejectors: Screening of turbulence modeling approaches. *Applied Thermal Engineering, 117,* 122–144.

Bouhanguel, A., Desevaux, P., & Gavignet, E. (2009). 3D CFD simulation of a supersonic air ejector. International seminar on ejector/jet-pump technology and application, Louvain-la-Neuve.

Brennen, C. (1995). *Cavitation and bubble dynamics.* s.l.: Oxford University Press.

Cardemil, J., & Colle, S. (2012). A general model for evaluation of vapor ejectors performance for application in refrigeration. *Energy Conversion and Management, 64,* 79–86.

Chunnanond, K., & Aphornratana, S. (2004). An experimental investigation of a steam ejector. *Applied Thermal Engineering, 24,* 311–322.

Dykas, S., & Wróblewski, W. (2011). Single- and two-fluid models for steam condensing flow modeling. *International Journal of Multiphase Flow, 37,* 1245–1253.

Dykas, S., & Wróblewski, W. (2013). Two-fluid model for prediction of wet steam transonic flow. *International Journal of Heat and Mass Transfer, 60,* 88–94.

ESDU. (1986). *Ejectors and jet pumps, data item 86030.* London: ESDU International Ltd.

García del Valle, J., Sierra-Pallares, J., Garcia Carrascal, P., & Castro Ruiz, F. (2015). An experimental and computational study of the flow pattern in a refrigerant ejector. Validation of turbulence models and real-gas effects. *Applied Thermal Engineering, 89,* 795–781.

Gerber, A. (2002). Two-phase Eulerian/Lagrangian model for nucleating steam flow. *ASME Journal of Fluids Engineering, 124,* 465–475.

Gerber, A. (2008). Inhomogeneous multifluid model for prediction of nonequilibrium phase transition and droplet dynamics. *Journal of Fluids Engineering, 130,* 031402-1–031402-11.

Gerber, A., & Kermani, M. (2004). A pressure based Eulerian-Eulerian multiphase model for non-equilibrium condensation in transonic steam flow. *International Journal of Heat and Mass Transfer, 47,* 2217–2231.

Gerber, A., et al. (2007). Predictions of non-equilibrium phase transition in a model low-pressure steam turbine. *Proceedings of the Institution of Mechanical Engineers, Part A: Journal of Power and Energy, 221,* 825–835.

Giacomelli, F., Biferi, G., Mazzelli, F., & Milazzo, A. (2016). CFD modeling of supersonic condensation inside a steam ejector. *Energy Procedia, 101,* 1224–1231.

Hemidi, A., et al. (2009a). CFD analysis of a supersonic air ejector. Part II: Relation between global operation and local flow features. *Applied Thermal Engineering, 29,* 2990–2998.

Hemidi, A., et al. (2009b). CFD analysis of a supersonic air ejector. Part I: Experimental validation of single-phase and two-phase operation. *Applied Thermal Engineering, 29,* 1523–1531.

Hill, P. G. (1966). Condensation of water vapour during supersonic expansion in nozzles. *Journal of Fluid Mechanics, 25*(3), 593–620.

Hirsch, C. (2007). *Numerical computation of internal and external flows* (Vol. 1, 2nd ed.). Amsterdam: Butterworth-Heinemann, Elsevier.

Hughes, F., Starzmann, J., White, A., & Young, J. B. (2016). A comparison of Modeling techniques for Polydispersed droplet spectra in steam turbines. *Journal of Engineering for Gas Turbines and Power, 138,* 042603.

IAPWS. (2011). *Thermophysical properties of metastable steam and homogeneous nucleation.* s.l.: International Assotiation for the Properties of Water and Steam.

Kantrovitz, A. (1951). Nucleation in very rapid vapour expansions. *Journal of Chemical Physics, 19,* 1097–1100.

Keshtiban, I., Belblidia, F. & Webster, M. (2004). *Compressible flow solvers for low mach number flows – A review*. s.l.: Institute of Non-Newtonian Fluid Mechanics, University of Wales, UK, Report No.: CSR 2-2004.

Launder, B., & Spalding, D. (1972). *Lectures in mathematical models of turbulence*. London: Academic Press.

Lemmon, E., Huber, M. & McLinden, M. (2013). *NIST standard reference database 23: Reference fluid thermodynamic and transport properties-REFPROP, Version 9.1*. s.l.: National Institute of Standards and Technology.

Mazzelli, F., & Milazzo, A. (2015). Performance analysis of a supersonic ejector cycle. *International Journal of Refrigeration, 49*, 79–92.

Mazzelli, F., Little, A. B., Garimella, S., & Bartosiewicz, Y. (2015). Computational and experimental analysis of supersonic air ejector: Turbulence modeling and assessment of 3D effects. *International Journal of Heat and Fluid Flow, 56*, 305–316.

Mazzelli, F., et al. (2016). *Condensation in supersonic steam ejectors: Comparison of theoretical and numerical models, International Conference on Multiphase Flow, ICMF*. Florence, s.n.

Menter, F. R. (1994). Two-equation Eddy-viscosity turbulence models for engineering applications. *AIAA Journal, 32*(8), 1598–1605.

Miettinen, A., & Siikonen, T. (2015). Application of pressure- and density-based methods for different flow speed. *International Journal for Numerical Methods in Fluids, 79*, 243–267.

Milazzo, A., & Mazzelli, F. (2017). Future perspectives in ejector refrigeration. *Applied Thermal Engineering, 121*, 344–350.

Moore, J. G., & Moore, J. (1999). Realizability in two-equation turbulence models. AIAA Paper 99-3779, 1–11.

Oberkampf, W., & Trucano, T. (2002). Verification and validation in computational fluid dynamics. *Progress in Aerospace Sciences, 38*, 209–272.

Patankar, S. (1980). *Numerical heat transfer and fluid flow*. s.l.:Taylor & Francis Group.

Pianthong, K., et al. (2007). Investigation and improvement of ejector refrigeration system using computational fluid dynamics technique. *Energy Conversion and Management, 48*(9), 2556–2564.

Pruppacher, H. R. (1995). A new look at homogeneous ice nucleation in supercooled water drops. *Journal of the Atmospheric Sciences, 52*, 1924–1933.

Reynolds, W. (1987). *Fundamentals of turbulence for turbulence modeling and simulation*. s.l.: Von Karman Institute.

Roache, P. (1997). Quantification of uncertainty in computational fluid dynamics. *Annual Review of Fluid Mechanics, 29*, 123–160.

Rodi, W. (1991). *Experience with two-layer models combining the k-E model with a one-equation model near the wall*, 29th Aerospace Sciences Meeting. Reno, AIAA.

Sarkar, S., & Balakrishnan, L. (1990). *Application of a Reynolds-stress turbulence model to the compressible shear layer*. s.l.: ICASE Report 90-18NASA CR 182002.

Sasao, Y., et al. (2013). *Eulerian-Lagrangian numerical simulation of wet steam flow through multi-stage steam turbine*. June 3–7, San Antonio, Texas, USA, Proceedings of ASME Turbo Expo 2013: Turbine Technical Conference and Exposition.

Shih, T.-H., et al. (1995). A new k-e Eddy viscosity model for high Reynolds number turbulent flows. *Computers Fluids, 24*(3), 227–238.

Shu, C.-W. (2009). High order weighted essentially non-oscillatory schemes for convection dominated problems. *SIAM Review, 51*, 82–126.

Sriveerakul, T., Aphornratana, S., & Chunnanond, K. (2007). Performance prediction of steam ejector using computational fluid: Part 1. Validation of the CFD results. *International Journal of Thermal Sciences, 46*, 812–822.

Starzmann, J. & Casey, M. (2010). *Non-equilibrium condensation effects on the flow field and the performance of a low pressure steam turbine*. June 14–18, Glasgow, UK, Proceedings of ASME Turbo Expo 2010: Power for Land, Sea and Air.

Starzmann, J., et al. (2016). *Results of the international wet steam modelling project*. Prague: s.n.

Taylor, J., Carrano, A., & Kandlikar, S. (2006). Characterization of the effect of surface roughness and texture on fluid flow—Past, present, and future. *International Journal of Thermal Sciences, 45*, 962–968.

Tennekes, H. & Lumley, J. (1972). *A first course in turbulence.* s.l.:MIT Press.

Toro, E. (2009). *Riemann solvers and numerical methods for fluid dynamics, a practical introduction.* Berlin/Heidelberg: Springer.

Van Leer, B. (1979). Towards the ultimate conservative difference scheme. V: A second order sequel to Godunov's Method. *Journal of Computational Physics, 32*, 101–136.

Wang, X., Lei, H. J., Dong, J. L., & Tu, J. Y. (2012). The spontaneously condensing phenomena in a steam-jet pump and its influence. *International Journal of Heat and Mass Transfer, 55*, 4682–4687.

White, A. (2003). A comparison of modelling methods for polydispersed wet-steam flow. *International Journal for Numerical Methods in Engineering, 57*, 819–834.

Wilcox, D. (1992). Dilatation–dissipation corrections for advanced turbulence models. *AIAA Journal, (11)*, 2639–2646.

Wilcox, D. (2006). *Turbulence Modeling for CFD.* La Canada: DCW Industries.

Wolfshtein, M. (1969). The velocity and temperature distribution in one-dimensional flow with turbulence augmentation and pressure gradient. *International Journal of Heat and Mass Transfer, 12*(3), 301–318.

Wölk, J., Wyslouzil, B., & Strey, R. (2013). Homogeneous nucleation of water: From vapor to supercooled droplets to ice. *AIP Conference Proceedings, 1527*, 55–62.

Wróblewski, W., Dykas, S., & Gepert, A. (2009). Steam condensing flow modeling in turbine channels. *International Journal of Multiphase Flow, 35*, 498–506.

Young, J. B. (1982). The spontaneous condensation in supersonic nozzles. *PhysicoChemical Hydrodynamics, 3*(1), 57–82.

Young, J. B. (1988). An equation of state for steam for Turbomachinery and other flow calculations. *Journal of Engineering for Gas Turbines and Power, 110*(1), 1–7.

Young, J. (1992). Two-dimensional, nonequilibrium, wet-steam calculations for nozzle and turbine cascades. *Journal of Turbomachinery, 114*, 569–579.

Zagarola, M., & Smits, A. (1998). Mean-flow scaling of turbulent pipe flow. *Journal of Fluid Mechanics, 373*, 33–79.

Zhu, Y., & Jiang, P. (2014). Experimental and numerical investigation of the effect of shock wave characteristics on the ejector performance. *International Journal of Refrigeration, 40*, 31–42.

Chapter 5
Experimental Activity

5.1 Testing of Ejectors

The first obvious citation concerning a relevant experimental study on ejectors must be acknowledged to the work of Keenan et al. (1950). Their "one-dimensional"[1] method for analysis of jet pumps was calibrated and validated against experimental data gathered from an air ejector, so that the assumption of ideal gas is perfectly acceptable in this case. A notable care was dedicated to guarantee a symmetrical, purely axial flow of the two streams at inlet. Five pressure taps along the ejector were used to acquire the longitudinal pressure distribution. The primary nozzle was mounted on a device allowing for axial movement. Various geometries were tested, obtaining area ratios between the primary and the mixer throat from 4 to 100. Calibrated orifices were used for mass flow rate measurements. Several experimental results and comparisons with calculations were reported, the best known result being the superior performance of the "constant-pressure mixing ejector" against the "constant-area" one.

Other experimental data are reported in the fundamental work by Munday and Bagster (1977), both coming from the previous literature and from direct measurements on an Ingersoll Rand industrial steam ejector. Unfortunately, the experimental apparatus is not described, and the results are expressed in terms of kJ (cooling) per kg of motive steam, as usual in the industrial reporting.

More recently, a steam ejector was tested by Chen and Sun (1997). A diffuser featuring a throat diameter of 17.8 mm was tested with three primary nozzles, ranging from 1.4 to 5.8 mm diameter of the nozzle throat (i.e., area ratio from 9.42 to 162). Complete maps of critical entrainment ratio vs. critical back pressure are reported for Mach number 2.7 and 4.35 at primary nozzle exit, showing maximum

---

[1] As stated in Chap. 3, a method that describes a finite number of sections within the ejector should be considered as zero dimensional, while the qualification of one dimensional should be reserved to those methods that describe the flow evolution continuously along the ejector axis.

© Springer International Publishing AG, part of Springer Nature 2018
G. Grazzini et al., *Ejectors for Efficient Refrigeration*,
https://doi.org/10.1007/978-3-319-75244-0_5

entrainment ratio above 0.5 and 1.3, respectively. Maximum pressure lift was 3.67 when the primary inlet pressure was 2.7 bar, and the Mach number at primary nozzle exit was 4.35. The results are compared to other experimental data from the literature. Very low generator pressures were tested (80–200 torr, corresponding to saturation temperatures of 47–67 °C), showing that operation is possible, but very low discharge pressures are allowed. Therefore, the authors proposed a two-stage ejector in order to increase the pressure lift. Even a holographic interferogram coming from a previous research is shown, and some indications on the flow structure within the mixing zone are gathered (flow visualization techniques will be dealt with in Sect. 5.2).

The same research group produced the well-known results published by Huang et al. (1999) and extensively described in Chap. 3. Switching to R141b avoided condensation. Eleven different ejectors were tested, combining two nozzles with eight different diffusers. A further paper from Huang and Chang (1999) reports results from 15 different configurations and claims an entrainment ratio as high as 0.54 when saturation temperatures at generator, condenser, and evaporator are 84, 28, and 8 °C, respectively, while at 90, 32, and 8 °C, the entrainment ratio is lowered to 0.45.

Roughly in the same period, Aphornratana and Eames (1997) tested a steam ejector with movable primary nozzle. The effect of the nozzle position was extensively tested, showing a well-defined optimum for most operating conditions. For a given operating condition, the decrease of the distance between the nozzle exit and the diffuser throat is seen to decrease the entrainment ratio but increase the critical pressure. A complete operating map for evaporator temperatures between 5 and 10 °C and generator temperature between 120 and 140 °C is reported. Maximum COP was 0.278, at 10 °C evaporator temperature and 130 °C generator temperature.

A selection of experimental results found in the literature is shown in Table 5.1. Note that in many cases, authors do not provide complete performance maps or charts where the critical conditions are properly highlighted. In some cases, it is also unclear whether the pump work is considered in the COP calculations. Hence, some data may not represent the real maximum performance point of the cycle.

The two last columns in the table deserve some explanation. The second-law efficiency is simply calculated as the ratio between the measured COP and an ideal reference evaluated on the basis of the saturation temperature (see Chap. 3 for a discussion about efficiency definitions). This is a rather oversimplified approach, as it implies that the level of overheating/subcooling at inlets and outlets of any heat exchanger is negligible. Hence some biases may arise when comparing chillers with different levels of overheating/subcooling.

Furthermore, a consistent source of thermodynamic losses is always found within any heat exchanger. This type of available energy destruction, totally neglected when calculating the ideal COP based on the saturation temperatures, may even be greater than the one occurring inside the supersonic ejector and throttling valve.

Contrarily to COP data, the second-law efficiency levels in Table 5.1 are comparable, despite the quite different working temperatures. This fact suggests that η_{II} is mostly influenced by differences in terms of system design while being unaffected

Table 5.1 Best performance of different ejector chillers taken from the literature

Paper	Fluid	Cooling capacity	Superheating	Temperature range [°C]			Other	Best operating point (temperatures in °C)					
				T_{eva}	T_{gen}	T_{cond}		T_{eva}	T_{gen}	T_{cond}	COP	COP_{ideal}	η_{II}
Aphornratana and Eames (1997)	Water	1 kW	Not used	5:10	120:140	26:36	By excluding heat losses COP results 30% higher on average	10	120	28	0.39	3.68	0.106
Nguyen et al. (2001)	Water	7 kW	No	1.7:5.1	73:79	26.9	Passive system with 7 meter gravity head, few conditions tested	1.5	76.7	26.9	0.32	1.54	0.21
Chunnanond and Aphornratana (2004)	Water	3 kW	Not used	5:15	120:140	24:39	No effect detected with superheating	10	120	28	0.49	3.68	0.133
Selvaraju and Mani (2006)	R134a	0.5 kW	No	2:12	65:90	27:37	Six geometry tested, NXP varies with different nozzles	12.5	70	30	0.49	1.90	0.258
Eames et al. (2007)	R245fa	4 kW	No	10:15	110:120	32:40	Recuperative heat exchanger, two nozzles tested, CRMC diffuser design	15	110	33.5	0.69	3.11	0.222
Yapici and Yetisen (2007)	Freon R11	1 kW	No	0:16	90:102	27:34	Difficult to identify critical conditions from results	9.5	102	32	0.2	2.34	0.085
Yapici et al. (2008)	R123	1.5 kW	No	8:15	80:105	32:37	Six different geometries with different area ratios	10	90	34	0.35	1.82	0.192
Ma et al. (2010)	Water	5 kW	10 °C	6:13	84:96	17:36	Primary flow control through movable spindle	13	90	36	0.47	1.85	0.254
Mazzelli and Milazzo (2015)	R245fa	40 kW	No	5:10	90:100	26:36	Tests made in industrial environment	10	90	29.4	0.54	2.44	0.220

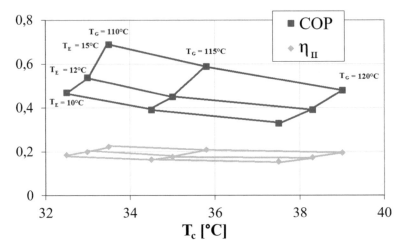

Fig. 5.1 Comparison between COP and η_{II} performance map, data from (Eames et al. 2007)

by changes in operating conditions. In particular, having neglected all the heat exchanger losses, this comparison gives information specifically on the quality of the supersonic ejector design, which is the main source of losses left.

To further prove this concept, Fig. 5.1 shows an experimental COP characteristics mapped into the equivalent second-law performance chart (data taken from (Eames et al. 2007)). The new map clearly shows a much attenuated dependence on the operating temperatures (the points almost lye on a horizontal line). Nevertheless, though lower, the sensitivity is still present, meaning that some conditions are more suited than other to a specific geometry (for instance, the primary nozzle is correctly expanded only for a specific set of pressure conditions; see Fig. 5.15, 5.16, and 5.17).

The last row in Table 5.1 refers to the ejector chiller built by Frigel S.p.A. on our design that will be described in Sect. 5.3. For a review of further experimental data, the reader may see the paper by Chen et al. (2013) or the recent paper by Besagni et al. (2016).

5.2 Flow Visualization

All experimental activities reported thus far can measure the ejector global performance in terms of flow rates and, at most, the distribution of static pressure along the ejector wall. The complex interaction between the motive and the entrained streams is only analyzed indirectly, on the basis of its effects on the global behavior. Flow visualization may give an insight on this interaction and on the detailed features of the flow. Obviously, the ejector must be extensively modified in order to allow optical access to the flow. For example, ejectors designed for flow visualization often have a rectangular cross section, with flat, transparent side walls. This fact may limit the applicability of the results gathered on such ejectors to the common axisymmetric geometry.

In addition to this, the visualization of compressible supersonic flow poses specific problems due to the high flow speeds and the presence of shock discontinuities. This is especially true for visualization techniques that use tracers, such as PIV or laser-scattering techniques. For these methods, the high speed and turbulence levels existing within ejectors require seeding the flow with very small particles that produce a poor signal intensity and contribute to the uncertainty of the measurement (Smits and Dussauge 2006). Moreover, the presence of shock waves introduces regions where particle tracers slip with respect to the surrounding flow (Scarano 2008); therefore, even with micron-sized particle, there is not a great assurance that the trajectory of the tracer does not depart from that of the flow, especially when the particles cross a shock-laden region.

The presence of severe variations of the local temperature and pressure may further result in unpredictable effects on the light emission properties of fluorescent tracers. For instance, it is known that acetone fluorescence depends on the local temperature and pressure of the flow. Disregarding these dependencies may introduce substantial errors and uncertainty on the quantitative data analysis (Karthick et al. 2017).

Finally, condensation of the air moisture content can produce additional difficulties due to the presence of a self-seeded tracer that must be distinguished from the one used for the measurement. On the other hand, this natural tracer can itself be used to analyze the mixing, shock, and turbulence structures of the flow, being the size of the water droplets ideal for tracking purposes (Bouhanguel et al. 2011).

The basic theory of flow visualization techniques will not be discussed here. For details the readers may refer to dedicated books like, for instance (Smits and Lim 2012; Settles 2001).

5.2.1 Shocks Structure

Thanks to the compressible nature of the flow, detailed studies of the shocks patterns within the ejector can be performed using diffraction techniques, i.e., shadowgraph or schlieren methods. These approaches are useful in that they are not intrusive and do not require the use of optical tracers. For example, the schlieren photographs (Carroll and Dutton 1990) reported in Sect. 2.2.2 are a good example of the capability of this technique when dealing with complex interactions between shock trains and boundary layers. Features like shock bifurcation, variable spacing between two consecutive shocks and many others, thanks to visual techniques become accessible for detailed analyses.

Early visualization studies of supersonic ejectors were carried out by Keenan et al. (1950) and Fabri and Siestrunck (1958) who have used schlieren methods to visualize the ejector shock patterns. More recent studies using the schlieren

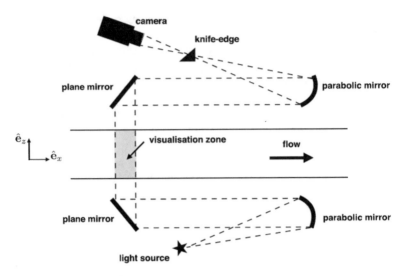

Fig. 5.2 Z-type schlieren setup (Lamberts et al. 2017)

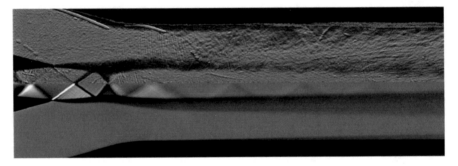

Fig. 5.3 Comparison between schlieren picture (top half) and numerical results (bottom half) (Reprinted with permission from Lamberts et al. (2017))

technique were performed, among others, by Matsuo et al. (1981 1982), Dvorak and Safarik (2005), Zhu and Jiang (2014a, b), and Rao and Jagadeesh (2014).

A recent work on schlieren visualization was carried out by Lamberts et al. (2017) who investigated the shock configuration as well as the momentum and exergy transfer inside an air ejector. The whole ejector was fitted with transparent walls and scanned by a halogen light source that was managed in a standard Z-type schlieren arrangement, as detailed in Fig. 5.2. The effect of scratches on the transparent wall was reduced by subtracting a "no-flow" image from the raw picture. A digital camera provided a spatial resolution of approximately 25 pixel/mm.

Figure 5.3 shows one result taken from Lamberts et al. (2017). The agreement between the density gradients highlighted by the schlieren image and those calculated by CFD seems quite satisfactory, with numerical results that reproduce well the position and inclination of the shocks.

Despite the general agreement, some slight discrepancies in the extension and intensity of the shock diamond pattern still exist as well as differences in the density gradient trends at wall. The authors suggest that these discrepancies may be partly due to differences in the type of averaging involved in the experimental and numerical approaches: the schlieren image superimposes the instantaneous flow features over the spanwise direction, while the CFD image is a time-averaged 2D slice.

Lamberts et al. (2017) further noted that for on-design operations, the shock structure at the nozzle exit was relatively stable, whereas the flow was found to be highly unstable with a jet flapping phenomenon at off-design conditions.

An interesting analysis of unsteady shock behavior was carried out by Rao and Jagadeesh (2014) who provided observations on the motion of the shock cells of an air ejector during off-design operations (referred by the author as mixed regime). A standard Z-type schlieren arrangement was used for this study together with a halogen light source and a high-speed camera with a frame rate of 2000 fps.

A typical result is shown in Fig. 5.4. The picture allows for the visualization of most of the flow characteristics features, including five shock cells and the mixing layer as well as the density fluctuations due to the turbulent structures downstream of the shock region.

To achieve a better understanding of the shock motion, Rao and Jagadeesh (2014) built a time trace of the density gradient intensity along the ejector midline. The resulting composite image is shown in Fig. 5.5. Clearly, five distinct bars can be observed, each showing an intensity peak moving back and forth. From the analysis of this trace, the authors could infer many interesting details about the nature of this oscillation:

- All shock cells move in the same direction at any given time, either forward or backward.
- The peak amplitude of the oscillation is about 5 mm, which is of the same order of the exit nozzle height, 6 mm.

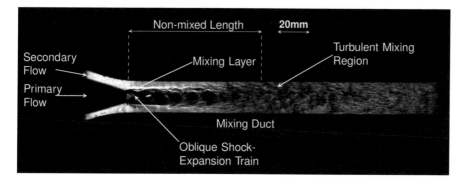

Fig. 5.4 An instantaneous schlieren image of the flow through the supersonic ejector (primary inlet pressure of 7.69 bar) (Reprinted with permission from Rao and Jagadeesh (2014))

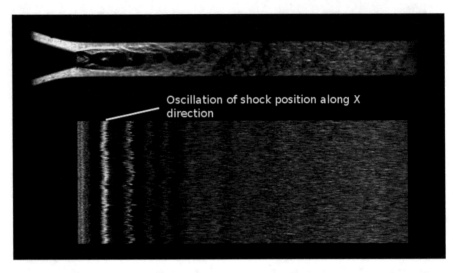

Fig. 5.5 A composite image showing the trace of the shock cell motion at the central line of the ejector (Reprinted with permission from Rao and Jagadeesh (2014))

- The amplitude of this oscillation is different for each peak and is lower close to the NXP and higher downstream. This means that there is an elongation-compression movement of the shock cells.
- The statistical correlation between the various peaks reveals that there is a strong correlation between different peaks oscillations, i.e., the shock cells move together.
- The analysis of the power spectrum of the time trace reveals discrete frequency peaks at around 50 Hz and 75 Hz. This is in partial agreement with the frequency peaks measured by the wall pressure sensors (peaks around 60 Hz).

The authors further argue that the pressure gradient induced by the wall may amplify the oscillation of the shocks through the possible manifestation of resonant modes. As a consequence, there may be optimal designs of the mixing duct with aeroacoustic characteristics that could be used for mixing enhancement (Rao and Jagadeesh 2014).

Finally, in their study the authors performed analysis of the mixing inside the ejector by exploiting laser-scattering visualization techniques. These types of analyses are illustrated in the next paragraph.

5.2.2 Mixing

A well-established technology uses scattering tracers in order to visualize the mixing process within the ejector. These tracers may be "natural" or artificially introduced into the flow. For example, an ejector using atmospheric air will surely have water vapor within the air stream, and this vapor may condense in the low-pressure zones of the ejector. Water droplets will scatter the light coming for a suitable source, i.e., a laser light sheet aligned with the ejector axis. This technique is a type of tomographic investigation that allows the visualization of a precise section of the flow. On the other hand, artificial tracers may be fluorescent substances that enhance the scattered light or confine it on specific wavelengths. Image processing may help the analysis of the flow, e.g., some unwanted features of the acquired images (e.g., scratches on the viewport) may be subtracted in order to highlight the desired flow details.

An example of these methods can be found in the paper by Bouhanguel et al. (2011) who investigated the flow inside an air ejector using three different techniques based on the laser sheet method (or laser tomography). Each of these methods allows qualitative observation of different phenomena, including the detection of the flow regime (single- or double-choked flow) and the analysis of the mixing of the two streams, as well as the visualization of the shock and turbulent flow structures.

The experimental apparatus adopted by Bouhanguel et al. (2011) is illustrated in Fig. 5.6. The setup includes a laser light source, an optical arrangement to generate a planar laser light, and an image acquisition system.

A first type of tomographic technique is based on the use of artificial tracers inserted within the suction stream. This approach was investigated by Desevaux et al. (1994) and Desevaux (2001) and allows the qualitative visualization of the mixing between the primary and secondary streams. Due to the presence of humidity in the air, water droplets form during the expansion of the primary stream (the process is similar to that analyzed in Chaps. 2 and 4 for pure steam). These droplets represent an additional scattering source which must be distinguished and separated from that emitted by the artificial tracer (this is done in different ways depending on the artificial tracer, for details see (Bouhanguel et al. 2011)).

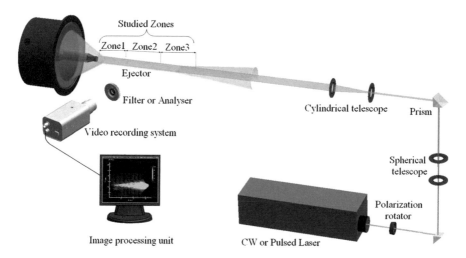

Fig. 5.6 Experimental setup for the laser tomography technique (Reprinted with permission from Bouhanguel et al. (2011))

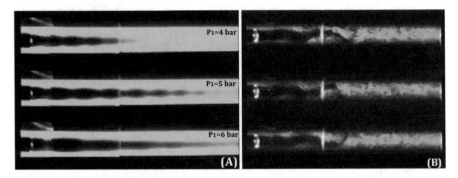

Fig. 5.7 Non-mixing region visualization using a fluorescent tracer for three different primary inlet pressures (A), non-mixing region at three different instants ($\Delta t = 0.1$ s) using a pulsed laser light source (B) (Reprinted with permission from Bouhanguel et al. (2011))

Figure 5.7a shows the visualization of the mixing inside the constant section duct of an air ejector for three different primary inlet pressures. The analysis is carried out using a fluorescent tracer that allows the visual assessment of the non-mixing length. The picture illustrates that the non-mixing region (the black area) increases with primary inlet pressure, probably due to a reduced level of overexpansion of the primary jet. Figure 5.7b illustrates the time evolution of the non-mixing length in three different instants. The images were recorded using the same laser-induced fluorescence technique coupled with a pulsed laser light source. This sequence of images reveals strong fluctuations of the non-mixing region due to the loss of stability of the primary jet when it flows out the motive nozzle into the mixing chamber (Bouhanguel et al. 2011).

Bouhanguel et al. (2011) also performed the analysis of the shock pattern inside the ejector by visualizing the scatter of water droplets formed after the condensation of the air humidity (natural tracer). Unfortunately, although providing qualitative insights on the flow regime, the acquired images did not result in very sharp contour of the shocks. This may be due to the time delay in droplet formation (i.e., the condensation time) that can smear the thickness of the shocks and move their location downstream the actual position. As a consequence, specific studies of the shock structures within ejectors should be better performed using diffraction techniques (shadowgraph or schlieren methods).

More recently, Karthick et al. (2017) attempted *quantitative* analyses of the mixing inside a planar air ejector exploiting the laser-induced fluorescence technique. The artificial tracer employed in the study was acetone vapors, whose emitted fluorescence intensity is directly proportional to the local concentration of the tracer (Lozano et al. 1992).

The authors define three different lengths connected to the mixing process, the mixing length, L_{MIX}; the non-mixed length, L_{NM}; and the potential core length, L_{PC}. These are illustrated in Fig. 5.8. In particular, Karthick et al. (2017) delineate the non-mixed length as the distance from the nozzle exit plane up to which the primary and the secondary flow are distinctly identifiable. In a time-averaged sense, this

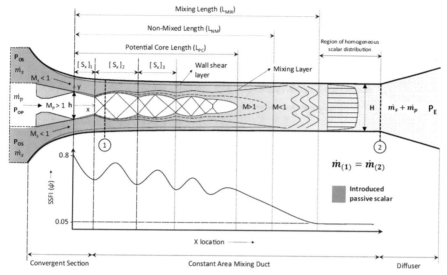

Pop-primary flow stagnation pressure; Pos-secondary flow stagnation pressure; P_E – Exit pressure; $[S_x]_n$ –shock cell spacing; n - shock cell number;

Fig. 5.8 Mixing characteristic lengths (Reprinted with permission from Karthick et al. (2017))

should coincide with the distance that is needed by the mixing layer to occupy the whole mixing chamber duct (in these terms, the laser-induced fluorescence method described by Bouhanguel et al. (2011) seems more suited to visualize the primary flow potential core length; however, for ejectors with narrow mixing chamber, the two quantities often coincide because the mixing layer occupies the secondary stream potential core prior than that of the primary flow).

In particular, the study of Karthick et al. (2017) focused on the assessment of the mixing length, L_{MIX}. In order to properly quantify this distance, the authors evaluated at each transverse cross section the Spatial Scalar Fluctuation Intensity (SSFI). This is defined as the root mean square of the spatial fluctuations intensity divided by the mean scalar variation in a given cross section (Karthick et al. 2017):

$$\psi = \frac{\sqrt{\overline{\Theta'}}}{\overline{\Theta}} \qquad \Theta = \frac{I}{I_{\max}} = \overline{\Theta} + \overline{\Theta'}$$

where the overbar represents a spatial averaged quantity and $I_{\max}$ is the maximum fluorescence intensity found in the flow field. The beginning of the fully mixed region was identified by the location where $\psi = 0.05$.

Figure 5.9 illustrates graphically the process to evaluate ψ and L_{MIX}. By using this technique, the authors investigated different operating conditions of the ejector, from highly over-expanded to highly under-expanded. In the first case, they found that the primary jet issuing from the nozzle was not centered due to the distortion

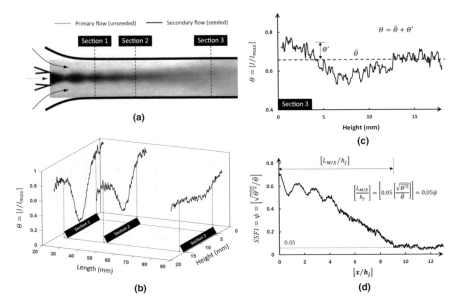

Fig. 5.9 Estimation procedure for the mixing length (L_{MIX}). (**a**) A typical time-averaged acetone planar laser-induced fluorescence image acquired by seeding the secondary flow; (**b**) normalized scalar profiles acquired at different transverse sections; (**c**) observation of fluctuating scalar field in a given section; (**d**) variations encountered in the Spatial Scalar Fluctuations Intensity (ψ) along the flow direction (Reprinted with permission from Karthick et al. (2017))

induced by the recirculation at the nozzle wall. Consequently, these cases were discarded from further investigations.

The analysis of the various results obtained by Karthick et al. (2017) revealed that the behavior of L_{MIX} is closely dependent on the nozzle operating conditions. The values of L_{MIX} were found to be reduced by 17.67% for the over-expanded flows and increased by 15.76% for the under-expanded flows from the perfectly expanded condition. Moreover, Karthick et al. (2017) found that the dependency of the mixing length on the nozzle expansion becomes linear after scaling this quantity with respect to the maximum primary jet height, as shown in Fig. 5.10 (in the figure, the scaled value of the mixing length is smaller for the under-expanded cases because the increase in jet height is higher than the corresponding increase in mixing length (Karthick 2017)).

Finally, the study also reports other supersonic confined jet characteristics like the potential core length (LPC) and the shock cell spacing (Sx) of the primary supersonic jet (for details see (Karthick et al. 2017)).

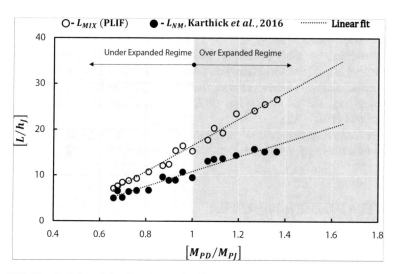

Fig. 5.10 Trend of the mixing length scaled with respect to the primary jet height and plotted against the Mach number ratio (design Mach number/fully expanded jet Mach number). The figure also shows the trends of the non-mixed length found in a previous work of the same authors (Karthick et al. 2016) (Reprinted with permission from Karthick et al. (2017))

5.2.3 Velocity Field

In the last decade, Particle Imaging Velocimetry (PIV) has become a standard tool for studying low-speed fluid flows (Smits and Dussauge 2006). Unfortunately, the application of this technique to compressible supersonic flows presents major difficulties with respect to the low-speed counterpart. These problems are mainly related to the non-ideal flow tracking, inhomogeneous seeding distribution, and optical distortions across shock waves (Scarano 2008). Additional difficulties are related to laser illumination and the speed and sensitivity of the cameras (Smits and Dussauge 2006).

To the authors' knowledge there exists only one attempt to measure the velocity field inside a supersonic ejector through PIV technique, namely, the preliminary study carried out by Bouhanguel et al. (2012). The authors, presented velocity measurements on a supersonic air ejector using PIV with two flow seeding methods (condensing water droplets and artificial scattering tracers) and compared the results with the velocity field obtained from CFD simulations.

The results showed that the PIV velocity profile compared well with the CFD results. The authors also found that the use of the natural seeding flow (condensing microdroplets) resulted in a better visual quality that may be explained by a higher number of natural tracers and by their very small size (mean diameter below 0.1 μm) that causes the tracer to scatter in the Rayleigh regime.

Unfortunately, no attempt was made to quantify the uncertainty connected with the various sources of errors (for instance, the polydispersed nature of the droplet population may introduce a different velocity lag for neighboring particles traveling across a shock (Scarano 2008)).

Nevertheless, the study proved the usefulness of PIV investigations, which can make available quantitative information that conventional diffraction or laser tomography visualizations cannot provide. Therefore, this method may represent an interesting tool for the validation of CFD simulations in supersonic ejectors (Bouhanguel et al. 2012).

A detailed discussion of PIV technique as applied to high-speed flow may be found in the nice reviews of (Scarano 2008; Havermann et al. 2008).

5.3 The DIEF Ejector Chiller

In 2010, DIEF started a cooperation with Frigel Firenze S.p.A. in order to open new market opportunities in the field of heat-driven chillers for industrial use.

The cooperation led to the project and construction of a new supersonic ejector chiller of relatively large size, suitable for experimental activity in an industrial environment. The plant has a nominal cooling power of 40 kW and is powered by low temperature heat (from 90 up to 100 °C or more). After an initial attempt to use R134a (which was the common choice at that time), the need to reduce the pressure at the generator suggested to revert to R245fa (Eames et al. 2013). This fluid was selected because of its relatively high critical temperature, low system pressure ratio, and, most importantly, positively sloped saturation vapor curve (i.e., dry-expansion fluid). Despite all the aforementioned qualities, R245fa has a relatively high GWP, posing potential restrictions with respect to expected regulations on fluorinated gases. However, the growing ORC market has stimulated the formulation of low-GWP alternatives matching the thermodynamic properties of R245fa, like HFO1233zd. Therefore, the results obtained with this peculiar fluid may retain their significance in the future (Mazzelli and Milazzo 2015).[2]

Figure 5.11 shows some pictures of the prototype. The prototype is conceived with a "ready to market" structure. During components placing, several features had to be taken into account to permit easy connection with external water circuits, allow user access for assembly and disassembly operations, and create a compact and moveable structure. A vertical arrangement was chosen for heat exchangers. This allowed reduction of space and mitigation of cavitation problems at the pump inlet. As a consequence, the main axis of ejector is vertical as well. The ejector is equipped with a movable primary nozzle, in order to optimize the axial position relative to the diffuser. At present the mechanism cannot be operated when the system is running, but in principle it could be modified for continuous adjustment during operation.

---

[2]Latest numerical simulations performed with HFO1233zd confirm that the performances are practically the same as those obtained with R245fa, as long as the same pressure conditions are imposed at the ejector inlets and outlets.

Fig. 5.11 Pictures of the DIEF chiller built by Frigel Firenze S.p.A

Fig. 5.12 Infrared image of
the first prototype ejector in
operation

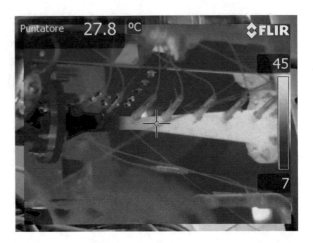

Nine static pressure probes are installed along the mixing chamber/diffuser duct in
order to analyze the internal pressure trends.

A Sporlan Electronic Expansion Valve is used to control the liquid level in the
evaporator based on the superheating at the evaporator exit. The superheating is
calculated by comparison between the measured temperature and the saturation
temperature calculated by the control system through a look-up Table. A pump
(SPECK Pumpen) with seven centrifugal stages has been selected because it is very
robust to cavitation.

It may be noted that the ejector depicted on the upper left picture in Fig. 5.11 is
different from that depicted on the right. The one on the left was the first prototype,
manufactured in carbon fiber. This ejector is now abandoned, but its manufacturing
process was very successful and will be hopefully used again in the future.

By the way, from an experimental point of view, an interesting feature of this
composite material is the good emissivity that facilitates infrared imaging. A first
result obtained with a FLIR infrared camera is shown in Fig. 5.12. The false colors
exhibited by the ejector witness the range of temperatures covered by the fluid (note
that carbon fiber has a relatively good thermal conductivity). Clearly this image has
only an indicative value, and it is not intended as a quantitative result.

The ejector shown on the right in Fig. 5.11 is instead the third prototype,
manufactured in aluminum.

Figure 5.13 shows the location of the transducers on the scheme. Temperature
measurements are obtained by resistance temperature detectors Pt100 whose precision class is 1/10 DIN. The probes are placed at the inlet and outlet connection of
each heat exchanger. Resistance values are read and converted by a National
Instruments cFP-RTD-124 module. Piezoresistive pressure transducers produced
by Keller are used to obtain pressure values. Water mass flow measurements in
the external circuit are carried out with Endress+Hauser Promag electromagnetic
flow meters. Electric power consumption of the feeding pump is measured by an
electronic wattmeter.

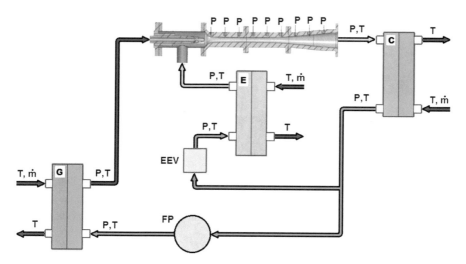

Fig. 5.13 Scheme of the DIEF chiller with transducers

Table 5.2 Instrumentation uncertainty

Instrument	Model	Equipped component	Total uncertainties
Piezoresistive pressure transducer	PA25HTT 0–30 bar	Diffuser	±(0.1% + 0.22% FS)
	PR23R 0.5–5 bar	Evaporator	±(0.1% + 0.22% FS)
	PA21Y 0–30 bar	Generator, condenser	±(0.08% + 1% FS)
Resistance temperature detectors	Pt100	Whole plant	±0.25 °C
Electromagnetic water flow meters	Promag 30F	Evaporator	±(0.22% + 0.06% FS)
	Promag 50P	Generator	±(0.5% + 0.04% FS)
	Promag 50 W	Condenser	±(0.5% + 0.04% FS)

Thus far, flow measurements inside the refrigeration plants are absent. Therefore, heat fluxes and refrigerant mass flow rates are measured by equivalence with the thermal fluxes flowing through the external water circuit. Due to this indirect method of measuring the mass fluxes, steady conditions are always sought to assure equality between water and refrigerant thermal fluxes. Nonetheless, the lack of direct mass flow measurement can lead to low accuracy of the experimental data. Hence, an extensive and detailed uncertainty analysis was performed to understand the level of confidence in the measurements.

Table 5.2 summarizes the final level of accuracy for each measured quantity.

The whole plant, including heat exchangers and piping, was designed using a numerical design tool based on a one-dimensional calculation method presented by Grazzini et al. (2012), as described in Chap. 3. Real fluid properties are used throughout the model by incorporating the NIST REFPROP subroutines (Lemmon et al. 2013).

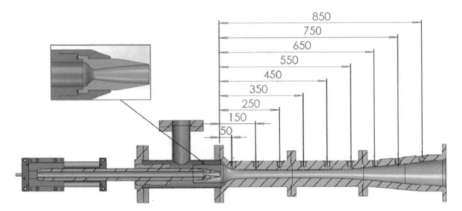

Fig. 5.14 Design of the ejector prototype currently installed inside the DIEF chiller (Mazzelli and Milazzo 2015)

Table 5.3 Main dimensions of the ejector prototype currently installed inside the DIEF chiller

	Nozzle	Diffuser
Throat diameter [mm]	10.2	31.8
Exit diameter [mm]	20.2	108.3
Length	66.4	950
Material	Aluminum	Aluminum

The design code implements a routine for the design of the ejector profile based on the CRMC (constant rate of momentum change) criterion by Eames (2002). The CRMC method is an attempt to reduce throat shock intensity by giving a prescribed momentum reduction rate throughout the mixing chamber and diffuser. The resulting mixing chamber/diffuser profile is continuous, and, from now on, it will be called "supersonic diffuser" or just "diffuser." Two different ejectors were designed according to this code and tested. However, the performances obtained with the first two configurations were unsatisfactory (Eames et al. 2013). Hence, the design concepts were reconsidered, and a new ejector was manufactured. The new supersonic diffuser follows the CRCM criterion but has an increased length to improve mixing and pressure recovery (Milazzo et al. 2014). Figure 5.14 shows the geometry of the ejector currently installed in the plant, while its main geometrical parameters are summarized in Table 5.3.

Numerical analyses on this third design showed that, for the set of operating conditions specified by the industrial partner, the primary nozzle was working under a high level of overexpansion. Therefore, two nozzles with a smaller exit area were simulated numerically, and the results are presented in Fig. 5.15 (the related CFD scheme is described in Chap. 4). As can be seen in the figure, a reduction of the nozzle exit area results in significant improvements both in terms of entrainment ratio and critical pressure. The improved performance is due to a reduction in the expansion level of the primary flow, which is attained by matching the nozzle exit pressure with that of the mixing chamber.

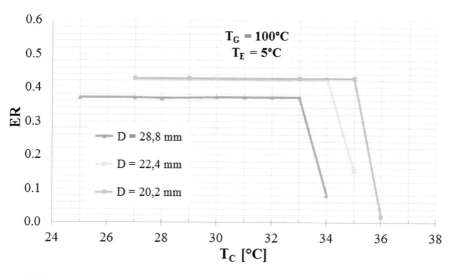

Fig. 5.15 Entrainment ratio for different values of the nozzle exit diameter

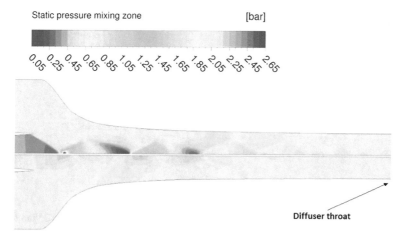

Fig. 5.16 Static pressure in the mixing region for two different nozzle exit diameters: 28.8 mm (above the axis) and 20.2 mm (below the axis) (Mazzelli and Milazzo 2015)

Figure 5.16 elucidates this concept by showing the static pressure field in the mixing chamber for two simulated cases. In the previous configuration (upper half of the figure), the primary flow expands down to very low exit pressures and shocks in order to reach the mixing chamber pressure. In the new configuration (lower half of the figure), the smaller pressure difference at the nozzle exit allows for a reduction in shock train intensities and pressure losses. This is also shown in Fig. 5.17 where the region with Mach number above unity is highlighted. The sonic line is much smoother in the new configuration and correctly follows the primary nozzle profile.

Mach Number

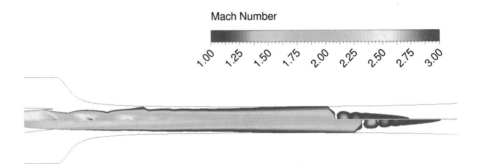

Fig. 5.17 Supersonic Mach field along the diffuser for two different nozzle exit diameters: 28.8 mm (above the axis) and 20.2 mm (below the axis) (Mazzelli and Milazzo 2015)

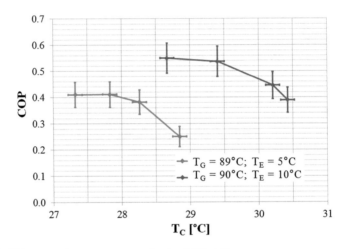

Fig. 5.18 COP vs. condenser temperature for the DIEF chiller

Based on these results, the nozzle design with exit diameter 20.2 mm was finally selected, manufactured, and inserted in the ejector. In agreement with numerical analyses, substitution of the primary nozzle produced significant improvements, and now chiller performance is aligned with or exceeds the results published by other authors (see Table 5.1).

A parameter that still needs to be improved is the pressure lift capacity, which at present makes this chiller inadequate for warm climates. Therefore, the chiller will evolve in the future toward this objective, accounting for what has already been said about the influence of a very good surface finish of the ejector wall on the critical pressure.

Figure 5.18 shows some experimental results of the refrigerator in terms of coefficient of performance (COP). The error bars for each evaluated operating point represent a confidence level of 95% and were calculated according to the

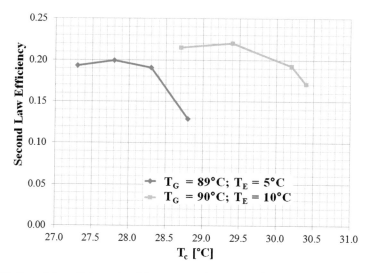

Fig. 5.19 Second-law efficiency trend for two operating curves of the DIEF refrigerator

procedure explained in (Mazzelli and Milazzo 2015). The generator temperature is around 90 °C which is a temperature suitable for solar cooling or waste heat recovery applications; 5 and 10 °C are imposed at the evaporator, which represent standard values for air conditioning. Finally, 5 °C superheating is set at the evaporator exit to avoid entrainment of liquid refrigerant inside the ejector.

Figure 5.18 shows that the COP reached by the DIEF chiller is above 0.4 when evaporating at 5 °C and around 0.55 when evaporating at 10 °C. These levels are not too far from those obtained by single-effect absorption chillers and match or exceed the performance obtained by other cycles in the literature.

By simply dividing each value of COP reported in Fig. 5.18 by the corresponding ideal COP (see Chap. 3), it is possible to reproduce the equivalent diagram for the second-law efficiency. Figure 5.19 shows the results of such operation.

Although the points are few, it is clear that the two curves present a definite maximum which occurs exactly at the critical operating conditions. This demonstrates once more that the best operating points are always at critical condition. Moreover, the maximum height of the two curves is very similar (with a difference of around 10%), meaning that this definition of efficiency accounts for the more or less favorable operating conditions.

Finally, the curves shown in Fig. 5.19 were obtained using an ideal COP based on the cycle saturation temperatures. Figure 5.20 shows a comparison between one efficiency curve from Fig. 5.19 and the corresponding curve based on the temperatures of the flow entering the heat exchangers from the external circuit side (i.e., the inlet temperature of the flow that cools the condenser and heat up the generator and evaporator). As can be clearly seen, the efficiency with the new formulation is less than a half of that calculated with the saturation temperatures. This analysis

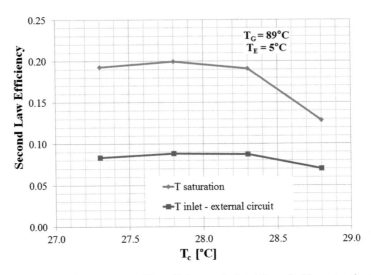

Fig. 5.20 Comparison between second-law efficiency calculated through either saturation temperatures or external circuit inlet temperatures

highlights the great importance of the thermodynamic losses resulting from a non-ideal heat exchange. These losses should always be accounted for in the view of a fair comparison between the performances of different cycles.

References

Aphornratana, S., & Eames, I. (1997). A small capacity steam-ejector refrigerator: Experimental investigation of a system using ejector with movable primary nozzle. *International Journal of Refrigeration, 20*, 352–358.

Besagni, G., Mereu, R., & Inzoli, F. (2016). Ejector refrigeration: A comprehensive review. *Renewable and Sustainable Energy Reviews, 53*, 373–407.

Bouhanguel, A., Desevaux, P., & Gavignet, E. (2011). Flow visualization in supersonic ejectors using laser tomography techniques. *International Journal of Refrigeration, 34*, 1633–1640.

Bouhanguel, A., Desevaux, P., Bailly, Y., & Girardot, L. (2012). *Particle image velocimetry in a supersonic ejector,* 15th International Symposium on Flow Visualization. June 25–28, Minsk, Belarus, s.n.

Carroll, B., & Dutton, J. (1990). Characteristics of multiple shock wave/turbulent boundary-layer interactions in rectangular ducts. *Journal of Propulsion, 6*, 186–193.

Chen, Y., & Sun, C. (1997). Experimental study of the performance characteristics of a steam-ejector refrigeration system. *Experimental Thermal and Fluid Science, 15*, 384–394.

Chen, X., Omer, S., Worall, M., & Riffat, S. (2013). Recent developmentsinejectorrefrigerationtechnologies. *Renewable and Sustainable Energy Reviews, 19*, 629–651.

Chunnanond, K., & Aphornratana, S. (2004). An experimental investigation of a steam ejector. *Applied Thermal Engineering, 24*, 311–322.

Desevaux, P. (2001). A method for visualizing the mixing zone between two co-axial flows in an ejector. *Optics and Lasers in Engineering, 35*, 317–323.

Desevaux, P., Prenel, J., & Hostache, G. (1994). An optical analysis of an induced flow ejector using light polarization properties. *Experiments in Fluids, 16*, 165–170.

Dvorak, V., & Safarik, P. (2005). Transonic instability in entrance part of mixing chamber of high-speed ejector. *Journal of Thermal Science, 14*, 258–264.

Eames, I. (2002). A new prescription for the design of supersonic jet-pumps: The constant rate of momentum change method. *Applied Thermal Engineering, 22*, 121–131.

Eames, I., Ablwaifa, A., & Petrenko, V. (2007). Results of an experimental study of an advanced jet-pump refrigerator operating with R245fa. *Applied Thermal Engineering, 27*, 2833–2284.

Eames, I., Milazzo, A., Paganini, D., & Livi, M. (2013). The design, manufacture and testing of a jet-pump chiller for air conditioning and industrial application. *Applied Thermal Engineering, 58*, 234–240.

Fabri, J., & Siestrunck, R. (1958). Supersonic air ejectors. In *Advances in applied mechanics* (Vol. 5). New York: Academic.

Grazzini, G., Milazzo, A., & Paganini, D. (2012). Design of an ejector cycle refrigeration system. *Energy Conversion and Management, 54*, 38–46.

Havermann, M., Haertig, J., Rey, C., & George, A. (2008). PIV measurements in shock tunnels and shock tubes. In C. W. A. Schroeder (Ed.), *Particle image velocimetry, Topics in applied physics* (Vol. 112, pp. 429–443). Berlin/Heidelberg: Springer.

Huang, B., & Chang, J. (1999). Empirical correlation for ejector design. *International Journal of Refrigeration, 22*, 379–388.

Huang, B., Chang, J., Wang, C., & Petrenko, V. (1999). A 1-D analysis of ejector performance. *International Journal of Refrigeration, 22*, 354–364.

Karthick, S. (2017). *Private communication.* s.l.:s.n.

Karthick, S., Rao, S., Jagadeesh, G., & Reddy, K. (2016). Parametric experimental studies on mixing characteristics within a low area ratio rectangular supersonic gaseous ejector. *Physics of Fluids, 28*, 1–26.

Karthick, S. K., Rao, S., Jagadeesh, G., & Reddy, K. (2017). Passive scalar mixing studies to identify the mixing length. *Experiments in Fluids, 58*(59), 1–20.

Keenan, J., Neumann, E., & Lustwerk, F. (1950). An investigation of ejector design by analysis and experiment. *Journal of Applied Mechanics, 17*, 299–309.

Lamberts, O., Chatelain, P., & Bartosiewicz, Y. (2017). New methods for analyzing transport phenomena in supersonic ejectors. *International Journal of Heat and Fluid Flow, 64*, 23–40.

Lemmon, E., Huber, M., & McLinden, M. (2013). *NIST standard reference database 23: Reference fluid thermodynamic and transport properties-REFPROP, Version 9.1*, s.l.: National Institute of Standards and Technology.

Lozano, A., Yip, B., & Hanson, R. (1992). Acetone: a tracer for concentration measurements in gaseous flows by planar laser-induced fluorescence. *Experiments in Fluids, 13*, 369–376.

Ma, X., Zhang, W., Omer, S., & Riffat, S. (2010). Experimental investigation of a novel steam ejector refrigerator suitable for solar energy applications. *Applied Thermal Engineering, 30*, 1320–1325.

Matsuo, K., Sasaguchi, K., Kiyotoki, Y., & Mochizuki, H. (1981). Investigation of supersonic air ejectors (Part 1, performance in the case of zero-secondary flow). *Bulletin of JSME, 24*, 2090–2097.

Matsuo, K., Sasaguchi, K., Kiyotoki, Y., & Mochizuki, H. (1982). Investigation of supersonic air ejectors (Part 2, effects of throat-area-ratio on ejector performance). *Bulletin of JSME, 25*, 1898–1905.

Mazzelli, F., & Milazzo, A. (2015). Performance analysis of a supersonic ejector cycle. *International Journal of Refrigeration, 49*, 79–92.

Milazzo, A., Rocchetti, A., & Eames, I. (2014). Theoretical and experimental activity on Ejector Refrigeration. *Energy Procedia, 45*, 1245–1254.

Munday, J. T., & Bagster, D. F. (1977). A new ejector theory applied to steam jet refrigeration. *Industrial and Engineering Chemistry Process Design and Development, 164*, 442–449.

Nguyen, V., Riffat, S., & Doherty, P. (2001). Development of a solar-powered passive ejector cooling system. *Applied Thermal Engineering, 21*, 157–168.

Rao, S. M. V., & Jagadeesh, G. (2014). Observations on the non-mixed length and unsteady shock motion in a two dimensional supersonic ejector. *Physics in Fluids, 26*(036103), 1–26.

Scarano, F. (2008). Overview of PIV in supersonic flows. In C. W. A. Schroeder (Ed.), *Particle image velocimetry, Topics in applied physics* (Vol. 112, pp. 445–463). Berlin/Heidelberg: Springer.

Selvaraju, A., & Mani, A. (2006). Experimental investigation on R134a vapor ejector refrigeration system. *International Journal of Refrigeration, 29*, 1160–1166.

Settles, G. (2001). *Schlieren and shadowgraph techniques: visualizing phenomena in transparent media*. Berlin: Springer.

Smits, A., & Dussauge, J.-P. (2006). *Turbulent shear layers in supersonic flow* (2nd ed.). New York: Springer.

Smits, A., & Lim, T. (2012). *Flow visualization, techniques and examples* (2nd ed.). London: Imperial College Press.

Yapici, R., & Yetisen, C. (2007). Experimental study on ejector refrigeration system powered by low grade heat. *Energy Conversion and Management, 48*, 1560–1568.

Yapici, R., et al. (2008). Experimental determination of the optimum performance of ejector refrigeration system depending on ejector area ratio. *International Journal of Refrigeration, 31*, 1183–1189.

Zhu, Y., & Jiang, P. (2014a). Experimental and numerical investigation of the effect of shock wave characteristics on the ejector performance. *International Journal of Refrigeration, 40*, 31–42.

Zhu, Y., & Jiang, P. (2014b). Experimental and analytical studies on the shock wave length in convergent and convergent–divergent nozzle ejectors. *Energy Conversion and Management, 88*, 907–914.

Chapter 6
Concluding Remarks

Many issues concerning ejectors and ejector chillers have been left out of this book. Some of them were cut on purpose. For example, our direct experience tells us that the noise produced by an ejector varies in a quite unpredictable fashion, and we decided to postpone any analysis on this point. This doesn't mean that the noise is not a relevant problem. Actually, it could prevent the diffusion of ejectors in some applications.

Another point that was neglected in the book is load control. Traditionally, ejectors are thought to be very rigid devices, even if some proposals have been made to control the cooling power. A straightforward approach would be to mount several ejectors in a plant and switch off one or more as the load request is reduced (clearly this increases the surface wetted by high-speed flows and eventually reduces the efficiency at full load).

We warmly encourage the interested readers to give us a feedback, and hopefully in the future, there will be a chance to improve our work.

A last outcome of our experimental analysis is a first indicative price list that, once sensors and other strictly experimental devices are eliminated, could give an idea of the cost for a medium-size ejector chiller. The approximate cost distribution is shown in the diagram below. Clearly, the heat exchangers constitute a big part of the total and, being off-the-shelf components, cannot be decreased even in case of mass production of the chiller. However, the single largest item in the list is the feed pump, which has a costly multistage structure, a magnetic coupling, and an inverter. Probably, a simpler and less expensive pump could be selected for normal production. Another significant point is the fluid, which, as specified in Chap. 5, is R245fa. This cost could decrease if a natural fluid were used or even increase for HFO. Finally, the ejector was rather complicated in this case, being equipped with static pressure ports, and its cost could be decreased by a careful optimization. The total cost of the chiller amounts to roughly 20,000 € for a cooling power of 50 KW.

For comparison, the market price of an absorption chiller of equivalent size may range between 25,000 and 30,000 €, which means that production cost should not be

© Springer International Publishing AG, part of Springer Nature 2018   175
G. Grazzini et al., *Ejectors for Efficient Refrigeration*,
https://doi.org/10.1007/978-3-319-75244-0_6

lower than 10,000 €. Therefore, the goal of making ejector chiller competitive in terms of costs per unit cooling power may be feasible in the near future.

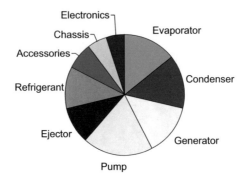

Ejector chiller prototype cost distribution

Index

© Springer International Publishing AG, part of Springer Nature 2018
G. Grazzini et al., *Ejectors for Efficient Refrigeration*,
https://doi.org/10.1007/978-3-319-75244-0

Printed in the United States
By Bookmasters